AF327054

Matrix Metalloproteinase Protocols

METHODS IN MOLECULAR BIOLOGY™

John M. Walker, SERIES EDITOR

176. **Steroid Receptor Methods:** *Protocols and Assays,* edited by *Benjamin A. Lieberman, 2001*

175. **Genomics Protocols**, edited by *Michael P. Starkey and Ramnath Elaswarapu, 2001*

174. **Epstein-Barr Virus Protocols**, edited by *Joanna B. Wilson and Gerhard H. W. May, 2001*

173. **Calcium-Binding Protein Protocols, Volume 2:** *Methods and Techniques,* edited by *Hans J. Vogel, 2001*

172. **Calcium-Binding Protein Protocols, Volume 1:** *Reviews and Case Histories,* edited by *Hans J. Vogel, 2001*

171. **Proteoglycan Protocols**, edited by *Renato V. Iozzo, 2001*

170. **DNA Arrays:** *Methods and Protocols,* edited by *Jang B. Rampal, 2001*

169. **Neurotrophin Protocols**, edited by *Robert A. Rush, 2001*

168. **Protein Structure, Stability, and Folding,** edited by *Kenneth P. Murphy, 2001*

167. **DNA Sequencing Protocols,** *Second Edition,* edited by *Colin A. Graham and Alison J. M. Hill, 2001*

166. **Immunotoxin Methods and Protocols**, edited by *Walter A. Hall, 2001*

165. **SV40 Protocols**, edited by *Leda Raptis, 2001*

164. **Kinesin Protocols**, edited by *Isabelle Vernos, 2001*

163. **Capillary Electrophoresis of Nucleic Acids, Volume 2:** *Practical Applications of Capillary Electrophoresis,* edited by *Keith R. Mitchelson and Jing Cheng, 2001*

162. **Capillary Electrophoresis of Nucleic Acids, Volume 1:** *The Capillary Electrophoresis System as an Analytical Tool,* edited by *Keith R. Mitchelson and Jing Cheng, 2001*

161. **Cytoskeleton Methods and Protocols**, edited by *Ray H. Gavin, 2001*

160. **Nuclease Methods and Protocols**, edited by *Catherine H. Schein, 2001*

159. **Amino Acid Analysis Protocols**, edited by *Catherine Cooper, Nicole Packer, and Keith Williams, 2001*

158. **Gene Knockoout Protocols**, edited by *Martin J. Tymms and Ismail Kola, 2001*

157. **Mycotoxin Protocols**, edited by *Mary W. Trucksess and Albert E. Pohland, 2001*

156. **Antigen Processing and Presentation Protocols**, edited by *Joyce C. Solheim, 2001*

155. **Adipose Tissue Protocols**, edited by *Gérard Ailhaud, 2000*

154. **Connexin Methods and Protocols**, edited by *Roberto Bruzzone and Christian Giaume, 2001*

153. **Neuropeptide Y Protocols**, edited by *Ambikaipakan Balasubramaniam, 2000*

152. **DNA Repair Protocols:** *Prokaryotic Systems,* edited by *Patrick Vaughan, 2000*

151. **Matrix Metalloproteinase Protocols**, edited by *Ian M. Clark, 2001*

150. **Complement Methods and Protocols**, edited by *B. Paul Morgan, 2000*

149. **The ELISA Guidebook**, edited by *John R. Crowther, 2000*

148. **DNA–Protein Interactions:** *Principles and Protocols* **(2nd ed.),** edited by *Tom Moss, 2001*

147. **Affinity Chromatography:** *Methods and Protocols,* edited by *Pascal Bailon, George K. Ehrlich, Wen-Jian Fung, and Wolfgang Berthold, 2000*

146. **Mass Spectrometry of Proteins and Peptides,** edited by *John R. Chapman, 2000*

145. **Bacterial Toxins:** *Methods and Protocols,* edited by *Otto Holst, 2000*

144. **Calpain Methods and Protocols**, edited by *John S. Elce, 2000*

143. **Protein Structure Prediction:** *Methods and Protocols,* edited by *David Webster, 2000*

142. **Transforming Growth Factor-Beta Protocols**, edited by *Philip H. Howe, 2000*

141. **Plant Hormone Protocols**, edited by *Gregory A. Tucker and Jeremy A. Roberts, 2000*

140. **Chaperonin Protocols**, edited by *Christine Schneider, 2000*

139. **Extracellular Matrix Protocols**, edited by *Charles Streuli and Michael Grant, 2000*

138. **Chemokine Protocols**, edited by *Amanda E. I. Proudfoot, Timothy N. C. Wells, and Christine Power, 2000*

137. **Developmental Biology Protocols, Volume III**, edited by *Rocky S. Tuan and Cecilia W. Lo, 2000*

136. **Developmental Biology Protocols, Volume II**, edited by *Rocky S. Tuan and Cecilia W. Lo, 2000*

135. **Developmental Biology Protocols, Volume I**, edited by *Rocky S. Tuan and Cecilia W. Lo, 2000*

134. **T Cell Protocols:** *Development and Activation,* edited by *Kelly P. Kearse, 2000*

133. **Gene Targeting Protocols**, edited by *Eric B. Kmiec, 2000*

132. **Bioinformatics Methods and Protocols**, edited by *Stephen Misener and Stephen A. Krawetz, 2000*

131. **Flavoprotein Protocols**, edited by *S. K. Chapman and G. A. Reid, 1999*

130. **Transcription Factor Protocols**, edited by *Martin J. Tymms, 2000*

129. **Integrin Protocols**, edited by *Anthony Howlett, 1999*

128. **NMDA Protocols**, edited by *Min Li, 1999*

127. **Molecular Methods in Developmental Biology:** *Xenopus and Zebrafish,* edited by *Matthew Guille, 1999*

126. **Adrenergic Receptor Protocols**, edited by *Curtis A. Machida, 2000*

125. **Glycoprotein Methods and Protocols:** *The Mucins,* edited by *Anthony P. Corfield, 2000*

124. **Protein Kinase Protocols**, edited by *Alastair D. Reith, 2001*

123. ***In Situ* Hybridization Protocols (2nd ed.),** edited by *Ian A. Darby, 2000*

122. **Confocal Microscopy Methods and Protocols**, edited by *Stephen W. Paddock, 1999*

121. **Natural Killer Cell Protocols:** *Cellular and Molecular Methods,* edited by *Kerry S. Campbell and Marco Colonna, 2000*

120. **Eicosanoid Protocols**, edited by *Elias A. Lianos, 1999*

119. **Chromatin Protocols**, edited by *Peter B. Becker, 1999*

118. **RNA–Protein Interaction Protocols**, edited by *Susan R. Haynes, 1999*

117. **Electron Microscopy Methods and Protocols**, edited by *M. A. Nasser Hajibagheri, 1999*

116. **Protein Lipidation Protocols**, edited by *Michael H. Gelb, 1999*

115. **Immunocytochemical Methods and Protocols (2nd ed.),** edited by *Lorette C. Javois, 1999*

114. **Calcium Signaling Protocols**, edited by *David G. Lambert, 1999*

METHODS IN MOLECULAR BIOLOGY™

Matrix Metalloproteinase Protocols

Edited by

Ian M. Clark

*University of East Anglia,
Norwich, UK*

Humana Press ✳ Totowa, New Jersey

Library of Congress Cataloging in Publication Data

Methods in molecular biology™.

Matrix metalloproteinase protocols / edited by Ian M. Clark.
    p. cm.—(Methods in molecular biology; 151)
  Includes bibliographical references and index.
  ISBN 0-89603-733-9 (alk. paper)
  1. Metalloproteinases—Laboratory manuals. 2. Extracellular matrix proteins—Laboratory manuals. I. Clark, Ian M. II.   Methods in molecular biology (Totowa, N.J.); v. 151.

  QP601.7.M38 2000
  572'.76—dc21   99-087618
                                                                                              CIP

Preface

Research in the matrix metalloproteinase field began with the demonstration by Gross and Lapière, in 1962, that resorbing tadpole tail expressed an enzyme that could degrade collagen gels. These humble beginnings have led us to the elucidation of around twenty distinct vertebrate MMPs, along with a variety of homologs from such diverse organisms as sea urchin, plants, nematode worm, and bacteria. This, coupled with four known specific inhibitors of MMPs, the TIMPs, gives a complex picture.

Part I of *Matrix Metalloproteinase Protocols* provides the reader with a selective overview of the MMP arena, and a chance to come to grips with where the field has been, where it is, and where it is going. I hope that this complements all of the methodology that comes later. Part II presents the reader with a diverse set of methods for the expression and purification of MMPs and TIMPs, bringing together the long and often hard-earned experience of a number of researchers. Part III allows the reader to detect MMPs and TIMPs at both the protein and mRNA level, whereas Part IV gives the ability to assay MMP and TIMP activities in a wide variety of circumstances.

For folks who are new to MMP research, I hope that this book will enable you to "hit the ground running," acquiring both some background and many useful laboratory techniques in one easy volume. For the seasoned MMP-er, I hope that it will add some new methods to your old favorites, or help the new people in your laboratory get going.

I'd like to thank all of the authors for the time and effort they put in. The demands of writing a book chapter are not trivial, especially against the background of heavy research and teaching commitments, and I have been overwhelmed with the number of eminent and senior people in MMP research who have contributed. I'd also like to acknowledge Kim Hoather-Potter, my Production Editor at Humana Press, for keeping the whole thing on track. Kim's hard work, interest, lively emails, and sense of humor helped make the difficult bits that much easier. Finally, I'd like to thank the Arthritis Research Campaign, UK, for keeping me in a job for long enough to make a commitment to editing this book; it's been an experience.

Ian M. Clark, PhD

Contents

Preface v

Contributors xi

PART I. MMPs AND TIMPs — AN OVERVIEW OF THE FIELD

1 MMPs and TIMPs:
   An Historical Perspective
   J. Frederick Woessner, Jr. *1*

2 Strategies for Cloning New MMPs and TIMPs
   Gloria Velasco and Carlos López-Otín 25

3 Structural Studies on MMPs and TIMPs
   Wolfram Bode and Klaus Maskos 45

4 Matrix Metalloproteinase Substrate Binding Domains,
   Modules and Exosites:
   Overview and Experimental Strategies
   Christopher M. Overall 79

5 The Matrix Metalloproteinase (MMP) and Tissue Inhibitor
   of Metalloproteinase (TIMP) Genes:
   *Transcriptional and Posttranscriptional Regulation,
   Signal Transduction and Cell-Type-Specific Expression*
   Matthew P. Vincenti 121

6 Models for Gain-of-Function and Loss-of-Function of MMPs:
   Transgenic and Gene Targeted Mice
   **Lisa M. Coussens, Steven D. Shapiro, Paul D. Soloway,
   and Zena Werb** 149

PART II. EXPRESSION AND PURIFICATION OF MMPs AND TIMPs

7 Expression of MMPs and TIMPs in Mammalian Cells
   **Karen M.-L. Yeow, Blaine W. Phillips, P. Paul Beaudry,
   Kevin J. Leco, Gillian Murphy, and Dylan R. Edwards** 181

8 Expression of Recombinant Matrix Metalloproteinases
   in *Escherichia coli*
   L. Jack Windsor and Darin L. Steele 191

9 Expression of Human Collagenase I (MMP-1) and TIMP-1
in a Baculovirus-Based Expression System
Rüdiger Vallon and Peter Angel *207*

10 Expression of Recombinant Matrix Metalloproteinases in Yeast
Glenn A. Doyle *219*

11 Expression of Recombinant Membrane-Type MMPs
**Michael J. Butler, Marie-Pia d'Ortho,
and Susan J. Atkinson** *239*

12 Refolding of TIMP-2 from *Escherichia coli* Inclusion Bodies
Richard A. Williamson *257*

13 Expression and Refolding of Full-Length Human TIMP-1
Deborah Davis *267*

14 Purification of MMPs and TIMPs
Ken-ichi Shimokawa and Hideaki Nagase *275*

PART III. DETECTION OF MMPS AND TIMPS

15 Monitoring MMP and TIMP mRNA Expression by RT-PCR
**Howard Wong, Huong Muzik, Lori L. Groft, Marc A. Lafleur,
Charles Matouk, Peter A. Forsyth, Gilbert A. Schultz,
Steven J. Wall, and Dylan R. Edwards** *305*

16 Measuring Transcription of Metalloproteinase Genes:
Nuclear Run-Off Assay vs Analysis of hnRNA
Anne M. Delany *321*

17 *In Situ* Hybridization for Metalloproteinases and Their Inhibitors
Tina L. Hurskainen and Suneel S. Apte *335*

18 Use of EIA to Measure MMPs and TIMPs
Noboru Fujimoto and Kazushi Iwata *347*

19 Immunohistochemistry of MMPs and TIMPs
Yasunori Okada *359*

20 Detecting Polymorphisms in MMP Genes
Shu Ye and Adriano M. Henney *367*

PART IV. ASSAY OF MMP AND TIMP ACTIVITIES

21 Methods for Studying Activation of Matrix Metalloproteinases
Vera Knäuper and Gillian Murphy *377*

22 Assay of Matrix Metalloproteinases Against Matrix Substrates
**Timothy E. Cawston, Paul Koshy,
and Andrew D. Rowan** *389*

23 Zymography and Reverse Zymography for Detecting
      MMPs and TIMPs
    Susan P. Hawkes, Hongxia Li, and Gary T. Taniguchi 399

24 *In Situ* Zymography
    Sarah J. George and Jason L. Johnson 411

25 Detection of Focal Proteolysis Using Texas-Red-Gelatin
    Rosalind M. Hembry 417

26 Antibodies to MMP-Cleaved Aggrecan
    **Amanda J. Fosang, Karena Last, David C. Jackson,
      and Lorena Brown** 425

27 Cartilage Proteoglycan Release Assay
    Caron J. Billington 451

28 Immunoassay for Collagenase-Mediated Cleavage
      of Types I and II Collagens
    **R. Clark Billinghurst, Mirela Ionescu,
      and A. Robin Poole** 457

29 Collagen Degradation Assays
    Anthony P. Hollander 473

30 Invasion Assays and Matrix Metalloproteinases:
      *Quantification of Cellular Invasion Using Propidium
      Iodide Labeling and Confocal Laser Scanning Microscopy*
    Ulrike Benbow, Kenneth A. Orndorff, and Alice L. Givan 485

31 Using Fluorogenic Peptide Substrates to Assay
      Matrix Metalloproteinases
    Gregg B. Fields 495

32 Kinetic Analysis of the Inhibition of Matrix Metalloproteinases
      by Tissue Inhibitor of Metalloproteinases (TIMP)
    Mike Hutton and Frances Willenbrock 519

33 Assaying Growth Factor Activity of Tissue Inhibitors
      of Metalloproteinases
    Kyoko Yamashita and Taro Hayakawa 533

Index 539

Contributors

PETER ANGEL • *German Cancer Research Center, Division of Signal Transduction and Growth Control, Deutsches Krebsforschungszentrum (DKFZ), Heidelberg, Germany*

SUNEEL S. APTE • *Department of Biomedical Engineering, Lerner Research Institute, Cleveland Clinic Foundation, Cleveland, OH*

SUSAN J. ATKINSON • *School of Biological Sciences, University of East Anglia, Norwich, UK*

P. PAUL BEAUDRY • *Department of Biochemistry and Molecular Biology, The University of Calgary, Alberta, Canada*

ULRIKE BENBOW • *Department of Medicine and Biochemistry, Dartmouth Medical School, Hanover, NH*

R. CLARK BILLINGHURST • *Department of Clinical Sciences, Colorado State University, Fort Collins, CO*

CARON J. BILLINGTON • *Department of Rheumatology, The Medical School, University of Newcastle, Framlington Place, Newcastle, UK*

WOLFRAM BODE • *Max-Planck-Institut für Biochemie, Abteilung Strukturforschung, Planegg-Martinsried, Germany*

LORENA BROWN • *Department of Microbiology and Immunology, University of Melbourne, Parkville, Australia*

MICHAEL J. BUTLER • *Department of Genetics, John Innes Centre, Norwich Research Park, Colney, Norwich, UK*

TIMOTHY E. CAWSTON • *Department of Rheumatology, School of Clinical and Medical Sciences, University of Newcastle, Newcastle, UK*

IAN M. CLARK • *School of Biological Sciences, University of East Anglia, Norwich, UK*

LISA M. COUSSENS • *Hormone Research Institute, University of California, San Francisco, CA*

DEBORAH DAVIS • *Rheumatology Research Unit, Addenbrookes Hospital, Cambridge/Roche Research Centre, Hertfordshire, UK*

ANNE M. DELANY • *Department of Research, St. Francis Hospital and Medical Center, Hartford/Department of Medicine, University of Connecticut School of Medicine, Farmington, CT*

MARIE-PIA D'ORTHO • *INSERM U296, Faculte de Medecine, Creteil, France*

GLENN A. DOYLE • *McArdle Laboratories for Cancer Research, University of Wisconsin-Madison, Madison, WI*

DYLAN R. EDWARDS • *School of Biological Sciences, University of East Anglia, Norwich, UK*

GREGG B. FIELDS • *Department of Chemistry and Biochemistry, Center for Molecular Biology and Biotechnology, Florida Atlantic University, Boca Raton, FL*

PETER A. FORSYTH • *Departments of Clinical Neuroscience and Medicine, Tom Baker Cancer Center, Foothills Hospital/Department of Pediatrics, Alberta Children's Hospital, University of Calgary, Canada*

AMANDA J. FOSANG • *Department of Paediatrics, University of Melbourne, Orthopaedic Research Unit, Royal Children's Hospital, Parkville, Australia*

NOBORU FUJIMOTO • *Biopharmaceutical Department, Fuji Chemical Industries, Toyama, Japan*

SARAH J. GEORGE • *Bristol Heart Institute, Bristol Royal Infirmary, University of Bristol, Bristol, UK*

ALICE L. GIVAN • *Englert Cell Analysis Laboratory of the Norris Cotton Cancer Center, Dartmouth Medical School, Lebanon, NH*

LORI L. GROFT • *Department of Medical Biochemistry, The University of Calgary, Calgary, Alberta, Canada*

SUSAN P. HAWKES • *Departments of Biopharmaceutical Sciences and Pharmaceutical Chemistry, University of California, San Francisco, CA*

TARO HAYAKAWA • *Department of Biochemistry, School of Dentistry, Aichi-Gakuin University, Nagoya, Japan*

ROSALIND M. HEMBRY • *School of Biological Sciences, University of East Anglia, Norwich, UK*

ADRIANO M. HENNEY • *Cardiovascular, Musculoskeletal and Metabolism Research Department, Zeneca Pharmaceuticals, Cheshire, UK*

ANTHONY P. HOLLANDER • *Department of Human Metabolism and Clinical Biochemistry, University of Sheffield Medical School, Sheffield, UK*

TINA L. HURSKAINEN • *Department of Dermatology, University of Oula, Oula, Finland*

MIKE HUTTON • *School of Biological Sciences, University of East Anglia, Norwich, UK*

MIRELA IONESCU • *Joint Diseases Laboratory, Shriners Hospital for Children, Montreal, Quebec, Canada*

KAZUSHI IWATA • *Biopharmaceutical Department, Fuji Chemical Industries, Toyama, Japan*

DAVID C. JACKSON • *Department of Microbiology and Immunology, University of Melbourne, Parkville, Australia*

JASON L. JOHNSON • *Bristol Heart Institute, Bristol Royal Infirmary, University of Bristol, Bristol, UK*

VERA KNÄUPER • *School of Biological Sciences, University of East Anglia, Norwich, UK*

PAUL KOSHY • *Department of Rheumatology, School of Clinical and Medical Sciences, University of Newcastle, Framlington Place, Newcastle, UK*

MARC A. LAFLEUR • *Department of Medical Biochemistry, The University of Calgary, Calgary, Alberta, Canada/School of Biological Sciences, University of East Anglia, Norwich, UK*

KARENA LAST • *Department of Paediatrics, University of Melbourne/Orthopaedic Research Unit, Royal Children's Hospital, Parkville, Australia*

HONGXIA LI • *Departments of Biopharmaceutical Sciences and Pharmaceutical Chemistry, University of California, San Francisco, CA*

KEVIN J. LECO • *Department of Biochemistry and Molecular Biology, The University of Calgary, Alberta, Canada*

CARLOS LÓPEZ-OTÍN • *Departamento de Bioquímica y Biologia Molecular, Facultad de Medicina, Universidad de Oviedo, Oviedo, Spain*

KLAUS MASKOS • *Max-Planck-Institut für Biochemie, Abteilung Strukturforschung, Planegg-Martinsried, Germany*

CHARLES MATOUK • *Departments of Clinical Neuroscience and Medicine, Tom Baker Cancer Center, Foothills Hospital/Department of Pediatrics, Alberta Children's Hospital, and the University of Calgary, Canada*

GILLIAN MURPHY • *School of Biological Sciences, University of East Anglia, Norwich, UK*

HUONG MUZIK • *Department of Medical Biochemistry, The University of Calgary/ Departments of Clinical Neurosciences and Medicine, Tom Baker Cancer Center, Foothills Hospital/Department of Pediatrics, Alberta Children's Hospital, Calgary, Alberta, Canada*

HIDEAKI NAGASE • *The Kennedy Institute of Rheumatology Division, Imperial College School of Medicine, Hammersmith, London, UK*

YASUNORI OKADA • *Department of Pathology, Keio University School of Medicine, Tokyo, Japan*

KENNETH A. ORNDORFF • *Englert Cell Analysis Laboratory of the Norris Cotton Cancer Center, Dartmouth Medical School, Lebanon, NH*

CHRISTOPHER M. OVERALL • *Department of Oral Biology, Faculty of Dentistry, and Biochemistry and Molecular Biology, Faculty of Medicine, University of British Columbia, Vancouver, Canada*

BLAINE W. PHILLIPS • *Department of Biochemistry and Molecular Biology, The University of Calgary, Alberta, Canada*

A. ROBIN POOLE • *Joint Diseases Laboratory, Shriners Hospital for Children, Departments of Surgery and Medicine, McGill University, Montreal, Quebec, Canada*

ANDREW D. ROWAN • *Department of Rheumatology, School of Clinical and Medical Sciences, University of Newcastle, Newcastle, UK*

STEVEN D. SHAPIRO • *Department of Pediatrics, Washington University School of Medicine, St. Louis, MO*

KEN-ICHI SHIMOKAWA • *Department of Biochemistry and Molecular Biology, University of Kansas Medical Center, Kansas City, KS*

GILBERT A. SCHULTZ • *Department of Medical Biochemistry, The University of Calgary, Calgary, Alberta, Canada*

PAUL D. SOLOWAY • *Department of Molecular and Cellular Biology, Roswell Park Cancer Institute, Buffalo, NY*

DARIN L. STEELE • *Department of Biochemistry and Molecular Genetics, University of Alabama, Birmingham, AL*

GARY T. TANIGUCHI • *Departments of Biopharmaceutical Sciences and Pharmaceutical Chemistry, University of California, San Francisco, CA*

RÜDIGER VALLON • *Forschungszentrum Karlsruhe, Institut für Genetik, Karlsruhe, Germany*

GLORIA VELASCO • *Departamento de Bioquímica y Biologia Molecular, Facultad de Medicina, Universidad de Oviedo, Oviedo, Spain*

MATTHEW P. VINCENTI • *Department of Medicine and Biochemistry, Dartmouth Medical School, Hanover, NH*

STEVEN J. WALL • *School of Biological Sciences, University of East Anglia, Norwich, UK*

ZENA WERB • *Department of Anatomy, University of California, San Francisco, CA*

FRANCES WILLENBROCK • *Laboratory of Structural and Mechanistic Enzymology, Department of Biochemistry, Queen Mary and Westfield College, University of London, London, UK*

RICHARD A. WILLIAMSON • *Department of Biosciences, University of Kent, Canterbury, Kent, UK*

L. JACK WINDSOR • *Department of Oral Biology, Indiana University School of Dentistry, Indianapolis, IN*

J. FREDERICK WOESSNER, JR. • *Department of Biochemistry and Molecular Biology, University of Miami, Miami, FL*

HOWARD WONG • *Department of Pathology, Calgary Laboratory Services, Calgary, Alberta, Canada*

KYOKO YAMASHITA • *Department of Biochemistry, School of Dentistry, Aichi-Gakuin University, Nagoya, Japan*

SHU YE • *Wessex Human Genetics Institute, Southampton University School of Medicine, Southampton General Hospital, Southampton, UK*

KAREN M.-L. YEOW • *Department of Biochemistry and Molecular Biology, University of Calgary, Alberta, Canada*

I

MMPs and TIMPs — An Overview of the Field

1

MMPs and TIMPs

An Historical Perspective

J. Frederick Woessner, Jr.

1. Introduction

The matrix metalloproteinase field has a clearly defined starting point, the seminal study of Jerome Gross and Charles Lapière in 1962 *(1)* in which fragments of resorbing tadpole tail were cultured on reconstituted collagen gels. A collagenolytic enzyme was recovered from the culture medium which could attack native collagen fibrils. Within a short time these studies were extended to show a similar activity in a wide variety of tissues, that the collagen molecule was cut into 3/4- and 1/4-length fragments, and that the enzyme activity was dependent on metal ions. The tissue inhibitors of metalloproteinases were discovered in the following decade in tissue culture *(2)* and serum *(3)*. It was soon recognized that the matrix metalloproteinases (MMPs) are involved in a great many processes of normal development and growth and in pathological processes such as arthritis and cancer metastasis as tabulated in Woessner, 1998 *(4)*. This has resulted in a tremendous outpouring of studies now numbering about 10,000 references with 1,000 appearing each year. There have been many recent reviews of the MMPs, including an entire book—Matrix Metalloproteinases—edited by Parks and Mecham *(5)* (1998) and detailed chapters in The Handbook of Proteolytic Enzymes *(6)* and Methods of Enzymology Vol. 248 *(7)*. These will be cited at appropriate points below.

There are now at least 18 distinct vertebrate MMPs. This group was originally defined on the basis of dependence on zinc for catalytic activity and the presence of a large propeptide responsible for latency. The recognition of other enzymes with such properties (e.g., reprolysins) has led to the definition of the group on the basis of DNA sequence. As outlined in the Handbook

From: *Methods in Molecular Biology, vol. 151: Matrix Metalloproteinase Protocols*
Edited by: I. Clark © Humana Press Inc., Totowa, NJ

of Proteolytic Enzymes *(6)* pp. 1144 – 1147, the matrixin subfamily and the serralysin subfamily are grouped together in family M10 of the metalloproteinase clan MB. The closely related family M12 contains the astacins and the reprolysins. The close relationship is not in the sequence but in the protein fold—the so-called metzincin fold *(8)*. Of these 4 subgroups, only the MMP subfamily is inhibited by tissue inhibitor of metalloproteinases (TIMPs), whereas all members are susceptible to hydroxamate inhibitors originally developed for the MMPs.

Table 1 lists the currently known MMPs and related forms from lower organisms. The MMPs have a very wide range of domains and segments that can be added to the basic core found in matrilysin: the propeptide and catalytic domains. These may be selected from the furin cleavage site RXKR, the fibronectin-like repeats, the hinge region, the hemopexin domain and the transmembrane domain. Chapter 4 provides more detail about the size and functions of the various domains and inserts. One consequence of these variations is that the MMPs may range in Mr from 105,000 for the largest proenzyme to 19,000 for the smallest active enzymes. There are three web sites that provide useful data about the MMPs. Neil Rawlings and Alan Barrett maintain a site named MEROPS that gives detailed information about each proteolytic enzyme; it is found at http://www.bi.bbsrc.ac.uk/MEROPS/. Those enzymes that are listed in the Enzyme Commission list (*see* **Table 1**) can be found at http://alpha.qmw.ac.uk/~ugca000/iubmb/enzyme/. Finally, each MMP that has been sequenced can be found by name and animal species in the SwissProt protease databank at http://expasy.hcuge.ch/sprot/sprot-top.html.

2. A Brief Historical Overview of the Individual MMPs

2.1. Interstitial Collagenase (MMP-1; Collagenase 1)

As mentioned previously the first discovery of a vertebrate collagenase was that of Gross and Lapière *(1)*. Although the major emphasis of their paper was on the frog enzyme, activity was also shown in tissues of rat, pig, chick, and mouse. The rat and mouse enzymes are likely, in retrospect, to have been MMP-13 whereas the frog enzyme could have been MMP-1 or MMP-18 or a mixture. The frog enzyme was shown to cleave the collagen molecule into a 3/4-length piece from the N-end and a 1/4-length piece from the C-end by the unusual method of reconstituting these fragments into Segment Long Spacing structures that could be viewed in the electron microscope *(9)*. A complete purification of skin MMP-1 was achieved in 1970 *(10)*. Oddly, it was not recognized until 1971 that the collagenases occurred as zymogens *(11)* that could be activated by trypsin *(12)* or organomercurials *(13)*. The first complete sequence, based on cDNA was reported in 1986 *(14)*. Reviews of collagenase 1 are found in *(15–17)*.

Table 1
The Matrix Metalloproteinases

MMP-No.	Name	E.C. number	Other names/notes
MMP-1	Collagenase 1	3.4.24.7	Interstititial collagenase
MMP-2	Gelatinase A	3.4.24.24	72 kDa gel'ase, type IV collagenase
MMP-3	Stromelysin 1	3.4.24.17	Transin, proteoglycanase, CAP
(MMP-4	Procollagen peptidase	—	Discontinued = MMP-3)
(MMP-5	3/4-Collagenase	—	Discontinued = MMP-2)
(MMP-6	Acid metalloprotease	—	Discontinued = MMP-3)
MMP-7	Matrilysin	3.4.24.23	Pump-1
MMP-8	Collagenase 2	3.4.24.34	Neutrophil collagenase
MMP-9	Gelatinase B	3.4.24.35	92 kDa gel'ase, type V collagenase
MMP-10	Stromelysin 2	3.4.24.22	Transin 2
MMP-11	Stromelysin 3		Furin motif
MMP-12	Macrophage elastase	3.4.24.65	Metalloelastase
MMP-13	Collagenase 3		Rodent intersititial collagenase
MMP-14	Membrane type matrix metalloproteinase 1		MT1-MMP, furin motif
MMP-15	Membrane type matrix metalloproteinase 2		MT2-MMP, furin motif
MMP-16	Membrane type matrix metalloproteinase 3		MT3-MMP, furin motif
MMP-17	Membrane type matrix metalloproteinase 4		MT4-MMP, furin motif
MMP-18	Collagenase 4		*Xenopus*
MMP-19	No trivial name		RASI-1
MMP-20	Enamelysin		
XMMP	No trivial name		*Xenopus*, furin motif
MMP-C31	(also H19, Y19)		*Caenorhabditis elegans*
—	Envelysin	3.4.24.12	Sea urchin
—	Soybean MMP		*Glycine max*, no cDNA
—	Fragilysin	3.4.24.74	*Bacteroides fragilis*

2.2. Gelatinase A (MMP-2, Type IV Collagenase)

Sellers et al. in 1978 *(18)* were the first to separate a gelatinase activity from collagenase and stromelysin in culture medium from rabbit bone. A similar enzyme, acting on basement membrane type IV collagen was reported by Liotta et al. the following year *(19)*. Gelatinase was purified from human skin *(20)*, mouse tumor cells *(21)*, rabbit bone *(22)*, and human gingiva *(23)*. The complete sequence of the human enzyme except for the signal peptide was reported by Collier et al. *(24)*. Gelatinase A has a triple repeat of fibronectin type II domains inserted in the catalytic domain; these participate in binding to the gelatin substrates of the enzyme *(25)*. Recent reviews of gelatinase A are found in *(26–28)*.

2.3. Stromelysin 1 (MMP-3)

A neutral proteinase distinct from collagenase was first noted in 1974 by Sapolsky et al. in human cartilage *(29)* and Werb and Reynolds in rabbit bone fibroblast culture *(30)*. The enzyme was metal-dependent and lost about 20,000 Daltons upon activation of the latent form *(31)*. Purification of the rabbit bone enzyme *(32)* permitted its detailed characterization and it was referred to as proteoglycanase. The name 'stromelysin' was introduced by Chin et al. *(33)* but the enzyme has also been named transin *(34)* and collagenase activating protein *(35)*. The first cDNA sequence was obtained for rat transin in 1985 *(34)*. Recent reviews are found in *(36–38)*.

2.4. The Missing Enzymes (MMP-4, -5, and -6)

These three enzymes were given numbers soon after their discovery, but it was subsequently determined that they were identical to other known MMPs. MMP-4 or collagen telopeptidase was described by Scott in 1983 in pig and human gingival extracts *(39,40)*. Nakano et al. *(41)* proposed the name MMP-4 for the human enzyme that removes the C-telopeptides from type I collagen. The subsequent demonstration that stromelysin cleaves the N-telopeptide, and probably the C-telopeptide *(42)*, from type II collagen suggested that MMP-4 was actually MMP-3. It is possible that MMP-13 may have been involved as well.

Overall and Sodek *(43)* described a 3/4-collagenase that acted on fragments of type I collagen produced by MMP-1. However, this was subsequently identified as MMP-2 *(44)*. Sapolsky et al. *(45)* extracted an aggrecan-degrading activity from human cartilage with an acid pH optimum of 5.5. This activity was purified *(46)* and shown to increase in osteoarthritis *(47)*. However, it was subsequently demonstrated *(48,49)* that this acid pH optimum was actually a property of MMP-3.

2.5. Matrilysin (MMP-7)

Matrilysin is the smallest of the MMPs, having only the propeptide and catalytic domains. It was first identified in the postpartum involuting uterus of the rat by Sellers and Woessner in 1980 *(50)*. However, it was not until 1988 that the enzyme was purified and characterized from rat uterus *(51)* and cloned and sequenced from a human library *(52)*. Expression of the human cDNA established that this enzyme, up until then known as putative matrix metalloproteinase or Pump-1, was a protease with properties similar to the rat enzyme *(53)*. The enzyme was also independently discovered as the kidney activator of urokinase *(54)*. Reviews of the enzyme and its properties may be found in *(55–57)*.

2.6. Neutrophil Collagenase (MMP-8, Collagenase 2)

A collagenase activity from human neutrophils was first reported by Lazarus et al. in 1968 *(58,59)*. This could be activated by a factor in synovial fluid *(60)*. The enzyme was located in the specific granules *(61)* and released upon phagocytosis *(62)*. The enzyme was first purified in 1983 *(63)* and shown to be a distinct gene product from MMP-1 by antibody reaction *(64)*. Cloning and sequencing was accomplished in 1990 *(65)*. Review articles are found in *(16,66,67)*. An early and lengthy series of papers in the 1970s purporting to describe neutrophil collagenase were finally shown by Ohlsson to refer to a serine enzyme and not to MMP-8 *(68)*.

2.7. Gelatinase B (MMP-9, Type V Collagenase)

Harris and Krane in 1972 *(69)* detected a gelatinase activity in rheumatoid synovial fluid. In retrospect this is likely to have been MMP-9. Sopata et al. *(70)* described a gelatinase from human polymorphonuclear leukocytes. This was shown to be latent and activated by organomercurials *(71)*. Rabbit macrophages produce a very similar enzyme which is able to digest type V collagen *(72)*. The neutrophil collagenase and gelatinase were resolved in 1980 *(73)*. Purification was achieved in 1983 *(74)* and sequencing of the cDNA in 1989 *(75)*. An interesting phenomenon, still not fully understood, is the binding of TIMP-1 to proMMP-9 to form a complex that further regulates the activation and subsequent activity of the enzyme *(76)*. Human neutrophil MMP-9 commonly occurs as a complex with lipocalin *(77)*. A series of papers concerning a 95 kDa protein in plasma that binds to gelatin culminated in the identification of this protein as MMP-9 *(78)*.

2.8. Stromelysin 2 (MMP-10, Transin 2)

This is the first of the MMPs to be discovered by cloning and sequencing, in this case as rat transin 2 *(79)*, which indicated an enzyme 71% identical to transin 1 (MMP-3) This was soon followed by the cloning of the human enzyme *(52)*. Both the rat and human forms were shown to have proteolytic activity when they were expressed *(80)*; they were inhibited by TIMP and activated latent proMMP-1. Review articles are found in *(36,38,81)*.

2.9. Stromelysin 3 (MMP-11)

The putative protease was first identified as a cDNA clone by Basset et al. in 1990 *(82)*. Several years went by before Murphy et al. *(83)* demonstrated that the expressed protein showed weak proteolytic activity. Pei and Weiss *(84)* first showed respectable catalytic action of the enzyme in digesting α1-proteinase inhibitor; $α_2$-macroglobulin was also attacked. This MMP contains an RKRR sequence at the end of the propeptide and it was shown by Pei *(85)* that there is indeed intracellular activation of the proenzyme by the action of furin. In this respect, stromelysin 3 is quite distinct from the other two stromelysins and the evolutionary tree also emphasizes this distinction *(4)*. The relative paucity of substrates has resulted in the lack of extensive reviews of this enzyme. Basset et al. *(86)* present a good review of the role of MMP-11 in cancer progression.

2.10. Macrophage Elastase (MMP-12)

Mouse macrophages were found to produce both serine and metallo proteases acting on elastin *(87)*. Banda and Werb *(88)* established the metalloprotease nature of the elastase and showed it was not blocked by α1-proteinase inhibitor, which was in fact a substrate. The mouse enzyme was purified in 1981 *(89)* and the studies of its specificity *(90)* were the earliest of any MMP except MMP-1. The mouse enzyme was cloned, sequenced, and assigned to chromosome 9 *(91)* followed in a few years by the human enzyme sequence and chromosome assignment to 11q22.2-22.3 *(92)*. Review articles are found in **refs.** *93,94*.

2.11. Collagenase 3 (MMP-13)

In the first paper of Gross and Lapière *(1)* collagenase was noted in the involuting rat uterus and in mouse and rat bone. It was assumed at that time that all the collagenases were the same and it was not recognized that the rat and mouse lacked the gene for MMP-1. When the rat uterine collagenase was sequenced *(95)* it was noted that the homology to human MMP-1 was unexpectedly remote. But it wasn't until the human MMP-13 was sequenced *(96)* that the explanation emerged; the rat enzyme was very much closer to the human MMP-13 than to MMP-1. There is a long history of the rat/mouse enzyme starting with purification from bone *(97)* and uterus *(98,99)*. The uterine enzyme is one of the earlier MMPs to have been extracted directly from tissue *(100)*. The human enzyme has not yet been characterized in much detail. Reviews may be found in *(16,101)*.

2.12. Membrane Type Matrix Metalloproteinase 1 (MT1-MMP)

Sato et al. in 1984 *(102)* cloned and sequenced a novel MMP that appeared to have a transmembrane domain close to its C-terminus and coming after the hemopexin domain; this enzyme could activate proMMP-2 also bound to the cell surface. Further cloning and sequencing of the human enzyme followed shortly in 3 labs *(103–105)*. The early reports used the nomenclature MT-MMP-1, however, this proved very confusing as many thought it meant MMP-1 (i.e., collagenase) inserted in the membrane. Reviews of this enzyme are found in *(106,107)*.

2.13. Membrane Type Matrix Metalloproteinases 2, 3, and 4 (MT2-MMP, MT3-MMP, MT4-MMP)

In view of the recent discovery of these enzymes, there is relatively little literature at the moment. There is a report of the cloning and sequencing of MT2-MMP by Takino et al. in 1995 *(108)*. However, they were unaware of the cloning of a second membrane type enzyme by Will and Hinzmann *(109)*. Therefore, the Will enzyme should be designated MT2 and the Takino enzyme is MT3. Proteolytic activity was demonstrated for expressed MT2-MMP *(110)*. MT3-MMP is subject to alternate splicing, yielding two sizes of enzyme *(111)*. MT4-MMP was sequenced by Puente et al. *(112)*. Reviews of these additional MT-MMPs are found as part of the articles cited for MT1-MMP.

2.14. MMP-18 and Higher

There is a single reference to MMP-18 or collagenase 4 *(113)*. The cDNA was isolated from a Xenopus library and represents the first amphibian MMP actually sequenced. The expressed protein is a collagenase. A brief review is presented in *(114)*.

The sequence of MMP-19 was first presented in 1996 *(115)*; the authors were unaware of the Xenopus enzyme and so used the name MMP-18 in their paper. However, it is more appropriately designated MMP-19; there is no trivial name assigned to this enzyme yet. Two further sequences were reported the following year *(116,117)*, the latter group worked with a rheumatoid arthritis synovial inflammation and so used the designation RASI-1 for their enzyme. The protein has been identified in tissues *(118)* but there is only sparse information on the enzyme activity *(116)*. The enzyme is similar to the other 55 kDa MMPs, but has a long hinge region between the catalytic and hemopexin domains.

Enamelysin (MMP-20) was noted as a metal-dependent activity in the developing enamel of pig and rat *(119,120)* and cow *(121)*. The pig enzyme was cloned and sequenced in 1996 *(122)* and the human, the following year *(123)*.

Finally, mention may be made of XMMP, a novel enzyme from Xenopus which has a long insert between the propeptide and the catalytic domain, including an RRKR motif indicative of furin processing *(124)*.

2.15. MMPs from Invertebrates, Plants, and Bacteria

The sea urchin contains a metalloproteinase active in the hatching process, this is now known by the name envelysin. Although a protease had been implicated in hatching for about 50 years, it was only in 1989 that a metalloprotease of about 54 kDa was purified *(125)*. The sequence followed in the next year *(126)* and the enzyme was found to be homologous to MMP-1. The propeptide was somewhat long at 148 amino acids and the catalytic plus hemopexin domain contained a more typical 421 residues. A second species was sequenced by Nomura et al. *(127)* and they introduced the current name. A review is available in *(128)*.

The nematode *Caenorhabditis elegans* has a number of sequences that appear to correspond to MMPs. However, there is only a single recent report on the nature of these enzymes, designated MMP-C31, MMP-H19, and MMP-Y19 *(129)*. The last two have the RXKR motif suggesting furin processing. The first two are inhibited by TIMP-1.

Soybean leaves were found to contain a metalloproteinase *(130)* that was purified in 1991 *(131)*. A sequence of the active form, obtained by classical amino acid sequencing, revealed a 20 kDa active domain related to MMP-1 *(132)*. It is inhibited by TIMP and hydroxamate compounds. A recent update is found in *(133)*.

The intestinal bacterium *Bacteroides fragilis* produces an enterotoxin that is important in human and animal disease. It was purified by van Tassell *(134)*, sequenced by Moncrief et al. *(135)* and shown to be a metalloproteinase related to MMP-1. The expressed protein was capable of causing disease in lamb, rat, and rabbit *(136)*. A recent review is found in *(137)*.

3. Tissue Inhibitors of Metalloproteinases (TIMPs)

The release of proteolytic enzymes into the extracellular matrix outside the cell poses certain risks of untrammeled breakdown. A number of mechanisms have evolved to help avoid uncontrolled proteolysis. First, the cells produce most MMPs only after receiving an appropriate signal requesting *de novo* synthesis. Second, the enzymes are in a proform that requires activation, presumably by controlled processes such as urokinase activation of plasminogen that in turn activates MMPs or by cell-surface MT-MMPs. Finally, there are general inhibitors in the matrix such as α_2-macroglobulin family members and the more specific TIMPs. There are currently 4 known TIMPs, although the reasons for this multiplicity are not yet clear. In addition to maintaining control over the activity of MMPs, TIMP-1 and -2 are able to bind directly to the hemopexin domain of MMP-9 and MMP-2, respectively, exerting further control over the activation process. There are also other roles of TIMPs that do not seem to be directly attributable to protease inhibition such as growth factor activity, steroido-genesis, and cell morphology modulation.

3.1. Tissue Inhibitor of Metalloproteinases 1 (TIMP-1)

It was first noted by Bauer et al. in 1975 (*2*) that cultured human fibroblasts produced an inhibitory protein. In the same year Woolley (*3*) noted a serum inhibitor dubbed β_1-anticollagenase. Cartilage was shown to have an extract-able inhibitor (*138*). A complete purification was achieved in 1979 (*139*) from skin fibroblasts, followed later by protein from rabbit bone (*140*) and cow aorta (*141*). An important feature of the inhibitor is its resistance to high tempera-ture, but sensitivity to reduction and alkylation (*141*). A sequence was obtained in 1985 (*142*). This provided the surprise that the protein was already known in a completely different context.

Work by Gauwerky et al. in 1980 (*143*) showed that human T-lymphoblasts produce a factor that potentiates erythroid factor. This erythroid potentiating factor was cloned and sequenced by Gasson et al. (*144*) and it is that sequence that TIMP was subsequently found to match (*142*). But TIMP has been redis-covered on a number of other occasions: fibroblasts were shown to express a gene upon stimulation by serum, the gene sequence matched EPF/TIMP (*145*). Phorbol stimulation of mouse cells produced a tumor promoting factor TPA-S1 which also had the sequence of TIMP (*146*); this appears in a number of literature reports under the name phorbin. Newcastle disease in mice induces a gene that matches TIMP (*147*) and bovine cumulus cells produce embryo-genin-1 which is also TIMP (*148*). Sertoli cells produce a stimulator of ste-roidogenesis that is a complex of TIMP-1 and cathepsin L (*149*). Some of these additional growth factor effects are reviewed in (*150,151*). Finally, TIMP-1 is able to bind to the hemopexin domain of proMMP-9 (*152*); here it is not react-ing with the active center, which is concealed by the propeptide, and so it is believed to have an important regulatory role.

3.2. Tissue Inhibitor of Metalloproteinases 2 (TIMP-2)

The second form of TIMP was first detected during the development of the technique of reverse zymography (*153*). Sheep cervix was noted to contain both a 28 kDa and a 20 kDa form of TIMP (*154*). De Clerck et al. (*155*) purified both forms from bovine endothelial cell culture and showed by sequence of the N-ter-minus that there was 51% identity. A partial sequence was also obtained by Goldberg et al. (*152*) who also observed the complex between TIMP-2 and proMMP-2. This complex is believed to have a regulatory role, particularly in the activation of proMMP-2 by cell-surface MMP-14 (*104*). A complete sequence was obtained by Edman degradation (*156*) followed shortly by cDNA sequencing *(157)*. As with TIMP-1, there have been several re-discoveries based on growth effects. Thus, a mouse factor that promoted survival of pituitary cells in culture proved to have the sequence of TIMP-2 *(158)* as did an autocrine factor that stimulated SV-40 transformed human fibroblasts *(159)*. A factor stimulating osteosarcoma cells also proved to be TIMP-2 *(160)*. As in the case of TIMP-1, it is unlikely that these effects can all be explained on the basis of blocking proteolysis. A review of TIMP-1 and TIMP-2, particularly with respect to their assay, is presented in *(161)*.

3.3. Tissue Inhibitor of Metalloproteinase 3 (TIMP-3)

Blenis and Hawkes in 1983 *(162)* found expression of a protein in the extracellular matrix produced by transformed chick embryo fibroblast cells. This protein of Mr 21,000 binds tightly to the matrix *(163)* but was not identified as a TIMP until 7 years later *(164)*. The name assigned to the protein based on cDNA sequence was ChiMP-3 *(165)*. The human and mouse TIMP-3 clones were sequenced by Apte and colleagues *(166,167)*. Considerable interest has been aroused by the discovery that Sorsby's fundus dystrophy involves mutations in this protein *(168)*.

3.4. Tissue Inhibitor of Metalloproteinases 4 (TIMP-4)

This inhibitor was first identified upon cloning *(169)*, mRNA was seen in heart but was low in kidney, placenta, colon, and testes and absent in many other tissues. TIMP-4 is similar to TIMP-2 in its ability to bind to proMMP-2 *(170)*. The mouse and rat proteins have also been sequenced *(171,172)*. Expressed TIMP-4 has been shown to inhibit at least MMP-1, -2, -3, -7, and -9. In general, most TIMPs appear to be able to inhibit most active MMPs, but there are considerable differences in binding affinities. An overview of homologies among 17 species of the 4 TIMPs is presented in *(173)*.

4. Conclusion

MMPs continue to be discovered. Highly specialized MMPs such as enamelysin, found only in the embryonic tooth enamel may have corresponding undiscovered counterparts in other tissues or organs. The human genome project will no doubt uncover further examples. In the case of TIMPs, there is an example from the *C. elegans* sequence of a TIMP that seems to consist solely of the N-terminal inhibitory domain. Similar inhibitors might exist in higher organisms. There have been several reports of small IMPs, below the size of TIMP, seen on zymograms *(174)*.

A great deal still remains to be learned about the MMPs and their inhibition. We know almost nothing about the natural substrates of the MMPs in vivo, about how the cell senses and regulates the level of enzymes and inhibitors out in the matrix, and about the ability of the cell to substitute one enzyme for another in knockouts. The clear involvement of the MMPs in the major disease processes of cancer, cardiovascular disease, and arthritis suggests that there will be significant funding for research on these enzymes. Attention will no doubt focus on ways to regulate the activities through gene manipulation and through specific inhibitors of enzyme expression or activity.

Acknowledgment

Supported by NIH Grant AR-16940.

References

1. Gross, J. and Lapière, C. M. (1962) Collagenolytic activity in amphibian tissues: a tissue culture assay. *Proc. Natl. Acad. Sci. USA* **48**, 1014–1022.
2. Bauer, E. A., Stricklin, G. P., Jeffrey, J. J., and Eisen, A. Z. (1975) Collagenase production by human skin fibroblasts. *Biochem. Biophys. Res. Commun.* **64**, 232–240.
3. Woolley, D. E., Roberts, D. R., and Evanson, J. M. (1975) Inhibition of human collagenase activity by a small molecular weight serum protein. *Biochem. Biophys. Res. Commun.* **66**, 747–754.
4. Woessner, J. F., Jr. (1998) The matrix metalloproteinase family, in *Matrix Metalloproteinases* (Parks, W. C. and Mecham, R. P., eds.) Academic Press, San Diego. 1–14.
5. Parks, W. C. and Mecham, R. P., eds. (1998) *Matrix Metalloproteinases*, Academic Press, San Diego.
6. Barrett, A. J., Rawlings, N. D., and Woessner, J. F., eds. (1998) *Handbook of Proteolytic Enzymes*, Academic Press, London.
7. Barrett, A. J., ed. (1995) *Methods in Enzymology. Vol. 248. Proteolytic Enzymes: Aspartic and Metallo Peptidases*, Academic Press, San Diego.
8. Bode, W., Gomis-Rüth, F. X., and Stöcker, W. (1993) Astacins, serralysins, snake venom and matrix metalloproteinases exhibit identical zinc-binding environments (HEXXHXXGXXH and Met- turn) and topologies and should be grouped into a common family, the 'metzincins'. *FEBS Lett.* **331**, 134–140.
9. Gross, J. and Nagai, Y. (1965) Specific degradation of the collagen molecule by tadpole collagenolytic enzyme. *Proc. Natl. Acad. Sci. USA* **54**, 1197–1204.
10. Bauer, E. A., Eisen, A. Z., and Jeffrey, J. J. (1970) Immunologic relationship of a purified human skin collagenase to other human and animal collagenases. *Biochim. Biophys. Acta* **206**, 152–160.
11. Harper, E., Bloch, K. J., and Gross, J. (1971) The zymogen of tadpole collagenase. *Biochemistry* **10**, 3035–3041.
12. Birkedal-Hansen, H., Cobb, C. M., Taylor, R. E., and Fullmer, H. M. (1975) Activation of latent bovine gingival collagenase. *Arch. Oral Biol.* **20**, 681–685.
13. Werb, Z. and Burleigh, M. C. (1974) A specific collagenase from rabbit fibroblasts in monolayer culture. *Biochem. J.* **137**, 373–385.

14. Goldberg, G. I., Wilhelm, S. M., Kronberger, A., Bauer, E. A., Grant, G. A., and Eisen, A. Z. (1986) Human fibroblast collagenase. Complete primary structure and homology to an oncogene transformation-induced rat protein. *J. Biol. Chem.* **261**, 6600–6605.

15. Dioszegi, M., Cannon, P., and Van Wart, H. E. (1995) Vertebrate collagenases. *Methods Enzymol.* **248**, 413–431.

16. Cawston, T. E. (1998) Interstitial collagenase, in *Handbook of Proteolytic Enzymes* (Barrett, A. J., Rawlings, N. D., and Woessner, J. F., eds.) Academic Press, London, pp. 1155–1162.

17. Jeffrey, J. J. (1998) Interstitial collagenases, in *Matrix Metalloproteinases* (Parks, W. C. and Mecham, R. P., eds.) Academic Press, San Diego, pp. 15–42.

18. Sellers, A., Reynolds, J. J., and Meikle, M. C. (1978) Neutral metallo-proteinases of rabbit bone. Separation in latent forms of distinct enzymes that when activated degrade collagen, gelatin and proteoglycans. *Biochem. J.* **171**, 493–496.

19. Liotta, L. A., Abe, S., Robey, P. G., and Martin, G. R. (1979) Preferential digestion of basement membrane collagen by an enzyme derived from a metastatic murine tumor. *Proc. Natl. Acad. Sci. USA* **76**, 2268–2272.

20. Seltzer, J. L., Eschbach, M. L., and Eisen, A. Z. (1985) Purification of gelatin-specific neutral protease from human skin by conventional and high-performance liquid chromatography. *J. Chromatogr.* **326**, 147–155.

21. Salo, T., Liotta, L. A., and Tryggvason, K. (1983) Purification and characterization of a murine basement membrane collagen-degrading enzyme secreted by metastatic tumor cells. *J. Biol. Chem.* **258**, 3058–3063.

22. Murphy, G., McAlpine, C. G., Poll, C. T., and Reynolds, J. J. (1985) Purification and characterization of a bone metalloproteinase that degrades gelatin and types IV and V collagen. *Biochim. Biophys. Acta* **831**, 49–58.

23. Nakano, T. and Scott, P. G. (1986) Purification and characterization of a gelatinase produced by fibroblasts from human gingiva. *Biochem. Cell. Biol.* **64**, 387–393.

24. Collier, I. E., Wilhelm, S. M., Eisen, A. Z., Marmer, B. L., Grant, G. A., Seltzer, J. L., Kronberger, A., He, C. S., Bauer, E. A., and Goldberg, G. I. (1988) H-ras oncogene-transformed human bronchial epithelial cells (TBE- 1) secrete a single metalloprotease capable of degrading basement membrane collagen. *J. Biol. Chem.* **263**, 6579–6587.

25. Collier, I. E., Krasnov, P. A., Strongin, A.Y., Birkedal-Hansen, H., and Goldberg, G. I. (1992) Alanine scanning mutagenesis and functional analysis of the fibronectin-like collagen-binding domain from human 92-kDa type IV collagenase. *J. Biol. Chem.* **267**, 6776–6781.

26. Murphy, G. and Crabbe, T. (1995). Gelatinases A and B. *Methods Enzymol.* **248**, 470–484.

27. Murphy, G. (1998) Gelatinase A, in *Handbook of Proteolytic Enzymes* (Barrett, A. J., Rawlings, N. D., and Woessner, J. F., eds.) Academic Press, London, pp. 1199–1205.

28. Yu, A. E., Murphy, A. N., and Stetler-Stevenson, W. G. (1998) 72-kDa gelatinase (gelatinase A): structure, activation, regulation and substrate specificity, in *Matrix Metalloproteinases* (Parks, W. C. and Mecham, R. P., eds.) Academic Press, San Diego, pp. 85–113.

29. Sapolsky, A. I., Howell, D. S., and Woessner, J. F., Jr. (1974) Neutral proteases and cathepsin D in human articular cartilage. *J. Clin. Invest.* **53**, 1044–1053.

30. Werb, Z. and Reynolds, J. J. (1974) Stimulation by endocytosis of the secretion of collagenase and neutral proteinase from rabbit synovial fibroblasts. *J. Exp. Med.* **140**, 1482–1497.

31. Werb, Z., Dingle, J. T., Reynolds, J. J., and Barrett, A. J. (1978) Proteoglycan-degrading enzymes of rabbit fibroblasts and granulocytes. *Biochem. J.* **173**, 949–958.

32. Galloway, W. A., Murphy, G., Sandy, J. D., Gavrilovic, J., Cawston, T. E., and Reynolds, J. J. (1983) Purification and characterization of a rabbit bone metalloproteinase that degrades proteoglycan and other connective- tissue components. *Biochem. J.* **209**, 741–752.

33. Chin, J. R., Murphy, G., and Werb, Z. (1985) Stromelysin, a connective tissue-degrading metalloendopeptidase secreted by stimulated rabbit synovial fibroblasts in parallel with collagenase. Biosynthesis, isolation, characterization, and substrates. *J. Biol. Chem.* **260**, 12,367–12,376.

34. Matrisian, L. M., Glaichenhaus, N., Gesnel, M. C., and Breathnach, R. (1985) Epidermal growth factor and oncogenes induce transcription of the same cellular mRNA in rat fibroblasts. *EMBO J.* **4**, 1435–1440.

35. Treadwell, B. V., Neidel, J., Pavia, M., Towle, C. A., Trice, M. E., and Mankin, H. J. (1986) Purification and characterization of collagenase activator protein synthesized by articular cartilage. *Arch. Biochem. Biophys.* **251**, 715–723.

36. Nagase, H. (1995) Human stromelysins 1 and 2. *Methods Enzymol.* **248**, 449–470.

37. Nagase, H. (1998) Stromelysin 1, in *Handbook of Proteolytic Enzymes* (Barrett, A. J., Rawlings, N. D., and Woessner, J. F., eds.) Academic Press, London, pp. 1172–1178.

38. Nagase, H. (1998) Stromelysins 1 and 2, in *Matrix Metalloproteinases* (Parks, W. C. and Mecham, R. P., eds.) Academic Press, San Diego, pp. 43–84.

39. Scott, P. G. and Goldberg, H. A. (1983) Cleavage of the carboxy-terminal cross-linking region of type I collagen by proteolytic activity from cultured porcine gingival explants. *Coll. Relat. Res.* **3**, 295–304.

40. Scott, P. G., Goldberg, H. A., and Dodd, C. M. (1983) A neutral proteinase from human gingival fibroblasts active against the C-terminal cross-linking region of type I collagen. *Biochem. Biophys. Res. Commun.* **114**, 1064–1070.

41. Nakano, T. and Scott, P. G. (1987) Partial purification and characterization of a neutral proteinase with collagen telopeptidase activity produced by human gingival fibroblasts. *Biochem. Cell. Biol.* **65**, 286–292.

42. Wu, J. J., Lark, M. W., Chun, L. E., and Eyre, D. R. (1991) Sites of stromelysin cleavage in collagen types II, IX, X, and XI of cartilage. *J. Biol. Chem.* **266**, 5625–5628.

43. Overall, C. M. and Sodek, J. (1987) Initial characterization of a neutral metalloproteinase, active on native 3/4-collagen fragments, synthesized by ROS 17/2.8 osteoblastic cells, periodontal fibroblasts, and identified in gingival crevicular fluid. *J. Dent. Res.* **66**, 1271–1282.

44. Nagase, H., Barrett, A. J., and Woessner, J. F., Jr. (1992) Nomenclature and glossary of the matrix metalloproteinases. *Matrix Suppl.* **1**, 421–424.

45. Sapolsky, A. I., Keiser, H., Howell, D. S., and Woessner, J. F., Jr. (1976) Metalloproteases of human articular cartilage that digest cartilage proteoglycan at neutral and acid pH. *J. Clin. Invest.* **58**, 1030–1041.

46. Azzo, W. and Woessner, J. F., Jr. (1986) Purification and characterization of an acid metalloproteinase from human articular cartilage. *J. Biol. Chem.* **261**, 5434–5441.

47. Dean, D. D., Martel-Pelletier, J., Pelletier, J. P., Howell, D. S., and Woessner, J. F., Jr. (1989) Evidence for metalloproteinase and metalloproteinase inhibitor imbalance in human osteoarthritic cartilage. *J. Clin. Invest.* **84**, 678–685.

48. Gunja-Smith, Z., Nagase, H., and Woessner, J. F., Jr. (1989) Purification of the neutral proteoglycan-degrading metalloproteinase from human articular cartilage tissue and its identification as stromelysin matrix metalloproteinase-3. *Biochem. J.* **258**, 115–119.

49. Wilhelm, S. M., Shao, Z. H., Housley, T. J., Seperack, P. K., Baumann, A. P., Gunja-Smith, Z., and Woessner, J. F., Jr. (1993) Matrix metalloproteinase-3 (stromelysin-1) Identification as the cartilage acid metalloprotease and effect of pH on catalytic properties and calcium affinity. *J. Biol. Chem.* **268**, 21,906–21,913.

50. Sellers, A. and Woessner, J. F., Jr. (1980) The extraction of a neutral metalloproteinase from the involuting rat uterus, and its action on cartilage proteoglycan. *Biochem. J.* **189**, 521–531.

51. Woessner, J. F., Jr. and Taplin, C. J. (1988) Purification and properties of a small latent matrix metalloproteinase of the rat uterus. *J. Biol. Chem.* **263**, 16,918–16,925.

52. Muller, D., Quantin, B., Gesnel, M. C., Millon-Collard, R., Abecassis, J., and Breathnach, R. (1988) The collagenase gene family in humans consists of at least four members. *Biochem. J.* **253**, 187–192.

53. Quantin, B., Murphy, G., and Breathnach, R. (1989) Pump-1 cDNA codes for a protein with characteristics similar to those of classical collagenase family members. *Biochemistry* **28**, 5327–5334.

54. Marcotte, P. A., Dudlak, D., Leski, M. L., Ryan, J., and Henkin, J. (1992) Characterization of a metalloprotease which cleaves with high site-specificity the Glu(143)-Leu(144) bond of urokinase. *Fibrinolysis* **6(Suppl. 1)**, 57–62.

55. Wilson, C. L. and Matrisian, L. M. (1996) Matrilysin: an epithelial matrix metalloproteinase with potentially novel functions. *Int. J. Biochem. Cell. Biol.* **28**, 123–136.

56. Woessner, J. F., Jr. (1995) Matrilysin. *Methods Enzymol.* **248**, 485–495.

57. Woessner, J. F. (1998) Matrilysin, in *Handbook of Proteolytic Enzymes* (Barrett, A. J., Rawlings, N. D., and Woessner, J. F., eds.) Academic Press, London, pp. 1183–1187.

58. Lazarus, G. S., Brown, R. S., Daniels, J. R., and Fullmer, H. M. (1968) Human granulocyte collagenase. *Science* **159**, 1483–1485.
59. Lazarus, G. S., Daniels, J. R., Brown, R. S., Bladen, H. A., and Fullmer, H. M. (1968) Degradation of collagen by a human granulocyte collagenolytic system. *J. Clin. Invest.* **47**, 2622–2629.
60. Kruze, D. and Wojtecka, E. (1972) Activation of leucocyte collagenase proenzyme by rheumatoid synovial fluid. *Biochim. Biophys. Acta* **285**, 436–446.
61. Murphy, G., Reynolds, J. J., Bretz, U., and Baggiolini, M. (1977) Collagenase is a component of the specific granules of human neutrophil leucocytes. *Biochem. J.* **162**, 195–197.
62. Oronsky, A. L., Perper, R. J., and Schroder, H. C. (1973) Phagocytic release and activation of human leukocyte procollagenase. *Nature* **246**, 417–419.
63. Macartney, H. W. and Tschesche, H. (1983) Latent and active human polymorphonuclear leukocyte collagenases. Isolation, purification and characterisation. *Eur. J. Biochem.* **130**, 71–78.
64. Hasty, K. A., Hibbs, M. S., Kang, A. H., and Mainardi, C. L. (1984) Heterogeneity among human collagenases demonstrated by monoclonal antibody that selectively recognizes and inhibits human neutrophil collagenase. *J. Exp. Med.* **159**, 1455–1463.
65. Hasty, K. A., Pourmotabbed, T. F., Goldberg, G. I., Thompson, J. P., Spinella, D. G., Stevens, R. M., and Mainardi, C. L. (1990) Human neutrophil collagenase. A distinct gene product with homology to other matrix metalloproteinases. *J. Biol. Chem.* **265**, 11,421–11,424.
66. Tschesche, H. (1995) Human neutrophil collagenase. *Methods Enzymol.* **248**, 431–449.
67. Tschesche, H. and Pieper, M. (1998) Neutrophil collagenase, in *Handbook of Proteolytic Enzymes* (Barrett, A. J., Rawlings, N. D., and Woessner, J. F., eds.) Academic Press, London, pp. 1162–1167.
68. Ohlsson, K. (1980) Polymorphonuclear leukocyte collagenase, in *Collagenase in Normal and Pathological Connective Tissues* (Woolley, D. E. and Evanson, J. M., eds.) Wiley, Chichester, England, pp. 209–222.
69. Harris, E. D., Jr. and Krane, S. M. (1972) An endopeptidase from rheumatoid synovial tissue culture. *Biochim. Biophys. Acta* **258**, 566–576.
70. Sopata, I. and Dancewicz, A. M. (1974) Presence of a gelatin-specific proteinase and its latent form in human leucocytes. *Biochim. Biophys. Acta* **370**, 510–523.
71. Sopata, I. and Wize, J. (1979) A latent gelatin specific proteinase of human leucocytes and its activation. *Biochim. Biophys. Acta* **571**, 305–312.
72. Mainardi, C. L., Seyer, J. M., and Kang, A. H. (1980) Type-specific collagenolysis: a type V collagen-degrading enzyme from macrophages. *Biochem. Biophys. Res. Commun.* **97**, 1108–1115.
73. Murphy, G., Bretz, U., Baggiolini, M., and Reynolds, J. J. (1980) The latent collagenase and gelatinase of human polymorphonuclear neutrophil leucocytes. *Biochem. J.* **192**, 517–525.

74. Rantala-Ryhanen, S., Ryhanen, L., Nowak, F. V., and Uitto, J. (1983) Proteinases in human polymorphonuclear leukocytes. Purification and characterization of an enzyme which cleaves denatured collagen and a synthetic peptide with a Gly-Ile sequence. *Eur. J. Biochem.* **134**, 129–137.

75. Wilhelm, S. M., Collier, I. E., Marmer, B. L., Eisen, A. Z., Grant, G. A., and Goldberg, G. I. (1989) SV40-transformed human lung fibroblasts secrete a 92-kDa type IV collagenase which is identical to that secreted by normal human macrophages. *J. Biol. Chem.* **264**, 17,213–17,221.

76. Goldberg, G. I., Strongin, A., Collier, I. E., Genrich, L. T., and Marmer, B. L. (1992) Interaction of 92-kDa type IV collagenase with the tissue inhibitor of metalloproteinases prevents dimerization, complex formation with interstitial collagenase, and activation of the proenzyme with stromelysin. *J. Biol. Chem.* **267**, 4583–4591.

77. Kjeldsen, L., Johnsen, A. H., Sengelov, H., and Borregaard, N. (1993) Isolation and primary structure of NGAL, a novel protein associated with human neutrophil gelatinase. *J. Biol. Chem.* **268**, 10,425–10,432.

78. Vartio, T. (1985) The 95000-Mr gelatin-binding protein in human serum and plasma. *Biochem. J.* **228**, 605–608.

79. Breathnach, R., Matrisian, L. M., Gesnel, M. C., Staub, A., and Leroy, P. (1987) Sequences coding for part of oncogene-induced transin are highly conserved in a related rat gene. *Nucleic Acids Res.* **15**, 1139–1151.

80. Nicholson, R., Murphy, G., and Breathnach, R. (1989) Human and rat malignant-tumor-associated mRNAs encode stromelysin-like metalloproteinases. *Biochemistry* **28**, 5195–5203.

81. Matrisian, L. M. (1998) Stromelysin 2, in *Handbook of Proteolytic Enzymes* (Barrett, A. J., Rawlings, N. D., and Woessner, J. F., eds.) Academic Press, London, pp. 1178–1180.

82. Basset, P., Bellocq, J. P., Wolf, C., Stoll, I., Hutin, P., Limacher, J. M., Podhajcer, O. L., Chenard, M. P., Rio, M. C., and Chambon, P. (1990) A novel metalloproteinase gene specifically expressed in stromal cells of breast carcinomas. *Nature* **348**, 699–704.

83. Murphy, G., Segain, J. P., O'Shea, M., Cockett, M., Ioannou, C., Lefebvre, O., Chambon, P., and Basset, P. (1993) The 28-kDa N-terminal domain of mouse stromelysin-3 has the general properties of a weak metalloproteinase. *J. Biol. Chem.* **268**, 15,435–15,441.

84. Pei, D., Majmudar, G., and Weiss, S. J. (1994) Hydrolytic inactivation of a breast carcinoma cell-derived serpin by human stromelysin-3. *J. Biol. Chem.* **269**, 25,849–25,855.

85. Pei, D. and Weiss, S. J. (1995) Furin-dependent intracellular activation of the human stromelysin-3 zymogen. *Nature* **375**, 244–247.

86. Basset, P., Bellocq, J. P., Lefebvre, O., Noel, A., Chenard, M. P., Wolf, C., Anglard, P., and Rio, M. C. (1997) Stromelysin-3: a paradigm for stroma-derived factors implicated in carcinoma progression. *Crit. Rev. Oncol. Hematol.* **26**, 43–53.

87. Werb, Z. and Gordon, S. (1975) Elastase secretion by stimulated macrophages. Characterization and regulation. *J. Exp. Med.* **142**, 361–377.

88. Banda, M. J., Clark, E. J., and Werb, Z. (1980) Limited proteolysis by macrophage elastase inactivates human alpha 1-proteinase inhibitor. *J. Exp. Med.* **152**, 1563–1570.

89. Banda, M. J. and Werb, Z. (1981) Mouse macrophage elastase. Purification and characterization as a metalloproteinase. *Biochem. J.* **193**, 589–605.

90. Kettner, C., Shaw, E., White, R., and Janoff, A. (1981) The specificity of macrophage elastase on the insulin B-chain. *Biochem. J.* **195**, 369–372.

91. Shapiro, S. D., Griffin, G. L., Gilbert, D. J., Jenkins, N. A., Copeland, N. G., Welgus, H. G., Senior, R. M., and Ley, T. J. (1992) Molecular cloning, chromosomal localization, and bacterial expression of a murine macrophage metalloelastase. *J. Biol. Chem.* **267**, 4664–4671.

92. Belaaouaj, A., Shipley, J. M., Kobayashi, D. K., Zimonjic, D. B., Popescu, N., Silverman, G. A., and Shapiro, S. D. (1995) Human macrophage metalloelastase. Genomic organization, chromosomal location, gene linkage, and tissue-specific expression. *J. Biol. Chem.* **270**, 14,568–14,575.

93. Senior, R. M. and Shapiro, S. D. (1998) Macrophage elastase, in *Handbook of Proteolytic Enzymes* (Barrett, A. J., Rawlings, N. D., and Woessner, J. F., eds.) Academic Press, London, pp. 1180–1183.

94. Shapiro, S. D. and Senior, R. M. (1998) Macrophage elastase (MMP-12), in *Matrix Metalloproteinases* (Parks, W. C. and Mecham, R. P., eds.) Academic Press, San Diego, pp. 185–197.

95. Quinn, C. O., Scott, D. K., Brinckerhoff, C. E., Matrisian, L. M., Jeffrey, J. J., and Partridge, N. C. (1990) Rat collagenase. Cloning, amino acid sequence comparison, and parathyroid hormone regulation in osteoblastic cells. *J. Biol. Chem.* **265**, 22,342–22,347.

96. Freije, J. M. P., Díez-Itza, I., Balbín, M., Sánchez, L. M., Blasco, R., Tolivia, J., and López-Otín, C. (1994) Molecular cloning and expression of collagenase-3, a novel human matrix metalloproteinase produced by breast carcinomas. *J. Biol. Chem.* **269**, 16,766–16,773.

97. Aer, J. (1971) Purification of rat bone collagenolytic enzymes. *Ann. Med. Exp. Biol. Fenn.* **49**, 1–8.

98. Jeffrey, J. J. and Gross, J. (1970) Collagenase from rat uterus. Isolation and partial characterization. *Biochemistry* **9**, 268–273.

99. Roswit, W. T., Halme, J., and Jeffrey, J. J. (1983) Purification and properties of rat uterine procollagenase. *Arch. Biochem. Biophys.* **225**, 285–295.

100. Weeks, J. G., Halme, J., and Woessner, J. F., Jr. (1976) Extraction of collagenase from the involuting rat uterus. *Biochim. Biophys. Acta* **445**, 205–214.

101. Jeffrey, J. J. (1998) Collagenase 3, in *Handbook of Proteolytic Enzymes* (Barrett, A. J., Rawlings, N. D., and Woessner, J. F., eds.) Academic Press, London, pp. 1167–1170.

102. Sato, H., Takino, T., Okada, Y., Cao, J., Shinagawa, A., Yamamoto, E., and Seiki, M. (1994) A matrix metalloproteinase expressed on the surface of invasive tumour cells. *Nature* **370**, 61–65.

103. Okada, A., Bellocq, J. P., Rouyer, N., Chenard, M. P., Rio, M. C., Chambon, P., and Basset, P. (1995) Membrane-type matrix metalloproteinase (MT-MMP) gene is expressed in stromal cells of human colon, breast, and head and neck carcinomas. *Proc. Natl. Acad. Sci. USA* **92**, 2730–2734.

104. Strongin, A. Y., Collier, I., Bannikov, G., Marmer, B. L., Grant, G. A., and Goldberg, G. I. (1995) Mechanism of cell surface activation of 72-kDa type IV collagenase. Isolation of the activated form of the membrane metalloprotease. *J. Biol. Chem.* **270**, 5331–5338.

105. Takino, T., Sato, H., Yamamoto, E., and Seiki, M. (1995) Cloning of a human gene potentially encoding a novel matrix metalloproteinase having a C-terminal transmembrane domain. *Gene* **155**, 293–298.

106. Seiki, M. (1998) Membrane-type matrix metalloproteinase, in *Handbook of Proteolytic Enzymes* (Barrett, A. J., Rawlings, N. D., and Woessner, J. F., eds.) Academic Press, London, pp. 1192–1195.

107. Knäuper, V. and Murphy, G. (1998) *Matrix Metalloproteinases* (Parks, W. C. and Mecham, R. P., eds.) Academic Press, San Diego, p. 199.

108. Takino, T., Sato, H., Shinagawa, A., and Seiki, M. (1995) Identification of the second membrane-type matrix metalloproteinase (MT-MMP-2) gene from a human placenta cDNA library. MT-MMPs form a unique membrane-type subclass in the MMP family. *J. Biol. Chem.* **270**, 23,013–23,020.

109. Will, H. and Hinzmann, B. (1995) cDNA sequence and mRNA tissue distribution of a novel human matrix metalloproteinase with a potential transmembrane segment. *Eur. J. Biochem.* **231**, 602–608.

110. d'Ortho, M. P., Will, H., Atkinson, S., Butler, G., Messent, A., Gavrilovic, J., Smith, B., Timpl, R., Zardi, L., and Murphy, G. (1997) Membrane-type matrix metalloproteinases 1 and 2 exhibit broad- spectrum proteolytic capacities comparable to many matrix metalloproteinases. *Eur. J. Biochem.* **250**, 751–757.

111. Matsumoto, S. I., Katoh, M., Saito, S., Watanabe, T., and Masuho, Y. (1997) Identification of soluble type of membrane-type matrix metalloproteinase-3 formed by alternatively spliced mRNA. *Biochim. Biophys. Acta. Gene Struct. Express.* **1354**, 159–170.

112. Puente, X. S., Pendás, A. M., Llano, E., Velasco, G., and López-Otín, C. (1996) Molecular cloning of a novel membrane-type matrix metalloproteinase from a human breast carcinoma. *Cancer Res.* **56**, 944–949.

113. Stolow, M. A., Bauzon, D. D., Li, J., Sedgwick, T., Liang, V. C., Sang, Q. A., and Shi, Y. B. (1996) Identification and characterization of a novel collagenase in *Xenopus laevis*: possible roles during frog development. *Mol. Biol. Cell.* **7**, 1471–1483.

114. Shi, Y.-B. and Sang, Q.-X. A. (1998) Collagenase 4, in *Handbook of Proteolytic Enzymes* (Barrett, A. J., Rawlings, N. D., and Woessner, J. F., eds.) Academic Press, London, pp. 1170–1172.

115. Cossins, J., Dudgeon, T. J., Catlin, G., Gearing, A. J., and Clements, J. M. (1996) Identification of MMP-18, a putative novel human matrix metalloproteinase. *Biochem. Biophys. Res. Commun.* **228**, 494–498.

116. Pendás, A. M., Knäuper, V., Puente, X. S., Llano, E., Mattei, M. G., Apte, S., Murphy, G., and López-Otín, C. (1997) Identification and characterization of a novel human matrix metalloproteinase with unique structural characteristics, chromosomal location, and tissue distribution. *J. Biol. Chem.* **272**, 4281–4286.

117. Kolb, C., Mauch, S., Peter, H. H., Krawinkel, U., and Sedlacek, R. (1997) The matrix metalloproteinase RASI-1 is expressed in synovial blood vessels of a rheumatoid arthritis patient. *Immunol. Lett.* **57**, 83–88.

118. Sedlacek, R., Mauch, S., Kolb, B., Schätzlein, C., Eibel, H., Peter, H.-H., Schmitt, J., and Krawinkel, U. (1998) Matrix metalloproteinase MMP-19 (RASI 1) is expressed on the surface of activated peripheral blood mononuclear cells and is detected as an autoantigen in rheumatoid arthritis. *Immunobiology* **198**, 408–423.

119. DenBesten, P. K. and Heffernan, L. M. (1989) Separation by polyacrylamide gel electrophoresis of multiple proteases in rat and bovine enamel. *Arch. Oral Biol.* **34**, 399–404.

120. Tanabe, T., Fukae, M., Uchida, T., and Shimizu, M. (1992) The localization and characterization of proteinases for the initial cleavage of porcine amelogenin. *Calcif. Tissue Int.* **51**, 213–217.

121. Moradian-Oldak, J., Simmer, J. P., Sarte, P. E., Zeichner-David, M., and Fincham, A. G. (1994) Specific cleavage of a recombinant murine amelogenin at the carboxy-terminal region by a proteinase fraction isolated from developing bovine tooth enamel. *Arch. Oral Biol.* **39**, 647–656.

122. Bartlett, J. D., Simmer, J. P., Xue, J., Margolis, H. C., and Moreno, E. C. (1996) Molecular cloning and mRNA tissue distribution of a novel matrix metalloproteinase isolated from porcine enamel organ. *Gene* **183**, 123–128.

123. Llano, E., Pendás, A. M., Knäuper, V., Sorsa, T., Salo, T., Salido, E., Murphy, G., Simmer, J. P., Bartlett, J. D., and López-Otín, C. (1997) Identification and structural and functional characterization of human enamelysin (MMP-20) *Biochemistry* **36**, 15,101–15,108.

124. Yang, M. Z., Murray, M. T., and Kurkinen, M. (1997) A novel matrix metalloproteinase gene (XMMP) encoding vitronectin-like motifs is transiently expressed in *Xenopus laevis* early embryo development. *J. Biol. Chem.* **272**, 13,527–13,533.

125. Lepage, T. and Gache, C. (1989) Purification and characterization of the sea urchin embryo hatching enzyme. *J. Biol. Chem.* **264**, 4787–4793.

126. Lepage, T. and Gache, C. (1990) Early expression of a collagenase-like hatching enzyme gene in the sea urchin embryo. *EMBO J.* **9**, 3003–3012.

127. Nomura, K., Shimizu, T., Kinoh, H., Sendai, Y., Inomata, M., and Suzuki, N. (1997) Sea urchin hatching enzyme (envelysin) - cDNA cloning and deprivation of protein substrate specificity by autolytic degradation. *Biochemistry* **36**, 7225–7238.

128. Gache, C., Lepage, T., Ghiglione, C., Emily-Fenouil, F., and Lhomond, G. (1998) Envelysin, in *Handbook of Proteolytic Enzymes* (Barrett, A. J., Rawlings, N. D., and Woessner, J. F., eds.) Academic Press, London, pp. 1195–1199.

129. Wada, K., Sato, H., Kinoh, H., Kajita, M., Yamamoto, H., and Seiki, M. (1998) Cloning of three *Caenorhabditis elegans* genes potentially encoding novel matrix metalloproteinases. *Gene* **211**, 57–62.

130. Ragster, L. and Chrispeels, M. J. (1979) Azocoll-digesting proteinases in soybean leaves: characteristics and changes during leaf maturation. *Plant Physiol.* **64**, 857–862.

131. Graham, J. S., Xiong, J., and Gillikin, J. W. (1991) Purification and developmental analysis of a metalloendoproteinase from the leaves of *Glycine max*. *Plant Physiol.* **97**, 786–792.

132. McGeehan, G., Burkhart, W., Anderegg, R., Becherer, J. D., Gillikin, J. W., and Graham, J. S. (1992) Sequencing and characterization of the soybean leaf metalloproteinase. Structural and functional similarity to the matrix metalloproteinase family. *Plant Physiol.* **99**, 1179–1183.

133. McGeehan, G. and Graham, J. S. (1998) Soybean metalloproteinase, in *Handbook of Proteolytic Enzymes* (Barrett, A. J., Rawlings, N. D., and Woessner, J. F., eds.) Academic Press, London, pp. 1190–1192.

134. van Tassell, R. L., Lyerly, D. M., and Wilkins, T. D. (1992) Purification and characterization of an enterotoxin from *Bacteroides fragilis*. *Infect. Immun.* **60**, 1343–1350.

135. Moncrief, J. S., Obiso, R., Jr., Barroso, L. A., Kling, J. J., Wright, R. L., van Tassell, R. L., Lyerly, D. M., and Wilkins, T. D. (1995) The enterotoxin of *Bacteroides fragilis* is a metalloprotease. *Infect. Immun.* **63**, 175–181.

136. Obiso, R. J., Lyerly, D. M., van Tassell, R. L., and Wilkins, T. D. (1995) Proteolytic activity of the *Bacteroides fragilis* enterotoxin causes fluid secretion and intestinal damage in vivo. *Infect. Immun.* **63**, 3820–3826.

137. Obiso, R. J., Jr. and Wilkins, T. D. (1998) Fragilysin, in *Handbook of Proteolytic Enzymes* (Barrett, A. J., Rawlings, N. D., and Woessner, J. F., eds.) Academic Press, London, pp. 1211–1213.

138. Kuettner, K. E., Hiti, J., Eisenstein, R., and Harper, E. (1976) Collagenase inhibition by cationic proteins derived from cartilage and aorta. *Biochem. Biophys. Res. Commun.* **72**, 40–46.

139. Welgus, H. G., Stricklin, G. P., Eisen, A. Z., Bauer, E. A., Cooney, R. V., and Jeffrey, J. J. (1979) A specific inhibitor of vertebrate collagenase produced by human skin fibroblasts. *J. Biol. Chem.* **254**, 1938–1943.

140. Cawston, T. E., Galloway, W. A., Mercer, E., Murphy, G., and Reynolds, J. J. (1981) Purification of rabbit bone inhibitor of collagenase. *Biochem. J.* **195**, 159–165.

141. Nolan, J. C., Ridge, S. C., Oronsky, A. L., and Kerwar, S. S. (1980) Purification and properties of a collagenase inhibitor from cultures of bovine aorta. *Atherosclerosis* **35**, 93–102.

142. Docherty, A. J. P., Lyons, A., Smith, B. J., Wright, E. M., Stephens, P. E., Harris, T. J. R., Murphy, G., and Reynolds, J. J. (1985) Sequence of human tissue inhibitor of metalloproteinases and its identity to erythroid-potentiating activity. *Nature* **318**, 66–69.

143. Gauwerky, C. E., Lusis, A. J., and Golde, D. W. (1980) Erythroid-potentiating activity: characterization and target cells. *Exp. Hematol.* **8**, 117–127.

144. Gasson, J. C., Golde, D. W., Kaufman, S. E., Westbrook, C. A., Hewick, R. M., Kaufman, R. J., Wong, G. G., Temple, P. A., Leary, A. C., Brown, E. L., et al. (1985) Molecular characterization and expression of the gene encoding human erythroid-potentiating activity. *Nature* **315**, 768–771.

145. Edwards, D. R., Parfett, C. L., and Denhardt, D. T. (1985) Transcriptional regulation of two serum-induced RNAs in mouse fibroblasts: equivalence of one species to B2 repetitive elements. *Mol. Cell. Biol.* **5**, 3280–3288.

146. Johnson, M. D., Housey, G. M., Kirschmeier, P. T., and Weinstein, I. B. (1987) Molecular cloning of gene sequences regulated by tumor promoters and mitogens through protein kinase C. *Mol. Cell. Biol.* **7**, 2821–2829.

147. Gewert, D. R., Coulombe, B., Castelino, M., Skup, D., and Williams, B. R. G. (1987) Characterization and expression of a murine gene homologous to human EPA/TIMP: a virus-induced gene in the mouse. *EMBO J.* **6**, 651–657.

148. Satoh, T., Kobayashi, K., Yamashita, S., Kikuchi, M., Sendai, Y., and Hoshi, H. (1994) Tissue inhibitor of metalloproteinases (TIMP-1) produced by granulosa and oviduct cells enhances in vitro development of bovine embryo. *Biol. Reprod.* **50**, 835–844.

149. Boujrad, N., Ogwuegbu, S. O., Garnier, M., Lee, C. H., Martin, B. M., and Papadopoulos, V. (1995) Identification of a stimulator of steroid hormone synthesis isolated from testis. *Science* **268**, 1609–1612.

150. Hayakawa, T. (1994) Tissue inhibitors of metalloproteinases and their cell growth- promoting activity. *Cell. Struct. Funct.* **19**, 109–114.

151. Gomez, D. E., Alonso, D. F., Yoshiji, H., and Thorgeirsson, U. P. (1997) Tissue inhibitors of metalloproteinases - structure, regulation and biological functions. *Eur. J. Cell. Biol.* **74**, 111–122.

152. Goldberg, G. I., Marmer, B. L., Grant, G. A., Eisen, A. Z., Wilhelm, S., and He, C. S. (1989) Human 72-kilodalton type IV collagenase forms a complex with a tissue inhibitor of metalloproteases designated TIMP-2. *Proc. Natl. Acad. Sci. USA* **86**, 8207–8211.

153. Herron, G. S., Banda, M. J., Clark, E. J., Gavrilovic, J., and Werb, Z. (1986) Secretion of metalloproteinases by stimulated capillary endothelial cells. II. Expression of collagenase and stromelysin activities is regulated by endogenous inhibitors. *J. Biol. Chem.* **261**, 2814–2818.

154. Raynes, J. G., Clarke, F. A., Anderson, J. C., Fitzpatrick, R. J., and Dobson, H. (1988) Collagenase inhibitor concentration in cultured cervical tissue of sheep is increased in late pregnancy. *J. Reprod. Fertil.* **83**, 893–900.

155. De Clerck, Y. A., Yean, T.-D., Ratzkin, B. J., Lu, H. S., and Langley, K. E. (1989) Purification and characterization of two related but distinct metalloproteinase inhibitors secreted by bovine aortic endothelial cells. *J. Biol. Chem.* **264**, 17,445–17,453.

156. Stetler-Stevenson, W. G., Krutzsch, H. C., and Liotta, L. A. (1989) Tissue inhibitor of metalloproteinase (TIMP-2) A new member of the metalloproteinase inhibitor family. *J. Biol. Chem.* **264**, 17,374–17,378.

157. Stetler-Stevenson, W. G., Brown, P. D., Onisto, M., Levy, A. T., and Liotta, L. A. (1990) Tissue inhibitor of metalloproteinases-2 (TIMP-2) mRNA expression in tumor cell lines and human tumor tissues. *J. Biol. Chem.* **265**, 13,933–13,938.

158. Matsumoto, H., Ishibashi, Y., Ohtaki, T., Hasegawa, Y., Koyama, C., and Inoue, K. (1993) Newly established murine pituitary folliculo-stellate-like cell line (TtT/GF) secretes potent pituitary glandular cell survival factors, one of which corresponds to metalloproteinase inhibitor. *Biochem. Biophys. Res. Commun.* **194**, 909–915.

159. Nemeth, J. A., Goolsby, C. L. (1993) TIMP-2, a growth-stimulatory protein from SV40 transformed human fibroblasts. *Exp. Cell Res.* **207**, 376–382.

160. Osawa, K., Shirai, T., Yanagawa, M., Yamada, K., Nishikawa, K., and Tanaka, H. (1994) Purification and characterization of growth and differentiation factors from human osteosarcoma cell line, OST-1-PF, in *Animal Cell Technology: Basic and Applied Aspects* (Kobayashi, T., Kitagawa, Y., and Okumura, K., eds.), Kluwer, Dordrecht, pp, 589–593.

161. Murphy, G. and Willenbrock, F. (1995) Tissue inhibitors of matrix metallo-endopeptidases. *Methods Enzymol.* **248**, 496–510.

162. Blenis, J. and Hawkes, S. P. (1983) Transformation-sensitive protein associated with the cell substratum of chicken embryo fibroblasts. *Proc. Natl. Acad. Sci. USA* **80**, 770–774.

163. Blenis, J. and Hawkes, S. P. (1984) Characterization of a transformation-sensitive protein in the extracellular matrix of chicken embryo fibroblasts. *J. Biol. Chem.* **259**, 11,563–11,570.

164. Staskus, P. W., Masiarz, F. R., Pallanck, L. J., and Hawkes, S. P. (1991) The 21-kDa protein is a transformation-sensitive metalloproteinase inhibitor of chicken fibroblasts. *J. Biol. Chem.* **266**, 449–454.

165. Pavloff, N., Staskus, P. W., Kishnani, N. S., and Hawkes, S. P. (1992) A new inhibitor of metalloproteinases from chicken: ChIMP-3. A third member of the TIMP family. *J. Biol. Chem.* **267**, 17,321–17,326.

166. Apte, S. S., Hayashi, K., Seldin, M. F., Mattei, M. G., Hayashi, M., and Olsen, B. R. (1994) Gene encoding a novel murine tissue inhibitor of metalloproteinases (TIMP), TIMP-3, is expressed in developing mouse epithelia, cartilage, and muscle, and is located on mouse chromosome 10. *Dev. Dyn.* **200**, 177–197.

167. Apte, S. S., Mattei, M. G., and Olsen, B. R. (1994) Cloning of the cDNA encoding human tissue inhibitor of metalloproteinases-3 (TIMP-3) and mapping of the TIMP3 gene to chromosome 22. *Genomics* **19**, 86–90.

168. Weber, B. H. F., Vogt, G., Pruett, R. C., Stohr, H., and Felbor, U. (1994) Mutations in the tissue inhibitor of metalloproteinases-3 (TIMP3) in patients with Sorsby's fundus dystrophy. *Nature Genet.* **8**, 352–356.

169. Greene, J., Wang, M., Liu, Y. E., Raymond, L. A., Rosen, C., and Shi, Y. E. (1996) Molecular cloning and characterization of human tissue inhibitor of metalloproteinase 4. *J. Biol. Chem.* **271**, 30,375–30,380.

170. Bigg, H. F., Shi, Y. E., Liu, Y. L. E., Steffensen, B., and Overall, C. M. (1997) Specific, high affinity binding of tissue inhibitor of metalloproteinases-4

(TIMP-4) to the COOH-terminal hemopexin-like domain of human gelatinase A: TIMP-4 binds progelatinase A and the COOH-terminal domain in a similar manner to TIMP-2. *J. Biol. Chem.* **272**, 15,496–15,500.

171. Leco, K. J., Apte, S. S., Taniguchi, G. T., Hawkes, S. P., Khokha, R., Schultz, G. A., and Edwards, D. R. (1997) Murine tissue inhibitor of metalloproteinases-4 (TIMP-4): cDNA isolation and expression in adult mouse tissues. *FEBS Lett.* **401**, 213–217.

172. Wu, I. M. and Moses, M. A. (1998) Molecular cloning and expression analysis of the cDNA encoding rat tissue inhibitor of metalloproteinase-4. *Matrix Biol.* **16**, 339–342.

173. Douglas, D. A., Shi, Y. E., and Sang, Q. X. A. (1997) Computational sequence analysis of the tissue inhibitor of metalloproteinase family. *J. Protein Chem.* **16**, 237–255.

174. Mott, J. D., Takahara, K., Rosenbach, M. T., Thomas, C. L., Lee, S., Greenspan, D. S., and Banda, M. J. (1998) A novel matrix metalloproteinase inhibitor in human brain tumors. *FASEB J.* **11**, A1255–A1255(Abstract).

2

Strategies for Cloning New MMPs and TIMPs

Gloria Velasco and Carlos López-Otín

1. Introduction

Matrix metalloproteinases (MMPs) and tissue inhibitors of metalloproteinases (TIMPs) play important roles in the remodeling of connective tissues associated with normal mammalian development and growth, and in the degradative processes accompanying diseases such as rheumatoid arthritis, pulmonary emphysema or tumor cell invasion and metastasis *(1)*. Because of the importance of these proteins in both normal and pathological conditions, over the last years many groups have tried to clone the diverse MMPs mediating these matrix remodeling events as well as the different TIMPs able to balance their proteolytic activities. The first evidence for the occurrence of MMPs was reported about 35 yr ago by Gross and Lapiere who described the presence of diffusible collagenolytic factors in tissue cultures of bullfrog tadpoles *(2)*. Some years after this finding, several groups independently reported the existence of naturally occurring metalloproteinase inhibitors known as TIMPs and active against most members of the MMP family *(3)*. The utilization of standard biochemical methods allowed the isolation of the first MMP and TIMP family members and their subsequent physico-chemical characterization. However, these studies were seriously hampered by the small amount of proteases and inhibitors usually found in normal conditions. The observation that these proteins were much more abundant in a series of pathological conditions such as inflammatory or tumor processes or during extracellular matrix remodeling events, facilitated the identification of additional members of both families and the molecular cloning of the first MMPs and TIMPs. More recently, the advent of more powerful molecular biology techniques and improved cloning strategies has made it possible to identify a large number of novel MMP and

From: *Methods in Molecular Biology, vol. 151: Matrix Metalloproteinase Protocols*
Edited by: I. Clark © Humana Press Inc., Totowa, NJ

TIMP family members. To date, 18 distinct MMPs have been identified, cloned, and characterized in vertebrates *(4)*. In addition, MMPs have been also cloned from embryonic sea urchin *(5)*, green alga *(6)* and soybean leaves *(7)*. The complexity of the TIMP family has also expanded during the last years and a total of 4 distinct inhibitors with ability to control the proteolytic activity of MMPs have been cloned and characterized at the molecular level *(8)*. These new additions to the growing list of MMPs and TIMPs have provided much more complexity to the field but have also opened new views on the role of these proteins in normal and pathological processes. Thus, evidence is accumulating that MMPs are not exclusively involved in the proteolytic degradation of extracellular matrix components, playing also direct roles in essential cellular processes such as differentiation, proliferation, angiogenesis and apoptosis *(9)*. Similarly, TIMPs appear to have additional roles other than their direct inhibition of MMP proteolytic activity, and a number of reports have described their involvement in cell growth *(10)*. The delineation of expanding roles for these proteins in a wide variety of biological processes has also reinforced previous observations indicating that misregulation of these proteases and inhibitors can have important pathological consequences. Nevertheless, it seems clear that most of this progress has been only possible by the cloning of an unexpected large number of these proteins. This chapter will give an overview of the different strategies used for cloning MMPs and TIMPs and their application to the identification and characterization of putative yet unknown members of these protein families that play essential roles in both normal and pathological conditions.

2. General Methods for Cloning MMPs and TIMPs

A survey of the literature on MMPs and TIMPs indicates that the strategies used for cloning the distinct members of these families can be summarized in four different groups (**Table 1**). However, it should be noticed that in some cases, a combination of several of these different approaches has been required to clone some specific family members. Furthermore, it is remarkable that some MMPs have been cloned by other procedures not involving an oriented search specifically aimed at cloning MMPs or TIMPs. This is the case for rat stromelysin-1 (MMP-3) first cloned as an oncogene-transformation induced gene in rat embryo fibroblasts *(11)*, or for human stromelysin-3 (MMP-11) cloned from a subtractive breast cancer cDNA library as a gene specifically expressed in stromal cells surrounding epithelial tumor cells *(12)*.

2.1. Methods Based on Purification and Biochemical Characterization of MMPs and TIMPs

The first MMP and TIMP family members were cloned by following approaches initially based on the isolation of the corresponding proteins from

Table 1
Strategies for Cloning MMPs and TIMPs

Name	Source	References
Purification and biochemical characterization of MMPs or TIMPs		
Collagenase-1 (MMP-1)	Human skin fibroblasts	*13*
Collagenase-2 (MMP-8)	Human neutrophils	*17,19*
Stromelysin-1 (MMP-3)	Human skin fibroblasts	*20*
Gelatinase A (MMP-2)	Human ras-transformed bronchial cells	*21*
Gelatinase B (MMP-9)	Human SV40-transformed fibroblasts	*22*
Envelysin	*Paracentrotus lividus* hatched embryos	*5*
Green alga MMP	*Chlamydomonas reinhardtii* vegetative cells	*6*
Soybean MMP	Soybean leaves	*7*
TIMP-1	Human fetal lung fibroblasts	*24*
TIMP-2	Human fetal heart cells	*28*
Low-stringency screening with MMP or TIMP probes		
Stromelysin-2 (MMP-10)	Rat genomic library screened with rat MMP-3	*29*
Matrilysin (MMP-7)	Human tumors cDNA library screened with h-MMP-3	*30*
Collagenase-4 (MMP-18)	Xenopus tadpole cDNA library screened with h-MMP-3	*32*
RT-PCR with degenerate oligonucleotides		
Collagenase-3 (MMP-13)	Human breast carcinoma	*35*
MT1-MMP (MMP-14)	Human placenta	*38*
MT2-MMP (MMP-15)	Human kidney carcinoma	*39*
MT3-MMP (MMP-16)	Human oral melanoma	*40*
MT4-MMP (MMP-17)	Human breast carcinoma	*41*
Metalloelastase (MMP-12)	Murine macrophages	*42*
Enamelysin (MMP-20)	Porcine enamel organ	*43*
X-MMP (MMP-21)	Xenopus embryos	*44*
TIMP-3	Human breast carcinoma	*45*
Screening of Expressed Sequence Tag-databases		
MMP-19	Human liver	*54*
TIMP-4	Human heart	*8*

different sources using standard biochemical techniques. After purification of the proteins of interest, partial amino acid sequences can be obtained by automatic Edman degradation of the intact protein or of peptides derived by enzymatic or chemical cleavage of the purified proteins. The obtained amino acid sequences can then be used for designing oligonucleotide probes encoding these protein fragments. Finally, the oligonucleotide probes are used as probes to screen cDNA libraries in which the specific MMPs or TIMPs are presumed to be expressed at high levels.

This approach was first successfully used for cloning human fibroblast collagenase (MMP-1 or collagenase-1) from skin fibroblasts *(13)*. To this purpose, procollagenase-1 was purified from conditioned medium of cultured adult skin fibroblasts (WUN 80547 cell strain) using cation-exchange and gel filtration chromatography. Since the N-terminal residue of the intact protein was blocked, a preparation of purified procollagenase-1 was subjected to cyanogen bromide cleavage and the resulting peptides were purified by reverse-phase high performance liquid chromatography (HPLC) and sequenced. Then, a mixture of 32 oligonucleotides, 17-bases long, encoding part of the amino acid sequence determined for a cyanogen bromide peptide was synthesized. The selected protein sequence (His-Phe-Asp-Glu-Asp-Glu) contained amino acids that had the lowest degree of degeneracy in the corresponding codons, thus diminishing the probe complexity. This oligomer was 5' end-labeled with polynucleotide kinase and hybridized to a cDNA library constructed from collagenase-producing human skin fibroblasts mRNA. Nucleotide sequencing of isolated clones hybridizing with the probe revealed the presence of an ORF encoding a protein of 469 amino acids with a predicted molecular weight of 51,929, which was called fibroblast collagenase. The sequence of this first cloned human MMP showed homology to an oncogene-induced rat protein of unknown function at that time and that we now know corresponds to rat stromelysin-1 *(11)*. Shortly after, other groups reported the cloning of collagenase-1 from different sources following a similar approach *(14,15)*. Since then, collagenase-1 has been the subject of a wide variety of biochemical, enzymatic, genetic, and clinical studies that have extended the knowledge of multiple normal and pathological aspects of this first human representative of the MMP family (reviewed in **ref. *16)***.

A similar strategy was subsequently used for the cloning of neutrophil collagenase (MMP-8 or collagenase-2) *(17)*. Partial amino acid sequences derived from the purified proenzyme were used to deduce 50 bases long synthetic oligonucleotides. In this case, a computer program was utilized to design "guessmers" based on codon usage, dinucleotide frequency, and potential probe self-complementarity *(18)*. The probes were used to screen a granulocyte cDNA library derived from mRNA of a patient with chronic granulocytic leukemia. Collagenase-2 was also cloned by Hasty et al. *(19)* from RNA of a granulocytic leukemia patient although they used as probe a 24-mer oligonucleotide derived from the Zn-binding region of collagenase-1. In both cases, the isolated cDNA clones encoded a 467-residue protein with about 58% identity to human collagenase-1 and displaying the same domain structure, including a signal peptide, an 80-residue propeptide, a catalytic domain with a Zn-binding site and a C-terminal hemopexin-like domain. Further studies revealed that recombinant collagenase-2 degraded type I collagen into the 3/4 and 1/4 fragments characteristic of mammalian interstitial collagenase activity. Thus, definitive evidence was provided that neutrophil collagenase is a member of the MMP family distinct from fibroblast collagenase, in both structural and enzymatic properties.

The cloning of human stromelysin-1 was also based on the same approach described above for cloning human collagenases-1 and -2 *(20)*. This enzyme was purified from human skin fibroblast (WUN 80547) conditioned medium by successive chromatographies in Zn-chelate-Sepharose, DEAE-Sepharose, and reactive red-agarose, and then subjected to Edman degradation. The N-terminal sequence of the 45-kDa active enzyme (Thr-Phe-Pro-Gly-Ile-Pro) was used to synthesize a 17-base long oligonucleotide probe. This probe contained 3 deoxyinosine residues at codon nucleotide positions of fourfold degeneracy in order to limit the number of oligonucleotides in each probe mixture. Then, a cDNA library from human fibroblasts (WUN 80547) was screened with this probe and the positive clones characterized by nucleotide sequencing. Structural analysis indicated that stromelysin-1 has 477 residues and the same overall organization as collagenases, confirming their evolutionary relationship. The sequence of stromelysin-1 from human skin fibroblasts was in agreement with that of a putative stromelysin cDNA clone isolated from a human gingival fibroblast cDNA library by cross-hybridization to a partial rabbit stromelysin-1 cDNA clone *(14)*.

The two members of the gelatinase subfamily of MMPs, 72K and 92K type IV collagenases or gelatinases A and B, were also originally cloned following the approach used for cloning of collagenases and stromelysins *(21,22)*. Gelatinase A was purified from H-*ras*-transformed bronchial epithelial cells by Zn chelate-Sepharose, reactive green-agarose and AcA-44 size-exclusion chromatography. The purified protein was digested with trypsin and the resulting fragments purified by reverse-phase HPLC and subjected to amino acid sequencing. The sequence of one of these peptides was used to construct a 21-mer oligonucleotide with 2 inosine residues at positions of fourfold codon degeneracy. The probe was used to screen a cDNA library of human skin fibroblast mRNA leading to the isolation of a number of positive clones. Structural analysis of these clones revealed that gelatinase A had a domain organization similar to that of collagenases and stromelysins. However, an additional domain was found consisting of 175 residues with homology to the type II motif of the collagen-binding region of fibronectin. Similarly, gelatinase B was purified from conditioned medium of SV40-transformed fibroblasts or TPA-differentiated monocytic leukemia U937 cells by successive chromatographies in reactive red-agarose, gelatin-Sepharose, Aca-44, and phenyl-Sepharose. The N-terminal sequence of the proenzyme was reverse-translated to generate a 48-nucleotide long probe containing 3 inosine residues. This probe was then used to screen a cDNA library from TPA-treated HT1080 fibrosarcoma cells with the finding of a series of positive clones encoding a 707 amino acid-long proenzyme of predicted Mr 78,426. Gelatinase B consists of a series of domains shared by most MMPs including the signal sequence, the propeptide, the catalytic region, and the C-terminal hemopexin-like domain. In addition, it contains the fibronectin-like domain also present in gelatinase A and a unique 54-amino acid long proline-rich domain homologous to the α2 chain of type V collagen *(21,22)*. The cloning of both human gelatinases has facilitated the subsequent cloning of the homologous enzymes in numerous species including mouse, rat, chicken, Xenopus, and newt. In addition, it has opened the possibility to perform a variety of biochemical studies which have demonstrated that gelatinases share broad overlapping substrate specificities, but differ considerably in terms of their transcriptional regulation, glycosylation patterns, and activation mechanisms. Finally, clinical studies have shown that gelatinases are usually overexpressed in a wide diversity of pathological conditions, suggesting that they may be general targets for future therapeutic intervention.

Furthermore, it is of interest that all MMPs from nonvertebrate origin characterized to date have been cloned through this general approach based on the previous purification of the corresponding enzymes. Thus, the sea-urchin MMP (also called hatching enzyme or envelysin) was cloned by immunoscreening of a blastula stage cDNA library from *Paracentrotus lividus* embryos with a polyclonal antibody raised against the enzyme purified from culture supernatants of hatched embryos *(5)*. Similarly, a green alga MMP involved in cell wall degradation to facilitate gamete fusion was cloned by screening of a cDNA library from *Chlamydomonas reinhardtii* vegetative cells with a 57-mer probe derived from the N-terminal sequence of the purified gamete lytic enzyme *(6)*. Finally, the first MMP identified in higher plants has been recently cloned by using degenerate oligonucleotides derived from the sequence of the protein purified from soybean leaves *(7)*. Although the biological significance of the presence of a MMP in plants remains unclear, it has been proposed that it may be involved in the tissue remodeling events which must occur during leaf expansion. Alternatively, an attractive albeit speculative hypothesis, is that the enzyme may play a defensive role against insects in plant leaves, through digestion of the collagen-like proteins within the midgut lining of insect pests, thus disrupting the normal digestive physiology *(7)*.

In addition to these efforts directed to cloning MMPs from different sources, several groups have tried to clone and characterize their naturally occurring inhibitors, known as TIMPs, and playing a critical role in controlling the matrix remodeling that takes place during normal development and in diseases such as cancer and arthritis. Based on structural and functional comparisons, these inhibitors constitute a protein family that in humans is composed at least by four different members which have been cloned by different methods. The first two members of this family, TIMP-1 and TIMP-2, have been cloned by following the classical approach involving the previous purification of the corresponding proteins. TIMP-1 is an ubiquitous glycoprotein that binds tightly to the active form of multiple MMPs and is also found associated specifically with the latent form of gelatinase B *(23,24)*. TIMP-1 was purified from human amniotic fluid and from the culture media of human fetal lung fibroblasts and subjected to N-terminal amino acid sequencing. Based on this information, a single 69-base oligonucleotide probe capable of encoding the 23 N-terminal amino acids of TIMP-1 was synthesized. To avoid codon degeneracy, the codons used for probe designing were those reported to appear most frequently in human genes. The 69-mer probe was used to identify positive colonies from a human fetal lung fibroblast cDNA library *(24)*. The isolated clones encoded a protein of 207 amino acids whose sequence was identical to a protein with erythroid-potentiating activity purified from medium of T-lymphoblast cells infected by HTLV-II virus *(25)*. This finding opened the possibility, subsequently confirmed, that TIMPs may have functions in cell growth distinct than those derived from their ability to inhibit MMPs *(3,10)*. This growth promoting activity of TIMP-1 is shared with TIMP-2, a 21 kDa nonglycosylated protein described in 1989 by several independent groups *(26,27)*. Human TIMP-2 was cloned from a fetal human heart library probed with a bovine TIMP-2 probe *(28)*. The bovine inhibitor had been previously cloned by using degenerate oligonucleotides derived from the N-terminal sequence of the protein purified from aortic endothelial cells. The isolated clones for TIMP-2 code for a mature protein of 194 residues and a signal peptide of 26 amino acids. This sequence is virtually identical to that of TIMP-2 cloned from a melanoma cell cDNA library screened with a 45-mer probe derived from the amino acid sequence of the inhibitor purified from cultured medium of melanoma cells. TIMP-2 shares about 40% identity with TIMP-1, including 12 conserved cysteine residues, and displays similar inhibitory properties against active MMPs. In addition, it binds with high affinity to the latent form of gelatinase A. However, both inhibitors differ in the regulation of their expression since TIMP-1 is responsive to a variety of external stimuli, such as phorbol esters, growth factors, cytokines, and serum, whereas TIMP-2 expression is for the most part constitutive (reviewed in **ref. 3**).

2.2. Methods Based on Low-Stringency Hybridization of Genomic or cDNA Libraries with MMP or TIMP Probes

The availability of MMP and TIMP cDNAs cloned by the RT-PCR strategy opened the possibility to clone new family members by using methods based on low-stringency hybridization of genomic or cDNA libraries with probes derived from the isolated MMP or TIMP cDNAs. This strategy was first successfully used by Breathnach et al. to clone rat stromelysin-2 through low-stringency screening of a rat genomic library with a cDNA probe for rat stromelysin-1 *(29)*. Subsequent cDNA cloning and structural analysis revealed that this second representative of the stromelysin subfamily of MMPs is 71% identical in sequence to stromelysin-1. However, their regulatory mechanisms are markedly different since stromelysin-1 is induced by a variety of cytokines, growth factors, and tumor promoters in fibroblast cells, whereas stromelysin-2 gene is not usually a target of any of these factors *(29)*. Shortly after, the same group used this approach to search for stromelysin-related genes that could be involved in tumor invasion and metastasis *(30)*. After hybridization of a cDNA library prepared from RNAs extracted from a pool of primary and metastatic tumors with a stromelysin-1 probe, a clone encoding human stromelysin-2 was isolated. Furthermore, this experiment led to the cloning of a novel member of the family originally named pump-1 and subsequently known as matrilysin (MMP-7). Structural analysis has revealed that matrilysin is synthesized without a hemopexin-like domain and thus it is comparably small in size. It can cleave a wide array of extracellular matrix substrates and can activate pro-collagenase-1, progelatinase A, and progelatinase B. The homologous rat enzyme has been cloned and its presence at very high levels in involuting rat uterus has been described, suggesting an essential participation of matrilysin in this reproductive process *(31)*.

An additional example of the usefulness of this method is represented by the recent cloning of *Xenopus* collagenase-4, the last identified member of the collagenase subfamily of MMPs *(32)*. Thus, this novel MMP has been cloned by screening a tadpole intestinal cDNA library with a human stromelysin-1 probe, under reduced stringency conditions. The finding of collagenase-4, a MMP whose expression is activated during amphibian metamorphosis, is a representative example of the presence of these enzymes during the extensive remodeling that occurs as larval tissues degenerate and adult organogenesis takes place, and emphasizes again the use of this model system as an excellent source for the cloning of novel family members.

In addition to its demonstrated validity for the cloning of novel MMPs, the strategy based on low stringency hybridization with previously cloned family members has been widely used for cloning orthologue enzymes from MMPs or

TIMPs originally isolated in other species. A representative example of this approach is the cloning of human macrophage metalloelastase (MMP-12) by using the murine cDNA as the probe to screen a human genomic DNA library. An isolated genomic fragment including two exons of the human gene was subsequently used as probe to isolate full-length cDNA clones from a human alveolar macrophage cDNA library *(33)*. Similarly, the human homologue of porcine enamelysin (MMP-20) has been recently cloned in our laboratory by screening of a human genomic library with a porcine enamelysin cDNA. The information deduced from isolated genomic clones was further used to PCR-amplify a full-length cDNA for human enamelysin *(34)*. Recombinant human enamelysin is able to degrade amelogenin, the major protein component of the enamel matrix. This fact together with its highly restricted expression to dental tissues strongly suggests that human enamelysin plays a central role in the process of tooth enamel formation *(34)*.

2.3. Methods Based on RT-PCR Amplification with Degenerate Oligonucleotides

Once several members of both MMP and TIMP families had been cloned, reverse-transcriptase polymerase chain reaction (RT-PCR) strategies using degenerate oligonucleotides encoding conserved sequences within the distinct proteins have provided an excellent choice to identify novel family members. The cloning of collagenase-3, MT-MMPs and TIMP-3 from human tissues are illustrative examples of the suitability of this procedure, but the technique has been widely applied by different groups to reveal additional members of these protein families in different species.

The cloning of human collagenase-3 was the result of studies directed to look for the presence of new proteases in tumor tissues *(35)*. Tumor specimens are an appropriate starting material to identify novel members of the MMP family because malignant processes require the combined action of over-expressed proteases that degrade the connective tissue to facilitate tumor growth and spread to distant sites. To clone collagenase-3, we utilized degenerate oligonucleotide primers spanning two highly conserved amino acid sequences in known MMPs and reverse-transcribed RNA from human breast carcinomas. The conserved structural motifs corresponded to the activation locus (Pro-Arg-Cys-Gly-Val-Pro-Asp) containing the Cys residue essential for the maintenance of the latency of these enzymes, and the Zn-binding site (Val-Ala-Ala-His-Glu-Phe-Gly-His) present in the catalytic domain of all MMPs. After synthesizing two degenerate oligonucleotides encoding these conserved motifs and performing RT of total RNA from a mammary carcinoma, a band of the expected size (about 400 bp) was obtained and cloned. Analysis of the nucleotide sequence of different clones revealed that some of them had a sequence similar to previously characterized human MMPs. Screening of a breast cancer cDNA library prepared from the same tumor employed for the RT experiment, and using the PCR generated fragment as probe, led to the identification of a positive clone. The isolated cDNA encoded a polypeptide of 471 amino acids with all structural features characteristic of MMPs. In addition, this novel human MMP contained in its amino acid sequence several residues specific to the collagenase subfamily of MMPs which led to the name of collagenase-3, since it represented the third member of this subfamily composed at that time of fibroblast and neutrophil collagenases. Further studies performed with recombinant collagenase-3 produced in different eukaryotic expression systems confirmed that it is a *bona fide* collagenase active against fibrillar collagens as anticipated from its amino acid sequence. Finally, it is of interest that collagenase-3 has also been cloned from IL-1 stimulated human chondrocytes by using this RT-PCR approach with degenerate oligonucleotides *(36)*. Further studies have extended these findings and revealed that collagenase-3 may play a critical role in the destruction of articular cartilage during arthritic processes, thus providing an additional dimension to the study of this MMP *(37)*.

The relevance of the PCR-based approach for MMP cloning is also demonstrated by the cloning of the different members of the MT subfamily of MMPs, whose discovery has represented a significant step to elucidate the mechanisms that underlie the pericellular activity of proteolytic enzymes *(38)*. In fact, all four distinct human MT-MMPs characterized to date have been cloned from different sources by RT-PCR with degenerate oligonucleotides *(38–41)*. Using primers corresponding to the above mentioned Cys-switch and Zn-binding sites, Sato et al. cloned in 1994 the first of these integral plasma membrane enzymes from a human placenta cDNA library. Since then, 3 additional MT-MMPs have been cloned using primers derived from the same regions and RNAs from different tumor tissues as templates: a kidney carcinoma for MT2-MMP, an oral melanoma for MT3-MMP, and a breast carcinoma for MT4-MMP. Full-length sequences for these enzymes were subsequently obtained by screening appropriate cDNA libraries. Structural comparisons between the four MT-MMPs have revealed that in addition to all typical domains shared with other MMPs, these four enzymes have a 10–12 residue insert with a furin-like recognition motif, and a final 75–105 residue insert containing a transmembrane sequence of about 24 residues and a short cytoplasmic C-terminal tail. Despite this similar structural organization, Northern blot analysis has shown that each MT-MMP shows a unique pattern of expression. Thus, MT1-MMP is expressed in numerous normal tissues, but is undetectable in brain and leukocytes. MT2-MMP is also undetectable in brain and leukocytes but is much more abundant than MT1-MMP in liver, heart, and skeletal muscle. MT3-MMP is mainly present in brain, placenta, lung, and heart, whereas MT4-MMP expression is very strong in leukocytes and brain. Functional analysis of MT-MMPs have revealed that these enzymes induce specific activation of pro-gelatinase A at the cell surface, thus being prime candidates to act as upstream activators in the gelatinase A proteolytic cascade involved in the tissue destruction accompanying tumor processes *(38)*.

Further examples of the application of the PCR-based homology cloning strategy with the universal MMP degenerate primers from Cys-switch and Zn-binding sites, include the identification of murine macrophage metalloelastase (MMP-12), porcine enamelysin (MMP-20) and amphibian X-MMP (MMP-21). The first of these enzymes was cloned by screening a murine macrophage cDNA library with a specific probe generated by two rounds of RT-PCR amplification *(42)*. The first round was carried out with the Cys-switch and Zn-binding derived primers and total RNA from P88D1 murine macrophages. The second round PCR was performed on an aliquot of the first round reaction mixture using the same Zn-reverse primer and a nested degenerate oligonucleotide based on the N-terminal sequence of the purified 22 kDa active enzyme. The deduced amino acid sequence of mouse MMP-12 predicts a 462 amino acid protein that is a distinct member of the MMP family. Enamelysin was cloned using RNA from porcine enamel organ as the template for RT-PCR *(43)*. The amplified 401 bp length fragment was then used for screening a cDNA library from porcine enamel, resulting in the isolation of a novel cDNA encoding a 483 amino acid residues protein similar in size to collagenases and stromelysin. However, enamelysin lacks distinctive residues of these two MMP subfamilies and thus cannot be classified into these MMP groupings. Similarly, X-MMP does not belong to any of the established MMP subfamilies. X-MMP is the most recently identified member of this protein family and has been cloned by RT-PCR using RNA from Xenopus embryos in early developmental stages and the two universal MMP-primers *(44)*. X-MMP is a 604 residue protein distantly related to other members of the family and unlike other MMPs, it has a vitronectin-like insert in its propeptide domain and lacks a proline-rich hinge region between its catalytic and C-terminal domains. A putative human homologue of X-MMP has not yet been reported.

The PCR-based homology cloning strategy has also been applied to search for novel members of the TIMP family. Thus, based on the fact that TIMPs are usually overexpressed in tumor tissues as part of a defensive response, we cloned TIMP-3 from human breast carcinomas using degenerate oligonucleotides encoding conserved sequences present in TIMP-1 and TIMP-2 *(45)*. The forward primer covered the central part of a conserved motif located in the N-terminal region of these inhibitors (Ile-His-Pro-Gln-Asp-Ala), whereas the sequence of the reverse primer was derived from a highly conserved region surrounding a Trp residue of the C-terminal of TIMPs (Glu-Cys-Leu-Trp-Thr-Asp). After RT-PCR amplification of total RNA isolated from a breast carcinoma, a band of the expected size (about 0.4 kb) was obtained and cloned. Nucleotide sequence analysis of isolated clones revealed that three of them were similar to the two human TIMPs previously isolated. To obtain a full-length cDNA for this putative novel TIMP, a cDNA library was prepared using as starting material polyA⁺ RNA from the same breast carcinoma used for the original RT-PCR experiment. Upon screening of this library using the PCR generated cDNA as probe, two positive clones were obtained. Nucleotide sequence analysis of the largest one revealed an ORF coding for a protein of 211 amino acids, and a predicted molecular mass of about 24 kDa. This novel member of the family was called TIMP-3 and shares 40% identity with TIMP-1 and 45% with TIMP-2, including the 12 cysteine residues which are conserved among all these proteins *(45)*. A similar strategy was also used by Silbiger et al. to clone human TIMP-3, although they used fetal kidney as the source of RNA and the forward oligonucleotide was derived from a distinct region (Val-Ile-Arg-Ala) and contained a restriction site to facilitate the subsequent cloning of the amplified products *(46)*. The sequences deduced for TIMP-3 cloned by following this approach are virtually identical to that reported by Apte et al. who cloned a cDNA for TIMP-3 by screening of a human placenta cDNA library with a probe specific for the murine inhibitor *(47)*. The cloning and sequencing of human TIMP-3 has also revealed that it is the homologue of a chicken inhibitor of metalloproteases called ChIMP-3, which was first isolated as a transiently expressed 21 kDa protein in the extracellular matrix of transforming chick fibroblasts *(48)*. Further studies have provided a biochemical characterization of this third member of the TIMP family with the finding that it is an effective inhibitor of MMPs, as well as of endothelial cell migration and angiogenesis. However, unlike the other TIMPs which are soluble, TIMP-3 is unique in being a component of extracellular matrix. TIMP-3 is produced by many epithelia and at high levels in cartilage and muscle, and its expression is induced in response to mitogenic stimulation, being subjected to a strict cell cycle regulation *(49)*. Finally, it is worthwhile mentioning that mutations in the human TIMP-3 gene cause Sorsby's fundus dystrophy, a dominantly inherited form of blindness, thus being the first member of the MMP and TIMP families in which genetic defects have been found *(50)*.

In summary, this approach has been of extraordinary value to clone novel members of the MMP and TIMP families. Several situations rather specific for MMPs and TIMPs have facilitated the success of this approach and may even favor its future application for the cloning of as yet unknown members of these families. Thus, both protein families have highly conserved sequences of enough length (6–7 amino acids) to design specific oligonucleotides. In addition, these regions are relatively devoid of amino acids such as Ser, Leu, or Arg displaying a high grade of degeneracy in their respective codons. The use of inosine at positions of fourfold degeneracy diminishes the complexity of the oligonucleotide mixture but in our experience good results can be equally obtained without the introduction of this base. Some groups have used oligonucleotides containing restriction sites at their ends to facilitate cloning of the PCR-amplified fragments, but this introduces additional mismatches into the primers which in some cases may lead to a low efficiency in the amplification process. It is also remarkable that the distance separating the diverse forward and reverse primers designed for cloning MMPs or TIMPs is about 300–400 bp, which is optimal because is easy to amplify and allows complete sequencing of the amplified fragments in a single sequencing run. Finally, an additional advantage of this strategy is that can be applied to small amounts of RNA extracted from a number of normal and pathological tissues in which these proteins may be overexpressed.

2.4. Methods Based on Screening of Expressed Sequence Tag-Databases

Expressed sequence tags (ESTs) are short nucleotide sequences derived by partial sequencing of the inserts of randomly selected cDNA clones *(51)*. More than 900,000 human ESTs generated from a large variety of tissue sources have now been deposited in a public database named dbEST. In addition, ESTs from model organisms are also being introduced in separate data bases, thus expanding the usefulness of this approach. Because the majority of the source cDNA libraries are normalized in vitro to reduce the high variation in abundance among the clones that represent individual mRNA species, it is likely that the dbEST already contains a significant percentage of all genes present in the human genome *(52)*. Consequently, the dbEST represents an invaluable and easily accessible tool to look for novel genes through automated computer search of ESTs with sequence similarity to genes of interest *(53)*. To this purpose, one of the most powerful methods is to conduct a homology search with the Basic Local Alignment Search Tool (BLAST), a program capable of rapidly detecting even very distant sequence similarities in a statistically rigorous manner. Where similarities are detected it is possible to make functional inferences concerning the encoded protein based on what is known about the function of the matched sequence. Recently, this vast source of data has also been used for cloning novel MMPs and TIMPs, MMP-19 and TIMP-4 being the first examples of the successful application of the EST-based approach to the identification and cloning of members of these protein families.

To clone human MMP-19, we first performed an extensive analysis of the dbEST with the BLAST algorithm, looking for sequences with homology to human MMPs. This search led to the identification of a short DNA sequence

(R55624, Merck EST project) which, when translated, showed a significant similarity to a region of the propeptide domain found in all previously characterized MMPs. A cDNA containing part of this EST was obtained by PCR amplification of total DNA prepared from a human liver cDNA library. Then, the amplified fragment was used as a probe to screen the same liver cDNA library, with the finding of several positive clones encoding a putative novel MMP which was tentatively called MMP-19. This protein exhibits the domain structure typical of MMPs but lacks specific features that would otherwise distinguish it as a collagenase, stromelysin, gelatinase or MT-MMP, suggesting that MMP-19 may represent the first member of a new MMP subfamily. Furthermore, in contrast to most MMPs, MMP-19 is expressed in a variety of normal human tissues, including placenta, lung, pancreas, ovary, spleen, and intestine suggesting that it may play a specialized role in these tissues *(54)*.

The same approach has also been used for cloning human TIMP-4, the most recently identified member of this family of proteinase inhibitors *(8)*. In this case, the automated search of the dbEST revealed an EST from a human brain cDNA library with high sequence similarity (about 45% identities) to TIMP-2. This EST was then used as a probe to examine the expression in a variety of human tissues by Northern blot analysis. Because the highest expression was found in heart, a cDNA library from human heart was screened with the EST probe. A full-length cDNA was obtained, coding for a protein of 224 amino acids, including a 29-residue signal sequence. After optimal alignment, TIMP-4 shows about 50% identity to TIMP-2 and TIMP-3, and 37% to TIMP-1. The predicted protein shares several essential features that are characteristic of the TIMP family including the consensus sequence Val-Ile-Arg-Ala-Lys present in the N-terminal region and proposed to serve as a hallmark of this family of inhibitors. The sequence deduced for TIMP-4 also displays the 12 conserved cysteine residues involved in the formation of the six disulfide bonds that fold these proteins in a two domain structure. In addition to these structural similarities, recombinant TIMP-4 produced in eukaryotic cell possesses an inhibitory activity against MMPs and is secreted extracellularly thus confirming that the cloned TIMP-4 is a new member of the TIMP family. In addition, recombinant TIMP-4 inhibits tumor cell invasion across reconstituted basement membranes. This finding together with the observation that this novel TIMP is expressed in stromal cells surrounding breast carcinomas suggests that TIMP-4 may reflect a host response to try to balance the local tissue degradation associated to tumor cell invasion. These studies open the possibility of TIMP-4 mediated antitumor and antimetastatic activities for controlling cancer progression *(55)*.

3. Conclusions and Perspectives

Over the last few years there has been a dramatic increase in the complexity of MMP and TIMP families. New proteases and inhibitors with unexpected novel functions have been cloned from human tissues as well as from a number of vertebrate species. Furthermore, the first MMPs from sea urchins, green algae and plants have been recently cloned and functionally characterized. The methods used for cloning these genes have clearly shifted from biochemical-based procedures, to molecular biology-based approaches, and more recently to computer search-based strategies. The classical approach to identifying MMPs and TIMPs started with the purification and biochemical characterization of a certain protein with an interesting biological activity as a protease or as a protease inhibitor. Subsequently, the gene encoding this protein of interest was cloned by using oligonucleotide probes derived from structural information of the purified protein, or less frequently, by immunoscreening of cDNA libraries with antibodies raised against the purified protein. After generalization of PCR-based techniques, this classical approach to cloning MMPs and their specific inhibitors has been largely replaced by RT-PCR methods with oligonucleotides derived from conserved sequences in these protein families. The successful application of this strategy by different research groups has considerably expanded the number of MMPs and at least eight new enzymes have been identified following this procedure. To our view, the RT-PCR approach still represents an alternative for the cloning of new members of these families, specially if it is applied to RNA from fetal tissues or other conditions involving extensive tissue remodeling in which MMPs or TIMPs are usually more abundant. Nevertheless, it is tempting to speculate that computer search of dbEST will be the method of choice for cloning novel MMPs and TIMPs. To date, only one member of each of these families has been cloned through this approach. However, it should be noted that most ESTs currently deposited on data bases are derived from genes expressed in normal tissues under physiological conditions in which MMPs or TIMPs may be largely under represented. The extension and generalization of EST projects to specific pathologic conditions, such as arthritic, inflammatory, or tumor processes in which these genes may be over expressed will provide an invaluable tool for the identification of new MMPs or TIMPs. Likewise, the conclusion of the large scale genome sequencing projects will provide a definitive view of the number and identity of MMPs and TIMPs present in human tissues as well as in model organisms. Their cloning and functional characterization will shed light on the contribution of these proteins to both normal and pathological processes. Finally, the targeted modulation of MMP and TIMP levels in vivo through genetic manipulation of the cloned genes will help to elucidate the biological significance of the apparent functional redundancy of these continuously growing protein families.

During editing of this chapter, three novel human MMPs: MMP-23, MT5-MMP, and MT6-MMP have been cloned *(56–59)*. The cloning strategies have involved the screening of EST-databases or the use of information derived from human genome sequencing projects.

Acknowledgments

This review is dedicated to the memory of Drs. Francisco Velasco and Pilar Negredo, and their daughter Pilar. The work in our laboratory is supported by grants from CICYT-Spain; Glaxo-Wellcome, UE-BiomedII, and Fuji-Chemical, Japan. We are also very grateful to all members of our laboratory for their continuous support and encouragement.

References

1. Birkedal-Hansel, H., Moore, W. G. I., Bodden, M. K., Windsor, L. J., Birkedal-Hansen, B., DeCarlo, A., and Engler, J. A. (1993) Matrix metalloproteinases: a review. *Crit. Rev. Oral Biol. Med.* **4**, 197–250.
2. Gross, J. and Lapiere, C. M. (1962) Collagenolytic activity in amphibian tissues: a tissue culture assay. *Proc. Natl. Acad. Sci. USA* **48**, 1014–1022.
3. Gomez, D. E., Alonso, D. F., Yoshiji, H., and Thorgeirsson, U. P. (1997) Tissue inhibitors of metalloproteinases: structure, regulation and biological functions. *Eur. J. Cell Biol.* **74**,11–122
4. Uría, J. A., Jiménez, M. J., Balbín, M., Freije, J. P. and López-Otín, C. (1998) Differential effects of TGFb on the expression of collagenase-1 and collagenase-3 in human fibroblasts. *J. Biol. Chem.* **273**, 9769–9777
5. Lepage, T. and Gache, C. (1990) Early expression of a collagenase-like hatching enzyme gene in the sea urchin embryo. EMBO J. **9**, 3003–3012.
6. Kinoshita, T., Fuzukawa, H., Shimada, T., Saito, T., and Matsuda, Y. (1992) Primary structure and expression of a gamete lytic enzyme in *Chlamydomonas reinhardtii*: similarity of functional domains to matrix metalloproteases. *Proc. Natl. Acad. Sci. USA* **89**, 4693–4697.
7. Pak, J. H., Liu, C. Y., Huangpu, J. and Graham, J. S. (1997) Construction and characterization of the soybean leaf metalloproteinase cDNA. *FEBS Lett.* **404**, 283–288.
8. Greene, J., Wang, M., Liu, Y. E., Raymond, L. A., Rosen, C., and Shi, Y. E. (1996) Molecular cloning and characterization of human tissue inhibitor of metalloproteinase 4. *J. Biol. Chem.* **271**, 30,375–30,380.
9. Werb, Z. (1997) ECM and cell surface proteolysis: Regulating cellular ecology. *Cell* **91**, 439–442.
10. Hayakawa, T., Yamashita, K., Tanzawa, K., Uchijima, E. and Iwata, K. (1992) Growth promoting activity of tissue inhibitor of metalloproteinase-1 (TIMP-1) for a wide range of cells. *FEBS Lett.* **298**, 29–32.
11. Matrisian, L. M., LeRoy, P., Ruhlmann, C., Gesnel, C., and Breathnach. (1986) Isolation of the oncogene and epidermal growth factor-induced transin gene:complex control in rat fibroblasts. *Mol. Cell. Biol.* **6**, 1679–1686.

12. Basset, P., Bellocq, J. P., Wolf, C., Stoll, I., Huntin, P., Limacher, J. M., Podhajcer, O. L., Chenard, M. P., Rio, M. C., and Chambon, P. (1990) A novel metalloproteinase gene specifically expressed in stromal cells of breast carcinomas. *Nature*, **348**, 699–704.

13. Goldberg, G. I., Wilhelm, S. M., Kronberger, A., Bauer, E. A., Grant, G. A., and Eisen, A. Z. (1986) Human fibroblast collagenase: Complete primary structure and homology to an oncogene transformation-induced rat protein. *J. Biol. Chem.* **261**, 6600–6605.

14. Whitham, S. E., Murphy, G., Angel, P., Rahmsdorf, H-J., Smith, B. J., Lyons, A., Harris, T. J. R., Reynolds, J. J., Herrlich, P., and Docherty, A. J. P. (1986) Comparison of human stromelysin and collagenase by cloning and sequence analysis. *Biochem. J.* **240**, 913–916.

15. Templeton, N. S., Brown, P. D., Levy, A. T., Margulies, I. M. K., Liotta, L. A., and Stetler-Stevenson, W. G. (1990) Cloning and characterization of human tumor cell interstitial collagenase. *Cancer Res.* **50**, 5431–5437.

16. Shingleton, W. D., Hodges, D. J., Brick, P., and Cawston, T. E. (1997) Collagenase: a key enzyme in collagen turnover. *Biochem. Cell. Biol.* **74**, 759–775.

17. Devarajan, P., Mookhtiar, K., Wart, H. V., and Berliner, N. (1991). Structure and expression of the cDNA encoding human neutrophil collagenase. *Blood* **77**, 2731–2738.

18. Dubnick, M., Lewis, L. K., and Mount, D. W. (1988) BIGPROBE: A complete program that predicts the sequence of long oligonucleotide probes with high reliability. *Nucleic. Acid. Res.* **16**, 1703.

19. Hasty, K. A., Pourmotabbed, T. F., Goldberg, G. I., Thompson, J. P., Spinella, D. G., Stevens, R. M., and Mainardi, C. L. (1990) Human neutrophil collagenase. A distinct gene product with homology to other matrix metalloproteinases. *J. Biol. Chem.* **265**, 11,421–11,424.

20. Wilhelm, S. M., Collier, I. E., Kronberger, A., Eisen, A. Z., Marmer, B. L., Grant, G. A., Bauer, E. A., and Goldberg, G. I. (1987) Human skin fibroblast stromelysin: structure, glycosylation, substrate specificity, and differential expression in normal and tumorigenic cells. *Proc. Natl. Acad. Sci. USA* **84**, 6725–6729.

21. Collier, I. E., Wilhelm, S. M., Eisen, A. Z., Marmer, B. L., Grant, G. A., Seltzer, J. L., Kronberger, A., He, C., Bauer, E. A., and Goldberg, G. I. (1988) H-ras oncogene-transformed human bronchial epithelial cells (TBE-1) secrete a single metalloprotease capable of degrading basement membrane collagen. *J. Biol. Chem.*, **263**, 6579–6587.

22. Wilhelm, S. M., Collier, I. E., Marmer, B. L., Eisen, A. Z., Grant, G. A., and Goldberg, G. I. (1989) SV40-transformed human lung fibroblasts secrete a 92-kDa type IV collagenase wich is identical to that secreted by normal human macrophages. *J. Biol. Chem.* **264**, 17,213–17,221.

23. Stricklin, G. P. and Welgus, H. G. (1983) Human skin fibroblast collagenase inhibitor. Purification and biochemical characterization. *J. Biol. Chem.* **258**, 12,252–12,258.

24. Docherty, A. J. P., Lyons, A., Smith, B. J., Wright, E. M., Stephens, P. E., and Harris, T. J. R. (1985) Sequence of human tissue inhibitor of metalloproteinases and its identity to erythroid-potentiating activity. *Nature* **318**, 66–69.

25. Gasson, J. C., Golde, D. W., Kaufman, S. E., Westbrook, C. A., Hewick, R. M., Kaufman, R. J., Wong, G. G., Temple, P. A., Leary, A., Brown, E. L., Orr, E. C., and Clark, S. C. (1985) Molecular characterization and expression of the gene encoding human erythroid-potentiating activity. *Nature* **315**, 768–771.

26. Goldberg, G. I., Marmer, B. L., Grant, G. A., Eisen, A. Z., Wilhelm, S., and He, C. (1989) Human 72-kDa type IV collagenase forms with a tissue inhibitor of metalloproteases designated TIMP-2. *Proc. Natl. Acad. Sci. USA* **86,** 8207–8211.

27. Stetler-Stevenson, W. G., Krutzsch, H. C., and Liotta, L. A. (1989) Tissue inhibitor of metalloproteinase (TIMP-2). A new member of the metalloproteinase inhibitor family. *J. Biol. Chem.* **264**, 17,374–17,378.

28. Boone, T. C., Johnson, M. J., De Clerck, Y. A., and Langley, K. E. (1990) cDNA cloning and expression of a metalloproteinase inhibitor related to tissue inhibitor of metalloproteinases. *Proc. Natl. Acad. Sci. U.S.A.* **87,** 2800–2804.

29. Breathnach, R. Matrisian, L. M., Gesnel, M. C. Staub, A., and Leroy, P. (1987) Sequences coding for part of oncogene-induced transin are higly conserved in a related rat gene. *Nucleic Acid Res.* **15,** 1139–1151.

30. Muller, D., Quantin, B., Gesnel, M-C., Millon-Collard, R., Abecassis, J., and Breathnach, R. (1988) The collagenase gene family in human consists of at least four members. *Biochem. J.* **253,**187-192.

31. Abramson, S. R., Conner, G. E., Nagase, H., Neuhaus, Y., and Woessner, J. F. (1995) Characterization of rat uterine matrilysin and its cDNA: relationship to human pump-1 and activation of procollagenases. *J. Biol. Chem.* **270**, 16,016–16,022.

32. Stolow, M. A., Bauzon, D. D., Li, J., Sedgwick, T., Liang, V.C-T., Sang, Q. A., and Shi, Y-B. (1996) Identification and characterization of a novel collagenase in Xenopus laevis: possible roles during frog development. *Mol. Biol. Cell* **7**, 1471–1483.

33. Shapiro, S. D., Kobayashi, D. K., and Ley, T. J. (1993) Cloning and characterization of a unique elastolytic metalloproteinase produced by human alveolar macrophages. *J. Biol. Chem.* **268**, 23,824–23,829.

34. Llano, E., Pendás, A. M., Knaüper, V., Sorsa, T., Salido, E., Murphy, G., Simmer, J. P., Bartlett, J. D., and López-Otín, C. (1997) Identification and structural and functional characterization of human enamelysin (MMP-20). *Biochemistry* **36,** 15,101–15,108.

35. Freije, J. M. P., Díez-Itza, I., Balbín, M., Sánchez, L. M., Blasco, R., Tolivia, J., and López-Otín, C. (1994) Molecular cloning and expression of collagenase-3, a novel human matrix metalloproteinase produced by breast carcinomas. *J. Biol. Chem.* **269,** 16,766–16,773.

36. Mitchell, P. G., Magna, H. A., Reeves, L. M., Lopresti-Morrow, L. L., Yocum, S. A., Rosner, P. J., Geoghegan, K. F., and Hambor, J.E. (1996) Cloning, expression, and type II collagenolytic activity of matrix metalloproteinase-13 from human osteoarthritic cartilage. *J. Clin. Invest.* **97,** 761–768.

37. Stahle-Bäckdahl, M., Sandsted, B., Bruce, K., Lindahl, A., Jiménez, M. G., Vega, J. A., and López-Otín, C. (1997) Collagenase-3 (MMP-13) is expressed during human fetal ossification and re-expressed in postnatal bone remodeling and in rheumatoid arthritis. *Lab. Invest.* **76**, 717–728.

38. Sato, H., Takino, T., Okada, Y., Cao, J., Shinagawa, A., Yamamoto, E., and Seiki, M. (1994) A matrix metalloproteinase expressed on the surface of invasive tumour cells. *Nature* **370,** 61–65.

39. Will, H. and Hinzmann, B. (1995) cDNA sequence and mRNA distribution of a novel human matrix metalloproteinase with a potential transmembrane domain. *Eur. J. Biochem.* **231,** 602–608.

40. Takino, T., Sato, H., Shinagawa, A., and Seiki, M. (1995) Identification of the second membrane-type matrix metalloproteinase (MT-MMP2) gene from a human placenta cDNA library. *J. Biol. Chem.,* **270,** 23,013–23,020.

41. Puente, X. S., Pendás, A. M., Llano, E., Velasco, G. and López-Otín, C. (1996) Molecular cloning of a novel membrane-type matrix metalloproteinase from a human breast carcinoma. *Cancer Res.* **56,** 944–949.

42. Shapiro, S. D., Griffin, G. L., Gilbert, D., Jenkins, N. A., Copeland, N. G., Welgus, H. G., Senior, R. M., and Ley, T. J. (1992) Molecular cloning, chromosomal localization, and bacterial expression of a murine macrophage metalloelastase. *J. Biol. Chem.* **267,** 4664–4671.

43. Bartlett, J. D., Simmer, J. P., Xue, J., Margolis, H. C., and Moreno, E. C. (1996) Molecular cloning and mRNA tissue distribution of a novel matrix metalloproteinase isolated from a porcine enamel organ. *Gene* **183,** 123–128.

44. Yang, M., Murray, M. T., and Kurkinen, M. (1997) A novel matrix metalloproteinase gene (XMMP) encoding vitronectin-like motifs is transiently expressed in Xenopus laevis early embryo development. *J. Biol. Chem.* **272,** 13,527–13,533.

45. Uría, J. A., Ferrando, A. A., Velasco, G., Freije, J. M. P., and López-Otín, C. (1994) Structure and expression in breast tumors of human TIMP-3, a new member of the metalloproteinase inhibitor family. *Cancer Res.* **54,** 2091–2094.

46. Silbiger, S. M., Jacobsen, V. L., Cupples, R. L., and Koski, R. A. (1994) Cloning of cDNAs encoding human TIMP-3, a novel member of the tissue inhibitor of metalloproteinase family. *Gene* **141,** 293–297.

47. Apte, S. S., Mattei, M. G., and Olsen, B. R. (1994) Cloning of the cDNA encoding human tissue inhibitor of metalloproteinases-3 (TIMP-3) and mapping of the TIMP3 gene to chromosome 22. *Genomics* **19,** 86–90.

48. Pavloff, N., Staskus, P. W., Kishnani, N. S., and Hawkes, S. P. (1992) A new inhibitor of metalloproteinases from chicken: ChIMP-3. *J. Biol. Chem.* **267,** 17,321–17,326.

49. Anand-Apte, B., Bao, L., Smith, R., Iwata, K., Olsen, B. R., Zetter, B. and Apte, S. S. (1997) A review of tissue inhibitor of metalloproteinases-3 (TIMP-3) and experimental analysis of its effect on primary tumor growth. *Biochem. Cell Biol.* **74,** 853–862.

50. Weber, B. H., Vogt, G., Pruett, R. C., Stohr, H., and Felbor, U. (1994) Mutations in the tissue inhibitor of metalloproteinase-3 (TIMP-3) in patients with Sorsby's fundus distrophy. *Nat. Genet.* **8,** 352–356.

51. Adams, M. D., Kelley, J. M., Gocayne, J. D., Dubnick, M., et al., (1991) Initial assessment of human gene diversity and expression pattern based upon 83 million nucleotides of cDNA sequence. *Nature* **377,** 3–17.

52. Soares, M. B., Bonaldo, M. F., Jelene, P., Su, L., Lawton, L., and Efstratiadis, A. (1994) Construction and characterization of a normalized cDNA library. *Proc. Natl. Acad. Sci. USA* **91**, 9228–9232.
53. Wolfsberg, T. G. and Landsman, D. (1997) A comparison of expressed sequence tags (ESTs) to human genomic sequences. *Nucleic Acids Res.* **25**, 1626–1632.
54. Pendás, A. M., Knäuper, V., Puente, X. S., Llano, E., Mattei, M. G., Apte, S., Murphy, G. and López-Otín, C. (1997) Identification and characterization of a novel matrix metalloproteinase with unique structural characteristics, chromosomal location and tissue distribution. *J. Biol. Chem. 272*: 4281–4286.
55. Liu, Y. E., Wang, M., Greene, J., Su, J., Ullrich, S., Li, H., Sheng, S., Alexander, P., Sang, Q. A., and Shi, Y. E. (1997) Preparation and characterization of recombinant tissue inhibitor of metalloproteinase 4 (TIMP-4). *J. Biol. Chem.* **272**, 20,479–20,483.
56. Guruajan, R. Grenet, J., Lahti, J. M., and Kidd, V. J. (1998) Isolation and characterization of two novel metalloproteinase genes linked to Cdc2L locus on human chromosome 1p36.3. *Genomics* **52**, 101–106.
57. Velasco, G., Pendas, A. M., Fueyo, A., Knauper, V., Murphy, G., and López-Otin, C. (1999) Cloning and characterization of human MMP-23, a new matrix metalloproteinase predominantly expressed in reproductive tissues and lacking conserved domains in other family members. *J. Biol. Chem.* **274**, 4570–4576.
58. Llano, F., Pendás, A. M., Freije, J. P., Nakano, A., Knauper, V., Murphy, G., and López-Otin, C. (1999) Identification and characterization of human MT5-MMP, a new membrane-bound activator of progelatinase A overexpressed in brain tumors. *Cancer Res.* **59**, 2570–2576.
59. Velasco, G., Cal, S., Merlos-Suarez, A., Ferrando, A. A., Alvarez, S., Nakano, A., Arribas, J., and López-Otin, C. (2000) Humana MT6-MMP identification progelatinase A activation, and expression in brain tumors. *Cancer Res.* **60**, 877–882.

3

Structural Studies on MMPs and TIMPs

Wolfram Bode and Klaus Maskos

1. Introduction

The proteolytic activity of the matrix metalloproteinases (MMPs) involved in extracellular matrix degradation must be precisely regulated by their endogenous protein inhibitors, the tissue inhibitors of metalloproteinases (TIMPs). Disruption of this balance results in serious diseases such as arthritis and tumor growth and metastasis. Knowledge of the tertiary structures of the proteins involved is crucial for understanding their functional properties and interference with associated dysfunctions. Within the last few years, several three-dimensional structures became available showing the domain organization, the polypeptide fold and the main specificity determinants of the MMPs. Complexes of the catalytic MMP domains with various synthetic inhibitors enabled the structure based design and improvement of high-affinity ligands, which might be elaborated into drugs. Very recently, structural information also became available for some TIMP structures and MMP-TIMP complexes, which allowed to derive the structural features governing the enzyme-inhibitor interaction. A multitude of reviews surveying work done on all aspects of MMPs and TIMPs appeared within recent years, but none of them concentrating on the three-dimensional structures. This review is considered to close this gap.

The matrix metalloproteinases (MMPs, matrixins) form a family of structurally and functionally related zinc endopeptidases. Collectively, these MMPs are capable in vitro and in vivo of degrading all kinds of extracellular matrix protein components such as interstitial and basement membrane collagens, proteoglycans, fibronectin, and laminin; they are thus implicated in connective tissue remodeling processes associated with embryonic development, pregnancy, growth, and wound healing *(1)*. Normally, the degenerative potential of the MMPs is held in check by the endogenous specific (the tissue inhibitors of

From: *Methods in Molecular Biology, vol. 151: Matrix Metalloproteinase Protocols*
Edited by: I. Clark © Humana Press Inc., Totowa, NJ

metalloproteinases, TIMPs) and nonspecific protein inhibitors (in particular α_2-macroglobulin). Disruption of this MMP-TIMP balance can result in pathologies such as rheumatoid and osteoarthritis, atherosclerosis, tumor growth, metastasis, and fibrosis (for recent overviews, *see* e.g., **refs. *2–6***). Therapeutic inhibition of MMPs is a promising approach for treatment of some of these diseases, and the MMP structures and their TIMP complexes are therefore attractive targets for rational inhibitor design (for recent literature, *see* **refs. *7,8***).

To date, at least 18 different human MMPs have been identified and/or cloned, which share significant sequence homology and a common multidomain organization; several counterparts have been found in other vertebrates, invertebrates, and from plant sources (*see* **ref. *9***), together forming the MMP or matrixin subfamily A of the metalloproteinase M10 family *(10)*. According to their structural and functional properties, the MMP family can be subdivided into five groups: (1) the collagenases (MMPs-1, -8 and -13); (2) the gelatinases A and B (MMPs-2 and -9); (3) the stromelysins 1 and 2 (MMPs-3 and 10); (4) a more heterogenous subgroup containing matrilysin (MMP-7), endometase (MMP-26), enamelysin (MMP-20), the *mmp20* gene product, macrophage metalloelastase (MMP-12), and MMP-19 (together making up the "classical" MMPs); and *(5)* the membrane-type-like MMPs (MT-MMPs-1 to -4 and stromelysin-3, MMP-11). These MMPs share a common multidomain structure, but are glycosylated to different extents and at different sites. According to sequence alignments *(9,11)*, the assembly of these domains might have been an early evolutionary event, followed by diversification *(12)*.

All MMPs are synthesized with an ~20 amino acid residue signal peptide and are (probably, except the MT-MMP-like furin-processed proteinases *[13–15]*) secreted as latent pro-forms; these proproteinases consist of an ~80 residue N-terminal prodomain followed by the ~170 residue catalytic domain (cd; *see* **Figs. 1** and **2**), which in turn (except for matrilysin) is covalently connected through a 10–70 residue Pro-rich linker to an ~195 residue C-terminal haemopexin-like domain (hld); in the MT-MMPs, the polypeptide chain possesses an additional

---

Fig. 1. (*opposite page*) Ribbon structure of the MMP catalytic domain (cd) shown in the standard orientation (**TOP**) cd of MMP-8 *(38)* shown together with the heptapeptide substrate. The catalytic and the structural zinc (*center and top*) and the calcium ions (flanking) are shown as dark gray and light gray spheres, and the three His liganding the catalytic zinc, the catalytic Glu in between, the characteristic Met, the Pro and the Tyr of the S1' wall forming segment, and the N-terminal Phe and the first Asp of the Asp pair forming the surface located salt bridge are shown with all atoms. (**BOTTOM**) Superposition of the MMP-8 cd (ribbon) with the cds of MMP-3, MMP-1, MMP-14, and MMP-7 (thin ropes), shown together with the heptapeptide substrate. The chain segment forming the extra domain of both gelatinases will be inserted in the sV-hB loop (*center, right*) and presumably extends to the right side. Figures made with SETOR *(84)*.

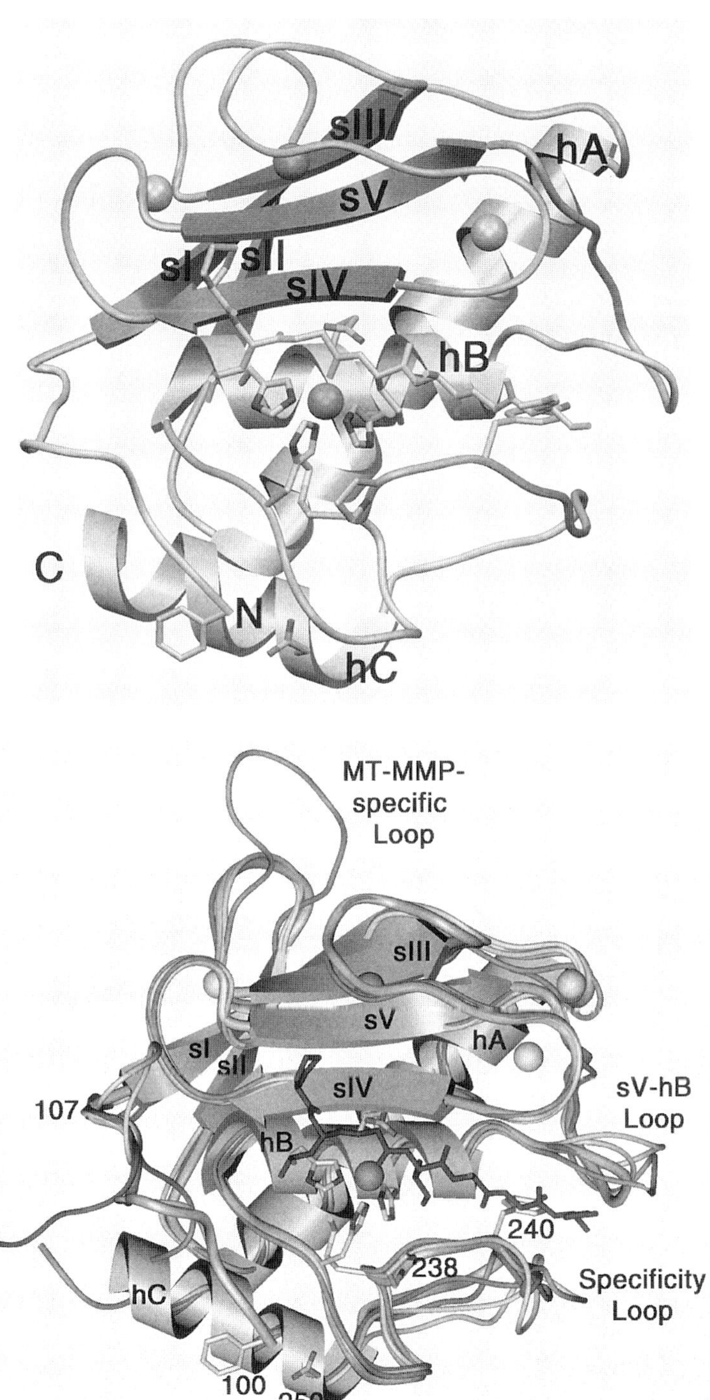
sIII
hA
sV
sI sII
sIV
hB
C
N
hC
MT-MMP-
specific
Loop
sIII
sV
hA
sI sII
sIV
sV-hB
Loop
107
hB
240
238
Specificity
Loop
hC
100 250

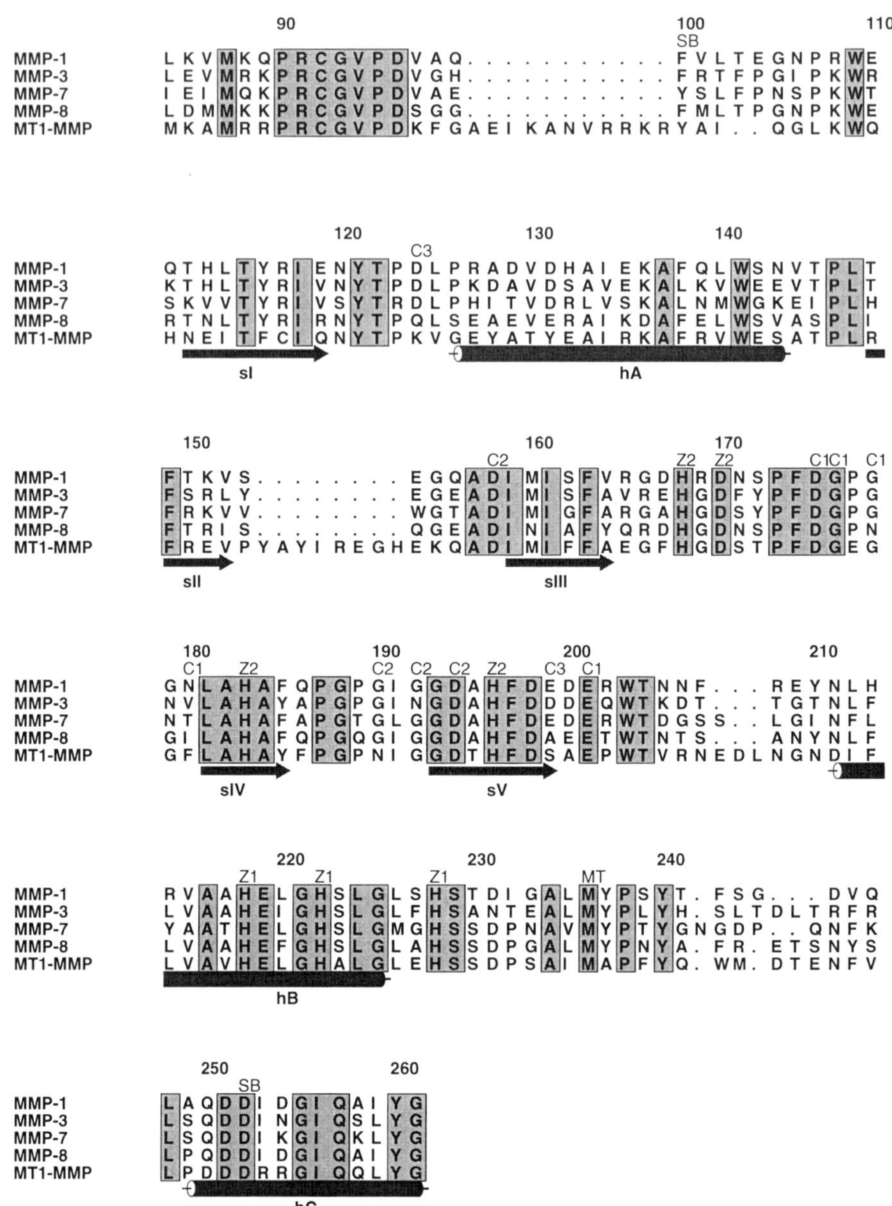

Fig. 2. Alignment of the catalytic domains of MMP-1, MMP-3, MMP-7, MMP-8, and MMP-14, made according to topological equivalencies *(21)*. Location and extent of α-helices and β-strands are given by cylinders and arrows, and symbols show residues involved in main and side chain interactions with the catalytic (Z1) and the structural zinc (Z2), the first (C1), second (C2) and third calcium ion (C3) of the cds. Figure made with ALSCRIPT *(85)*.

75–100 residue extension, which presumably forms a transmembrane helix and a small cytoplasmic domain *(14)*. Removal of this hld in the collagenases eliminates their characteristic capability to cleave triple-helical collagen, but does not significantly affect hydrolytic activity toward gelatin, casein or synthetic substrates (*see* **ref. *16***). In both gelatinases, the catalytic domains have an additional 175 amino acid residue insert comprising three fibronectin-related type II modules confering gelatin and collagen binding.

The TIMP family currently includes 4 different members (TIMPs-1 to 4; *[17–20]*), which after optimal topological superposition exhibit 41–52% sequence identity (*see* **Fig. 3**; *see* **refs. *21,22***). Besides their inhibitory role, these TIMPs seem to have other functions such as growth factor-like and anti-angiogenic activity (*see* e.g., **refs. *23,24***). The TIMP cDNAs encode an ~25 residue leader peptide, followed by the 184 to 194 amino acid residue mature inhibitor (*see* **Figs. 3** and **4**). Virtually all TIMPs form tight 1:1 complexes with MMPs. Except for the rather weak interaction between TIMP-1 and MT1- and 2-MMPs *(14,25–27)*, the TIMPs do not seem to differentiate much between the various MMPs (*see* **ref. *28***). TIMPs-1 and 2 are unique in that they also bind to the pro-forms of gelatinase B and A, respectively *(29)*; the complex between MT1-MMP and TIMP-2 seems to act as a cell-surface-bound "receptor" for progelatinase A activation in vivo, using these noninhibitory interactions between TIMP-2 and progelatinase A *(26,30,31)*. Removal of the C-terminal 1/3 of the TIMP polypeptide chain gives rise to so-called N-terminal TIMP domains (N-TIMPs), which retain most of their reactivity toward their target MMPs *(32,33)*.

Only in early 1994, the first X-ray crystal structures of the catalytic domains (blocked by various synthetic inhibitors, *see* **Table 1**) of human fibroblast collagenase/MMP-1 *(34–36)* and human neutrophil collagenase/MMP-8 *(36–38)*, and an NMR structure of the catalytic domain of stromelysin-1/MMP-3 *(39)* became available, which were later complemented by additional cd structures of MMP-1 *(40,41)*, matrilysin/MMP-7 *(42)*, MMP-3 *(43–47)*, MMP-8 *(48,49)*, and MT1-MMP *(21)*. In 1995, the first X ray structure of an MMP pro-form, the C-terminally truncated prostromelysin-1, was published *(43,46)*, and the first structure of a mature full-length MMP, namely of porcine fibroblast collagenase/MMP-1 *(50)*, was described. At that time, structures of the isolated haemopexin-like domains from human gelatinase A *(51,52)* and from collagenase-3/MMP-13 *(53)* were also reported *(54)*. More recently, the crystal structures of the cds from human *(54a)* and mouse collagenase-3 (MMP-13) *(54b)* and of full-length pro-gelatinase A (MMP-2) *(54c)* were published, besides a number of other MMP cds containing synthetic inhibitors (*see* **Table 1**). Of the TIMPs, a first preliminary NMR model of human N-TIMP-2 was presented in 1994 *(55)*, which showed that the polypeptide framework of the N-terminal part of the TIMPs resem-

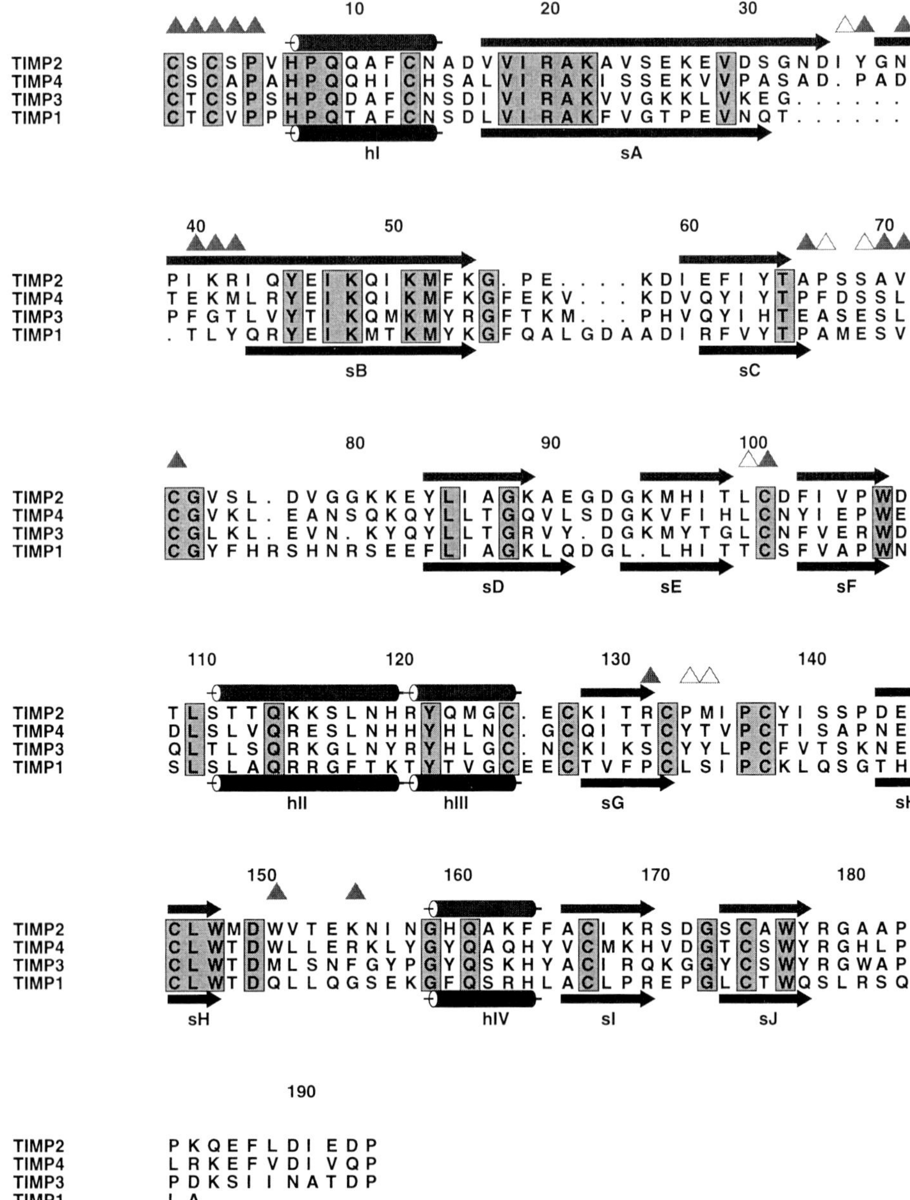

Fig. 3. Alignment of TIMPs-1 to -4 *(17–20)*, made according to topological equivalencies *(21)*. The numbering is that of mature TIMP-2. Location and extent of α-helices and β-strands are given by cylinders and arrows, and symbols show residues involved in main and side chain interactions with MT1-MMP *(21)*. Figure made with ALSCRIPT *(85)*.

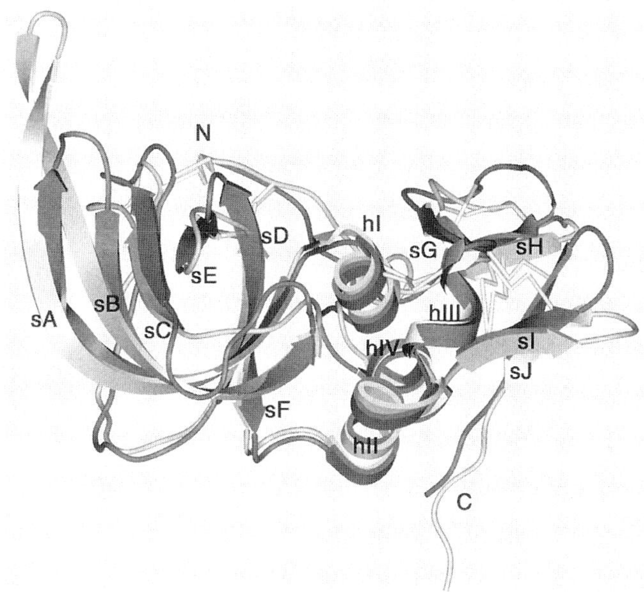

Fig. 4. The TIMP structure. The wedge-like TIMP-1 (dark gray ribbon) and TIMP-2 molecules (light gray ribbon) are superimposed upon minimizing the deviations between all equivalent α-carbon atoms *(21,47)*. This front view is related to the standard view *(see* **Fig. 1**) by a 90 degree rotation about a horizontal. The polypeptide chains start with their N-terminal segment (marked with N, centre top), then build up the N-terminal (left) and the C-terminal subdomain (right), before terminating in the flexible tails (marked with C, bottom right). The strands sA to sJ and the helices hI to hIV are labeled *(47)*. Figure made with SETOR *(84)*.

bles so-called OB-fold proteins; a refined N-TIMP-2 has recently been published, describing some enhanced mobility of contacting inhibitor segments *(56,57)*. Only in 1997, the first structure of a complete TIMP, human deglycosylated TIMP-1, in complex with the catalytic domain of human MMP-3, was published *(47)*, followed by the X-ray structure of TIMP-2 in complex with the catalytic domain of MT1-MMP *(21)*. More recently, also the crystal structure of isolated human TIMP-2 *(57a)* and the NMR structure of N-TIMP-1 *(57b)* were published.

In the following, the structures of the MMPs and TIMPs and their detailed interactions will be presented. On the basis of their topological equivalencies, for both the MMPs *(see* **Fig. 2**) and the TIMPs *(see* **Fig. 3**) structure-based sequence alignments will be given *(21)*. The MMP nomenclature used is based on the cDNA sequence of (human) fibroblast collagenase/MMP-1 as the refer-

Table 1
3D Structures of MMPs and TIMPs as Deposited at the PDB

Protein	Domain	Inhibitor	Method	Reference	PDB
HFC	cd	Carboxyalkyl	X-ray	*34a*	1CGL
HFC	cd	N-term neighbour	X-ray	*40*	1CGE
HFC	cd	N-term neighbour	X-ray	*40*	1CGF
HFC	cd	Hydroxamate	X-ray	*35*	2TCL
HFC	cd	Hydroxamate	X-ray	*41*	1HFC
HFC	cd	Free	NMR	*91*	2AYK
HNC	cd	Hydroxamate	X-ray	*54a*	966C
HNC	cd	Hydroxamate	X-ray	*37*	1JAP
HNC	cd	Hydroxamate	X-ray	*38*	1JAN
HNC	cd	Hydroxamate	X-ray	*48*	1JAO
HNC	cd	Hydroxamate	X-ray	*48*	1JAQ
HNC	cd	Hydroxamate	X-ray	*49*	MMB
HNC	cd	Hydroxamate	X-ray	*65*	1KBC
HNC	cd	Hydroxamate	X-ray	*90*	1A85
HNC	cd	Carboxylate	X-ray	*54h*	1BZS
HNC	cd	Hydroxamate	X-ray	*36*	1MNC
SL1	cd	Carboxyalkyl	NMR	*39*	2SRT
SL1	cd	Carboxyalkyl	X-ray	*43*	1SLN
SL1	cd	Carboxyalkyl	X-ray		1HFS
SL1	cd	Hydroxamate	NMR	*92*	1UMT
SL1	cd	Carboxyalkyl	X-ray	*43*	1SLM
SL1	cd	Hydroxamate	X-ray	*54e*	1BIW
SL1	cd	Carboxamide	X-ray	*54f*	1BQO
SL1	cd	Hydroxamate	X-ray	*54g*	1B3D
SL1	cd	Nonpeptidic	X-ray	*54i*	1CAQ
SL1	cd	Nonpeptidic	X-ray	*54i*	1CIZ
SL1	cd	Thiadiazole	X-ray	*54j*	1USN
SL1	cd	Thiadiazole	X-ray	*54j*	2USN
MMP-7	cd	Carboxylate	X-ray	*42*	1MMP
MMP-7	cd	Hydroxamate	X-ray	*42*	1MMQ
MMP-7	cd	Sulfodiimine	X-ray	*42*	1MMR
MMP-2	prod+fn+cd+hld		X-ray	*54c*	1CK7
MMP-2	cd lacking fn	Hydroxamate	X-ray	*54d*	1QIB
MMP-13	cd	Hydroxamate	X-ray	*54a*	456C
MMP-13	cd	Hydroxamate	X-ray	*54a*	830C
proSL1	prod+cd	Pro	X-ray	*43*	1SLM
MMP-1	cd+hld	Hydroxamate	X-ray	*50*	1FBL
MMP-2	hld		X-ray	*51*	1GEN
MMP-2	hld		X-ray	*52*	1RTG
MMP13	hld		X-ray	*53*	1PEX
MMP3-TIMP1	cd+timp-1		X-ray	*47*	1UEA
MMP14-TIMP2	cd+timp-2		X-ray	*21*	1BUV
N-TIMP-2			NMR	*57*	2TMP
TIMP-2			X-ray	*57a*	1BR9
N-TIMP-2			NMR	*57b*	1D2B

ence MMP (*see* **Fig. 2**). For the assignment of peptide substrate residues and substrate recognition sites on the proteinase, the nomenclature of Schechter and Berger *(58)* will be used: P1, P2 and so on, and P1', P2' and so on, indicate the residues in N- and C-terminal direction of the scissile peptide bond of a bound peptide substrate (analogue); and S1, S2 and so on, and S1', S2' and so on, the opposite binding sites on the enzyme. TIMP residues will be given with TIMP-1/TIMP-2 numbers.

2. Structures and Mechanisms

2.1. MMPs

2.1.1. Catalytic Domain

The catalytic domains of the MMPs exhibit the shape of an oblate ellipsoid. In the "standard" orientation, which in this article as well as in most other MMP papers is preferred for the display of MMPs, a small active-site cleft notched into the flat ellipsoid surface extends horizontally across the domain to bind peptide substrates from left to right (*see* **Fig. 5**). This cleft harboring the "catalytic zinc" separates the smaller "lower subdomain" from the larger "upper subdomain."

This upper subdomain formed by the first three quarters of the polypeptide chain (up to Gly225) consists of a five-stranded β-pleated sheet, flanked by three surface loops on its convex side and by two long regular α-helices on its concave side embracing a large hydrophobic core (*see* **Fig. 1**). The polypeptide chain starts on the molecular surface of the lower subdomain, passes β-strand sI, the amphipathic α-helix hA, and β-strands sII, sIII, sIV, and sV, before entering the "active-site helix" hB (for nomenclature, *see* **Fig. 1**). In the classical MMPs, strands sII and sIII are connected by a relatively short loop bridging sI; in the MT-MMPs, however, this loop is expanded into the spur-like, solvent exposed "MT-MMP-specific loop" of hitherto unknown function (*see* **Fig. 1B**). In all MMPs, strands sIII and sIV are linked via an "S-shaped double loop," which is clamped via the "structural zinc" and the first of two to three bound calcium ions to the β-sheet. This S-loop extends into the cleft-sided "bulge" continuing in the antiparallel "edge strand" sIV; this bulge-edge segment is of prime importance for binding of peptidic substrates and inhibitors (*see* **Fig. 6A**). The sIV-sV connecting loop together with the sII-sIII bridge sandwiches the second bound calcium. After strand sV, the chain passes the large open sV-hB loop before entering the active-site helix hB; this helix provides the first His (218) and the second His (222) which ligand the catalytic zinc, and the "catalytic Glu219" in between, all of them representing the N-terminal part of the "zinc binding consensus sequence" HEXGHXXGXXH (*see* **Fig. 2**) characteristic of the metzincin superfamily *(59,60)*.

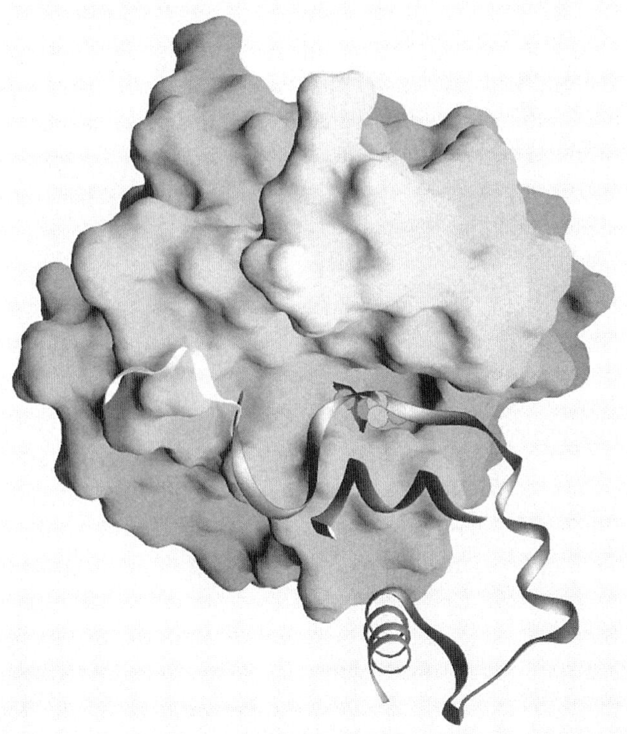

Fig. 5. The pro-form of the cd of Pro-MMP-3 *(43)* shown in standard orientation. The cd from the hinge Pro107 onwards is given as a solid surface centered around the catalytic zinc (half sphere, center) located in the active-site cleft. The pro-segment is shown as a ribbon with its ordered segments only, i.e., with the first (apparently separate) helix, the second helix, the connecting loop, the third helix, the switch loop (running across the catalytic zinc, liganding it via the conserved Cys side chain, shown as a CPK model) and the rocker arm (white color), which swings down after activation cleavage. Figure made with GRASP *(76)*.

This active-site helix abruptly stops at Gly225, where the peptide chain bends down, descends (presenting the third zinc-liganding histidine, His228, and the following MMP-invariant Ser229, the function of which is not evident judged from the cd structure alone) and runs through a wide right-handed spiral (catalytic domain's "chin") terminating in the 1,4-tight "Met-turn" (of the strongly conserved sequence Ala234-Leu235-Met236-Tyr237, with an obligatory methionine residue at turn position 3). The chain then turns back to the molecular surface to an (except in human stromelysin-3) invariant Pro238, forms with a conserved Pro238-Xaa-Tyr240 segment (the "S1' wall forming segment") the outer wall of the S1' pocket, runs through another wide loop of slightly variable length and conformation (which due to fencing round the most important S1' pocket and codetermining its extension might also be called the "specificity loop,"), before it passes the C-terminal α-helix hC, which ends with the conserved Tyr260-Gly261 residue pair.

The overall structures of all MMP cds known so far are very similar, with the collagenase structures resembling one another most, the MMP-7 and the MT1-MMP structures deviating most (**Fig. 1B**), and the MMP-2 exhibiting the three well-separated fibronectin-like domains (not shown in Fig. 1). Larger main chain differences occur (1) in the N-terminal segment up to Pro107 (depending on the length of the N-terminus and the presence of a TIMP); (2) in the sII-sIII bridge (with the elongated and more exposed MT loop in the MT-MMPs); (3) in the sV-hB loop; and (4) in the specificity loop. With two (compared to MMP-7) and three (MMPs-1, -3, and -8) additional residues, the open sV-hB loop of MT1-MMP deviates most, followed by MMP-7. In progelatinase A (and certainly also in MMP-9), the four residues between Leu205 and Tyr210 are replaced by a 187 residue insert, which forms a large cloverleaf-like extra domain (fn) *(57c)* consisting of three tandem copies of fibronectin type II-like modules *(61)*; furthermore remarkable is the unusual cis-conformation of the peptide bond preceding Tyr210 in MMPs-1 and -8, giving the possibility to form a favourable hydrogen bond with an adjacent strand *(41,48)*. The specificity loop is shortest in MMP-1 and MMP-2; those of MMPs-3, -8 and -14 (with three and two additional residues) resemble one another, although that of MMP-7 (two) deviates most (**Figs. 1B** and **2**).

Besides the catalytic zinc, all MMP catalytic domains possess another zinc ion, the structural zinc, and two (MMP-8, MT1-MMP) or three bound calcium ions (MMP-1, MMP-3, MMP-7) (**Figs. 1** and **2**). The structural zinc and the first (probably most tightly bound) calcium are sandwiched between the double-S loop and the outer face of the β-sheet. In the vast majority of the classical and the furin-activatable MMPs, the structural zinc is coordinated by the imidazoyl Nε2 or Nδ1 atoms, respectively, of three His residues (provided by the S-loop, sIV and sV) and by one carboxylate oxygen of the Asp residue in the S-loop next but one to the liganding His residue (*see* **Fig. 2**). This zinc is completely buried in the protein matrix; the impossibility to exchange it in the MMP-8 crystals *(37)* suggested its extremely tight binding. The second part of the S-loop encircles the adjacent calcium ion and packs it against the side chain carboxylate groups of an invariant Asp-Glu couple protruding from strand sV; these carboxyl oxygens, together with three carbonyl and one carboxyl oxygen of the S-loop, coordinate this calcium in a nearly octahedral manner. The second calcium ion, sandwiched between the sIV-sV loop and sIII, is octahedrally coordinated by three carbonyl groups, a bulk solvent and one carboxylate oxygen of the invariant Asp which starts strand sV. The loop immediately follow-

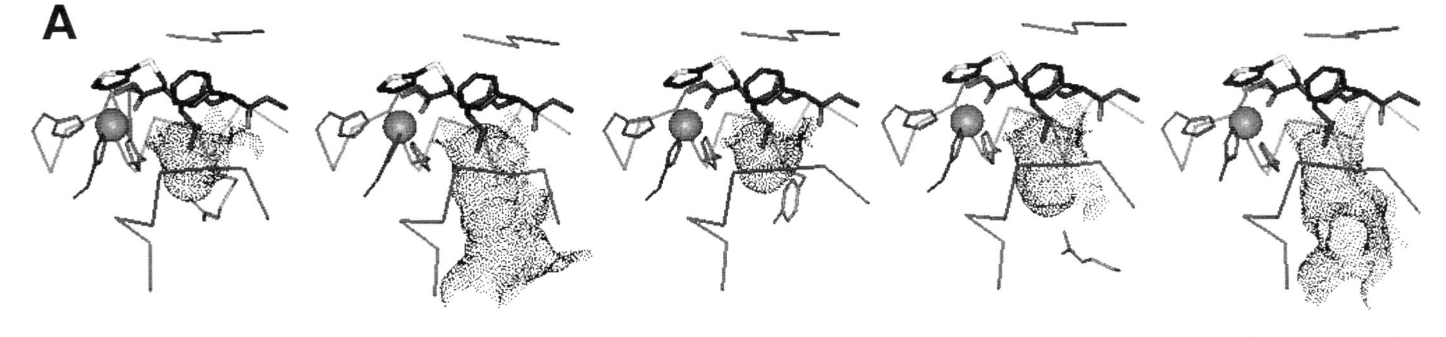

Fig. 6. Peptide substrate and inhibitor interaction and specificity. (**A**) Comparison of the S1' pockets (dot surface) of (from left to right) MMP-1, MMP-3, MMP-7, MMP-8 and MMP14. Besides the Cα plots of the bulge-edge strand (dark gray, top), of the active-site helix (light gray, center) and the segment comprising the Met-turn and the S1' wall forming segment (dark gray, bottom), the full structure of the inhibitor lead Batimastat (4-(*N*-hydroxyamino)-2(R)-isobutyl-3(S)-([2-thienyl-thiomethyl]succinyl)-L-phenylalanine-*N*-methylamide) alias BB-94 (as bound to MMP-8 *[49]*), the catalytic zinc (sphere), the side chains of the three zinc liganding His residues, and the side chains of residues Arg214, Tyr214, and Arg243 restricting the S1' pockets of MMP-1, MMP-7 and MMP-8, respectively, in size are shown (Figure courtesy F. Grams) (**Fig. 6B** *on next page*).

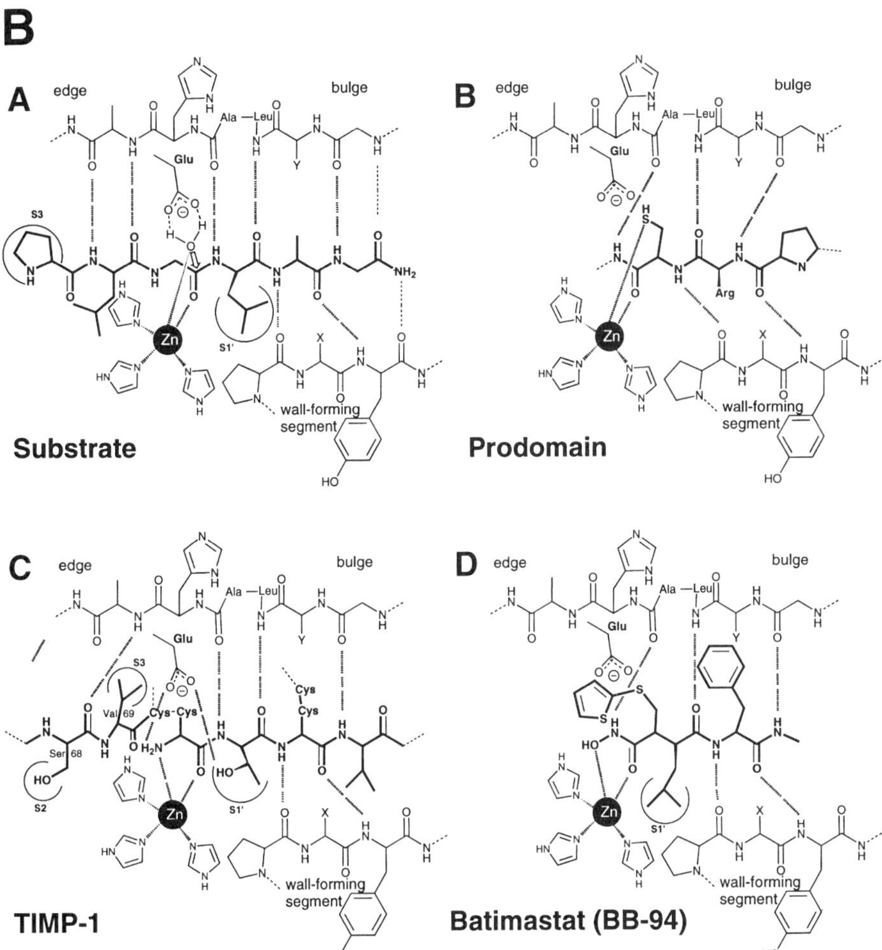

Fig. 6 **(B)** Schematic drawing of the putative encounter complex between (A) a modeled Pro-Leu-Gly-Leu-Ala-Gly-amide hexapeptide substrate *(48)*, (B) the prodomain switch peptide *(43)*, (C) the N-terminal segment and the C-connector loop of TIMP-1 *(47)*, and (D) Batimastat *(49)* and the MMP active site. The substrate polypeptide chain, the switch peptide, the N-terminal segment, and the inhibitor peptide mimic (bold connections) run antiparallel to the bulge-edge strand (top) and parallel to the S1' wall forming segment (bottom), forming several inter-main-chain hydrogen bonds (dashed lines). Dominant hydrophobic interactions are made through the substrate's P1 and P3 side chains with the S1' pocket and the S3 subsite (emphasized through two troughs). In the enzyme-substrate encounter, the catalytic water activated by the catalytic Glu residue is suitably placed and activated to attack the carbonyl group of the scissile P1–P1' peptide bond.

ing sV encircles the third calcium and coordinates it through two carbonyl groups (of residues 199 and 201; *see* **Fig. 2**) and one carboxylate oxygen of Asp124, whose presence seems to correlate with the existence of a third calcium site.

For the fibroblast and the neutrophil collagenases, a several-fold larger activity has been demonstrated for active enzymes starting with a highly conserved Phe100 residue compared with species truncated by one or two residues (a phenomenon also called "superactivity" or "superactivation,") *(62–64)*. Position and fixation of the N-terminus of the mature classical MMPs indeed seem to depend on the presence of the N-terminal amino acid Phe/Tyr100, i.e., on the accurate processing/tailoring of the MMP precursor *(37,38)*. In cases that it starts with (the highly conserved) Phe100 (*see* **Fig. 2**), the N-terminal heptameric segment preceding the conserved Pro107-Lys/Arg-Trp109 triple is tightly packed against a hydrophobic surface groove made by the C-terminal helix hC and the descending segment centering around the third His ligand of the catalytic zinc; the N-terminal Phe100 ammonium group makes a surface located salt bridge with the side chain carboxylate of an Asp250, which is the first residue of a strictly invariant helix hC based Asp250-Asp251 pair *(38)*. The side chain of the second of both Asp residues is buried in a solvent filled protein cavity and hydrogen bonded via the Met-turn to the first zinc-liganding His. In this way, formation of the Phe100 . . . Asp250 salt bridge might be signaled to the active center. In the absence of an N-terminal Phe/Tyr100 (a state associated with a lower catalytic activity), the N-terminal (hexa)peptide preceding Pro107 is disordered and might interfere with substrate binding, offering an alternative explanation for the reduced activity of such N-terminally truncated MMPs.

2.1.2. Specificity Determinants

Bounded at the upper rim by the bulge-edge segment and the second part of the S-loop, and at the lower side by the third zinc-liganding imidazole and the S1' wall-forming segment, the active-site cleft of all MMPs is relatively flat at the left ("nonprimed") side, but carves into the molecular surface at the catalytic zinc and to the right ("primed") side leveling-off again to the surface further to the right (*see* **Fig. 5**). In its center and to the right side, this cleft exhibits a slightly negative potential. In unliganded MMPs, the catalytic zinc residing in its center is coordinated by the three imidazole Nε2 atoms of the three histidines (His218, 222, and 228) and by a fixed water molecule, which simultaneously is in hydrogen bond distance to the carboxylate group of the catalytic Glu219. In case of MMP complexes with bidentate inhibitors (such as those with a hydroxamic acid function) (*see* **Figs. 6 A,B**), this water is replaced by two oxygen atoms, which together with the three imidazoles, ligand the catalytic zinc in a trigonal-bipyrimidal (penta-coordinate) manner *(37)*. As in all other metzincins *(59,60)*, the zinc-imidazole ensemble of the MMPs is placed above the distal ε-methyl-sulfur moiety of the strictly conserved Met236 of the Met-turn, which forms a hydrophobic base of still unclear function (*see* **Figs. 1A,B**). An MMP-8 catalytic domain with this Met replaced by a selenomethionine retained its catalytic activity, and exhibited slightly decreased kinetic and thermal properties but (except for a slight local disturbance) virtually no conformational differences compared to the wild-type domain *(65)*.

Immediately to the right of the catalytic zinc the S1' specificity pocket invaginates, which in size and shape differs considerably among the various MMPs (**Fig. 6A**). This pocket is mainly formed (seen in standard orientation) by (1) the initial part of the active-site helix hB ("back side"); (2) the somewhat mobile *(44)* phenolic side chain of Tyr240 ("right-hand flank"); (3) the main chain atoms of the underlying wall forming segment Pro-X-Tyr ("front side"); (4) the flat side of the first zinc liganding His218 imidazole ("left side"); and (5) the Leu(Ile/Val)235 residue of the Met-turn, which together with the Leu/Tyr/Arg214 or the Arg243 side chain (if present) form its "bottom" or line it toward the second exit opening at the lower molecular surface, respectively. In all MMPs, the interior of the pocket has direct water mediated connections to the bulk water; this pocket is, however, of quite different size and shape depending on the presence of a Leu, or a Tyr, or an Arg at position 214.

Nearly all of the synthetic inhibitors analyzed so far in MMP complexes contain a chelating group (such as a hydroxamic acid, a carboxylate, or a thiol group) for zinc ion ligation, and a peptidic or peptidomimetic moiety mimicking peptide substrate binding to the substrate recognition site (*see* **Table 1**). Of the synthetic inhibitors published in complex with an MMP, only the Pro-Leu-Gly-hydroxamic acid inhibitor *(37,38)* binds to the left-hand subsites (the nonprimed subsites S3 to S1) alone, antiparallel to the edge strand ("left-side inhibitor"). A few synthetic inhibitors bind across the active site, whereas in the majority of synthetic inhibitors studied so far this peptidic moiety interacts in an extended manner with the primed right-hand subsites ("right-side inhibitors"), inserting between the (antiparallel) bulge-edge segment and the (parallel) S1' wall forming segment of the cognate MMP under formation of a three-stranded mixed β-sheet (*see* **Fig. 6B**).

An L-configurated P1'-like side chain is perfectly arranged to extend into the hydrophobic bottleneck of the S1' pocket. This P1'–S1' interaction is the main determinant for the affinity of inhibitors and the cleavage position of peptide substrates. Depending in particular on the length and character of residue 214 harboured in the N-terminal part of the active-site helix hB, the size of the S1' pocket differs considerably among the MMPs (**Fig. 6A**). In MMP-1 and MMP-7, the side chains of Arg214 and Tyr214, respectively, extend into the

S1' opening limiting it to a size and shape still compatible with the accommodation of medium-sized P1' residues, but less for very large side chains, in agreement with peptide cleavage studies on model peptides *(66–68)*. Some more recent MMP-1 structures show, however, that the Arg214 side chain can swing out of its normal site thus allowing also binding of synthetic inhibitors with larger P1' side chains (M. Browner, personal communications). The smaller Leu214 residues of MMP-2, MMP-3 and MMP-14 do not bar the internal S1' "pore," which extends right through the molecule to the lower surface, i.e., equals more a long solvent-filled "tube" (**Fig. 6A**). In spite of a small Leu214 residue, however, the S1' pocket of MMP-8 is of medium size and is closed at the bottom, due to the Arg243 side chain extending into the S1' space from the specificity loop *(37)*.

Second in importance for substrate specificity seems to be the interaction of the P3 residue (in collagen cleavage sites always a Pro residue) with the mainly hydrophobic S3 pocket (**Fig. 6B**). The S2 site is a shallow depression extending on top of the imidazole ring of the second zinc-liganding His; its polarity character might be influenced by residue 227 preceding the third zinc-liganding His, His228. Longer side chains of P1 residues (in collagen cleavage sites mostly a Gly, for references *see [69]*) are placed in the surface groove lined by the His183 side chain of the edge strand together with the last bulge residue 180 (marked with a "Y" in **Fig. 6B**); depending on this latter side chain, one or the other P1 side chain might be preferred. P2' side chains extend away from the surface, squeezed between the bulge rim and the side chain of the middle residue of the Pro-X239-Tyr wall forming segment; the quality of interaction will be determined particularly by the nature of bulge residue180 and residue 239 (marked with an "X" in **Fig. 6B**), which in the MMPs-14, -15, and -16 and MMP-11 is an exposed Phe *(21)*. Further to the right side, the molecular surface again has a hydrophobic/polar depression, which could accomodate P3' side chains of differing nature (*see* the accommodation of residue 4 of bound TIMPs in **Fig. 6B**).

By replacing the zinc-chelating groups of such peptidic left- and right-side inhibitors by a normal peptide bond, a contiguous peptide substrate was constructed indicating the probable binding geometry of a normal substrate-MMP encounter complex (**Fig. 6B**) *(48)*. Accordingly, the peptide substrate chain is aligned in an extended manner to the continuous bulge-edge segment, under formation of an antiparallel two-stranded β-pleated sheet, which expands on the right-hand side into a three-stranded mixed parallel-antiparallel sheet due to additional alignment with the S1' wall forming segment. The bound peptide substrate (such as the hexapeptide shown in **Fig. 6B**) forms five and two inter main-chain hydrogen bonds, respectively, to both crossing-over MMP segments. Similar to the reaction mechanism previously suggested for the more distantly related zinc endopeptidase thermolysin *(70)*, the MMP catalyzed cleavage of the scissile peptide bond will probably proceed via a general-base mechanism *(48)*. The carbonyl group of the scissile bond (Gly-Phe in **Fig. 6B**) is directed nearly toward the catalytic zinc and strongly polarized. The zinc-bound water molecule is activated by the carboxylate/carboxylic acid of the catalytic Glu219, the pK of which might (in particular after complete shielding from the bulk water upon substrate/inhibitor binding) be shifted to higher values, due to packing in the protein matrix without charge stabilizing internal hydrogen bonds (the importance of this Glu219 for proteolytic activity in MMPs has been demonstrated through replacement with Asp, Ala, and Gln residues *(71,72)*, resulting in mutants with lowered or extremely low catalytic activity, respectively). This activated water molecule squeezed between the carboxylate group of the catalytic Glu and the scissile peptide bond carbonyl group is properly oriented to attack via its lone pair orbital the electrophilic carbonyl carbon. The tetrahedral intermediate is presumably stabilized by both the zinc and the carbonyl group of the first Ala residue of the edge strand sIV. Simultaneously, one water proton could be transferred via the Glu carboxylate (acting as a proton shuttle) to the amino group, which after cleavage of the peptide bond and transfer of a second proton could leave the enzyme-substrate complex together with the N-terminal substrate fragment. Remarkably, there is no other electrophile (such as His231 in thermolysin *[70]* or Tyr149 in astacin *[73,74]*) in the catalytic zinc environment of the MMPs, which could further stabilize the carboxy anion of the presumed tetrahedral intermediate; a frequently observed water molecule suggested to be activated by the carbonyl group of Pro238 *(42)* does not seem to be placed suitably for this purpose.

2.1.3. The MMP Prodomain

Currently, the only pro-MMP structures known are those of the C-terminally truncated MMP-3 *(43,46)* and of full-length pro-MMP-2 *(54c)*. The propeptide has an egg-like shape, attached with its rounded-off side to the active site of the catalytic domain. It essentially consists of three mutually perpendicularly packed α-helices and a segment connecting it with the cd (*see* **Fig. 5**). The ~15 residue long solvent exposed N-terminal segment is flexible in pro-MMP3, but fixed in pro-MMP-2, due to specific interactions with the third fn type-II module; the polypeptide chain then continues with helix 1, which it connects through a mostly disordered surface located (in pro-MMP-2 disulfide-bridged) segment with the 4-turn α-helix 2 directed toward the "chin" of the catalytic domain; the prochain then deviates from the catalytic domain surface through a defined multiple-turn loop, turns back through helix 3 to the conserved Pro90, where it kinks and enters the active-site cleft as the almost invariant Pro90-Arg-Cys-Gly-Val-Pro-Asp96 "switch segment"; this loop (with the first three residues and the Asp strictly conserved, and with only a very few variations in the latter three residues) is similarly arranged, but with opposite direction, to a bound P3' to P2 peptidic substrate down to the Val residue, i.e., it aligns parallel to the bulge-edge segment and antiparallel to the S1' wall forming segment under formation of five inter-main chain hydrogen bonds (**Fig. 6B**). At the conserved Val residue, the propeptide chain kinks away from the cd surface, turns back and runs via the final X99-Phe/Tyr100 cleavage site (which in this static structure is neither accessible nor in a suitable proteinase binding conformation) into the "rocker arm" (*see* **Fig. 5**), which terminates at the strongly conserved Pro107. From this Pro onward the polypeptide chain of both pro-MMPs is in register with that of the mature i.e. activated MMPs (*see* **Fig. 6B, panel B**). In MMP-11 and in the four MT-MMPs, up to 11 residues are inserted between the switch loop of the prochain and the cd, and the prochain terminates in an Arg-X-Arg/Lys-Arg sequence typical for cleavage by furin-like convertases *(14,75)*; intracellular cleavage of the MT-MMPs and MMP-11 by Golgi-associated furin-like enzymes results in expression of the active enzymes *(15)*.

The bent switch segment is bridged by the side chains of the strictly conserved residues Arg91 and Asp96, whose side chains are turned toward one another under formation of a symmetric 2N...2O salt bridge located on top of the third zinc-liganding His228; from the propeptide side, this salt bridge is sandwiched by the phenol side chains of (up to) three strongly conserved Tyr/Phe residues presented by helices 1 and 2, which shield it against the bulk water and strengthen the prodomain-cd interaction as long as the prodomain is intact. Somewhat similar to P2' and P1' side chains of bound peptide substrates, the Arg and the Cys side chains are directed away from the active-site cleft and toward the S1' pocket, respectively, with the Sγ of the latter turned to the catalytic zinc, acting as the fourth ligand in a tetrahedral coordination sphere (*see* **Fig. 6B**). In spite of the bound prodomain, the entire active-site region of the catalytic domain of this pro-form is remarkably similar in structure to that in the mature enzyme.

Pro-MMP activation by (other) proteinases (*see* **ref. *64***) seems to proceed via a stepwise mechanism: Some early cleavages occurring in the flexible, exposed helix1-helix2 loop (the "bait") not only might expose the hydrophobic core of the prochain domain, but in particular also destabilize this domain, thus exposing other (downstream) cleavage sites and weakening and finally destroying the Cys-catalytic zinc interaction, eventually leading to a liberation and 'flexibilization' of the X99-Phe/Tyr100 activation-cleaved peptide bond (as originally predicted by the "cysteine switch hypothesis" *[77]*); this segment then might adapt and bind to the substrate binding site of an approaching cleaving proteinase. In case of liberation of a Phe/Tyr100 N-terminus (in the classical MMPs), the activation cleavage seems to be accompanied by substantial rearrangement of the N-terminal 100–107 rocker arm (*see* **Fig. 5**) into the surface crevice described above; the N-terminal residue moves for about 17 Å to form the above mentioned surface located salt bridge with the carboxylate of Asp250 of the helix C based Asp250/Asp251 pair; in this movement, the Pro107 residue acts as a flexible joint, with its main chain angles changing from an α-helix-like (in the pro-form) to a polyproline-II-like conformation (in the mature form).

It might be noteworthy that the MMP propeptides are not required for proper (re)folding of the catalytic domains, in contrast to some other proproteinases *(46)*.

2.1.4. The Haemopexin-Like MMP Domain

Except MMP-7, all vertebrate/human MMPs are expressed with a C-terminal haemopexin-like domain (hld). Some of these hlds have been shown to be involved in substrate recognition and to confer substrate specificity, most dramatically in the collagenase subfamily, where the capability to cleave native triple-helical collagen is associated with the covalently bound hld (*see* **ref. *16***). In the MT-MMPs, these hlds have a 75–100 residue extension, which seems to constitute a connecting peptide, a transmembrane segment, and a short cytoplasmatic fragment *(14)*.

The hlds of the classical MMPs exhibit the shape of an oblate ellipsoidal disc and exhibit very similar structures *(50–53)*. The hld polypeptide chain (with an early recognized sequence similarity with haemopexin, a plasma haem-binding and transporting protein, *see* **ref. *78***) is essentially organized in four β-sheets (blades) I–IV, which are arranged almost symmetrically around a central axis in consecutive order, giving rise to the formation of a four-bladed propeller of pseudo fourfold symmetry (*see* **Fig. 7**). Each propeller blade is made by four antiparallel β-strands connected in a W-like topology, and is strongly twisted. The first innermost strands in all four blades enter the propeller at one site (the "entrance side") and run almost parallel to one another along the propeller axis, forming a central funnel-shaped tunnel, which opens slightly toward the exit. The fourth strands of blades II and III are interrupted by characteristic β-bulges, which allow these strands to keep in phase with the antiparallel third strands in spite of the overall sheet curvature. In all four blades, the outer segments loop around the periphery of the disc and end up in short helical segments. The C-terminus of the blade IV helix is tethered to the entering strand of blade I via the single disulfide bridge, stabilizing the whole domain. In spite of the strong volume increase in radial direction, the hld ellipsoid is evenly filled with protein mass, mainly achieved by a general increase in the size of the amino acid residues when going from the center to the periphery.

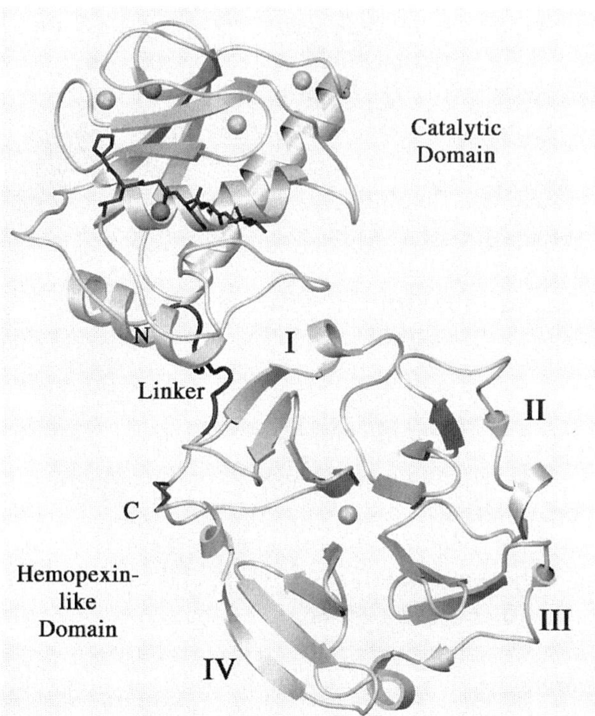

Fig. 7. Ribbon drawing of the full-length active MMP-1 *(50)*. The cd is shown in approximate standard orientation. The heptapeptide substrate (dark) has been modeled into the active-site cleft of the catalytic domain (top), which contains the two zinc ions (dark gray spheres) and three calcium ions (light gray spheres). A linker peptide (dark connection, center) connects this cd with the four-blade C-terminal haemopexin-like domain (hld, bottom), which is viewed at almost along the central tunnel toward its exit side. In all hlds known so far, a calcium ion (light sphere) is bound to the entry side of the tunnel; in some structures, additional solvent molecules located in the central part of the tunnel have been assigned as a chlorine-calcium pair and another chlorine ion, respectively. Also shown is the single disulfide bridge connecting the N- and the C-termini of the hld. Figure made with SETOR *(84)*.

Within the central tunnel, up to four heavier ions have been identified. At the entrance to this tunnel a calcium ion is placed (*see* **Fig. 7**), which is tetragonally surrounded by the first carbonyl group of each of the four inner strands. The center of the tunnel harbors an ion pair, which has been interpreted as a chloride-calcium couple. The calcium ion is octahedrally surrounded by the four carbonyl oxygens of the four residues positioned third in the innermost strands, and by the chlorine and a second chlorine/water molecule on both sides. The chlorine ion, situated between both calcium ions, is tetrahedrally coordinated by the nitrogen atoms of the same amide groups which also ligand the second calcium *(52)*. The function of these ions is not yet clear. In the three different hld structures known so far the charge patterns differ more than the surface contours *(53)*. In this respect, the hlds from the collagenases show particularly strong similarities to one another. In all hlds, (up to four) charge uncompensated Asp residues are arranged around the tunnel entrance. On the exit side of the MMP-2 hld, a surface patch of positively charged residues is noteworthy, which might be involved in binding to TIMP-2 *(52)*; however, some novel mutagenesis studies locate the TIMP-2 binding site to the junction of blades III and IV on the peripheral rim of the MMP-2 hld (C. Overall, personal communication).

In the full-length MMP structures *(50,54c)*, the catalytic domain and the hemopexin-like domain make noncovalent contacts only along small domain edges, with the outermost strand of the first blade of the hld propeller contacting the C-terminal helix hC of the cd (*see* **Fig. 7**). The 17 (or 12) amino acid residue "linker" between both domains bulges backwards and runs in a loose manner antiparallel to helix hC, before it joins (about four residues before the first Cys residue) the hld moiety. This linker is rich in Pro residues, but is quite flexible and thus a preferred target for hydrolytic cleavage *(50)*. Chimeric constructs made to define the structural features encoding the triple-helicase specificity of collagenases *(79–81)* show that both the cd and the hld have important determinants, and that both must be suitably linked to act in concert to confer helicase specificity. It has been suggested that after binding of the triple-helical substrate to the MMP-1 and 8 cds, the fully stretched linker (acting like a spacer) would just allow the hld to fold over the cd, sandwiching, trapping and (possibly) destabilizing the substrate *(53)*. However, MT1-MMP, also exhibiting triple helicase activity *(82,83)*, exhibits a much longer linker *(16)*. Alanine scanning experiments with MMP-8 *(83)* indeed show that also the linker sequence motif, with special emphasis on the Pro residues, is important to retain collagenolytic activity. Hopefully, structures of full-length MMPs in complex with heterotrimeric collagen-like peptides, which are preferentially cleaved by the full-length collagenases *(86)*, will help to solve this controversial issue in the future. Modeling experiments have shown that collagen-like triple-helical structures are far too bulky and that each constituent single-strand is not in an appropriate conformation, to allow a productive cleavage interaction with the MMP catalytic site, and that the scissile strand of triple-helical collagen must be freed for productive binding *(37)*.

2.2. TIMPs

2.2.1. TIMP Structure

The TIMPs have the shape of an elongated contiguous wedge consisting of an N-terminal segment (Cys1 to Pro5), an all-β-structure left-hand part, an all-helical center, and a β-turn structure to the right (according to the "front view" in **Fig. 4**) *(47)*. The N- and the C-terminal halves of the polypeptide chain form two opposing subdomains. The N-terminal subdomain exhibits a so-called **OB**-fold, known for a number of **o**ligosaccharide/**o**ligonucleotide **b**inding proteins *(55)*. This region consists of a five-stranded β-pleated sheet of Greek-key topology rolled into a closed β-barrel of elliptical cross-section. The narrower opening of this barrel is bounded by the sB-sC loop, whereas its wider exit is, in contrast to other OB-fold proteins, covered by an extended segment connecting strands sC and sD, designated as "connector" *(47)*. After leaving the barrel, the polypeptide chain passes two helices, forms a two-stranded β-sheet, runs through a wide multiple-turn loop, and terminates in a β-hairpin sheet. The last three (TIMP-1) to 10 (TIMP-2) C-terminal residues do not exhibit a defined conformation and presumably form a flexible tail on the TIMP surface *(21,47)*.

The TIMP edge is formed by five sequentially separate chain segments, namely the extended N-terminal segment Cys1-Pro5 flanked by the sA-sB loop and the sC-connector loop on the left-hand side, and by the sG-sH loop and the multiple-turn loop on the right-hand side (*see* **Fig. 4**). The N-terminal segment is tightly connected to the adjacent sC-connector and to the underlying sE-sF loop via disulfide bridges Cys1-Cys70/72 and Cys3-Cys99/101 (*see* **Fig. 3**). Particularly remarkable features of TIMP-2 are the quite elongated sA-sB β-hairpin loop, which does not follow the OB-barrel curvature but is twisted and extends away, and the much longer negatively charged flexible C-terminal tail. TIMP-4 seems to share these features with TIMP-2, whereas TIMP-3 exhibits a short sA-sB loop and a long but not negatively charged tail *(21,22)*. All together, the TIMP topologies differ considerably, in spite of 40% overall sequence identities.

2.2.2. The TIMP-MMP Inhibition Mechanism

In complexes with MMPs, the wedged-shaped TIMPs bind with their edge into the entire length of the active-site cleft of their cognate MMPs *(21,47)* (*see* **Fig. 8**), with removal of about 1300 Å^2 surfaces of each molecule from contact with bulk water and some 'rigidification' of the participating loops *(57)* upon complex formation. The majority of all intermolecular contacts are made by the N-terminal segment Cys1-Pro5, the sC-connector loop and the connecting disulfide bridge; in the case of TIMP-2 (and probably also TIMP-4), the par-

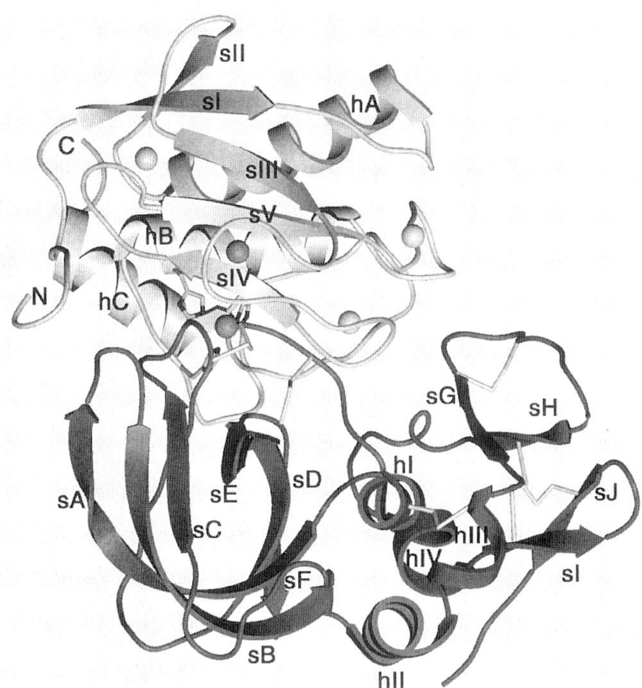

Fig. 8. MMP-TIMP complex. The complex formed by the MMP-3 catalytic domain (top, light colored ribbon) and TIMP-1 (bottom, dark colored ribbon) *(47)* is shown in front view. All disulfide bridges are given, and the zinc and calcium ions are represented as dark gray and light gray spheres. The wedge-shaped inhibitor binds with its edge made by six segments into the entire active-site cleft of the MMP-3, which in this view is directed toward the bottom. The N-terminal Cys1 of TIMP-1 is located on top of the catalytic zinc, and the first five N-terminal residues bind to the active site in a substrate-like manner. Figure made with SETOR *(84)*.

ticipation of the sA-sB loop in intermolecular contacts is considerable. The first five TIMP residues Cys1 to Pro5 bind to the MMP active-site cleft in a substrate- or product-like manner, i.e., similar as P1, P1', P2', P3', and P4' peptide substrate residues insert between the bulge and the wall forming segments, forming five intermolecular inter-main chain hydrogen bonds (*see* **Fig. 6B, panel C**). The sC-connector loop, in particular residues Ser68/Ala70 and Val69/Leu71, in contrast, interacts in a somewhat substrate-inverse manner with the left-hand subsites S2 and S3 *(21,47)*.

Cys1 is located directly above the catalytic zinc, with its N-terminal α-amino nitrogen and its carbonyl oxygen atoms placed directly above the catalytic zinc, coordinating it together with the three imidazole rings from the cognate MMP. The α-amino group of Cys1 approximately occupies the site of the bound "attacking" water molecule and forms a hydrogen bond to one carboxylate oxygen atom of the catalytic Glu219. In spite of this close interaction, the Glu219Asp mutation in MMP-2 does not significantly affect TIMP-1 binding *(71)*. The Thr/Ser side chain of the second TIMP residue extends into the S1' pocket of the cognate MMP, similar to the side chain of a P1' peptide substrate residue (*see* **Fig. 6B**), without filling this pocket properly; the Thr/Ser2 side chain oxygen is hydrogen bonded to the catalytic Glu219 carboxylate in the target MT1-MMP; mutagenesis experiments with N-TIMP-1 have shown that replacement of Thr2 by other natural amino acids can lead to TIMP species, which are able to discriminate much more between different MMPs *(32)*. The Cys3, Val/Ser4, and Pro5 side chains touch subsites S2', S3', and S4' in a manner expected for substrate P2', P3', and P4' side chains (*see* **Fig. 6B**).

It is noteworthy that a very similar "substrate-like" interaction has been found in two crystal forms of "noninhibited" MMP-1 cds *(40)*, where the opened-out N-terminal segment (Leu102-Thr-Glu-Gly105) of one molecule inserts into the S1 to S3' subsites of a symmetry-related molecule, utilizing identical subsite and intermolecular hydrogen bond interactions as observed in the MMP-TIMP complexes. A similar peptide-metalloproteinase interaction has, furthermore, been observed in complexes between serralysins and endogenous inhibitors produced by various *Serratia*-like bacteria; these ~100 amino acid residue inhibitors interact via a flexible N-terminal strand (of sequence Ser1-Ser2-Leu3-Arg4) with the S1 to S3' subsites of their target serralysin, in a roughly similar manner as the N-terminal TIMP tetrapeptide does in the MMP-3-TIMP-1 complex *(87)*; neither the (large) rest of the inhibitor molecules nor the other interactions bear any further resemblance with the TIMPs, however.

In the MT1-MMP-TIMP-2 complex, the quite long sA-sB hairpin loop of TIMP-2 folds alongside the S-loop over the rim of the active-site cleft and reaches up to the β-sheet of its cognate MMP *(21)*; in spite of the relatively large overall interface between the sA-sB loop and the molecular MMP surface, most of the intermolecular contacts do not seem to be designed for optimal complementarity, however. The other edge loops of TIMP (namely the sE-sF loop, the sG-sH loop, and the multiple-turn loop, **Fig. 4**) are involved in a relatively small number of intermolecular contacts, with both C-terminal edge loops (in particular in TIMP-1) showing relatively high flexibility. Binding data indicate, however, that the interaction of the C-terminal TIMP subdomains with a cognate MMP might be of much more importance for the TIMP inhibition of the gelatinases and MMP-13 *(16)*, probably correlating with tighter intermolecular contacts.

Superposition experiments with the structure of full-length porcine MMP-1 *(50)* show that the C-terminal hld of full-length MMPs as positioned in that structure would be compatible with TIMP binding *(47)*. In such a TIMP-full-length MMP complex both domains, the TIMP and the hld, would just touch one another, in agreement with kinetic binding studies showing (except for gelatinases A and B and MMP-13 *(16)* that the C-terminal MMP domains contribute relatively little to TIMP binding *(28,32)*. The negatively charged C-terminal tail of TIMP-2 seems to facilitate the noninhibitory binding to the progelatinase A hld mainly via electrostatic interactions. This specific TIMP-2 tail-MMP-2 hld interaction is important for formation of the MT1-MMP-TIMP-2-progelatinase complex implicated in progelatinase activation *(88,89)*.

Acknowledgment

The valuable contributions by Drs. H. Nagase, H. Tschesche, and K. Brew and their groups as well as by our colleagues Drs. F. Grams, F.-X. Gomis-Rueth, and C. Fernandez-Catalan to structures described here are greatly acknowledged.

References

1. Woessner, J. F., Jr. (1991) Matrix metalloproteinases and their inhibitors in connective tissue remodeling. *FASEB J.* **5**, 2145–2155.
2. Nagase, H., Das, S. K., Dey, S. K., Fowlkes, J. L., Huang, W., and Brew, K. (1997) in *Inhibitors of metalloproteinases in development and disease* (Hawkes, S. P., Edwards, D. R., and Khokha, R., eds.) Harwood Academic Publ., Lausanne, Switzerland.
3. Johnson, L. L., Dyer, R., and Hupe, D. J. (1998) Matrix metalloproteinases. *Curr. Opin. Chem. Biol.* **2**, 466–471.
4. Yong, V. W., Krekoski, C. A., Forsyth, P. A., Bell, R., and Edwards, D. R. (1998) Matrix metalloproteinases and diseases of the CNS. *Trends Neurosci.* **21**, 75–80.
5. Coussens, L. M., Werb, Z. (1996) Matrix metalloproteinases and the development of cancer. *Chem. Biol.* **3**, 895–904.
6. Chambers, A. F. and Matrisian, L. M. (1997) Changing views of the role of matrix metalloproteinases in metastasis. *J. Natl. Cancer Inst.* **89**, 1260–1270.
7. Beckett, R. and Whittaker, M. (1998) Matrix metalloproteinase inhibitors. *Exp. Opin. Ther. Patents* **8**, 259–282.
8. Bottomley, K. M., Johnson, W. H., and Walter, D. S. (1998) Matrix metalloproteinase inhibitors in arthritis. *J. Enz. Inhib.* **13**, 79–101.
9. Massova, I., Kotra, L. P, Fridman, R., and Mobashery, S. (1998) Matrix metalloproteinases: structures, evolution, and diversification. *FASEB J.* **12**, 1075–1095.
10. Rawlings, N. D. and Barrett, A. J. (1995) Evolutionary families of metallopeptidases. *Meth. Enzymol.* **248**, 183–229.

11. Sang, Q. A. and Douglas, D. A. (1996) Computational sequence analysis of matrix metalloproteinases. *J. Prot. Chem.* **15**, 137–160.

12. Murphy, G. J., Murphy, G., and Reynolds, J. J. (1991) The origin of matrix metalloproteinases and their familial relationships. *FEBS lett.* **289**, 4–7.

13. Sato, H., Takino, T., Okada, Y., Cao, J., Shinagawa, A., Yamamoto, E., and Seiki, M. (1994) A matrix metalloproteinase expressed on the surface of invasive tumor cells. *Nature* **370**, 61–65.

14. Sato, H., Kinoshita, T., Takino, T., Nakayama, K., and Seiki, M. (1996) Activation of a recombinant membrane type 1-matrix metalloproteinase (MT1-MMP) by furin and its interaction with tissue inhibitor of metalloproteinases (TIMP)-2. *FEBS Lett.* **393**, 101–104.

15. Pei, D. and Weiss, S. J. (1996) Transmembrane-deletion mutants of the membrane-type matrix metalloproteinase-1 process progelatinase A and express intrinsic matrix-degrading activity. *J. Biol. Chem.* **271**, 9135–9140.

16. Murphy, G. and Knäuper, V. (1997) Relating matrix metalloproteinase structure to function: why the "hemopexin" domain? *Matrix Biol.* **15**, 511–518.

17. Docherty, A. J. P., Lyons, A., Smith, B. J., Wright, E. M., Stephens, P. E., Harris, T. J. R., Murphy, G., and Reynolds, J. J. (1985) Sequence of human tissue inhibitor of metalloproteinases and its identity to erythroid potentiating activity. *Nature* **318**, 65–69

18. Stetler-Stevenson, W. G., Krutzsch, H. C., and Liotta, L. A. (1989) Tissue inhibitor of metalloproteinase (TIMP-2). *J. Biol. Chem.* **264**, 17,374–17,378.

19. Apte, S. S., Olsen, B. R., and Murphy, G. (1995) The gene structure of tissue inhibitor of metalloproteinases (TIMP)-3 and its inhibitory activities define the distinct TIMP gene family. *J. Biol. Chem.* **270**, 14,313–14,318.

20. Greene, J., Wang, M., Liu, Y. E., Raymond, L. A., Rosen, C., and Shi, Y. E. (1996) Molecular cloning and characterisation of human tissue inhibitor of metalloproteinase 4. *J. Biol. Chem.* **271**, 30,375–30,380.

21. Fernandez-Catalan, C., Bode, W., Huber, R., Turk, D., Calvete, J. J., Lichte, A., Tschesche., and Maskos, K. (1998) Crystal structure of the complex formed by the membrane type 1-matrix metalloproteinase with the tissue inhibitor of metalloproteinases-2, the soluble progelatinase A receptor. The EMBO J. **17**, 5238–5248.

22. Douglas, D. A., Shi, Y. E., and Sang Q. A. (1997) Computational sequence analysis of the tissue inhibitor of metalloproteinase family. *J. Prot. Chem.* **16**, 237–255.

23. Gomez, D. E., Alonso, D. F., Yoshiji, H., and Thorgeirsson, U. P. (1997) Tissue inhibitors of metalloproteinases: structure, regulation and biological functions. *Eur. J. Cell Biol.* **74**, 111–122.

24. Cawston, T. (1998) Matrix metalloproteinases and TIMPs: properties and implications for the rheumatic diseases. *Mol. Med. Today* **4**, 130–137.

25. Butler, G. S., Will, H., Atkinson, S.J., and Murphy, G. (1997) Membrane-type-2 matrix metalloproteinase can initiate the processing of progelatinase A and is regulated by the tissue inhibitors of metalloproeinases. *Eur. J. Biochem.* **244**, 653–657.

26. Will, H., Atkinson, S. J., Butler, G. S., Smyth, B., and Murphy, G. (1996) The soluble catalytic domain of membrane type 1 matrix metalloproteinase cleaves the propeptide of progelatinase A and initiates autocatalytic activation. *J. Biol. Chem.* **271**, 17,119–17,123.

27. Zucker, S., Drews, M., Conner, C., Foda, H. D., DeClerck, Y. A., Langley, K. E., Bahou, W. F., Docherty, A. J. P., and Cao, J. (1998) Tissue inhibitor of metalloproteinase-2 (TIMP-2) binds to the catalytic domain of the cell surface receptor, membrane type 1-matrix metalloproteinase 1 (MT1-MMP). *J. Biol. Chem.* **273**, 1216–1222.

28. Murphy, G. and Willenbrock, F. (1995) Tissue inhibitors of matrix metallo-endopeptidases. *Meth. Enzym.* **248**, 496–510.

29. Strongin, A. Y., Collier, I. E., Bannikov, U., Marmer, B. L., Grant, G. A., and Goldberg, G. I. (1995) Mechanism of cell surface activation of 72-kDa type IV collagenase. *J. Biol. Chem.* **270**, 5331–5338.

30. Strongin, A. Y., Marmer, B. L., Grant, G. A., and Goldberg, G. I. (1993) Plasma membrane-dependent activation of the 72-kDa type IV collagenase is prevented by complex formation with TIMP-2. *J. Biol. Chem.* **268**, 14,033–14,039.

31. Kinoshita, T., Sato, H., Takino, T., Itoh, M., Akizawa, T., and Seiki, M. (1996) Processing of a precursor of 72-kilodalton type IV collagenase/gelatinase A by a recombinant membrane-type 1 matrix metalloproteinase. *Cancer Res.* **56**, 2535–2538.

32. Huang, W., Meng, Q., Suzuki, K., Nagase, H., and Brew, K. (1997) Mutational study of the amino-terminal domain of human tissue inhibitor of metallopro-teinases 1 (TIMP-1) locates an inhibitory region for matrix metalloproteinases. *J. Biol. Chem.* **272**, 22,086–22,091.

33. Murphy, G., Houbrechts, A., Cockett, M. I., Williamson, R. A., O'Shea, M., and Docherty, A. J. P. (1991) The N-terminal domain of human tissue inhibitor of metalloproteinases retains metalloproteinase inhibitory activity. *Biochemistry* **30**, 8097–8102.

34. Lovejoy, B., Cleasby, A., Hassell, A. M., Longley, K., Luther, M. A., Weigl, D., McGeehan, G., McElroy, A. B., Drewry, D., Lambert, M. H., and Jordan, S. R. (1994) Structure of the catalytic domain of fibroblast collagenase complexed with an inhibitor. *Science* **263**, 375–377.

35. Borkakoti, N., Winkler, F. K., Williams, D. H., D'Arcy, A., Broadhurst, M. J., Brown, P. A., Johnson, W. H., and Murray, E. J. (1994) Structure of the catalytic domain of human fibroblast collagenase complexed with an inhibitor. *Nature Struct. Biol.* **1**, 106–110.

36. Stams, T., Spurlino, J. C., Smith, D. L., Wahl, R. C., Ho, T. F., Qoronfleh, M. W., Banks, T. M., and Rubin, B. (1994) Structure of human neutrophil collagenase reveals large S1' specificity pocket. *Nature Struct. Biol.* **1**, 119–123.

37. Bode, W., Reinemer, P., Huber, R., Kleine, T., Schnierer, S., and Tschesche, H. (1994) The X-ray crystal structure of the catalytic domain of human neutrophil collagenase inhibited by a substrate analogue reveals the essentials for catalysis and specificity. *EMBO J.* **13**, 1263–1269.

38. Reinemer, P., Grams, F., Huber, R., Kleine, T., Schnierer, S., Pieper, M., Tschesche, H., and Bode, W. (1994) Structural implications for the role of the N-terminus in the 'superactivation' of collagenases — a crystallographic study. FEBS Lett. **338**, 227–233.

39. Gooley, P. R., O'Connell, J. F., Marcy, A. I., Cuca, G. C., Salowe, S .P., Bush, B. L., Hermes, J. D., Esser, C. K., Hagmann, W. K., Springer, J. P., and Johnson, B. A. (1994) NMR structure of inhibited catalytic domain of human stromelysin-1. *Nature Struct. Biol.* **1**, 111–118.

40. Lovejoy, B., Hassell, A. M., Luther, M. A., Weigl, D., and Jordan, S. R. (1994) Crystal structures of recombinant 19-kDa human fibroblast collagenase complexed to itself. *Biochemistry* **33**, 8207–8217.

41. Spurlino, J. C., Smallwood, A. M., Carlton, D. D., Banks, T. M., Vavra, K. J., Johnson, J. S., Cook, E. R., Falvo, J., Wahl, R. C., Pulvino, T. A., Wendoloski, J. J., and Smith, D. L. (1994) 1.56Å structure of mature truncated human fibroblast collagenase. Proteins: *Str. Fct. Gen.* **19**, 98–109.

42. Browner, M. F., Smith, W. W., and Castelhano, A. L. (1995) Matrilysin-inhibitor complexes: common themes among metalloproteases. *Biochemistry* **34**, 6602–6610.

43. Becker, J. W., Marcy, A .I., Rokosz, L. L., Axel, M. G., Burbaum, J. J., Fitzgerald, P. M. D., Cameron, P. M., Esser, C. K., Hagmann, W. K., Hermes, J. D., and Springer, J. P. (1995) Stromelysin-1: Three-dimensional structure of the inhibited catalytic domain and of the C-truncated proenzyme. *Prot. Sci.* **4**, 1966–1976.

44. Dhanaraj, V., Ye, Q-Z, Johnson, L. L., Hupe, D. J., Ortwine, D. F., Dunbar, J. B., Rubin, J. R., Pavlovsky, A., Humblet, C., and Blundell, T. L. (1996) X-ray structure of a hydroxamate inhibitor complex of stromelysin catalytic domain and its comparison with members of the zinc metalloproteinase superfamily. *Structure* **4**, 375–386.

45. vanDoren, S. R., Kurochkin, A. V., Hu, W., Ye, Q. Z., Johnson, L. L., Hupe, D. J., and Zuiderweg, E. R. (1995) Solution structure of the catalytic domain of human stromelysin complexed with a hydrophobic inhibitor. *Protein Sci.* **4**, 2487–2498.

46. Wetmore, D. R. and Hardman, K. D. (1996) Roles of the propeptide and metal ions in the folding and stability of the catalytic domain of stromelysin (matrix metalloproteinase 3). *Biochemistry* **35**, 6549–6558.

47. Gomis-Rüth, F. X., Maskos, K., Betz, M., Bergner, A., Huber, R., Suzuki, K., Yoshida, N., Nagase, H., Brew, K., Bourenkov, G. P., Bartunik, H., and Bode, W. (1997) Mechanism of inhibition of the human matrix metalloproteinase stromelysin-1 by TIMP-1. *Nature* **389**, 77–81.

48. Grams, F., Reinemer, P., Powers, J. C., Kleine, T., Pieper, M., Tschesche, H., Huber, R., and Bode, W. (1995) X-ray structures of human neutrophil collagenase complexed with peptide hydroxamate and peptide thiol inhibitors. Implications for substrate binding and rational drug design. *Eur. J. Biochem.* **228**, 830–841.

49. Grams, F., Crimmin, M., Hinnes, L., Huxley, P., Pieper, M., Tschesche, H., and Bode, W. (1995) Structure determination and analysis of human neutrophil collagenase complexed with a hydroxamate inhibitor. *Biochemistry* **34**, 14,012–14,020.

50. Li, J.-Y., Brick, P., O'Hare, M. C., Skarzynski, T., Lloyd, L. F., Curry, V. A., Clark, I. M., Bigg, H. F., Hazleman, B. L., Cawston, T. E., and Blow, D. M. (1995) Structure of full-length porcine synovial collagenase reveals a C-terminal domain containing a calcium-linked, four-bladed b-propeller. *Structure* **3**, 541–549.

51. Libson, A., Gittis, A., Collier, I., Marmer, B., Goldberg, G., and Lattman, E. E. (1995) Crystal structure of the hemopexin-like C-terminal domain of gelatinase A. *Nature Struct. Biol.* **2**, 938–942.

52. Gohlke, U., Gomis-Rüth, F.-X., Crabbe, T., Murphy, G., Docherty, A. J. P., and Bode, W. (1996) The C-terminal (haemopexin-like) domain structure of human gelatinase A (MMP2): structural implications for its function. *FEBS Lett.* **378**, 126–130.

53. Gomis-Rüth, F. X., Gohlke, U., Betz, M., Knäuper, V., Murphy, G, López-Otín, C., and Bode, W. (1996) The helping hand of collagenase-3 (MMP-13): 2.7Å crystal structure of its C-terminal haemopexin-like domain. *J. Mol. Biol.* **264**, 556–566.

54. Bode, W. (1995) A helping hand for collagenases: the haemopexin-like domain. *Structure* **3**, 527–530.

54a. Lovejoy, B., Welch, A. R., Carr, S., Luong, C., Broka, C., Hendricks, R. T., Campbell, J. A., Walker, K. A. M., Martin, R., Van Wart, H., and Browner, M. F. (1999) Crystal structures of MMP-1 and –13 reveal the structural basis for selectivity of collagenase inhibitors. *Nature Str. Biol.* **6**, 217–221.

54b. Botos, I., Meyer, E., Swanson, S. M., Lemaitre, V., Eeckhout, Y., and Meyer, E. F. (1999) Structure of recombinant mouse collagenase-3 (MMP-13). *J. Mol. Biol.* **292**, 837–844.

54c. Morgunova, E., Tuuttila, A., Bergmann, U., Isupov, M., Lindquist, Y., Schneider, G., and Tryggvason, K. (1999) Structure of human pro-matrix metalloproteinase-2: activation mechanism revealed. *Science* **284**, 1667–1670.

54d. Dhanaraj, V., Williams, M. G., Ye, Q.-Z., Molina, F., Johnson, L. L., Ortwine, D. F., Pavlovsky, A., Rubin, J. R., Skeenan, R. W., White, A. D., Humblet, C., Hupe, D. J., and Blundell, T. L. (1999) X-ray structure of gelatinase A catalytic domain complexed with a hydroxamate inhibitor. *Croatia Chem. Acta* **72**, 575.

54e. Natchus, M. G., Cheng, M., Wahl, C. T., Pikul, S., Almstead, N. G., Bradley, R. S., Taiwo, Y. O., Mieling, G. E., Dunaway, C. M., Snider, C. E., Mciver, J. M., Barnetts, B. L., McPhail, S. J., Anastasio, M. V., and De, B. (1998) Design and synthesis of conformationally-constrained MMP inhibitors. *Bioorg. Med. Chem. LETT.* **8**, 2077.

54f. Pikul, S., McDowDunham, K. L., Almstead, N. G., De, B., Natchus, M. G., Anastasio, M. V., McPhail, S. J., Snider, C. E., Taiwo, Y. O., Rydel, T., Dunaway, C. M., Gu, F., and Mieling, G. E. (1998) Discovery of potent, achiral matrix metalloproteinase inhibitors. *J. Med. Chem.* **41**, 3568.

54g. Chen, L., Rydel, T. J., Dunaways, C. M., Pikul, S., Dunham, K. M., Gu, F., and Barnett, B. L. (1998) Crystal structure of the stromelysin catalytic domain at 2.0 Å resolution: inhibitor-induced conformational changes. *J. Mol. Biol.* **293**, 545.

54h. Matter, H., Schwab, W., Barbier, D., Billen, G., Haase, B., Neises, B., Schudok, M., Torwart, W., Schreuder, H., Brachvogel, V., and Loenze, P. (1999) Quantitative structure-activity relationship of human neutrophil collagenase (MMP-8) inhibitors using comparative molecular field analysis and X-ray structure analysis. *J. Med. Chem.* **42**, 908.

54i. Pavlovsky, A. G., Williams, M. G., Ye, Q.-Z.m, Ortwinw, F., Purchase II, C. F., White, A. D., Dhanaraj, V., Roth, B. D. Johnson, L. L., Hupe, D., Humblet, C., and Blundell, T. L. (1999) X-ray structure of human stromelysin catalytic domain complexed with non-peptide inhibitors: implication for inhibitor selectivity. *Prot. Sci.* **8**, 1455.

54j. Finzel, B. C., Baldwin, E. T., Bryant, Jr., G. L., Hess, G. F., Wilks, J. W., Trepod, C. M., Mott, J. E., Marshall, V. P., Petzold, G. L., Poorman, R. A., O'Sullivan, T. J., Schostarez, H. J., and Mitchell, M. A. (1998) Structural characterizations of nonpeptidic thiadiazole inhibitors of matrix metalloproteinases reveal the basis for stromelysin selectivity. *Prot. Sci.* **7**, 2118.

55. Williamson, R. A., Martorell, G., Carr, M. D., Murphy, G., Docherty, A. J., Freedman, R. B., and Feeney, J. (1994) Solution structure of the active domain of tissue inhibitor of metalloproteinases-2. A new member of the OB fold protein family. *Biochemistry* **33**, 11,745–11,759.

56. Williamson, R. A., Carr, M. D., Frenkiel, T. A., Feeney, J., and Freedman, R. B. (1997) Mapping the binding site for matrix metalloproteinase on the N-terminal domain of the tissue inhibitor of metalloproteinases-2 by NMR chemical shift perturbation. *Biochemistry* **36**, 13,882–13,889.

57. Muskett, F. W., Frenkiel, T. A., Feeney, J., Freedman, R. B., Carr, M. D., and Williamson, R. (1998) High resolution structure of the N-terminal domain of tissue inhibitor of metalloproteinases-2 and characterization of its interaction site with matrix metalloproteinase-3. *J. Biol. Chem.* **273**, 21,736–21,743.

57a. Tuuttila, A., Morgunova, E., Bergmann, U., Lindquist, Y., Maskos, K., Fernandez-Catalan, C., Bode, W., Tryggvason, K., and Schneider, G. (1998) Three-dimensional structure of human tissue inhibitor of metalloproteinases-2 at 2.1 Å resolution. *J. Mol. Biol.* **284**, 1133–1140.

57b. Wu, B., Arumugam, S., Gao, G., Le, G., Semenchenko, V., Huang, W., Brew, K., and VanDoren, S.R. (2000) NMR structure of tissue inhibitor of metalloproteinases-1 implicates localized induced fit in recognition of matrix metalloproteinases. *J. Mol. Biol.* **295**, 257–268.

58. Schechter, I. and Berger, A. (1967) On the size of the active site in proteases. *I. Papain. Biochem. Biophys. Res. Commun.* **27**, 157–162.

59. Bode, W., Gomis-Rüth, F.-X., and Stöcker, W. (1993) Astacins, serralysins, snake venom and matrix metalloproteinases exhibit identical zinc-binding environments (HEXXHXXGXXH and Met-turn) and topologies and should be grouped into a common family, the 'metzincins'. *FEBS lett.* **331**, 134–140.

60. Stöcker, W., Grams, F., Baumann, U., Reinemer, P., Gomis-Rüth, F. X., McKay, D. B., and Bode, W. (1995) The metzincins - topological and sequential rela-

tions between the astacins, adamalysins, serralysins, and matrixins (collagenases) define a superfamily of zinc-peptidases. *Protein Science* **4**, 823–840.

61. Pickford, A. R., Potts, J. R., Bright, J. R., Phan, I., and Campbell, I. D. (1997) Solution structure of a type 2 module from fibronectin: implications for the structure and function of the gelatin-binding domain. *Structure* **5**, 359–370.

62. Knäuper, V., Murphy, G., and Tschesche, H. (1996) Activation of human neutrophil procollagenase by stromelysin 2. *Eur. J. Biochem.* **235**, 187–191.

63. Suzuki, K., Enghild, J. J., Morodomi, T., Salvesen, G., and Nagase, H. (1990) Mechanisms of activation of tissue procollagenase by matrix metalloproteinase 3 (stromelysin). *Biochemistry* **29**, 10,261–10,270.

64. Nagase, H. (1997) Activation mechanisms of matrix metalloproteinases. *Biol. Chem.* **378**, 151–160.

65. Pieper, M., Betz, M., Budisa, N., Gomis-Rüth, F.-X., Bode, W., and Tschesche, H. (1997) Expression, purification, characterization, and X-ray analysis of selenomethionine 215 variant of leukocyte collagenase. *J. Prot. Chem.* **16**, 637–650.

66. Netzel-Arnett, S., Fields, G. B., Birkedal-Hansen, H., and van Wart, H. E. (1991) Sequence specificities of human fibroblast and neutrophil collagenase. *J. Biol. Chem.* **266**, 6747–6755.

67. Netzel-Arnett, S., Sang, Q. X., Moore, W. G. I., Navre, M., Birkedal-Hansen, H., and van Wart, H. E. (1993) Comparative sequence specificities of human 72- and 92-kDa gelatinases (type IV collagenases) and PUMP (matrilysin) *Biochemistry* **32**, 6427–6432.

68. Niedzwiecki, L., Teahan, J., Harrison, R. K., and Stein, R. L. (1992) Substrate specificity of the human matrix metalloproteinase stromelysin and the development of continuous fluoremetric assays. *Biochemistry* **31**, 12,618–12,623.

69. Birkedal-Hansen, H., Moore, W. G. I., Bodden, M. K., Windsor, L. J., Birkedal-Hansen, B., DeCarlo, A., and Engler, J. A. (1993) Matrix metalloproteinases: a review. *Crit. Rev. Oral Biol. Med.* **4 (2)**, 197–250.

70. Matthews, B. W. (1988) Structural basis of the action of thermolysin and related zinc peptidases. *Acc. Chem. Res.* **21**, 333–340.

71. Crabbe, T., Zucker, S., Cockett, M. I., Willenbrock, F., Tickle, S., O'Connell, J. P., Scothern, J. M., Murphy, G., and Docherty, A. J. P. (1994) Mutation of the active site glutamic acid of human gelatinase A: effects on latency, catalysis, and the binding of tissue inhibitor of metalloproteinases-1. *Biochemistry* **33**, 6684–6690.

72. Windsor, L. J., Bodden, M. K., Birkedal-Hansen, B., Engler, J. A., and Birkedal-Hansen, H. (1994) Mutational analysis of residues in and around the active site of human fibroblast-type collagenase. *J. Biol. Chem.* **269**, 26,201–26,207.

73. Bode, W., Gomis-Rüth, F. X., Huber, R., Zwilling, R., and Stöcker, W. (1992) Structure of astacin and implications for activation of astacins and zinc-ligation of collagenases. *Nature* **358**, 164–166.

74. Grams, F., Dive, V., Yiotakis, A., Yiallouros, I., Vassiliou, S., Zwilling, R., Bode, W., and Stöcker, W. (1996) Structure of astacin with a transition-state analogue inhibitor. *Nature Struct. Biol.* **3**, 671–675.

75. Noel, A., Santavicca, M., Stoll, I., L'Hoir, C., Staub, A., Murphy, G., Rio, M. C., and Basset, P. (1995) Identification of structuural determinants controlling human and mouse stromelysin-3 proteolytic activities. *J. Biol. Chem.* **270**, 22,866–22,872.

76. Nicholls, A., Bharadwaj, R., and Honig, B. (1993) Grasp - graphical representation and analysis of surface properties. *Biophys. J.* **64**, A166.

77. van Wart, H. E. and Birkedal-Hansen, H. (1990) The cysteine switch: a principle of regulation of metalloproteinase activity with potential applicability to the entire matrix metalloproteinase gene family. *Proc. Natl. Acad. Sci. USA* **87**, 5578–5582.

78. Faber, H. R., Groom, C. R., Baker, H. M., Morgan, W. T., Smith, A., and Baker, E. N. (1995) 1.8 Å crystal structure of the C-terminal domain of rabbit serum haemopexin. *Structure* **3**, 551–559.

79. Murphy, G., Allan, J. A., Willenbrock, F., Cockett, M. I., O'Connell, J. P., and Docherty, A. J. P. (1992) The role of the C-terminal domain in collagenase and stromelysin specificity. *J. Biol. Chem.* **267**, 9612–9618.

80. Sanchez-Lopez, R., Alexander, C. M., Behrendtsen, O., Breathnach, R., and Werb, Z. (1993) Role of zinc-binding- and hemopexin domain-encoded sequences in the substrate specificity of collagenase and stromelysin-2 as revealed by chimeric proteins. *J. Biol. Chem.* **268**, 7238–7247.

81. Hirose, T., Patterson, C., Pourmotabbed, T., Mainardi, C. L., and Hasty, K. A. (1993) Structure-function relationship of human neutrophil collagenase: identification of regions responsible for substrate specificity and general proteinase activity. *Proc. Natl. Acad. Sci. USA* **90**, 2569–2573.

82. Ohuchi, E., Imai, K., Fujii, Y., Sato, H., Seiki, M., and Okada, Y. (1997) Membrane type 1 matrix metalloproteinase digests interstitial collagens and other extracellular matrix macromolecules. *J. Biol. Chem.* **272**, 2446–2451.

83. Knäuper, V., Docherty, A. J. P., Smith, B., Tschesche, H., and Murphy, G. (1997) Analysis of the contribution of the hinge region of human neutrophil collagenase (HNC, MMP-8) to stability and collagenolytic activity by alanine scanning mutagenesis. *FEBS lett.* **405**, 60–64.

84. Evans, S. V. (1993) SETOR: hardware lighted three-dimensional solid model representations of macromolecules. *J. Mol. Graph.* **11**, 134–138.

85. Barton, G. J. (1993) ALSCRIPT: a tool to format multiple sequence alignments. *Protein Eng.* **6**, 37–40.

86. Ottl, J., Battistuta, R., Pieper, M., Tschesche, H., Bode, W., Kühn, K., and Moroder, L. (1996) Design and synthesis of heterotrimeric collagen peptides with a built-in cystine knot. *FEBS Lett.* **398**, 31–36.

87. Baumann, U., Bauer, M., Letoffe, S., Delepelaire, P., and Wandersman, C. (1995) Crystal Structure of a complex between *Serratia marcescens* metalloprotease and an inhibitor from *Erwinia chrysanthemi*. *J. Mol. Biol.* **248**, 653–661.

88. Cao, J., Sato, H., Takino, T., and Seiki, M. (1995) The C-terminal region of membrane type matrix metalloproteinase is a functional transmembrane domain required for progelatinase A activation. *J. Biol. Chem.* **270**, 801–805.

89. Butler, G. S., Butler, M. J., Atkinson, S. J., Will, H., Tamura, T., van Westrum, S. S., Crabbe, T., Clements, J., d'Ortho, M. P., and Murphy, G. (1998) The TIMP2 membrane type 1 metalloproteinase "receptor" regulates the concentration and efficient activation of progelatinase A. *J. Biol. Chem.* **273**, 871–880.
90. Brandstetter, H., Engh, R. A., Graf von Roedern, E., Moroder, L., Huber, R., Bode, W., and Grams, F. (1998) Structure of malonic acid-based inhibitors bound to human neutrophil collagenase. A new binding mode explains apparently anomalous data. *Prot. Sci.* **7**, 1303–1309.
91. Moy, F. J., Chandra, P. K., Cosmi, S., Pisano, M. R., Urbano, C., Wilhelm, J., and Powers, R. (1998) High-resolution solution structure of the inhibitor free catalytic fragment of human fibroblast collagenase, determined by multidimensional NMR. *Biochemistry* **37**, 1495–1504.
92. VanDoren, S. R., Kurochkin, A. V., Hu, W., Ye, Q.-Z., Johnsson, L. L., Hupe, D. J., and Zuiderweg, E. R. P. (1995) Solution structure of the catalytic domain of human stromelysin complexed with a hydrophobic inhibitor. *Prot. Sci.* **4**, 2487–2498.

4

Matrix Metalloproteinase Substrate Binding Domains, Modules and Exosites

Overview and Experimental Strategies

Christopher M. Overall

1. Introduction

Connective tissue degradation occurs in chronic inflammatory diseases such as arthritis and periodontitis to adversely affect life quality. More importantly, disturbances in connective tissue homeostasis may be life threatening in various lung, neurological, and cardiovascular diseases and is pivotal in tumor metastasis. Matrix metalloproteinases (MMPs) form one of the most important families of proteinases that participate in the degradative aspects of these diseases. The specific inhibition of MMPs by the four members of the tissue inhibitor of metalloproteinase (TIMP) family can regulate the extracellular activity of MMPs *(1)*. Not surprisingly, altered TIMP expression is also known to occur in many disease processes. Activation of MMP zymogens is another critical aspect of the regulation of connective tissue matrix composition, structure, and function. For some soluble MMPs, activation occurs at the cell surface following proteolytic cleavage by membrane type MMPs (MT-MMPs), often in a TIMP-dependent pathway. For other MMPs, activation occurs in the extracellular environment in an activation cascade initiated by tissue proteinases, such as plasmin, kallikrein, and tryptase, a process that is often amplified by the activated MMPs themselves functioning as proMMP activators. Therefore, understanding the structural basis of MMP function, in particular substrate recognition and cleavage, MMP inhibition by TIMPs and synthetic inhibitors, and the domain:domain interactions that occur in the activation and association of MMPs and TIMPs with the cell membrane and in the matrix, may point to new avenues of therapeutic intervention or refine existing MMP inhibitor strategies.

From: *Methods in Molecular Biology, vol. 151: Matrix Metalloproteinase Protocols*
Edited by: I. Clark © Humana Press Inc., Totowa, NJ

Most normal tissue and tumor cells synthesize MMPs, which together have the capacity to initiate and complete the degradation of connective tissue matrices. To date the MMP gene family encodes 21 homologous proteinases (MMPs 1–3, 7–24; *see* Woessner, Chapter 1) that carry out related functions. Like many extracellular proteins, MMPs are multidomain mosaic proteins that share similar primary, secondary, and tertiary structures having most likely evolved from a common ancestor through gene duplication, exon shuffling, and subsequent divergence by amino acid substitutions to generate new MMPs with altered specificities, regulation properties, and other characteristics.

Functionally, the most important domain of MMPs is the catalytic domain, the structure of which has been covered extensively in Chapter 3 by Bode and Maskos. Like many proteinases MMPs are secreted as inactive zymogens. Responsible for enzyme latency is an 80 residue propeptide that extends from the catalytic domain to the N-terminus of the enzyme. At the C-terminus of the catalytic domain a potentially flexible proline-rich linker connects to a hemopexin C-terminal domain (C domain) present in most MMPs except for matrilysin (MMP-7) *(2)*, and MMP-23 *(3)* which lack a hemopexin domain. MMP-21 *(4)* has a hemopexin domain linked to the catalytic domain without a linker. The carboxy-terminus of soluble secreted MMPs is at, or just after, the last cysteine in the hemopexin C domain *(5,6)*. This and another absolutely conserved cysteine at the start of the hemopexin domain, form the single disulfide-bridge found in most MMPs except for gelatinases A and B. The gelatinases contain an additional six disulfide bridges in a unique 174-residue domain composed entirely of three fibronectin type II modules inserted at the N terminal side of the active site α-helix B in the catalytic domain *(7,8)*. This was originally termed the "gelatin binding domain," but in view of the additional *native* collagen binding properties of the domain *(9,10)* the term "collagen binding domain" (CBD) is more appropriate. In the MT-MMPs, departing from the side of the hemopexin C domain after the disulphide bridge, a flexible linker connects to a transmembrane putative α-helix, which is followed by a small cytoplasmic domain. These function as a plasma membrane anchor and potentially interact with intracellular proteins. **Figure 1** depicts the archetypal domain structure of the MMPs.

MMPs are unified by a common global fold and mechanism of catalysis, but distinguished within the family by selectivity for different physical and chemical determinants on substrates. A significant challenge is understanding the molecular determinants of MMP substrate specificity. Despite the great interest in MMPs, the more recently reported enzymes have not yet been screened extensively for substrates, nor have many of the earlier described MMPs been reexamined to discover novel substrates from the large number of newly described extracellular matrix or cell-associated proteins. This is surprising

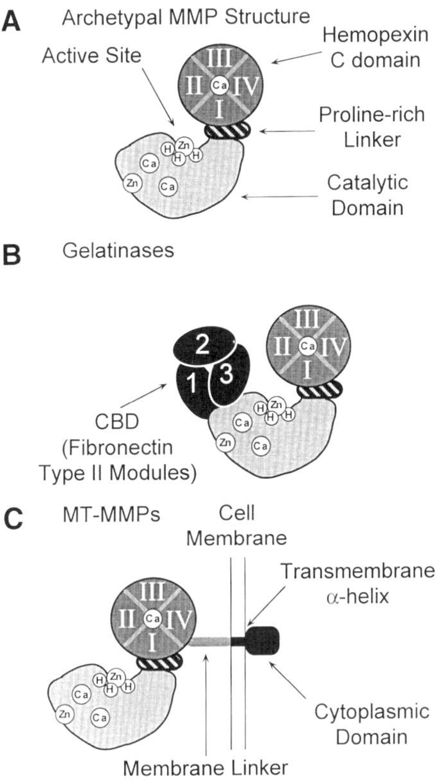

Fig. 1. Archetypal Domain Organization of Soluble MMPs and MT-MMPs. In Panel **(A)** MMP features and domains are labeled. In panels **(B)** and **(C)** only additional features unique to gelatinases (B) and MT-MMPs (C) are labeled. Hemopexin modules I-IV are labeled in approximately their correct orientation according to Li et al. *(24)*. All soluble MMPs except matrilysin (MMP-7) and MMP-23 have a hemopexin C domain. Gelatinase B has a longer proline-rich linker than other MMPs and this is homologous to the type V collagen α-chain. In the MT-MMPs (C) the "membrane linker" connecting the C terminus of the hemopexin C domain to the transmembrane helix varies considerably in length and commences from the C-terminal side of the distal cysteine residue that forms the intradomain disulfide crosslink. Topographically, this linker therefore starts near the terminus of the proline-rich linker at the "top" of the domain. Therefore, in schematic representations of these molecules the membrane linker should be drawn at the "side" of the molecule rather than at its base. *CBD*, collagen binding domain.

since it is reasonable to anticipate that most, if not all, extracellular matrix proteins will be degraded by one or more MMPs with varying degrees of proficiency. In the MMP family, the ability to cleave native collagen is a signature property of the collagenases, several MT-MMPs, and the gelatinases. Mechanistically, native collagen cleavage is a three-step process that first necessitates collagen binding. This is followed by triple helicase activity whereby the native collagen triple helix is locally disrupted to provide access to the scissile bonds. The last step is peptide bond cleavage, occurring one α-chain at a time. The collagenases and MT-MMPs *(11,12)* specifically cleave interstitial native types I, II and III collagens. Although several MMPs including the stromelysins and matrilysin can cleave denatured collagen (gelatin), gelatinolysis is most efficiently performed by the gelatinases (reviewed in **ref.** *13*). Gelatinase A (MMP-2, 72-kDa gelatinase) and gelatinase B (MMP-9, 92-kDa gelatinase) also degrade *native* types IV, V, VII, X, and XI collagens and so must also have a triple helicase mechanism.

The cell membrane-associated MMPs (MT-MMPs and gelatinase A) would be expected to have profoundly different regulation properties and in vivo roles in connective tissue degradation compared with their soluble counterparts. In tissue degradative events requiring the bulk removal of matrix such as in inflammation, post-partum uterus involution, and tissue repair, the overlapping substrate specificities of the soluble MMPs and their ability to act at a distance from cells facilitates the rapid and complete degradation of connective tissue. In contrast, the subtler remodeling of normal tissue must occur in a highly regulated manner in order to maintain the integrity of connective tissue structure and function. Accordingly, these fastidious requirements for regulated degradation are likely to be best achieved by cell-directed focal proteolysis that the MT-MMPs and cell membrane localized soluble MMPs such as gelatinase A appear well suited for. Indeed, this may provide the enzymic mechanism for the fibroblastic intracellular degradation pathway of phagocytosed collagenous matrix proposed to account for much of the normal turnover of connective tissue matrices *(14)*. In addition, it is now becoming clear that many cell surface receptors, prohormones and growth factors can be activated by proteolytic cleavage at the cell surface initiated by MMP activity. Recent work in our laboratory also shows an important role for MMPs in the proteolytic inactivation of various chemokines (McQuibban and Overall, submitted). Thus, MMP activity can regulate cell function not just through alterations of the extracellular environment that is registered by the cell through integrins *(15)*, but also directly through cleavage of cell surface receptors and their cognate ligands. The regulated cleavage of such a diverse range of MMP substrates partially explains the need for a large gene family of proteinases to accomplish extracellular matrix remodeling and cell regulation in a variety of diseases and conditions, in tissue and organ development, and wound repair.

In this chapter, I will overview the essential structural features of MMPs that has lead to the evolutionary radiation of the large number of susceptible proteins cleaved by MMPs. In view of the many different methods that could be utilized to investigate protein:protein interactions and the role of MMP domains in substrate and TIMP recognition, this chapter will consider experimental strategies and approaches for their study rather than specific methodologies. The 3D structure of MMPs and the atomic details of the active site have been dealt with extensively in the preceding chapter in this volume and will not be dwelt upon here. Therefore, this chapter will emphasize the simplicity of generating diverse proteinase specificities through the modular design of MMPs. In particular, I will focus on the properties, evolutionary aspects, and kinetic implications of MMP substrate binding "exosites" located on specialized protein domains and modules. A theoretical consideration of the impact of exosites on enzyme kinetics will be presented which predicts that in substrate-depleted matrices exosites modify enzymic activity to function as if still under zero order conditions, thus increasing the catalytic efficiency of these enzymes in remodeling. Another related, but distinct function of exosites that will also be discussed is the collagen binding and triple helicase mechanisms essential for native collagen degradation. Hypotheses will be presented for the role of exosites in what we term the "molecular tectonics" responsible for the collagen triple helicase activities of the collagenases and gelatinases. Lastly, MMP domains outside the catalytic domain also play important roles in TIMP binding and zymogen activation. This will be discussed in relation to substrate binding exosites that we have recently found influence gelatinase A activation. A new model will be presented for the homeostatic control of cellular collagen density that is regulated in part by collagen interactions with MT-MMPs and gelatinase A.

2. Domains, Modules, and Motifs

Throughout evolution as organisms became more complex, the number of functions required of a cell increased. Hence, functional genetic units were duplicated and recombined to generate gene families of proteins that performed related tasks, and more functions were added to existing proteins. The modification of proteins by the addition of new structural or protein binding modules to alter recognition and binding specificities is an evolutionarily rapid and selectable route to gene family diversity. In the recipient mosaic protein the function of the imported module is then often refined by amino acid substitutions. In proteinases this results in substrate specificity diversity, and new kinetic, inhibitory, and cell or tissue localization properties.

2.1. Domains

Complex mosaic proteins such as MMPs are composed of a hierarchy of structural units: domains, modules, and motifs. A protein domain is a subregion of the entire protein molecule that is functionally and structurally autonomous, but does not have to be contiguous in sequence *(16)*. For example, MMP catalytic and hemopexin C domains are stable and can function as isolated proteins in vitro and in vivo. Indeed, in the natural enzymes, these two domains are linked by a flexible spacer peptide that is predicted to allow separate movement of the two domains.

2.2. Modules

Modules are a distinct subset of protein domains that, like domains, are also homologous structural and functional units that have been repeatedly used throughout evolution as building blocks in functionally diverse proteins. In this way, modules could be added to the "core" protein to import new properties. There are now over a hundred examples recognized of protein modules that are components of both extracellular and intracellular proteins (for a compendium *see* **ref. *16***). Extracellular modules occur frequently in cell surface receptors, in members of the coagulation and fibrinolytic pathways, and extracellular matrix proteins—with intracellular modules found in proteins such as transcription factors, signaling molecules, and cytoskeletal proteins. Examples include the epidermal growth factor (EGF), immunoglobulin, fibronectin type I, II, and III modules, SH_2 and SH_3 domains, and von Willebrand factor type C domain. Modules fold independently to form compact structures, but are typically smaller than domains being composed of from between 40–100 amino acid residues. The smaller modules are generally stabilized by one or more disulfide bonds due to the lower stability imparted by the limited elements of secondary structure that can be present. As the size of modules increases, the raised stability of the more extensive secondary structure dominates and usually requires less disulfide use. Thus, due to the reducing intracellular environment, intracellular modules tend to be larger than many extracellular ones.

Protein modules often correspond to single exons *(17,18)*. Being encoded by single exons, modules are contiguous in sequence. Throughout evolution this has facilitated production of mosaic proteins composed of several modules by exon shuffling and duplication *(19–21)*. Here the function of the peptide chain of the module encoded for by the exon DNA drives evolutionary selection and successful shuffling of the exon through the genome as a discrete functional unit. Thus, module shuffling in mosaic proteins, as exemplified by the MMP family, generates increased diversity of protein function in a simple manner. In this process, the introns act as buffers to prevent gene disruption and to facilitate recombination. However, following integration and a successful alteration in protein properties of the recipient protein, intron mobility in the genomic DNA can then mask the original integration event at the sequence level *(17)*. Thus, only recent evolutionary shuffling events show modules with exon borders that exactly match the functional protein module. Nonetheless, recognition of exon boundaries is a useful tool for the first approximation of the boundaries of functionally discrete parts of a protein.

Topographically, modules may string together to form beaded-like protein structures as typified by large segments of fibronectin and fibrillin 1 *(22)*. Alternatively, a protein domain may contain inserted modules or be entirely composed of several modules. The 3D positioning of the N- and C-termini of the module in part dictates these arrangements. Termini at opposite ends of the module will in general favor linear arrays. Termini near one another allows packing with other modules to form a compact globular domain or facilitates insertion into a globular domain with minimal disruption of the domain fold. In the gelatinases three fibronectin type II modules are inserted in the two-lobed globular catalytic domain before the zinc binding sequence coded by exon 8 *(23)*. Therefore, these smaller modules, each separately encoded by an exon (exons 5, 6, and 7), break up the contiguity of the catalytic domain DNA and protein sequences. Other sequences have also been inserted in the stromelysin-3, MT-MMP, and the *Xenopus* MMP, XMMP (MMP-21) prodomains or catalytic domains *(6)*.

Although most modules can fold and function independently, this is not always the case. In some proteins, homophilic combinations of modules form higher order structures composed of the module repeats with extensive internal inter-modular packing interfaces—the whole domain now representing the stable functional and selectable unit. An example of this is the four bladed ß-propeller hemopexin C domain of MMPs. Its constituent building blocks are four hemopexin modules, each of which forms a blade of the ß-propeller structure. Although composed of four individual modules, the hemopexin C domain is further stabilized by an intradomain, intermodular disulfide bridge between hemopexin modules I and IV after the proline-rich linker at the start of the domain and by central structural Ca^{2+} ions *(24–28)*.

2.3. Motifs

At a finer level of structural organization, amino acid substitutions on module and domain backbones alter binding specificities and therefore function. Indeed, the biological function of a protein module or domain typically depends on only a small number of functional residues, a motif, displayed in a functionally permissive orientation by the polypeptide backbone. Identification of functional residues is critical if the biological activity of the protein is to be understood in molecular detail. Motifs are often used for protein:protein interactions or to bind cofactors or prosthetic groups. The classic familiar motif in the zincin proteinase clan is the HExxH Zn^{2+} binding site *(29)*. In the metzincins, this forms part of an extended Zn^{2+} binding motif HExxHxxGxxH *(30; see* Chapter 3). Another motif common to all MMPs (except MMP-23) *(3)* is the zinc switch sequence PRCG(N/V)D in the prodomain that is responsible for maintaining enzyme latency *(31)*.

Molecular determinants of function are often located in the flexible loops connecting ß-strands or α-helices in modules or domains. Here, amino acid substitutions, insertions or deletions can occur within a family of proteins or modules, usually without affecting the structural stability of the domain or modular fold. Such variation can substantially modify the properties of the module or domain. For example, the RGD (Arg-Gly-Asp) integrin-binding motif on fibronectin module III$_{10}$ is on a connecting loop between two ß-strands *(32)*. In the MMP prodomain of the convertase-sensitive MMPs (two soluble MMPs [stromelysin-3 MMP-11, XMMP MMP-21] and five MT-MMPs [MMP-14-17, MMP-24]) a dibasic recognition sequence (RXRR, RRKR, RRRR) has been inserted that is recognized and cleaved by furin-like convertases in the Golgi *(33)*. In contrast, some motifs such as the HExxH Zn^{2+} binding site occur displayed on one face of an α-helix. This is not so much because of the structural considerations of the zinc binding properties of the motif (although this is clearly important), but because higher order protein organization is also required to bind substrate and generate different substrate specificities. This is accomplished through positioning other strands or loops of the catalytic domain nearby which then can modulate substrate binding and preference. Possibly as a consequence of forming such an integral part of the catalytic domain, the catalytic zinc also makes a significant contribution to the structural stability of the catalytic domain *(34)*.

3. Experimental Strategies

The study of multidomain protein structure and function has been greatly advanced by protein engineering techniques to generate isolated domains, domain deletions, and domain swap and chimera variants. **Table 1** lists these approaches and compares their respective advantages and disadvantages. Several groups have studied the structural biology and catalytic activity of various MMPs through investigation of isolated recombinant catalytic domains. Thus, many MMPs have now been expressed only as the fully active form of the catalytic domain lacking the prodomain, the linker, and hemopexin C domain. Although considerable information is lost by these deletions, this approach greatly simplifies protein expression and purification procedures, crystallization, and a protein of this size is amenable to NMR studies. Such isolated catalytic domains are particularly useful for the study of peptide cleavage, synthetic MMP inhibitor binding, and structural studies.

Often the study of substrate binding sites on individual domains provides valuable complementary data to domain deletion recombinant proteins. Domain:domain or module:module interactions with either flanking or distant sites in the protein sequence may influence the functional properties of the domain or module. For example, a CBD deletion mutant of gelatinase A lost type IV collagen binding properties *(35)*. However, the isolated CBD binds denatured, but not native type IV collagen *(9)*. Together, this indicates that the type IV collagen-binding site requires surface contributions from both the CBD and the catalytic domain—most likely, at or around the S_3' subsite on α-helix B. The gelatin-binding site in fibronectin also requires intermodular interactions between the two type II and the flanking type I modules within the 42-kDa gelatin-binding fragment of fibronectin *(36–38)*.

Full-length MMPs can be expressed in eukaryotic systems quite successfully, but yields are typically much lower than achieved in *E. coli*. Although attempts have been made at expression of full-length MMPs in *E. coli* these can be difficult systems to optimize. A frequent exacerbating problem is localization of the recombinant protein in insoluble *E. coli* inclusion bodies. This necessitates devising dissolution and refolding procedures to recover soluble, functional protein. In general, a denaturation step is included to achieve this in which the inclusion bodies are broken up and their contents dissolved in 6 *M* guanidinium, 8 *M* urea, 2 *M* arginine, or 1% sarcosyl. However, in the renaturation steps, an open prodomain can render the refolded catalytic domain active. Thus, MMP autocatalytic activity during refolding of proMMPs often results in variable cleavage of the prodomain or the linker with consequent loss of the hemopexin C domain. This makes it difficult to obtain homogenous preparations of full-length, latent recombinant enzyme that includes the prodomain, or hemopexin domain, or both. Incorporation of synthetic MMP inhibitors into buffers at these steps can reduce this problem or a catalytically inactive mutant that does not autodegrade can be made by replacement of the active site Glu with an Ala.

Correct refolding of recombinant MMPs can be further complicated due to the presence of disulfide bridges. Since the MMP catalytic domain does not contain disulfide crosslinks (other than the gelatinases, which contain 6 in the three-fibronectin type II modules) and the hemopexin C domain only contains a single disulfide bridge, this is further justification for expression of only the catalytic domain in the first instance. These problems are exacerbated in some MMPs such as MMP-9 *(8)*, rat MMP-8 *(39)* and MMP-21 *(5)* which contain one or two additional unpaired cysteine residues in the hemopexin C or catalytic domains, respectively. Nonetheless, since substrate binding exosites reside on the hemopexin (*see* **Subheading 5.2.**) and fibronectin type II modules (*see* **Subheading 5.1.**), and TIMP stabilization secondary sites also occur on the hemopexin C domain, a comprehensive understanding of MMP catalysis of macromolecular substrates, TIMP inhibition, and activation warrants the attempt to express and study full-length enzymes.

Chimeric proteins are intended to replace and thereby allow the assessment of function of a contiguous segment of surface exposed strands or helices of a

Table 1
Experimental Approaches to Study Module and Domain Function

1. Deletion mutants

Advantages
- Allows for assessment of the role of the domain on enzyme function by loss of function.
- Provides indirect evidence for the properties of the deleted domain.
- N- or C-terminal deletions will typically be stable if an entire domain or module is removed.
- Removal of disordered flexible N- or C-terminal "ends" from the folded organized core may significantly improve crystallization success without disturbing the global fold of the protein.
- Removal of N- or C-terminal ends may reduce the protein size into the range suitable for NMR studies.

Pitfalls
- Internal deletions have great potential to structurally perturb the remnant protein unless a compensating linker or spacer is left or an appropriate linker is introduced.
- Progressive N- or C-terminal deletion mutants, particularly those based on convenient restriction sites in the cDNA or those also made without consideration of the 3-D structure, can significantly disrupt the remnant protein structure also leading to loss of function. For example, deletions that break an essential disulfide bridge or which sever an α-helix, can drastically destabilize the protein. In these cases, loss of function cannot be ascribed with certainty to the deleted protein segment.
- Exposure of the domain packing face on the remnant protein may introduce new binding properties including aggregation, precipitation, or destabilization of the protein.

2. Recombinant domains

Advantages
- Simplifies data interpretation by removal of other potential substrate or protein binding sites, including the active site.
- Can allow unequivocal assignment of binding sites or other properties to a particular domain.
- Recombinant domains are useful in competition experiments with the wild-type proteinase. Adding domains to in vitro assays may inhibit the function of the enzyme in a dominant negative manner.

Pitfalls
- May not fold autonomously. This is generally not a problem if the domain boundaries have been carefully selected with consideration of the 3-D structure, or in the absence of structural information, according to exon boundaries.
- Loss of binding sites formed with contiguous tertiary structures. That is, loss of cooperativity from other parts of the protein.
- Exposure of a domain packing face that normally would be buried may introduce new properties, such as self binding leading to aggregation or precipitation.

3. Domain swap mutants and domain chimeras

Advantages

- Confer domain-specific properties on the recipient protein.
- May remove domain-specific properties from the donor protein.
- May alter proteinase substrate specificity.
- May allow for the identification of binding sites.
- Can be used to determine the role of natural amino acid substitutions in homologous domains.

Pitfalls

- Incorrect domain-domain interactions and spacing between the donor and recipient domains may alter function or stability. Therefore, loss of function cannot necessarily be ascribed to the domain swap.
- Chimeric substitutions within a domain may significantly perturb the protein backbone and function. This is an important consideration when large packing interfaces are present between the donor protein and the recipient.
- Only gain-of-function chimeras are unequivocally useful indicators of the functional importance of the new chimeric protein segment.

4. Site-directed mutagenesis

Advantages

- Site specific.
- Resolves in molecular detail determinants of protein function.
- Compared with other protein engineering methods, has the potential for the least structural perturbations possible.
- Active site mutations that inactivate the proteinase (E->A) can be used to study substrate binding in the absence of cleavage of the protein or autocatalysis.

Pitfalls

- May perturb structure.
- Single mutations of important residues in high affinity binding sites may not have readily detectable effects in the absence of changes to other residues.
- In the absence of a rational strategy, can be laborious.

domain. However, these inserted segments must still pack onto the domain core through side chain and main chain interactions. Well-ordered internal amino acids contribute significantly to the stability of proteins. In general, the native structures of globular domains exhibit efficient packing of internal non-polar side chains into a dense hydrophobic core with few internal defects. Indeed, single amino acid replacements in the internal core can increase or decrease the melting temperature of these proteins by 10°C *(40,41)*. Therefore, the effect of several substitutions may significantly disrupt the efficient, well ordered internal packing of side chains. Protein chimeras that replace large segments of the protein core with the corresponding part of another protein, particularly with nonconserved amino acid residues at the chimera interface, will therefore likely perturb the internal stability of the domain. Care must be exercised in the design of chimeras to reduce or minimize such disruptions. However, very little biophysical characterization of MMP chimeric domains has been reported. It would be expected that minor to considerable shifts in the backbone of the domain and alterations in stability would have resulted to affect function in some way. Nonetheless, the fact that chimeras replacing large segments of the hemopexin or MMP catalytic domains work at all indicates that considerable tolerance exists in these domains for quite extensive substitutions. This concern does not invalidate the chimeric approach, indeed chimeras are an extremely useful tool when gain of function occurs, but loss of function results always need to be qualified as such and interpreted with caution. The temptation to use convenient DNA restriction enzyme sites in the cDNA to define the boundaries of the chimera domain should be restrained. Rather, domains, modules and chimeras thereof should be constructed at rationally selected sites by use of polymerase chain reaction (PCR) to introduce appropriate restriction sites at the defined ends of the cDNA of interest.

Interpretation of domain chimera binding studies should be performed cognisant of the caveats listed in **Table 1**. On the basis of loss of TIMP-2 binding from gelatinase A/stromelysin-1 hemopexin C domain chimeras it was concluded that the gelatinase A TIMP-2 binding site was located on hemopexin modules III and IV, commencing from a cationic triad at Lys566/Lys567/Lys568 *(42)*. However, an extensive site directed mutagenesis study found that the Lys566/Lys567/Lys568 were not directly involved in TIMP-2 interaction *(43)*. Moreover, since several important TIMP-2 contact residues were also located as far N-terminal as position 547, these data indicated that the chimera interface in the preceding study had not replaced the TIMP-2 binding site with a nonfunctional segment from stromelysin, but had in fact split the binding site. Indeed, the 3D-structure shows the cationic cluster at 547/549/550/561 to be topographically between the lysine triad and residues in hemopexin module IV that were also found to be pivotal for the TIMP-2 binding interaction *(43)*.

4. Determinants of MMP Catalytic Function

The catalytic power of MMPs is encoded in four essential structural elements: (1) binding of the main chain of the substrate in the active site cleft; (2) the specificity subsite pockets that define the active site, in particular the S_1' subsite; (3) the zinc ion-binding histidine triad and catalytic glutamate; and (4) binding to substrate binding "exosites" outside the active site. Main chain binding of protein or peptide substrates to the active site occurs through formation of an antiparallel ß-sheet structure, with hydrogen bonds formed between the substrate and enzyme. A polarized H_2O molecule bound to the active site Zn^{2+} ion and general base glutamate residue is essential in catalyzing peptide bond cleavage. In MMPs, the Zn^{2+} ion is coordinated by the three histidines in the HExxHxxGxxH metzincin signature motif *(44)*. Mutagenesis of any one of these His residues ablates catalytic activity *(34,45)*. By analogy with the bacterial thermolysin collagenases, nucleophilic attack by the catalytic water molecule in MMPs occurs on the carbonyl of the substrate scissile peptide bond to form a tetrahedral transition-state intermediate in which the negatively charged oxygen is stabilized by coordination with the catalytic Zn^{2+} ion.

The variety of known and potential MMP substrates requires different molecular strategies to be used in their cleavage, often by the same MMP. To accomplish this, critical functional elements of the catalytic domain have been adapted to suit diverse substrates. Thus, the active site specificity subsites differ between MMPs to accommodate different peptide backbones of substrates around the scissile bond. A crucial molecular determinant of MMP substrate specificity is the side chain of the amino acid residue (P_1') immediately after the scissile bond *(46–50)*. The S_1' specificity pocket of MMPs accommodates this side chain and so the size and chemical characteristics of the S_1' subsite are important in determining peptide bond preference for cleavage. For example, the small S_1' pocket of MMP-1 and MMP-7 restrains substrate preference to small hydrophobic residues at P_1' *(50–52)*. In contrast, other MMPs such as MMP-2, MMP-3, MMP-8, MMP-9, and MMP-13 have large S_1' pockets *(50,53–55)* and can accommodate a more diverse range of amino acids at P_1'. However, little is currently known of the nature, degree and extent of conformational changes induced in and around the specificity subsites of the MMPs by substrate or inhibitor binding. Any such torsional entropic cost of binding would be expected to be more than counteracted by favorable interactions with the substrate or inhibitor. Having accurate structures of the free protein and bound complexes with information on structural shifts and dihedral angle perturbation is essential for designing new inhibitors more specific toward one or more MMPs and away from others.

5. Exosites

In addition to the active center subsites, the second dominant strategy to facilitate substrate binding and subsequent cleavage by MMPs is through specialized secondary substrate binding sites on discrete substrate binding domains, or smaller functional modules, located outside the active site. Adding protein-binding domains and modules increases the affinity of the proteinase for particular substrates and can modify the specificity of the main function of the MMP catalytic domain, which is to cleave scissile bonds. At the Conference *"Inhibitors of Metalloproteinases in Development and Disease,"* Banff 1996, these secondary specificity sites were termed "ectodomains" *(56)* or "exosites" *(10)* after our previous adoption of this term *(9)*. Since one domain or module can display multiple binding sites for the same or different substrates we favor the term *exosites*.

Substrate interactions with exosites can influence the behavior of a proteinase in a number of ways. Exosites modulate and broaden the substrate specificity profile of MMPs by providing an additional contact area not influenced by the primary specificity subsites. Variation in the substrate binding properties of these modules or domains can alter the substrate preference of the MMP, and as part of the MMP degradative system, may add competitive advantage to the proteinase to degrade particular substrates. In this way, the function of the proteinase is refined and can be made, in general, more specific or efficient. Substrate binding is often the main function of specialized modules. Thus, simplicity through modular design is an attractive concept where the addition of modules and domains to the proteinase catalytic domain generates new diversity in substrate preference. In addition to tethering substrates to the enzyme to potentiate cleavage, exosites may be involved in essential "substrate preparation" prior to cleavage. For example, the localized "unwinding" of native collagen substrates by MMPs has been termed triple helicase activity. Exosites can also target the enzyme to substrate in tissues *(9,57,58)* or to cell associated substrates *(59,60)*. Thus, the identification of substrate exosites and the development of specific drugs designed to bind and block these sites potentially offers new pathways to highly selective anti-MMP therapeutics which are selective for the degradation of specific substrates by that MMP. This promises a novel therapeutic approach with reduced side effects.

In MMPs, exosites are found on the hemopexin C domain and, in the gelatinases, on the three-fibronectin type II modules. The novel insert in stromelysin-3 *(61)* near the S_3 subsite may also bind specific substrates. The vitronectin-insert sequence in the prodomain of XMMP *(5)* would not be defined as an exosite since the prodomain is removed upon activation and so cannot influence subsequent catalytic activity. However, targeting of proXMMP to specific substrate locations in the matrix prior to activation would undoubtedly influence the actions of this MMP in tissues. In collagenases 1–3, stromelysin-1, and MT1-MMP the hemopexin C domain binds native collagen (*62–66*; Tam and Overall, unpublished data), but the hemopexin C domain of gelatinase A does not (*10,28,67*). Instead the triple fibronectin type II repeat in the gelatinases forms an alternative collagen binding domain which lies proximal to the S_3' subsite, 16–19 Å from the active site Zn^{2+} ion. The insertion site of this CBD is in a loop of the catalytic domain that connects the active site α-helix B with ß-strand V on the top of the protein. As discussed above, protein loops that connect elements of secondary structure are generally structurally plastic to insertions, deletions, or amino acid substitutions. Functionally, the insertion of a 22-kDa fibronectin type II module triple repeat in such a connecting loop in the gelatinases does not appear to have perturbed the global fold of the enzyme.

5.1. Fibronectin Type II Modules

In fibronectin (*36,38,68*) and bovine seminal fluid protein PDC-109 (*69*) fibronectin type II modules are involved in binding denatured collagen. A similar function for these modules in the CBDs of gelatinase A (*9,35,70,71*) and gelatinase B (*72*) was subsequently reported. In addition to binding denatured type I collagen (gelatin), our characterization of the recombinant human gelatinase A CBD demonstrated that exosites on this domain also account for most, if not all, of the native type I, V, and X collagen binding properties of gelatinase A (*9,10,73*). The CBD also provides the enzyme with binding specificity to denatured types II, IV, V, and X collagens, and unexpectedly, elastin (*9*). As an isolated domain the CBD does not bind native type IV collagen (*9,73*). However, CBD-deletion mutants of gelatinase A show reduced native type IV collagen binding and degradation (*35*) indicating the importance of juxtaposed segments of the catalytic domain in forming the native type IV collagen binding site. A 90% reduction in gelatinolysis (*35*) and a total loss of elastinolytic activity (*74*) occurs following deletion of the CBD. This confirms the functional importance of the CBD in binding these substrates. A similar reduction in gelatinolysis occurs in CBD deletion mutants of gelatinase B (*75*). In addition, type XI collagen degradation is abrogated by this deletion. Nonetheless, for reasons unknown, Ye et al. (*76*) found little effect on gelatin degradation by a CBD deletion of recombinant gelatinase A. This may be explained by a lower specific activity of the purified wild-type protein prepared in their system.

Binding differences are apparent between the CBDs of gelatinases A and B that may reflect differential substrate specificity properties of these enzymes. Dimethyl sulfoxide (DMSO) used at a concentration that disrupts gelatin binding did not affect gelatinase B activity (*72*), whereas dimethize sulfoxide (DMSO) reduces gelatinase A activity by 33% (*see* **Fig. 2**) in agreement with

Murphy et al. *(35)* who reported a 50% reduction in gelatinolysis upon addition of DMSO. However, a common problem with this and domain deletion approaches is structural perturbation. DMSO may disrupt hydrophobic substrate interactions in the active cleft or packing in the hydrophobic protein core and so also perturb enzyme structure or function. Similarly, the internal deletion of a 22-kDa domain from a 45-kDa gelatinase A catalytic:CBD recombinant protein without inserting a compensating spacer peptide may induce structural and functional defects that alter the activity of the enzyme against macromolecular substrates, but not the activity of the active site against short septapeptide substrates. In competition experiments the importance of the CBD for gelatinolysis can be readily demonstrated without perturbing structure. Upon addition of recombinant CBD, but not the hemopexin C domain, a concentration dependent reduction in gelatinolysis by recombinant full-length gelatinase A is seen, with the CBD functioning in a dominant negative manner (*see* **Fig. 2**).

The matrix binding properties of the CBD also have the potential to localize the enzyme to collagen either in the extracellular matrix *(9,35,58)* or on the cell surface linked to ß1-integrins *(59)*. When expressed as fusion proteins, the individual fibronectin type II modules of gelatinase A and B can also bind gelatin *(70,72)*. However, we have found the individual gelatinase A CBD modules (CBD1, CBD2, CBD3), expressed as isolated recombinant proteins without a fusion protein partner, to be unstable in solution. This may be the result of exposure of the hydrophobic intermodule packing faces on the isolated modules. Protein stability was markedly increased when two CBD modules were expressed together as CBD12 (lacking the third module) or CBD23 (lacking the first module). The deletion of either the first or third module reduces the binding affinity for denatured type I collagen by an order of magnitude to 1.4×10^{-6} *M* for CBD12 and 1.1×10^{-6} *M* for CBD23, from 1.2×10^{-7} *M* for the full-length CBD123 *(73)*. A similar effect was found for the binding to native type I collagen. These CBD deletion mutants also bind different substrates indicating different substrate binding specificities for each module. Thus, CBD12 does not bind *denatured* type II, IV or V collagens whereas CBD23 does *(73)*. This result indicates that either CBD3 is responsible for binding these specific collagen types or that a cooperative binding site is formed only between CBD2 and CBD3 and not between CBD1 and CBD2. Other data in support of the formation of cooperative binding sites includes the observation that *native* type V collagen binds CBD23 but not CBD12. However, an essential role for CBD1 is apparent for elastin and type X collagen binding. Compared with the CBD123, which binds elastin *(9)* and native and denatured type X collagen, neither CBD12 nor CBD23 bind these substrates *(73)*. This points to coopera-

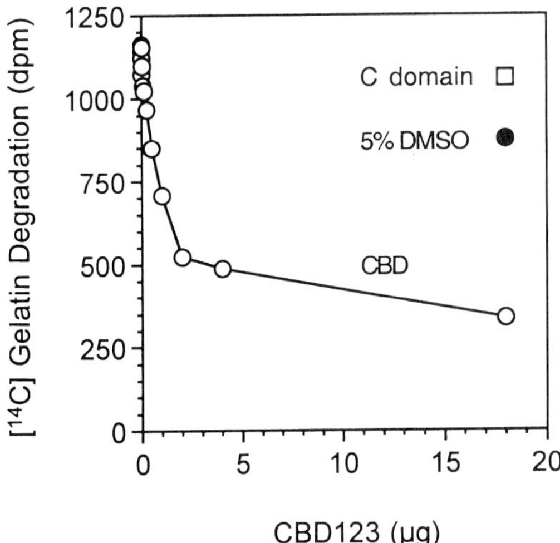

Fig. 2. The importance of the gelatinase A CBD for gelatinolysis. Recombinant human gelatinase A expressed from CHO cells was added to metabolically labeled 14C-glycine denatured type I collagen for 2 h at 37°C. Gelatin degradation was quantitated by measuring the radioactivity of 10% TCA/1% tannic acid-soluble gelatin fragments *(107)*. In competition experiments, recombinant gelatinase A CBD was added to the reaction in the amounts indicated (*open circles*). A concentration-dependent decrease in gelatinolytic activity was observed that was not apparent upon addition of recombinant C domain of gelatinase A (*open square*) or 5% DMSO (*closed circle*), as indicated.

tive interactions occurring between the three modules to form a complex binding site for some proteins.

Our experiments analyzing the potential of the CBD to bind simultaneously several substrate molecules also support the concept of module-module interactions being essential for substrate binding. Whereas the gelatinase A CBD123 can bind two molecules of gelatin or native collagen at the same time *(9)* the single module deletion forms of the CBD triple repeat (CBD12 and CBD23) do not *(73)*. Thus, two cooperative binding pockets may form per CBD123 (*see* **Fig. 3**), with one being lost upon deletion of a CBD module and with it, the ability to bind two molecules of substrate simultaneously. In the CBD123 protein, substrate interaction may occur in one pocket to the same homologous binding site on each of the two modules to more effectively bind the substrate. The remaining module may form the second binding site alone, or bound substrate may be stabilized through an alternative interaction with another face of one of the other two modules. Indeed the location of an effective lysine mutant (Lys263Ala) that disrupts collagen binding of the gelatinase A CBD is on the

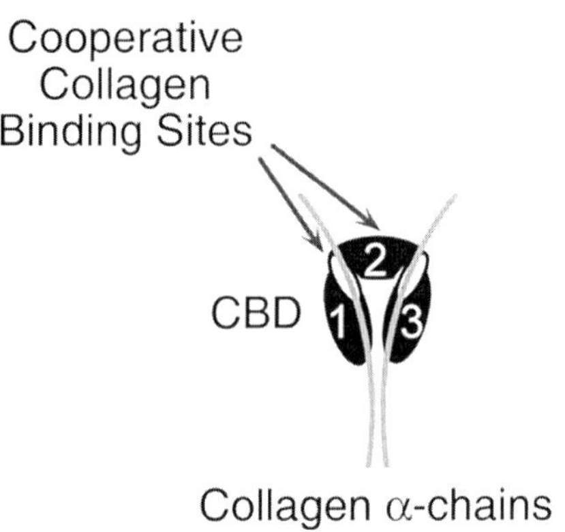

Fig. 3. The gelatinase A CBD may form two binding pockets through cooperative interactions involving two CBD modules each. The cooperative binding sites are schematically drawn as white ovals between two fibronectin type II modules.

"top" surface of the module (Steffensen and Overall, unpublished data). This is on the opposite side of the module away from the concave cup shaped collagen binding surface (*72*; Moore, Conner and Overall, unpublished data).

The structure of the CBD and catalytic domain from either gelatinase remains unreported. However, modeling studies have been published of the structure of the gelatinase A catalytic domain *(55)* and the catalytic domain with the CBD *(77)* based on the nuclear magnetic resonance (NMR) structure of the fibronectin type II module of PDC 109 *(78)*. Whereas the structure of an individual module can be predicted by threading algorithms with a reasonable degree of confidence *(16)* the packing interactions between modules in any modular protein cannot yet be accurately predicted. This casts doubt on the usefulness of models derived from predicting module-module interactions. Indeed, consideration of the cooperative interactions required of the different CBD modules to bind substrates indicates that the CBD structure is likely to be complex. It is unclear how much intermodular flexibility is allowed in the domain by hinging at the linkers between the start and finish of each module. Are all three modules packed together in one stable orientation or is intermodular movement permitted in the structure? Since the N- and C-termini of the type II modules of fibronectin are positioned near each other *(79,80)*, this indicates the likelihood of a globular domain structure for the CBD rather than an extended array. Indeed, the CBD termini must insert near one another in the catalytic domain in order not to perturb the MMP global fold. If independent movement is permitted, this may allow a molecular treadwheel action to "feed" new substrate into the active site of the enzyme. Independent movement may also facilitate a novel triple helicase function for the CBD on those collagens cleaved by the gelatinases (*see* **Subheading 6.3.3.**).

5.2. The Hemopexin C Domain

The second major location of MMP exosites is on the hemopexin C domain. This domain has a beautiful common fold. The overall shape is a squat cylinder comprised of four ß-sheets each representing a hemopexin module and each forming a blade of the four bladed ß-propeller structure *(24–27)*. Each blade is a twisted four stranded ß-sheet that links to the adjacent blades by small connecting strands on the rim of the domain and from a longer loop at the center. Extensive packing faces on the blade sides between adjacent modules arrange one next to another in a stable configuration. Packing of the wedge shaped modules also forms a central pore in the domain, in which are bound either one *(24,25)* or two *(26,27)* Ca^{2+} ions. The central Ca^{2+} ions in the gelatinase A hemopexin C domain are structurally important—their chelation eliminates heparin and fibronectin interactions *(28)*. Although X-ray crystallography also showed a potential Zn^{2+} ion binding site on the gelatinase A hemopexin C domain *(25)*, inductive coupled plasma mass spectrometry analysis did not measure any significant zinc molar content in a recombinant human gelatinase A hemopexin C domain *(28)*.

The hemopexin C domain has homology to the collagen-binding domain of vitronectin (which therefore will also likely have domains with a four bladed ß-propeller structure). Notably, unconfirmed studies from one group report the binding of the gelatinase A hemopexin C domain to $\alpha_v\beta_3$ integrin *(81)*, a recognized vitronectin receptor. The hemopexin C domain of collagenase-1 is known to bind native collagen *(62,63,82)*. Hemopexin C domain deletion ablates collagenolysis but not catalytic competence—that is, truncated collagenase-1 lacking the C domain still degrades synthetic peptide substrates and casein, but not native type I collagen *(62,64)*. In addition, substitution of the two cysteine residues disrupts the hemopexin C domain structure and causes the loss of type I collagenolysis, but still allows casein degradation *(63)*. Other collagenases such as neutrophil collagenase (MMP-8) *(39,65)*, collagenase-3 (MMP-13) *(66)*, and MT1-MMP (Tam and Overall, unpublished data) also bind native collagen on their hemopexin C domains. However, hemopexin C domain deletion mutants of collagenase-3 cleaved type IV, IX, X, and XIV collagens, fibronectin, and tenascin *(66)* indicating the absence of important hemopexin domain exosites for these substrates. Thus, collagen binding by exosites on the hemopexin C domain of collagenases plays an essential role in triple helicase activity for some fibrillar collagen types and also likely potentiates cleavage of the bound, and now locally unwound, collagen substrate. However, it is important to note that these are two separate, but linked processes in the mechanistic aspects of native collagen cleavage.

Unlike the collagenases, removal of the hemopexin C domain from gelatinase A *(67)* or gelatinase B *(83)* does not modify native or denatured collagen substrate specificity. Consistent with this is the demonstration that recombinant human gelatinase A hemopexin C domain does not bind native or denatured collagen type IV, V, or X substrates of gelatinase A *(10,28)*, possibly as a result of the acquisition of binding sites for TIMP-2 *(43,84,85)* and TIMP-4 *(82)*. This is important because it shows that the molecular tectonics of triple helicase activity must be fundamentally different between the collagenases and gelatinases *(28)* (*see* **Subheading 6.3.**).

In general, substrate-binding properties of the hemopexin C domain of MMPs have not been examined comprehensively. Indeed the initial reports comparing substrate specificity of hemopexin C domain deletion mutants of stromelysin-1 and the gelatinases with the wild-type enzymes found no significant differences in substrate specificity or cleavage patterns *(64,67,83,86,87)*. This has prompted the suggestion by several authors that the primary role of the hemopexin C domain is to stabilize the interaction of TIMPs with active MMPs. Not only is there a stabilization site for the TIMP inhibitory interaction in the gelatinases on the hemopexin C domain, but also a separate high affinity site that can bind TIMPs *(43,84,85)*. Nonetheless, the sea urchin MMP envelysin is absolutely dependent on the hemopexin C domain for fertilization envelope protein cleavage *(88)*. In addition, several substrates of gelatinase A including fibronectin, fetuin (α2-HS) (Greene, Overall and Ochieng, unpublished data), and monocyte chemotactic protein-3 (MCP-3) (McQuibban and Overall, unpublished data), also bind the hemopexin C domain at different sites. In enzyme assays, recombinant hemopexin C domain functions as a competitive inhibitor for these substrates. Interestingly, the identification of MCP-3 as a novel substrate for gelatinase A was made using the yeast 2 hybrid screen. This is the first example of a proteinase substrate being identified using this analysis. Thus, it is likely that other MMP substrates will be identified that bind the hemopexin C domains in interactions essential for their effective cleavage.

6. MMP Exosite Function

6.1. Exosites Target MMPs to Substrate in Tissues or on Cells

Exosites can influence the behavior of MMPs in a number of important ways. First, by increasing the affinity of MMPs for substrate, exosites are a highly efficient adaptation that enables proteinase targeting to macromolecular assemblies of substrate such as fibers in the tissue or on cells. In the extracellular environment substrate distribution is not uniform, as it is experimentally in vitro. For example, microenvironments occur in the immediate vicinity of the cell, between cells, and at different locations in the extracellular matrix. At one extreme, tendons represent essentially an infinite concentration of collagen substrate, *albeit* mostly insoluble. Elsewhere, the tissue collagen concentration is considerably lower, but unevenly distributed around the cell and in the fibrils and fibers present in the tissue. Since substrate concentration differs across tissues, the kinetic properties of MMPs are location dependent. MMPs may function under substrate excess conditions (zero order kinetics) at one site and under substrate-depleted conditions (first order kinetics) elsewhere. Persistence of binding to substrate prevents diffusion away with the enzyme no longer in equilibrium with the aqueous phase (tissue fluid), thus favoring enzyme action under zero-order conditions.

Paradoxically, a large accumulation of substrate such as collagen fibrils, and packing of fibrils into fibers, also reduces substrate accessibility. Since only the surface layer of collagen in fibrils is initially accessible to the enzyme, collagenase degradation of type I collagen fibrils is about one-third less than for soluble collagen *(57)*. Indeed, only ~12% of the total number of molecules present in a 50-nm diameter collagen fibril are surface exposed. As the fibril is degraded the subjacent deeper layers of collagen now become accessible to the enzyme which delays proteinase entry to substrate-depleted conditions with reaction rates less than V_{max} *(57)*. Thus, collagenolysis of fibrils saturated with enzyme continues linearly with time, and does not decline due to substrate depletion, until degradation has reached the point at which the fibril is depolymerized. Only then is the enzyme forced into first order conditions.

Gelatinase A may also be targeted to native collagen, a poor substrate of the enzyme *(89,90)*, by interactions involving the CBD. One consequence of gelatinase A binding native type I collagen in the telopeptide regions *(9)* is that the enzyme is localized on collagen fibrils poised for later activity during tissue remodeling. In this way, gelatinase A is precisely located on a potential substrate that will be generated following collagenase cleavage of the native collagen. At that time the bound, cleaved and denatured collagen may feed into the active site of the enzyme as gelatinase A "moves" along the denatured collagen degrading the cleaved collagen α-chains to small fragments. Enzyme may also remain bound to native collagen in the fibril but still be readily accessible for cleavage of nearby gelatin chains generated by collagenase action on the fibril. Indeed, gelatinase A can simultaneously bind two molecules of collagen or gelatin on the CBD favoring actions of this type *(9)*. Gelatinases A *(59)* and B *(60)* also bind to the cell surface through interactions involving the CBD. For gelatinase A this was shown to be through the CBD interacting with ß1-integrin-bound cell surface native collagen *(59)*. Olsen et al. (1998) found that gelatinase B binds cell surface denatured α2 IV collagen chains.

6.2. Impact of Exosites on Enzyme Kinetics

In addition to spatially targeting MMPs to substrate, exosite binding also impacts on the kinetic properties of the proteinase in other ways. Whereas the molecular environment of the Zn^{2+} ion is critical in determining the specificity of peptide bond cleavage and the rate of product formation k_{cat} (where k_{cat} is the number of moles of substrate degraded per unit time per enzyme molecule), exosites impact on substrate binding affinity and therefore K_m (1/affinity). Second, in substrate-depleted conditions, we would predict that exosites kinetically enhance substrate degradation. Typical enzyme kinetic considerations of model proteinases utilize simple substrates, often peptide substrate mimics. In these experimental conditions substrate binding to the active site, as reflected by the K_m, is a relatively simple consideration. However, calculating the kinetic parameters of the cleavage of macromolecular substrates that bind at multiple sites on the enzyme and discerning the contribution of this binding K_D to the K_m is not trivial. Here, the overall K_D of the enzyme for the substrate is a reflection of binding to both the active site and to exosites. By tethering substrates to the catalytic domain, exosites promote subsequent productive interactions of the substrate with the active site where otherwise the dissociation rate (k_{-1}) is large in comparison to the rate of product formation k_{cat}. This consideration predicts that substrate binding to exosites would decrease the K_m and therefore increase the specificity constant k_{cat}/K_m to favor degradation. Indeed, a CBD-deletion mutant of gelatinase A showed approximately a 10-fold reduction in the specificity constant for gelatin through increases in K_m *(35)*. Septapeptide cleavage was not affected since the binding properties of the active site would be expected to be relatively unaltered.

Our studies have led us to appreciate that by binding macromolecular substrates, exosites also increase the local concentration of substrate in the vicinity of the enzyme at the active site. Thus, there are in effect two concentrations of substrate to consider at a particular tissue site: the tissue total $[S]_{total}$, and secondly, the local "tethered" substrate concentration $[S]_{local}$. For maximal proteolytic activity per proteinase molecule, the zero order reaction rate, the substrate concentration must be in excess relative to the proteinase. That is, the reaction rate must be independent of substrate concentration, with any change in the amount of product formed at a site of remodeling over time dependent upon the amount of active proteinase present. As substrate is depleted in the environment of the enzyme, the proteinase active site is no longer saturated, substrate concentration becomes rate limiting, the cleavage reaction becomes first order (proportional to the first power of substrate concentration), and the rate of substrate cleavage slows due to substrate exhaustion. However, substrate binding to exosites would tether potential substrates to the enzyme independent of binding to the active site. For substrates such as gelatin, which are cleaved at multiple sites along the α-chain, exosite binding of even one gelatin chain increases $[S]_{local}$ to generate substrate excess conditions of the gelatin scissile bonds. This should enable the proteinase to be converted to the enzyme substrate complex [ES] for a longer period of time, and at a lower $[S]_{total}$ than predicted to be limiting. Thus, V_{max} (the maximum reaction velocity occurring when the entire available enzyme has been converted to [ES]) may be more readily achieved. In other words, the enzyme now requires less total tissue substrate to be saturated and so the maximal velocity is reached at a lower $[S]_{total}$. Of course, the exosites themselves need to be saturated for the V_{max} to be achieved under these conditions. When exosites are not bound with substrate the enzyme functions at $[S]_{total}$.

In a clear example of the potential for exosites to increase the $[S]_{local}$ independent of substrate bound to the active site we have shown that the CBD of gelatinase A can simultaneously bind two molecules of native or denatured collagen *(9)*. Since only one α-chain can be cleaved at a time, the binding of a second molecule of substrate by the CBD may therefore favor its subsequent cleavage following expulsion of the cleaved products of the first α-chain from the active site.

Generating conditions in which proteinases function with zero order kinetics has a number of advantages for a cell. This is the most economical way for the cell to synthesize the minimum amount of proteinase required to degrade a specified amount of substrate. Since removal of substrate is proportional to the amount of active enzyme generated, the cell can more effectively regulate enzyme expression to accomplish a given amount of degradation. Because less proteinase is expressed under these conditions, regulation through enzyme inhibition can be more readily achieved by TIMPs and $\alpha2$-macroglobulin. In these ways, proteinase exosites may facilitate more precise control of extracellular proteolysis.

As discussed in **Subheading 6.3.**, triple helicase activity of collagenase is absolutely dependent upon native collagen binding by the hemopexin C domain of the enzyme. However, because of this action the effect of the hemopexin C domain collagen binding exosite on the subsequent cleavage step of α-chains is difficult to measure. The collagenase hemopexin C domain does not bind gelatin; therefore collagenase cleavage of gelatin is only dependent upon binding at the active site. Although triple helical collagen is cleaved one chain at a time this is not the same as gelatin cleavage. Part of the triple helical segment of the collagen is bound to the hemopexin C domain thus favoring entry of the locally denatured α-chains into the active site. Since removal of the hemopexin C domain of collagenases completely ablates triple helicase activity, the effect of hemopexin C domain deletion on the actual collagen chain cleavage rate cannot be measured by these deletion mutants.

6.3. Collagen Triple Helicase Activity

For some substrates, exosite binding is an absolute requirement for degradation. By tertiary or quaternary structure disruption exosites may expose scissile bonds for subsequent cleavage. Exosite "substrate preparation" is exemplified by collagen triple helicase activity. The collagen triple helical structure, stabilized in part by its high proline and hydroxyproline content, presents unique problems for enzyme cleavage. Triple helical collagen is ~1.5 nm in diameter. Since the width of the active site cleft of collagenases is ~0.5 nm, only a single collagen α-chain can be bound and cleaved at a time. Therefore, the collagen triple helix must unfold at the cleavage site. Of the many potential Gly-Ile, Gly-Leu scissile bonds on each α-chain only those at position 775-776 in native type I collagen are cleaved. Thus, these proteinase susceptible sites are masked by the triple helical conformation of the molecule and so are not cleaved by collagenase or gelatinase. A lower Pro and Hyp content in the 775-776-cleavage region results in a localized looser helix *(91)* that may facilitate cleavage here.

There is a large entropic cost of collagen α-chain desolvation upon local denaturation. Indeed, only about 22 molecules of collagen are degraded per molecule of collagenase per hour (k_{cat}) *(57)*. That is, one α-chain is cleaved per 45 s. This turnover rate is one of the slowest observed for an enzyme-catalyzed reaction. The difficulty in cleaving collagen is also shown by the activation energy *(EA)* values calculated by Welgus et al. *(92)*. The *EA* of human collagenase cleavage of different collagenous substrates is very temperature dependent. Arrhenius energy plots for denatured α1-chains (13,900 calories) are similar to other enzyme-catalyzed reactions. However, for native collagen in solution the value is 49,200 calories and for fibrillar collagen it is 101,050 calories, revealing the influence of the triple helix and higher order structures on cleavage rates. The kinetic implication is that for every 2°C increase, collagenase activity will increase ~3-fold compared with more typical reaction rates which increase 1.7–3.0-fold for a 10°C increase in temperature. At sites of inflammation where the temperature can rise 5°C, a great enhancement in collagenase rate would therefore be expected.

6.3.1. Collagen Clamping and Intercollation Triple Helicase Actions

Although the mechanism of triple helicase activity remains enigmatic, collagenase hemopexin C domain and proline-rich linker exosites play critical roles in disrupting the tertiary and quaternary structure of interstitial collagens. Clamping native collagen to the collagenase active site by the hemopexin C domain may splay apart the collagen α-chains *(27,30)* (*see* **Fig. 4A**). Other mechanisms are also possible: intercollation of the proline-rich linker with the collagen triple helix may separate the 3 α-chains *(93)*. In so doing, this is proposed to displace one α-chain that can then bind the active site for cleavage (*see* **Fig. 4B**). Mutagenesis and linker swap chimera studies have both shown a role for the linker, but also failed to confirm its pivotal importance in triple helicase activity *(65,94–96)*. However, incorrect spacing or orientation of the hemopexin C domain may have resulted from these types of experiments to reduce the triple helicase actions of the chimeric enzymes.

6.3.2. Triple Helicase "Molecular Tectonics"

Collagen triple helicase activity exemplifies a precise allosteric mechanism of molecular binding and release. We have considered all force vectors that could potentially be applied to a triple helix to induce an unwinding action. From this consideration, several new triple helicase mechanisms are hypothesized in **Fig. 4**. It is most likely that at a minimum, two-point binding is required to physically induce distortion of the triple helix since one point binding cannot "move" the bound protein. Simultaneous binding of collagen to both the catalytic domain and the hemopexin C domain may cause a bend in the collagen that distorts the helix to enable cleavage (*see* **Fig. 4C**). In Figure 4C this is drawn as a bend "up" around the catalytic domain. Equally possible is a bend "down" to wrap around the hemopexin C domain. Topographically, the most likely sites for collagen binding on the catalytic and hemopexin C domains are on the rim of the hemopexin domain that is closest to the catalytic cleft. Thus, the edge of hemopexin module II and the area around the S_3' subsite may be the two collagen binding sites. The 3D structure of the full-length collagenase molecule *(24)* shows that this region forms a distinct cleft that is of a size that could bind a protein with the diameter of a collagen triple helix. In gelatinase A, the S_3' subsite is near the insertion site of the three fibronectin type II modules which, as discussed in **Subheading 6.3.3.**, we also hypothesize are involved in the triple helicase actions of the gelatinases. Indeed, in our current studies, site-directed mutations at this locus in MMP-8 disrupt catalysis *(39)*.

As an alternative mechanism to the localized melting of the collagen triple helix induced on binding or intercollation, active molecular mechanisms may also play a role. Movement of the hemopexin C domain relative to the catalytic domain may provide the motive force required for collagen distortion and triple helicase activity. If so, what drives such molecular tectonics? Single point binding cannot exert a relative motive force. After anchoring collagen to both the catalytic domain and the hemopexin C domain (two-point contact), subsequent relative movement of the hemopexin C domain or linker may disrupt the triple helix. Since this is clearly a non-ATP dependent reaction, the energy for this

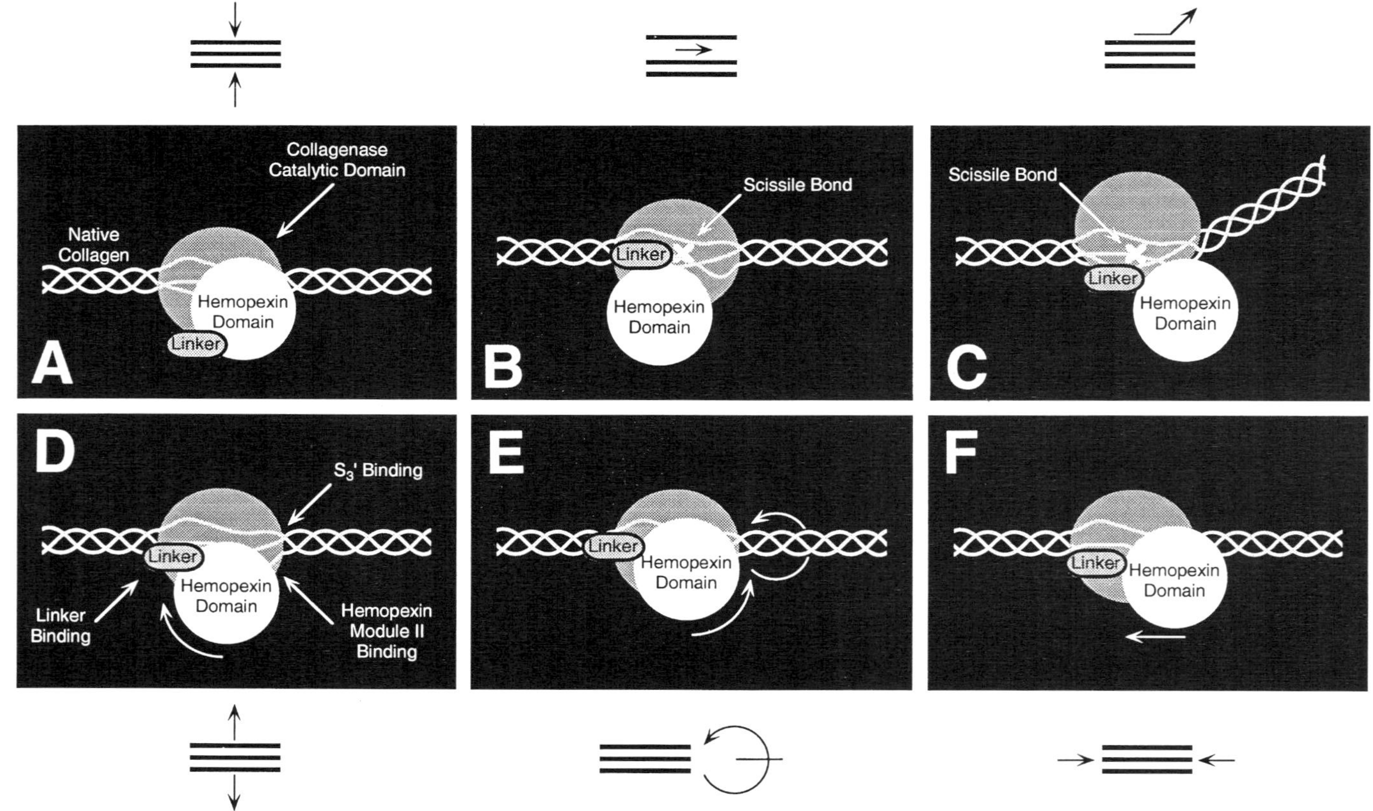

Fig. 4. Potential collagenase triple helicase molecular tectonics. (**A**) Lateral compression or clamping; (**B**) α-chain intercollation; (**C**) bending; (**D**) lateral tension; (**E**) axial rotation; (**F**) axial compression bulge.

movement may be derived from satisfying a third collagen binding interaction. For strain to be exerted on the collagen triple helix, such a proposed third binding interaction would only occur after physical movement of the hemopexin C domain (with the bound collagen) to make this contact. This movement would then induce distortion in the bound collagen that would separate the α-chains.

What molecular movements might satisfy three-point collagen contact? An opening action of the hemopexin C domain from the clamp position shown if **Fig. 4A** may physically tease the triple helix apart. In this scenario the third and last point of contact might be the linker. After first clamping to the collagen, hemopexin C domain movement may then occur to allow the linker to bind or to intercollate with the α-chains. This may laterally pull the collagen α-chains apart after a hinge-like action (**Fig. 4D**) or more likely by a twist of the hemopexin C domain to axially rotate the triple helix open (**Fig. 4E**). (Rotation in the opposite direction would tighten the right-handed triple helix). Compression force exerted axially along the collagen triple helix might also cause distension and localized unraveling of the triple helix at the compression bulge (**Fig. 4F**). Innovative experiments will be required to establish the validity of these proposed mechanisms, but as reviewed in **Subheading 6.1.**, collagen binding has already been shown to the catalytic domain, linker and hemopexin domain. The challenge now is to map their precise locations on these domains and to determine their differential roles in triple helicase activity.

6.3.3. Gelatinase Triple Helicase Molecular Tectonics

As discussed previously *(28)* the molecular tectonics of triple helicase actions must be fundamentally different between the collagenases and gelatinases because the gelatinase A C domain cannot bind native collagen. Alternatively, different mechanisms may be necessary because of structural distinctions between the classes of native collagens cleaved by these two groups of enzymes. The gelatinase CBD is likely to compensate for the absence of any collagen binding properties by the hemopexin C domain, to facilitate native collagen type IV, V, X, and XI triple helicase activities and cleavage by the gelatinases. The CBD may clamp native collagen against the active site to splay or bend open the triple helix in an analogous mechanism to that proposed for collagenase by Bode *(30)* and Gomis-Rüth et al. *(27)* and shown in **Fig. 4A**. Alternatively, we favor a model of "strand intercollation" (**Fig. 5**). We hypothesize that the three-fibronectin type II modules bind and interdigitate with the collagen triple helix to splay the three α-chains apart. This would allow a single α-chain to bind and occupy the active site cleft. The CBD binds to denatured collagen with higher affinity than to native collagen *(9)*. Therefore, to satisfy a binding preference for single α-chains rather than the native collagen helix, independent modular movement might facilitate penetration of the gelatinase

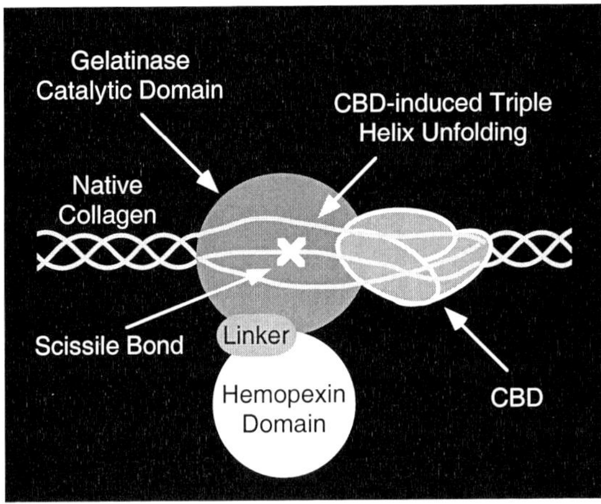

Fig. 5. Proposed triple helicase activity by the gelatinase CBD.

type II modules into regions of loosened collagen triple helix that otherwise masks potential cleavage sites. This binding propensity for separate α-chains could mechanistically drive the local distortion of the triple helical collagen structure to further loosen the helix to allow for substrate α-chain binding in the active site. This may be one reason for having three fibronectin type II modules in the gelatinases located near the S_3' subsite, whereas other fibronectin type II module-containing proteins, such as factor XII or fibronectin, have only one or two. Moreover, since the CBD appears to only bind two α-chains at a time (*73, see* **Subheading 5.1.**), this may allow the noninteracting α-chain the steric freedom to bind the active site cleft.

6.4. Exosites Modulate Gelatinase A Activation: Implications in the Control of Pericellular Collagen Homeostasis.

Concanavalin A-induced cell surface activation of progelatinase A *(97)* occurs in a quaternary complex of progelatinase A linked to cell surface MT-MMPs via a TIMP-2 bridge *(98)*. The TIMP-2 inhibitory N domain binds and inhibits the active site of MT-MMP *(99)*. The C domain of TIMP-2 binds to the gelatinase A hemopexin C domain *(67,85)* at the junction of hemopexin modules III and IV on the outer rim of the domain *(43)*. Clustering with a second MT-MMP molecule completes the quaternary complex to cleave the prodomain of the tethered progelatinase A at Asn[37]-Leu[38] *(10,42,99)*. Intermolecular autocatalytic cleavage then occurs to produce fully active gelatinase A with a N-terminus of Tyr[81]. This may be either soluble active gelatinase A or cell membrane bound active gelatinase A. Heparin binding to the hemopexin C domain of gelatinase A *(28,42,100)*, but not to the MT1-MMP C domain (Tam and Overall, unpublished data), favors clustering which increases activation. Addition of recombinant gelatinase A hemopexin C domain, but not the MT1-MMP hemopexin C domain, to concanavalin A treated cells blocks the cellular activation process in a dominant negative manner *(101)*.

Recent work from our laboratory has shown that substrate-binding exosites modulate gelatinase A activation. Gelatinase A has the potential to bind cell surface fibronectin via the hemopexin C domain *(28)*, but only in the absence of TIMP-2 *(10,101)*. This may represent a method of localizing TIMP-2-free progelatinase A to cell surfaces that subsequently can be liberated and allowed to enter the MT-MMP activation pathway upon competition with TIMP-2 *(10)*. In view of the critical TIMP-2 dependence of gelatinase A activation *(42)*, low levels of TIMP-2 that do not favor binding to cell surface MT-MMP or subsequent activation, would still allow cell surface localization of the enzyme to fibronectin. We have also shown that exosites on the CBD localize progelatinase A to cell surface collagen in a β_1-integrin dependent manner *(59)*. This forms a distinct pool of cell-localized enzyme that may be one mechanism for targeting stromal-derived enzyme to tumor cell surfaces *in trans*. When bound to β_1-integrin cell-associated collagen, progelatinase A is recalcitrant to cellular activation by MT-MMPs *(59)*. Recombinant CBD added to unstimulated cells releases progelatinase to the medium. Following concanavalin A treatment of cells to initiate the cellular activation of gelatinase A *(97)*, the addition of recombinant CBD further increases the activation of gelatinase A *(59)*. Thus, recombinant CBD competes progelatinase A from the collagen, facilitating entry of the zymogen into the activation complex. This indicates that progelatinase A bound to collagen via the CBD of the enzyme resists entry to the hemopexin C domain-dependent activation pathway and that intact cellular collagen reduces the activation of progelatinase A (*see* **Fig. 6**). However, cell growth in a three-dimensional collagen matrix also induces MT-MMP expression and gelatinase A activation *(102–103)*.

In an elaboration of our model proposed in Steffensen et al. *(59)*, we feel these opposite actions of collagen on gelatinase A activation may therefore form the basis for a homeostatic mechanism to maintain a normal pericellular collagen network. Collagen binding to cell surface integrins may signal an increase in transcriptional activation of MT-MMP expression *(104)* (M. S. Stack, personal communication). Native collagen is a natural substrate for MT-MMPs *(11,33)*. When present on the cell surface, the MT-MMPs may therefore degrade the pericellular collagen and in so doing release bound progelatinase A from the cell (**Fig. 6B**). The gelatinized collagen fragments may remain bound to the enzyme at the CBD. Progelatinase A could then complex with MT-MMPs via TIMP-2 for activation (**Fig. 6C**). This would generate active gelatinase precisely at the time and site required to complete the

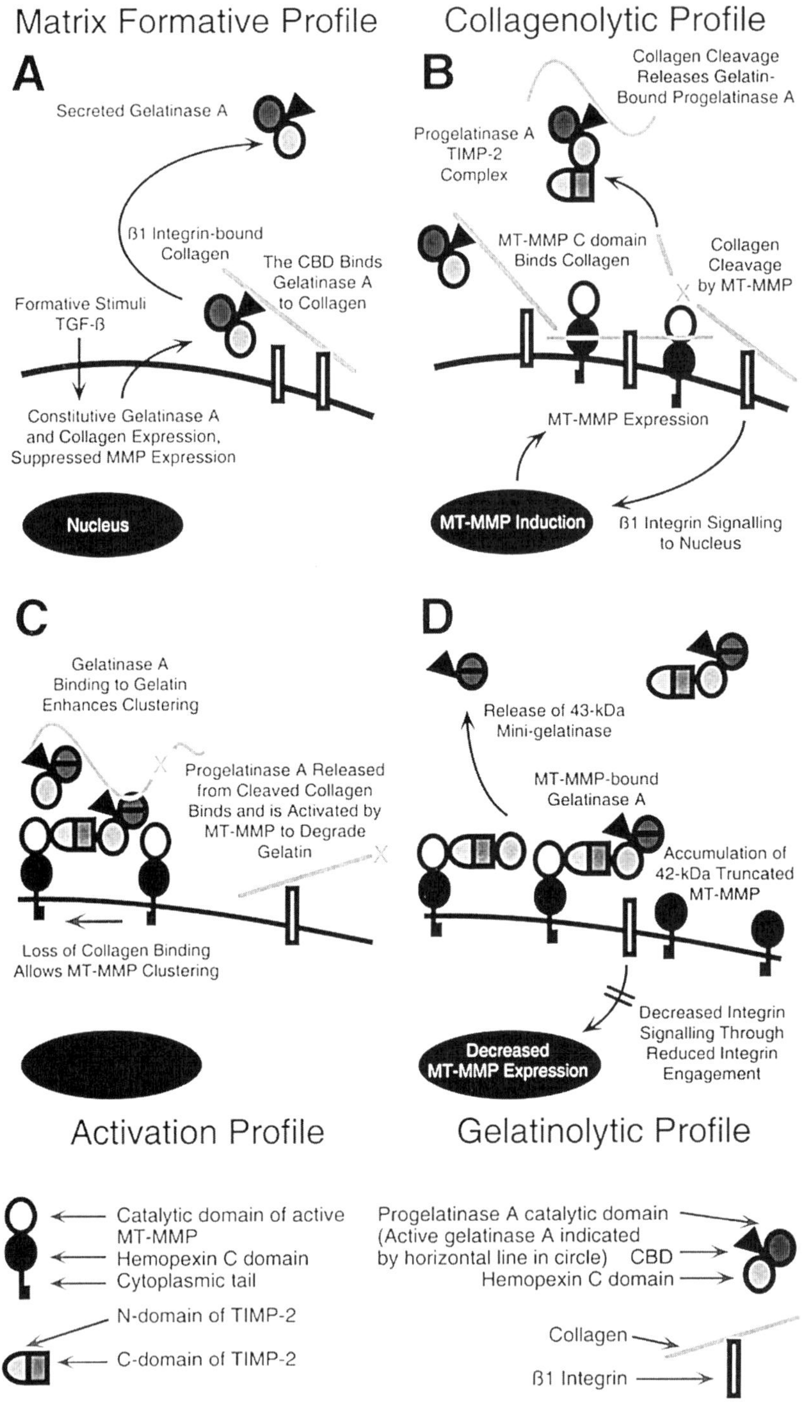

Matrix Formative Profile
A
Secreted Gelatinase A
ß1 Integrin-bound Collagen
The CBD Binds Gelatinase A to Collagen
Formative Stimuli TGF-ß
Constitutive Gelatinase A and Collagen Expression, Suppressed MMP Expression
Nucleus
Collagenolytic Profile
B
Collagen Cleavage Releases Gelatin-Bound Progelatinase A
Progelatinase A TIMP-2 Complex
MT-MMP C domain Binds Collagen
Collagen Cleavage by MT-MMP
MT-MMP Expression
MT-MMP Induction
ß1 Integrin Signalling to Nucleus
C
Gelatinase A Binding to Gelatin Enhances Clustering
Progelatinase A Released from Cleaved Collagen Binds and is Activated by MT-MMP to Degrade Gelatin
Loss of Collagen Binding Allows MT-MMP Clustering
Activation Profile
D
Release of 43-kDa Mini-gelatinase
MT-MMP-bound Gelatinase A
Accumulation of 42-kDa Truncated MT-MMP
Decreased MT-MMP Expression
Decreased Integrin Signalling Through Reduced Integrin Engagement
Gelatinolytic Profile
Catalytic domain of active MT-MMP
Hemopexin C domain
Cytoplasmic tail
N-domain of TIMP-2
C-domain of TIMP-2
Progelatinase A catalytic domain (Active gelatinase A indicated by horizontal line in circle)
CBD
Hemopexin C domain
Collagen
ß1 Integrin

Fig. 6. (*opposite page*) Role of collagen-binding exosites in the cellular activation of gelatinase A. (**A**) Matrix formative cell profile. Constitutively expressed progelatinase A is secreted from the cell, some of which binds the pericellular collagen layer coupled to ß1-integrin. (**B**) Collagenolytic cell profile. Signaling from ß1-integrin engagement with native collagen induces MT-MMP transcription and expression. MT-MMPs bind cell surface collagen through hemopexin C domain interactions. This restricts membrane trafficking and clustering of MT-MMPs, and hence gelatinase A activation. Degradation of MT-MMP-bound cell collagen releases progelatinase A from the cell coupled to the gelatinized degraded collagen. TIMP-2 complexes with progelatinase A. (**C**) Activation cell profile. Gelatinase A complexed with TIMP-2 binds MT-MMPs. Degradation and removal of the cellular collagen by MT-MMPs, which otherwise restricts membrane trafficking of the bound MT-MMPs, now allows MT-MMPs to cluster. Gelatinase A tethered either to the gelatin formed from the degraded collagen or to integral membrane heparan-sulphate proteoglycans colocalizes enzyme to form cell membrane rafts of MT-MMPs and progelatinase A. Together this accelerates gelatinase A activation by unoccupied MT-MMPs. MT-MMP cleaves the gelatinase A zymogen propeptide at Asn^{37}-Leu^{38}, which is followed by autolytic inter-molecular cleavage at Tyr^{81}. The clustering of free or MT-MMP bound active gelatinase A molecules via gelatin or heparin sulphate potentiates the final activation cleavage. Activated gelatinase A cleaves gelatin and releases bound enzyme. (**D**) Gelatinolytic cell profile. With the reduced levels of integrin-bound collagen, integrin signaling is reduced leading to decreased MT-MMP gene transcription. As activation proceeds, the 42-kDa inactive cleaved form of MT-MMP accumulates on the cell surface. Active gelatinase A remains bound on the TIMP-2 inhibited-MT-MMP. Minigelatinase A, generated by autolytic cleavage of the linker *(97)*, may be shed from the activation complexes leaving gelatinase A hemopexin C domain bound to the MT-MMP-TIMP-2 complex. Notably, the proteolytic profile of the cell has now not only switched to being gelatinolytic, but this process also actively reduces cellular collagenolytic activity. The combined effect allows collagen to accumulate, particulary when under the influence of TGF-ß.

---

degradation of the cleaved and denatured collagen still bound to the enzyme and in the vicinity. Use of cyanogen bromide fragments of collagen has shown that gelatinase A can bind at multiple sites on collagen α-chains *(9)*. Thus, binding of several molecules of gelatinase A on each gelatin chain would serve to colocalize the enzyme with MT-MMP activation complexes to favor the final autolytic activation cleavage of the gelatinase. In addition, the binding of TIMP-2/gelatinase A complexes to MT-MMPs would inhibit the collagenolytic potential of the cell and change the cells proteolytic profile from collagenolytic to gelatinolytic following the acquisition of bound active gelatinase A (**Fig. 6D**). Moreover, the cleavage of the activating MT-MMP to the 42-kDa inactive form *(105,106)* associated with this process would further reduce the collagenolytic potential of the cell (**Fig. 6D**). The function of this degraded form of MT-MMP, which represents the hemopexin C domain, is unclear. However, it still retains collegen binding properties and so can compete with the collagenase activities of full length MT-MMP. The cytoplasmic tail may also continue to bind cytoplasmic proteins and so function in a dominant negative manner to disrupt the accumulation of MT-MMP clusters. Thus, the degradation of cellular collagen would reduce the integrin-activated expression of the MT-MMPs, allowing native collagen to accumulate (**Fig. 6A**). Resorptive stimuli would also trigger this cascade by increasing MT-MMPs, similar to that shown in **Fig. 6B**, with resolution being most likely directed by transforming growth factor ß (TGF-ß.) Notably, TGF-ß1 increases collagen and gelatinase A expression in human fibroblasts *(107)*, but suppresses other MMPs. Gelatinase A expressed by a cell with a formative cell profile has been suggested to remove misfolded or abnormal collagen generated during rapid TGF-ß-induced synthesis *(108)*.

Although collagen binding to hemopexin C domain exosites on the MT-MMPs may favor clustering of MT-MMPs to enhance gelatinase activation (**Fig. 7A**) we feel it more likely that when bound to collagen this would serve to keep MT-MMPs *apart* by reducing membrane trafficking (*see* **Fig. 7B**). Thus, MT-MMP binding to rigid collagen molecules would reduce clustering mediated by cytoplasmic couplers binding the MT-MMP cytoplasmic tails. In contrast, we also hypothesize that the flexible nature of gelatin, when bound to the gelatinase A CBD may allow gelatinase A molecules to remain tethered to one another and thus to colocalize more readily. During the collagenolytic and gelatinolytic phases (*see* **Fig. 6C, D**) this may assist in the formation of rafts of MT-MMP and gelatinase A activation complexes on the cell membrane. This interaction at the CBD would be analogous to the clustering effect of heparin sulphate chains binding the hemopexin C domain of gelatinase A *(28,42,100)* to facilitate the final autolytic activation cleavage at Tyr[81] *(99)*. Thus, native collagen binding to exosites may play a role both on gelatinase A and MT-MMPs to reduce the cellular activation of gelatinase A, whereas denatured collagen may physically enhance the activation process. Notwithstanding the TIMP-2 dependence of this system *(42)*, the presence of normal levels of native collagen on cells may explain why any constitutive MT-MMP expression does not lead to widespread gelatinase A activation in the absence of inducing agents such as Con A or TPA. Indeed, Con A treatment of cells leads to the degradation of all secreted collagen, fibronectin and SPARC *(97)*.

7. Conclusion

Substrate interactions with MMP exosites play an essential role in the biological function of these enzymes at many different levels. Productive binding of substrate by a proteinase prior to cleavage is a relatively rare event. Exosites

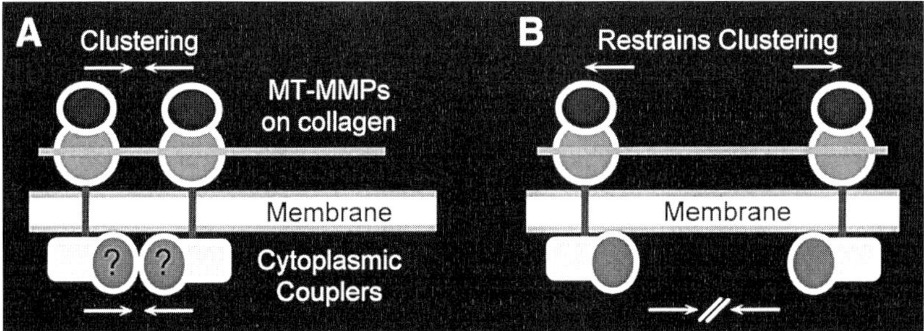

Fig. 7. The influence of collagen bound to MT-MMP hemopexin C domain prior to cleavage. **(A)** Native collagen may cluster MT-MMPs on the cell surface. Part of this driving force may be through cytoplasmic couplers or cytoskeleton binding the MT-MMP cytoplasmic domain. **(B)** More likely, MT-MMP binding to cellular collagen may restrict membrane movement of the enzyme with the rigid collagen molecules serving as "spokes" to keep MT-MMPs apart. This would prevent colocalization of MT-MMPs by cytoplasmic couplers. Only through subsequent degradation of the bound collagen would membrane traffic of MT-MMPs occur, resulting in the formation of rafts of clustered MT-MMP. The denatured collagen so formed may then serve to tether gelatinase A to these rafts facilitating the autolytic activation cleavage of cell surface gelatinase A.

bind substrates and facilitate the presentation of substrate scissile bonds in the optimal orientation to the active site. In addition to favoring binding and cleavage of substrate, exosites may assist in substrate preparation to expose scissile bonds prior to cleavage. At the cell surface, the diverse pericellular molecular environment allows soluble MMPs many potential routes to bind and localize at or near the cell membrane. Binding to cell associated matrix can form pools of progelatinase A poised for activation. Although stimulating MT-MMP expression, native collagen at the cell surface has the potential to reduce the activation of collagen-bound gelatinase A and also to reduce the clustering of MT-MMPs, to further impair activation. New investigations of MMP exosites will undoubtedly lead to a greater understanding of these complex issues. Future protein engineering studies should also reveal additional novel functions for these sites. The challenge is to map in molecular detail substrate exosites to understand further MMP behavior and to devise new therapeutic approaches that target exosites for highly selective MMP control in various diseases.

Determining the MMP degradome, that is the entire substrate repertoire of a proteinase in a tissue or cell, is a significant challenge. An important outcome of a "degradomics" approach is the identification of novel and important new substrates that will point to intersections of the MMP system with signaling and structural protein networks in health and disease. The utilization of exosites as baits in the yeast-two-hybrid or other screening techniques, offers a promising approach to identify new MMP substrates to complete a degradomics analysis of MMPs.

Acknowledgments

This work was supported by grants from the Canadian Medical Research Council and the National Cancer Institute of Canada with funds raised by the Canadian Cancer Society.

References

1. Overall, C. M. (1994) Regulation of tissue inhibitor of matrix metalloproteinase expression. *Ann. N.Y. Acad. Sci.* **732,** 51–64.
2. Muller, D., Quantin, B., Gesnel, M. C., Millon-Collard, R., Abecassis, J., and Breathnach, R. (1988) The collagenase gene family in humans consists of at least four members. *Biochem. J.* **253,** 187–192.
3. Velasco, G., Pendas, A. M., Fueyo, A., Knauper, V., Murphy, G., and Lopez-Otin, C. (1999) Cloning and characterization of human MMP-23, a new matrix metalloproteinase predominantly expressed in reproductive tissues and lacking conserved domains in other family members. *J. Biol. Chem.* **274,** 4570–4576.
4. Yang, M., Murray, M. T., and Kurkinen, M. (1997) A novel matrix metalloproteinase gene (XMMP) encoding vitronectin-like motifs is transiently expressed in Xenopus laevis early embryo development. *J. Biol. Chem.* **272,** 13,527–13,533.
5. Sang, Q. A. and Douglas, D. A. (1996) Computational sequence analysis of matrix metalloproteinases. *J. Protein Chem.* **15,** 137–160
6. Massova, I., Kotra, L. P., Fridman, R., and Mobashery, S. (1998) Matrix metalloproteinases: structure, evolution, and diversification. *FASEB J.* **12,** 1075–1095.
7. Collier, I. E., Wilhelm, S. M., Eisen, A. Z., Marme,r B. L., Gran,t G. A., Seltzer, J. L., Kronberger, A, He, C. S., Bauer, E. A., and Goldberg, G. I. (1988) H-ras oncogene-transformed human bronchial epithelial cells (TBE-1) secrete a single metalloprotease capable of degrading basement membrane collagen. *J. Biol. Chem.* **263,** 6579–6587.
8. Wilhelm, S. M., Collier, I. E., Marmer, B. L., Eisen, A. Z., Grant, G. A., and Goldberg, G. I. (1989) SV40-transformed human lung fibroblasts secrete a 92-kDa type IV collagenase which is identical to that secreted by normal human macrophages. *J. Biol. Chem.* **264,** 17,213–17,221.
9. Steffensen, B. J., Wallon, U. M., and Overall, C. M. (1995) Extracellular matrix binding properties of recombinant fibronectin type II-like modules of human 72-kDa gelatinase/type IV collagenase. High affinity binding to native type I collagen but not native type IV collagen. *J. Biol. Chem.* **270,** 11,555–11,566.
10. Overall, C. M., Wallon, U. M., Steffensen, B., De Clerk, Y., Tschesche, H., and Abbey, R. (2000) In *Inhibitors of matrix metalloproteinases in development and*

disease (Edwards, D., Hawkes, S., and Kokha, R., eds.) Gordon and Breach, Amsterdam, Holland, pp. 57–69.

11. Ohuchi, E., Imai, K., Fujii, Y., Sato, H., Seiki, M., and Okada, Y. (1997) Membrane type 1 matrix metalloproteinase digests interstitial collagens and other extracellular matrix macromolecules. *J. Biol. Chem.* **272,** 2446–2451.

12. Pei, D. and Weiss, S. J. (1996) Transmembrane-deletion mutants of the membrane-type matrix metalloproteinase-1 process progelatinase A and express intrinsic matrix-degrading activity. *J. Biol. Chem.* **271,** 9135–9140.

13. Hewitt, R. E., Corcoran, M. L., and Stetler-Stevenson, W. G. (1996) The Activation, Expression and Purification of Gelatinase A (MMP2). *Trends Glycosci. Glycotechnol.* **8,** 23–36.

14. Sodek, J. and Overall, C. M. (1988) In *The Biological Mechanisms of Tooth Eruption and Root Resorption,* Edited by Davidovitch, Z., Published by EBSCO Media, Birmingham, AL, pp. 303–311.

15. Werb, Z., Tremble, P. M., Behrendtsen, E., Crowley, E., and Damsky, C. H. (1989) Signal transduction through the fibronectin receptor induces collagenase and stromelysin gene expression. *J. Cell Biol.* **109,** 877–889.

16. Bork, P., Downing, A. K., Kieffer, B., and Campbell, I. D. (1996) Structure and distribution of modules in extracellular proteins. *Quat Rev. Biophys.* **29,** 119–167.

17. Patthy, L. (1991) Exons—original building blocks of proteins? *Bioessays* **13,** 187–192.

18. Patthy, L. (1996) Exon shuffling and other ways of module exchange. *Matrix Biol.* **15,** 301–310.

19. Gilbert, W. (1978) Why genes in pieces? *Nature* **271,** (5645), 501.

20. Blake, C. (1979) Exons encode protein functional units. *Nature* **277,** (5698), 598.

21. Dorit, R. L., Schoenbacher, L., and Gilbert, W. (1990) How big is the universe of exons? *Science* **250,** 1377–1382.

22. Campbell, I. D. and Downing, A. K. (1998) NMR of modular proteins. *Nature Struct. Biol. NMR Supp.* 496–499.

23. Collier I. E., Bruns, G. A. P., Goldberg, G. I., and Gerhard, D. S. (1991) On the structure and chromosome location of the 72- and 92-kDa human type IV collagenase genes. *Genomics* **9,** 429–434.

24. Li, J., Brick, P., O'Hare, M. C., Skarzynski, T., Lloyd, L. F., Curry, V. A., Clark, I. M., Bigg, H. F., Hazleman, B. L., Cawston, T. E., and Blow, D. M. (1995) Structure of full-length porcine synovial collagenase reveals a C-terminal domain containing a calcium-linked, four-bladed beta-propeller. *Structure* **3,** 541–549.

25. Libson, A. M., Gittis, A. G., Collier, I. E., Marmer, B. L., Goldberg, G. I., and Lattman E. E. (1995) Crystal structure of the haemopexin-like C-terminal domain of gelatinase A. *Nature Struct. Biol.* **2,** 938–942.

26. Gohlke, U., Gomis-Rüth, F-Z., Crabbe, T., Murphy, G., Docherty, A. J. P., and Bode, W. (1996) The C-terminal (haemopexin-like) domain structure of human gelatinase A (MMP2): structural implications for its function. *FEBS Letters* **378,** 126–130.

27. Gomis-Rüth F. X., Gohlke, U., Betz, M., Knauper, V., Murphy, G., López-Otín, C., and Bode, W. (1996) The helping hand of collagenase-3 (MMP-13): 2.7 A crystal structure of its C-terminal haemopexin-like domain. *J. Mol. Biol.* **264,** 556–566.

28. Wallon, U. M. and Overall, C. M. (1997) The hemopexin-like domain (C domain) of human gelatinase A (matrix metalloproteinase-2) requires Ca^{2+} for fibronectin and heparin binding. Binding properties of recombinant gelatinase A C domain to extracellular matrix and basement membrane components. *J. Biol. Chem.* **272,** 7473–7481.

29. Rawlings, N. D. and Barrett, A. J. (1995) Evolutionary families of metallopeptidases. *Meth Enzymol.* **248,** 183–228.

30. Bode, W. (1995) A helping hand for collagenases: the hemopexin-like domain. *Structure* **2,** 527–530.

31. Springman, E. B., Angleton, E. L., Birkedal-Hansen, H., and Van Wart, H. E. (1990) Multiple modes of activation of latent human fibroblast collagenase: evidence for the role of a Cys73 active-site zinc complex in latency and a "cysteine switch" mechanism for activation. *Proc. Natl. Acad. Sci. USA* **87,** 364–368.

32. Leahy, D. J., Aukhil, I., and Erickson, H. P. (1996) 2.0 A crystal structure of a four-domain segment of human fibronectin encompassing the RGD loop and synergy region. *Cell* **84,** 155–164.

33. Pei, D. and Weiss, S. J. (1995) Furin-dependent intracellular activation of the human stromelysin-3 zymogen. *Nature* **375,** 244–247.

34. Windsor, L. J., Bodden, M. K., Birkedal-Hansen, B., Engler, J. A., and Birkedal-Hansen, H. (1994) Mutational analysis of residues in and around the active site of human fibroblast-type collagenase. *J. Biol. Chem.* **269,** 26,201–26,207.

35. Murphy, G., Nguyen, Q., Cockett, M. I., Atkinson, S. J., Allan, J. A., Knight, C. G., Willenbrock, F., and Docherty, A. J. P. (1994) Assessment of the role of the fibronectin-like domain of gelatinase A by analysis of a deletion mutant. *J. Biol. Chem.* **269,** 6632–6636.

36. Owens, R. J. and Baralle, F. E. (1986) Mapping the collagen-binding site of human fibronectin by expression in Escherichia coli. *EMBO. J.* **5,** 2825–2830.

37. Litvinovich, S. V., Strickland, D. K., Medved, L. V., and Ingham, K. C. (1991) Domain structure and interactions of the type I and type II modules in the gelatin-binding region of fibronectin. All six modules are independently folded. *J. Mol. Biol.* **217,** 563–575.

38. Skorstengaard, K., Holtet, T. L., Etzerodt, M., and Thogersen, H. C. (1994) Collagen-binding recombinant fibronectin fragments containing type II domains. *FEBS Lett.* **343,** 47–50.

39. Overall, C. M., Lowne, D., Wells, G., Burel, S., McCullouch, C. A. G., and Clements, J. M. (1999) Cloning, CHO Cell Expression, and Activation of Rat Collagenase-2 (MMP-8). *J. Dent. Res.* **78,** (IADR Abstracts), 458.

40. Shirley, B. A., Stanswsens, P., Hahn, U., and Pace, N. C. (1992) Contribution of hydrogen bonding to the conformational stability of ribonuclease T1. *Biochemistry* **31,** 725–732.

41. Maurus, R., Overall, C. M., Bogumil, R., Luo, Y. L., Mauk, G., Smith, M., and Brayer, G. (1997) A myoglobin variant with a polar substitution in a conserved hydrophobic cluster in the heme binding pocket. *Biochim. Biophys. Acta* **1341,** 1–13.

42. Butler, G. S., Butler, M. J., Atkinson, S. J., Will, H., Tamura, T., van Westrum, S. S., Crabbe, T., Clements, J., d'Ortho, M.-P., and Murphy, G. The TIMP-2 membrane type I metalloproteinase "receptor" regulates the concentration and efficient activation of progelatinase A (1998) *J. Biol. Chem.* **273,** 871–880.

43. Overall, C. M., King, A. E., Sam, D. K., Ong, A. D., Lau, T. T. Y., Wallon, U. M., DeClerck, Y. A., and Atherstone, J. (1999) Identification of the tissue inhibitor of metalloproteinases-2 (TIMP-2) binding site on the hemopexin carboxyl domain of human gelatinase A by site-directed mutagenesis. The hierarchical role in binding TIMP-2 of the unique cationic clusters of hemopexin modules III and IV. *J. Biol. Chem.* **274,** 4421–4429.

44. Bode, W., Gomis-Ruth, F.-X., and Stockler, W. (1993) Astacins, serralysins, snake venom and matrix metalloproteinases exhibit identical zinc-binding environments (HEXXHXXGXXH and Met-turn) and topologies and should be grouped into a common family, the 'metzincins'. *FEBS Lett.* **331,** 134–140.

45. Pourmotabbed T., Solomon, T. L., Hasty, K. A., and Mainardi, C. L. (1994) Characteristics of 92 kDa type IV collagenase/gelatinase produced by granulocytic leukemia cells: structure, expression of cDNA in E. coli and enzymic properties. *Biochim. Biophys. Acta.* **1204,** 97–107.

46. Fields, B. B., Van Wart H. E., and Birkedal-Hansen, H. (1987) Sequence specificity of human skin fibroblast collagenase. Evidence for the role of collagen structure in determining the collagenase cleavage site. *J. Biol. Chem.* **262,** 6221–6226.

47. Netzel-Arnett, S., Fields, G. B., Birkedal-Hansen, H., and Van Wart, H. E. (1991) Sequence specificities of human fibroblast and neutrophil collagenases. *J. Biol. Chem.* **266,** 6747–6755.

48. Netzel-Arnett, S., Sang, Q. X., Moore, W. G., Navre, M., Birkedal-Hansen, H., and Van Wart, H. E. (1993) Comparative sequence specificities of human 72- and 92-kDa gelatinases (type IV collagenases) and PUMP (matrilysin). *Biochemistry* **32,** 6427–6432.

49. McGeeham, G. M., Bickett, D. M., Green, M., Kassel, D., Wiseman, J. S., and Berman, J. (1994) Characterization of the peptide substrate specificities of interstitial collagenase and 92-kDa gelatinase. Implications for substrate optimization. *J. Biol. Chem.* **269,** 32,814–32,820.

50. Welch, A. R., Holman, C. M., Huber, M., Brenner, M. C., Browner, M. F., and Van Wart, H.E. (1996) Understanding the P1' specificity of the matrix metalloproteinases: effect of S1' pocket mutations in matrilysin and stromelysin-1. *Biochemistry* **35,** 10,103–10,109.

51. Borkakoti, N., Winkler, F. K., Williams, D. H., D'Arcy, A., Broadhurst, M. J., Brown, P. A., Johnson, W. H., and Murray, E. J. (1994) Structure of human fibroblast collagenase complexed with an inhibitor. *Nature Struct. Biol.* **1,** 106–110.

52. Lovejoy, B., Cleasby, A., Hassell, A. M., Longley, K., Luther, M. A., Weigl, D., McGeehan, G., McElroy, A. B., Drewry, D., Lambert, M. H., and Jordan, S. R. (1994) Structure of the catalytic domain of fibroblast collagenase complexed with an inhibitor. *Science* **263,** 375–377.

53. Stams, T., Spurlino, J. C., Smith, D. L., Wahl, R. C., Ho, T. F., Qoronfleh, M. W., Banks, T. M., and Rubin, B. (1994) Structure of human neutrophil collagenase reveals large S1' specificity pocket. *Nature Struct. Biol.* **1,** 119–123.

54. Bode, W., Reinemer, A., Huber, R., Kleine, T., Schnierer, S., and Tschesche, H. (1994) The X-ray crystal structure of the catalytic domain of human neutrophil collagenase inhibited by a substrate analogue reveals the essentials for catalysis and specificity. *EMBO J.* **13,** 1263–1269.

55. Massova, I., Fridman, R., and Mobashery, S. (1997) Structural insights into the catalytic domains of human matrix metalloproteinaase-2 and human matrix metalloproteinase-9: Implications for substrate specificites. *J. Mol. Model.* **3,** 17–30.

56. Docherty, A. J. P. (2000) In *Inhibitors of Matrix Metalloproteinases in Development and Disease* (Edwards, D., Hawkes, S., and Kokha, R., eds.) Gordon and Breach, Amsterdam, Holland, in press.

57. Welgus, H. G., Jeffrey, J. J., Stricklin, G. P., Roswit, W. T., and Eisen, A. Z. (1980) Characteristics of the action of human skin fibroblast collagenase on fibrillar collagen. *J. Biol. Chem.* **255,** 6806–6813.

58. Allan, J. A., Docherty, A. J., Barker, P. J., Huskisson, N. S., Reynolds, J. J., and Murphy, G. (1995) Binding of gelatinases A and B to type-I collagen and other matrix components. *Biochem. J.* **309,** 299–306.

59. Steffensen, B., Bigg, H. F., and Overall, C. M. (1998) The involvement of the fibronectin type II-like modules of human gelatinase A in cell surface localization and activation. *J. Biol. Chem.* **273,** 20,622–20,628.

60. Olsen, M. W., Toth, M., Gervasi, D. C., Sado, Y., Ninomiya, Y., and Fridman, R. (1998) High affinity binding of latent matrix metalloproteinase-9 to the alpha 2 (IV) chain of collagen IV. *J. Biol. Chem.* **273,** 10,672–10,681.

61. Basset, P., Bellocq, J. P., Wolf, C., Stoll, I., Hutin, P., Limacher, J. M., Podhajcer, O. L., Chenard, M. P., Rio, M. C., and Chambon, P. (1990) A novel metalloproteinase gene specifically expressed in stromal cells of breast carcinomas. *Nature* **348,** 699–704.

62. Clark, I. M. and Cawston, T. E. (1989) Fragments of human fibroblast collagenase. Purification and characterization. *Biochem. J.* **263,** 201–206.

63. Windsor, L. J., Birkedal-Hansen, H., Birkedal-Hansen, B., and Engler, J. A. (1991) An internal cysteine plays a role in the maintenance of the latency of human fibroblast collagenase. *Biochem.* **30,** 641–647.

64. Murphy G., Allan, J. A., Willenbrock, F., Cockett, M. I., O'Connell, J. P., and Docherty, A. J. P. (1992) The role of the C-terminal domain in collagenase and stromelysin specificity. *J. Biol. Chem.* **267,** 9612–9618.

65. Hirose, T., Patterson, C., Pourmotabbed, T., Mainardi, C. L., and Hasty, K. A. (1993) Structure-function relationship of human neutrophil collagenase: identifi-

cation of regions responsible for substrate specificity and general proteinase activity. *Proc. Natl. Acad. Sci. USA* **90,** 2569–2573.

66. Knauper, V., Cowell, S., Smith, B., López-Otin, C., O'Shea, M., Morris, H., Zardi, L., and Murphy, G. (1997) The role of the C-terminal domain of human collagenase-3 (MMP-13) in the activation of procollagenase-3, substrate specificity, and tissue inhibitor of metalloproteinase interaction. *J. Biol. Chem.* **272,** 7608–7616.

67. Murphy, G., Willenbrock, F., Ward, R. V., Cockett, M. I., Eaton, D., and Docherty, A. J. P. (1992) The C-terminal domain of 72 kDa gelatinase A is not required for catalysis, but is essential for membrane activation and modulates interactions with tissue inhibitors of metalloproteinases. *Biochem. J.* **283,** 637–641.

68. Ingham, K. C., Brew, S. A., and Migliorini, M. M. (1989) Further localization of the gelatin-binding determinants within fibronectin. Active fragments devoid of type II homologous repeat modules. *J. Biol. Chem.* **264,** 16,977–16,980.

69. Banyai, L., Trexler, M., Koncz, S., Gyenes, M., Sipos, G., and Patthy, L. (1990) The collagen-binding site of type-II units of bovine seminal fluid protein PDC-109 and fibronectin. *Eur. J. Biochem.* **193,** 801–806.

70. Banyai, L. and Patthy, L. (1991) Evidence for the involvement of type II domains in collagen binding by 72 kDa type IV procollagenase. *FEBS Lett.* **282,** 23–25.

71. Banyai, L., Tordai, H., and Patthy, L. (1994) The gelatin-binding site of human 72 kDa type IV collagenase (gelatinase A). *Biochem. J.* **298,** 403–407.

72. Collier, E. E., Krasnov, P. A., Strongin, A. Y., Birkedal-Hansen, H., and Goldberg, G. I. (1992) Alanine scanning mutagenesis and functional analysis of the fibronectin-like collagen-binding domain from human 92-kDa type IV collagenase. *J. Biol. Chem.* **267,** 6776–6781.

73. Abbey, R., Steffensen, B., and Overall, C. M. (1999) Differential substrate binding to the fibronectin type II modules of human gelatinase A. Evidence for cooperative binding sites (In preparation).

74. Shipley, J. M., Doyle, G. A. R., Fliszar, C. J., Ye, Q.-Z., Johnson, L. J., Shapiro, S. D., Welgus, H. G., and Senior, R. M. (1996) The structural basis for the elastinolytic activity of the 92-kDa and 72-kDa gelatinases. Role of the fibronectin type II-like repeats. *J. Biol. Chem.* **271,** 4335–4341.

75. O'Farrell, T. J. and Pourmotabbed, T. (1998) The fibronectin-like domain is required for the type V and XI collagenolytic activity of gelatinase B. *Arch. Biochem. Biophys.* **354,** 24–30.

76. Ye, Q. Z., Johnson, L. L., Yu, A. E., and Hupe, D. (1995) Reconstructed 19 kDa catalytic domain of gelatinase A is an active proteinase. *Biochemistry* **34,** 4702–4708.

77. Banyai, L., Tordai, H., and Patthy, L. (1996) Structure and domain-domain interactions of the gelatin binding site of human 72-kilodalton type IV collagenase (gelatinase A, matrix metalloproteinase 2). *J. Biol. Chem.* **271,** 12,003–12,008.

78. Constantine, K. L., Madrid, M., Banyai, L., Trexler, M., Patthy, L., and Llinas, M. (1992) Refined solution structure and ligand-binding properties of PDC-109 domain b. A collagen-binding type II domain. *J. Mol. Biol.* **223,** 281–298.

79. Sticht, H., Pickford, A. R., Potts, J. R., and Campbell, I. D. (1998) Solution structure of the glycosylated second type 2 module of fibronectin. *J. Mol. Biol.* **276,** 177–187.
80. Pickford, A. R., Potts, J. R., Bright, J. R., Phan, I., and Campbell, I. D. (1997) Solution structure of a type 2 module from fibronectin: Implications for the structure and function of the gelatin-binding domain. *Structure* **5,** 359–370.
81. Brooks, P. C., Stromblad, S., Sanders, L. C., von Schalscha, T. L., Aimes, R. T., Stetler-Stevenson, W. G., Quigley, J. P., and Cheresh, D. A. (1996) Localization of matrix metalloproteinase MMP-2 to the surface of invasive cells by interaction with integrin alpha v beta 3. *Cell* **85,** 683–693.
82. Bigg, H. F., Shi, Y. E., Liu, Y. E., Steffensen, B., and Overall, C. M. (1997) Specific, high affinity binding of tissue inhibitor of metalloproteinases-4 (TIMP-4) to the COOH-terminal hemopexin-like domain of human gelatinase A. TIMP-4 binds progelatinase A and the COOH-terminal domain in a similar manner to TIMP-2. *J. Biol. Chem.* **272,** 15,496–15,500.
83. O'Connell, J. P., Willenbrock, F., Docherty, A. J. P., Eaton, D., and Murphy, G. (1994) Analysis of the role of the COOH-terminal domain in the activation, proteolytic activity, and tissue inhibitor of metalloproteinase interactions of gelatinase B. *J. Biol. Chem.* **269,** 14,967–14,973.
84. Stetler-Stevenson, W. G., Krutzsch, H. C., and Liotta, L. A. (1989) Tissue inhibitor of metalloproteinase (TIMP-2). A new member of the metalloproteinase inhibitor family. *J. Biol. Chem.* **264,** 17,374–17,378.
85. Howard, E. W. and Banda, M. J. (1991) Binding of tissue inhibitor of metalloproteinases 2 to two distinct sites on human 72-kDa gelatinase. Identification of a stabilization site. *J. Biol. Chem.* **266,** 17,972–17,977.
86. Koklitis, P. A., Murphy, G., Sutton, C., Angal, S. (1991) Purification of recombinant human prostromelysin. Studies on heat activation to give high-Mr and low-Mr active forms, and a comparison of recombinant with natural stromelysin activities. *Biochem. J.* **276,** 217–221.
87. Marcy, A. I., Eiberger, L. L., Harrison, R., Chan, H. K., Hutchinson, N. I., Hagmann, W. K., Cameron, P. M., Boulton, D. A., and Hermes, J. D. (1991) Human fibroblast stromelysin catalytic domain: expression, purification, and characterization of a C-terminally truncated form. *Biochemistry* **30,** 6476–6483.
88. Nomura, K., Shimizu, T., Kinoh, H., Sendai, Y., Inomata, M., and Suzuki, N. (1997) Sea urchin hatching enzyme (envelysin): cDNA cloning and deprivation of protein substrate specificity by autolytic degradation. *Biochemistry* **36,** 7225–7238.
89. Sodek, J. and Overall, C. M. (1992) Matrix metalloproteinases in periodontal tissue remodelling. *Matrix Supplement* **1,** 352–362.
90. Aimes, R. T. and Quigley, J. P. (1995) Matrix metalloproteinase-2 is an interstitial collagenase. Inhibitor-free enzyme catalyzes the cleavage of collagen fibrils and soluble native type I collagen generating the specific 3/4- and 1/4-length fragments. *J. Biol. Chem.* **270,** 5872–5876.
91. Highberger, J. H., Corbett, C., and Gross, J. (1979) Isolation and characterization of a peptide containing the site of cleavage of the chick skin collagen alpha 1[I] chain by animal collagenases. *Biochem. Biophys. Res. Comm.* **89,** 202–208.

92. Welgus, H. G., Jeffrey, J. J., and Eisen, A. Z. (1981) Human Skin fibroblast collagenase. Assessment of activation energy and deuterium isotope effect with collagenous substrates. *J. Biol. Chem.* **256,** 9516–9521.

93. De Souza, S. J., Pereira, H. M., Jacchieri, S., and Brentani, R. R. (1996) Collagen/collagenase interaction: Does the enzyme mimic the conformation of its own substrate? *FASEB J.* **10,** 927–930.

94. De Souza, S. J. and Brentani, R. (1992) Collagen binding site in collagenase can be determined using the concept of sense-antisense peptide interactions. *J. Biol. Chem.* **267,** 13,763–13,767.

95. Sanchez-López, R., Alexander, C. M., Behrendtsen, O., Breathnach, R., and Werb, Z. (1993) Role of zinc-binding and hemopexin domain encoded sequences in the substrate specificity of collagenase and stromelysin-2 as revealed by chimeric proteins. *J. Biol. Chem.* **268,** 7238–7247.

96. Knauper, V., Docherty, A. J. P., Smith, B., Tschesche, H., and Murphy, G. (1997) Analysis of the contribution of the hinge region of human neutrophil collagenase (HNC, MMP-8) to stability and collagenolytic activity by alanine scanning mutagenesis. *FEBS Letters* **405,** 60–64.

97. Overall, C. M. and Sodek, J. (1990) Concanavalin A produces a matrix-degradative phenotype in human fibroblasts. Induction and endogenous activation of collagenase, 72-kDa gelatinase, and Pump-1 is accompanied by the suppression of the tissue inhibitor of matrix metalloproteinases. *J. Biol. Chem.* **265,** 21,141–21,151.

98. Strongin, A. Y., Collier, I., Bannikow, G., Marmer, B. L., Grant, G. A., and Goldberg, G. I. (1995) Mechanism of cell surface activation of 72-kDa type IV collagenase. *J. Biol. Chem.* **270,** 5331–5338.

99. Will, H., Atkinson, S. J., Butler, G. S., Smith, B., and Murphy, G. (1996) The soluble catalytic domain of membrane type I matrix metalloproteinase cleaves the propeptide of progelatinase A and initiates autoproteolytic activation. Regulation by TIMP-2 and TIMP-3. *J. Biol. Chem.* **271,** 17,119–17,123.

100. Crabbe, T., Joannou, C., and Docherty, A. J. P. (1993) Human progelatinase A can be activated by autolysis at a rate that is concentration-dependent and enhanced by heparin bound to the C-terminal domain. *Eur. J. Biochem.* **218,** 431–438.

101. Overall, C. M., Wallon, U. M. W., McQuibban, G. A., Tam, E., Bigg, H. F., Morrison, C. M., DeClerck, Y. A., King, A., and Overall, C. M. (2000) Domain binding studies of gelatinase A activation by membrane type-1 matrix metalloproteinase and TIMP-2. Analysis of the potential interactions of the 43-KDa form of MT1-MMP (submitted).

102. Gilles, C., Polette, M., Seiki, M., Birembaut, P., and Thompson, E. W. (1997) Implication of collagen type I-induced membrane-type 1-matrix metalloproteinase expression and matrix metalloproteinase-2 activation in the metastatic progression of breast carcinoma. *Lab Invest.* **76,** 651–660.

103. Haas, T. L., Davis, S. J., and Madri, J. A. (1998) Three-dimensional type I collagen lattices induce coordinate expression of matrix metalloproteinases MT1-MMP and MMP-2 in microvascular endothelial cells. *J. Biol. Chem.* **273,** 3604–3610.

104. Seftor, R. E. B., Seftor, E. A., Stetlet-Stevenson, W. G., and Hendrix, M. J. C. (1993) The 72 kDa type IV collagenase is modulated via differential expression of αvß3 and α5ß1 integrins during human melanoma cell invasion. *Cancer Res.* **53,** 3411–3415.

105. Lohi, J. Lehti, K., Westermarck, J., Kahari, V., and Keski-Oja, J. Regulation of membrane-type matrix metalloproteinase-1 expression by growth factors and phorbol 12-myristate 13-acetate. *Eur. J. Biochem.* **239,** 239–247.

106. Lehti, K., Lohi, J., Valtanen, H., and Keski-Oja, J. Proteolytic processing of membrane-type-1 mtrix metalloproteinase is associated with gelatinase A activation at the cell surface. *Biochem. J.* (1998) **334,** 345–353.

107. Overall, C. M., Wrana, J. L., and Sodek, J. (1989) Independent Regulation of Collagenase, 72-kDa Progelatinase, and Metalloendoproteinase Inhibitor Expression in Human Fibroblasts by Transforming Growth Factor-β. *J. Biol. Chem.* **264,** 1860–1869.

108. Overall, C. M., Wrana, J. L., and Sodek, J. (1991) Transcriptional and Post-transcriptional Regulation of 72-kDa Gelatinase/Type IV Collagenase by Transforming Growth Factor-β1 in Human Fibroblasts. Comparisons with Collagenase and Tissue Inhibitor of Matrix Metalloproteinase Gene Expression. *J. Biol. Chem.* **266,** 14,064–14,071.

5

The Matrix Metalloproteinase (MMP) and Tissue Inhibitor of Metalloproteinase (TIMP) Genes

Transcriptional and Posttranscriptional Regulation, Signal Transduction and Cell-Type-Specific Expression

Matthew P. Vincenti

1. Introduction

Developmental and homeostatic remodeling of the extracellular matrix (ECM) is a highly regulated process orchestrated by a family of zinc-containing, calcium-dependent neutral proteases known as the matrix metalloproteinases (MMP). This family of enzymes, which now contains twenty members, can collectively degrade all structural proteins of the ECM including interstitial collagens (I, II, III, and V), basement membrane collagens (IV), fibronectin, laminin, proteoglycan, and elastin (**Table 1**). The enzymatic activity of MMP family members is controlled by a group of inhibitor proteins known as the Tissue Inhibitors of Metalloproteinases (TIMPs), which consist of four family members (**Table 2**). Whereas all four TIMPs can inhibit all MMPs in vitro, preferential TIMP-MMP interactions and tissue-restricted TIMP expression suggest that each TIMP has a specific function (**Table 2**).

The relative balance of MMPs and TIMPs is thought to determine the rate of ECM turnover, so that even slight deviations in gene expression can contribute to the progression of some diseases. For instance, elevated expression of collagenase-1 (MMP-1), collagenase-2 (MMP-8) collagenase-3 (MMP-13), and stromelysin-1 (MMP-3), and TIMP-1 have been documented connective tissue diseases such as rheumatoid and osteoarthritis *(1–6)*. The ability of cancer cells to invade the ECM and metastasize is dependent on MMPs produced either by

From: *Methods in Molecular Biology, vol. 151: Matrix Metalloproteinase Protocols*
Edited by: I. Clark © Humana Press Inc., Totowa, NJ

Table 1
The Matrix Metalloproteinases (MMPs)

MMP family member	Common names	Substrates	Selected references
MMP-1	Collagenase-1	Collagens I, II, III, VI, X, gelatins, aggrecan, entactin	*(154–158)*
MMP-2	72 kDa Gelatinase, Gelatinase A	Gelatins, collagens I,IV, V, VII, X, XI, fibronectin, laminin, aggrecan, elastin, large tenascin C, vitronectin, ß-amyloid protein precursor	*(157–159)*
MMP-3	Stromelysin-1	Aggrecan, gelatins, fibronectin, laminin, collagen III, IV, IX, X, large tenascin C, vitronectin	*(28,157,158, 160,161)*
MMP-7	Matrilysin, Pump	Aggrecan, fibronectin, laminin, collagen IV, elastin, entactin, small tenascin C, vitronectin	*(157,158, 162,163)*
MMP-8	Collagenase-2, Neutrophil Collagenase	Collagens I, II, III, aggrecan	*(157,158,164)*
MMP-9	92 kDa Gelatinase, Gelatinase B	Gelatins, collagens IV, V, XIV, aggrecan, elastin, entactin, vitronectin	*(157,158, 165,166)*
MMP-10	Stromelysin-2	Aggrecan, fibronectin, laminin, collagen IV	*(157,158,167)*
MMP-11	Stromelysin-3	Fibronectin, laminin, collagen IV, aggrecan, gelatins	*(141,158)*
MMP-12	Metalloelastase	Elastin	*(147,158)*
MMP-13	Collagenase-3	Collagens I, II, III	*(88,89,124, 158,168,169)*
MMP-14	Membrane-type-1-MMP	Collagens I, II, III, fibronectin, laminin, vitronectin, proteoglycan, ProMMP-2, ProMMP-13	*(158,170)*
MMP-15	Membrane-type-2-MMP	Not known	*(158,171)*
MMP-16	Membrane-type-3-MMP	ProMMP-2	*(158,172)*
MMP-17	Membrane-type-4-MMP	Not known	*(158,173)*
MMP-18	Collagenase-4, *Xenopus* Collagenase	Collagen I	*(158,174,175)*
MMP-19		Not Known	*(158,176,177)*
MMP-20	Enamelysin	Amelogenin (The major tooth enamel matrix protein)	*(178)*

Table 2
Tissue Inhibitors of Metalloproteinases (TIMPs)

Family member	Reported functions	Selected References
TIMP-1	Inhibition of all known MMP family members. Associates with proMMP-9. Inhibits angiogenesis. Erythroid-potentiating actvity.	*(179–181)*
TIMP-2	Inhibitions all known MMP family members. Associates with MT1-MMP and MMP-2 at cell surface and regulates MMP-2 activation.	*(181–184)*
TIMP-3	Inhibition of all known MMP family members. Extracellular matrix-associated. Mutation associated with Sorsby's fundus dystrophy.	*(55,181,185–188)*
TIMP-4	Inhibition of all known MMP family members. Restricted expression suggests tissue specific TIMP function.	*(181,189,190)*

the tumors themselves, or the surrounding stromal cells *(7,8)*. Finally, the rupture of atherosclerotic plaques involves MMP expression by foam cells and smooth muscle endothelial cells, resulting in myocardial infarcts *(9,10)*. Clearly, inappropriate expression of MMPs in specific tissues, and disruption of the MMP/TIMP balance can have dire consequences, and therefore has been the subject of considerable study.

Connective tissue cells express the genes coding for MMP and TIMP family members either constitutively, or transiently during specific developmental or pathological events. For the inducible MMP and TIMP genes, the mechanisms of gene activation or suppression are often complex, involving changes in the rates of transcription and mRNA processing. Recent work has begun to define signal transduction pathways that lead to MMP and TIMP gene activation. It is important to note that these mechanisms of gene regulation are not global, and significant differences have been described between cell types. Ultimately, understanding the regulation of these genes as it applies to disease progression may someday lead to important new therapies for arthritis, cancer, and cardiovascular disease.

2. Transcriptional Regulation of MMP and TIMP Genes

Many of the MMP and TIMP genes are transcribed at low to undetectable levels in resting connective tissue cells, but transcription rates rise dramatically in response to various extracellular stimuli. To assess transcriptional activation of MMP and TIMP genes, investigators have relied primarily on three assays. In the Nuclear Run-on Assay, nuclei are isolated from stimulated cells and radiolabeled uridine triphosphate (UTP) is incorporated into nascent transcripts. The nascent transcripts are then hybridized to specific cDNA probes to measure the transcription rate of the MMP or TIMP of interest. More recently, heterogeneous nuclear RNA (hnRNA), which is the pool of nascent, unspliced transcripts, has been detected by Reverse Transcriptase Polymerase Chain Reaction (RT-PCR). In a third assay, fragments of promoter, exons or introns are linked to a reporter gene, and the resulting construct is transiently or stably transfected into cells. The ability of MMP or TIMP sequences to drive expression of the reporter gene is then measured under various experimental conditions to identify transcriptional regulatory sequences. Taken together, this work has demonstrated that MMP-1 *(11–24)*, MMP-2 *(25)*, MMP-3 *(14,16,26–36)*, MMP-7 *(37,38)* MMP-9 *(39–41)*, MMP-12 *(42)*, MMP-13 *(43–46)*, TIMP-1 *(47–53)*, TIMP-2 *(54)* and TIMP-3 *(55,56)* are all transcriptionally activated by extracellular signals such as growth factors, phorbol esters, matrix proteins, and inflammatory cytokines, as well as by the expression of transforming oncogenes. In contrast, stromelysin-3 (MMP-11) is only transcriptionally activated by the differentiating agent retinoic acid *(57–59)*, or by stromal cell interactions with tumor cells *(60)*. The remaining family members (MMP-8, MMP-10, MMP-14 to MMP-20, and TIMP-4) have not yet been characterized at the transcriptional level.

2.1. Role of PEA3 and AP-1

Activating Protein-1 (AP-1) sites, which bind the Fos and Jun families of transcription factors, and Polyoma Enhancer A-binding Protein-3 (PEA-3) sites, which bind the Ets family of transcription factors, are typically present in the promoters of inducible MMP and TIMP genes, and are essential elements for transcriptional activation. Individual AP-1 and PEA-3 elements often play distinct roles in basal and induced transcription, and this may be a function of the specific transcription factors that bind to these sites. The first MMP family member to be described, MMP-1, contains adjacent AP-1 and PEA3 sites that are proximal to the transcriptional start site, and these sites are required for activation of the promoter by transforming oncogenes and tumor promoting phorbol esters *(11,12,22,61,62)*. The proximal AP-1 site contributes to basal transcription of MMP-1, whereas a second AP-1 site, which resides further 5' in the promoter, mediates fold transcriptional induction by phorbol esters *(20)*. The AP-1 site in the stromelysin-1 promoter (MMP-3) plays a similar role in basal transcription, since it is not necessary for fold induction of transcription *(31,34)*. Instead, two PEA3 sites that reside 5' of AP-1 are the essential phorbol ester-responsive elements *(29,33)*. Interestingly, differential activation of MMP-1 and MMP-3 may depend upon the c-Ets family members that are activated in cells. Ets-2 binds to the PEA3 sites in both genes and activates transcription, whereas Erg-1 preferential activates MMP-1 and represses MMP-3 *(35)*. This suggests that the location of individual AP-1 and PEA-3 sites, and the promoter sequences which surround them, may determine the function of these elements in transcriptional regulation.

2.2. Role of Cytokine and Growth Factor-Activated Transcription Factors

Although factors that bind to AP-1 and PEA-3 sites are essential for inducible MMP transcription, other transcription factors often contribute, particularly in response to growth factors and cytokines. The inflammatory cytokine interleukin-1 (IL-1) transactivates the rabbit MMP-1 promoter through an NF-κB-binding site located approx 3 kilobases from the start of transcription, which cooperates with the proximal AP-1 site *(63)*. This cooperativity between NF-κB and AP-1 also activates the 92 kDa gelatinase (MMP-9) promoter in response to tumor necrosis factor alpha (TNF) *(39)* or IL-1 *(41)*. In the human MMP-3 gene, the AP-1 site is not sufficient to drive the IL-1 response, but requires sequences that flank the AP-1 site, and that bind IL-1-inducible nuclear proteins *(30,34)*. Both PEA-3 and AP-1 "cross-talk" with a Platelet-derived Growth Factor (PDGF)-responsive element (SPRE) in the MMP-3 promoter to regulate transcription in PDGF-stimulated cells *(36)*. Thus, while transcription factors that bind to PEA-3 and AP-1 sites are often essential to drive MMP transcription in response to growth factors and cytokines, these factors often require interaction with other transcription factors to recruit the RNA polymerase II complex to the transcriptional start site.

Four TIMP genes have been cloned to date, but transcriptional regulation has been described primarily for human and murine TIMP-1. Like the inducible MMPs, PEA-3 and AP-1 sites in the TIMP-1 promoter *(50,51,64,65)* can cooperate with additional elements, such as CCAAT boxes *(51,64)*, to drive basal, serum and virus-induced transcription. Regulation of TIMP-1 transcription may have an additional layer of complexity that has not been described for the MMP genes. Regulatory sequences have been found 3' of the TIMP-1 transcriptional start site *(51,64)*, and an element that spans the border of exon 1 and intron 1 (nt +21 to +58) has been implicated in basal transcription *(51)*. These differences in TIMP-1 and MMP transcriptional regulation may contribute to the fact that agents such as TGF-ß and retinoic acid, which are potent inhibitors of MMP gene expression, enhance expression of the TIMP-1 gene *(55,66,67)*.

2.3. Transcriptional Repression of MMP Genes

Elevated MMP gene expression in connective tissue diseases such as in rheumatoid arthritis has been, and continues to be, the target of several therapies. Steroid hormones have been used clinically to impede disease progression, and these compounds work in part by repressing MMP transcription. The corticosteroids dexamethazone and prednisone inhibit MMP-1 *(68,69)* and MMP-3 *(26)* transcription by direct association between the glucocorticoid receptor and AP-1 proteins *(69)*. Retinoic acid and its derivatives (the retinoids) are less commonly used to treat arthritis because of toxic side effects, but are routinely used to impede the invasive potential of some head and neck carcinomas *(70,71)*. Like corticosteroids, the retinoids activate cellular receptors (Retinoic Acid Receptors, RAR and Retinoid X Receptors, RXR) to associate with AP-1-binding proteins and impede MMP-1 and MMP-3 transactivation by AP-1 *(72–75)*. In contrast to most MMP genes, the MMP-11 promoter does not contain an AP-1 site. Instead, this promoter contains a Retinoic Acid Response Element (RARE), which binds RAR directly and positively regulates MMP-11 transcription in retinoid-treated cells *(58)*. This suggests that the effects of retinoids on MMP transcription are determined by the presence or absence of AP-1 sites and/or RAREs in MMP promoters. However, RARE are also present in the promoters of genes that control cell growth and differentiation, events that also effect MMP gene expression. Therefore, such a model of retinoid-mediated MMP transcription may be too simplistic.

Like steroid hormones, transforming growth factor-beta (TGF-β) suppresses MMP-1 *(76)* MMP-3 *(77)* and MMP-13 *(78)* gene expression in normal fibroblasts. This, coupled with the ability of TGF-β to activate transcription of the genes for type I collagen, fibronectin *(25)*, and TIMP-1 *(25,76)* by these cells can promote ECM deposition in many stromal tissues. A TGF-β-inhibitory Element (TIE) has been described in the MMP-3 promoter, which mediates TGF-β repression by binding c-Fos *(77)*. Homologous elements have been identified in the MMP-1 *(79)* and MMP-13 *(44)* promoters, and may regulate TGF-β repression in these genes as well. However, TGF-β also activates transcription of the epidermal growth factor (EGF) receptor gene *(80)*, and EGF in turn activates transcription of MMPs *(81)* and TIMPs *(76)*. Therefore, the transcriptional effect of TGF-β on MMP and TIMP genes may be a balance between increased sensitivity to growth factor stimulation and repression by induced transcription factors.

3. Posttranscriptional Regulation of MMP Genes

While transcriptional activation can be demonstrated for inducible MMP genes, the increase of steady-state mRNA is often much greater than can be explained by increases of transcription rate alone. These observations have prompted several laboratories to investigate whether the rate of mRNA degradation changes in response to stimulus. To address this, the rate of mRNA decay can be assayed in pulse-chase experiments, in the presence of inhibitors of transcription, or using transfected genes under the transcriptional control of constitutive promoters. Such studies have demonstrated that the MMP-1 mRNA is stabilized in cells treated with IL-1 *(18,79)*, EGF *(81)*, and interferon-gamma (IFN-γ) *(82)*. Mutational analysis of the 3' untranslated region (UTR) has shown that AU-rich sequences, which control mRNA turnover in cytokine genes *(83)* and proto-oncogenes *(84)*, are also required for regulated decay of MMP-1 transcripts *(18)*. Recent work on the rat interstitial collagenase gene, which is homologous to human MMP-13, has shown that this transcript is stabilized by cortisol *(85)* and PDGF *(86)*, and destabilized by TGF-β *(87)*, in osteoblast-enriched cells. Although the 3' UTRs of the human *(88)* and rabbit *(89)* MMP-13 genes contain AU-rich sequences, it is not yet known if the human MMP-13 gene is also regulated at the posttranscriptional level.

Complex posttranscriptional mechanisms also control TIMP-1 gene expression in phorbol ester and serum-stimulated cells. The TIMP-1 transcript is stabilized following phorbol 12-myristate 13-acetate (PMA)-differentiation of macrophages, possibly due to an increase of poly-A tail length *(52)*. In fibroblasts, the TIMP-1 gene also produces transcripts of multiple lengths, and serum treatment increases the abundance of the shortest species *(90)*. Interestingly, as these shorter TIMP-1 transcripts accumulate, they are preferentially translated *(91)*, suggesting an additional level of serum-induced gene amplification. This phenomenon does not appear to occur in PMA-differentiated macrophages, where increased TIMP-1 transcript stability is undermined by reduced translational efficiency *(52)*. Thus, following *de novo* transcription, there is a strict balance between the rate of mRNA degradation and the rate of translational initiation and elongation. This level of gene regulation may be necessary for rapid changes in MMP and TIMP biosynthesis in response to a continuously changing environment.

4. Signaling Intermediates Involved in Transcriptional Activation

The thorough characterization of MMP and TIMP gene regulation has coincided with an explosion in the field of signal transduction. In particular, several protein kinase cascades have been described that are responsible for activating specific transcription factors. The incorporation of knowledge from the signaling field is now allowing a more complete picture of MMP and TIMP gene expression to be drawn.

As has been previously discussed, the transcription factor NF-κB activates MMP-1 and MMP-9 promoters in response to inflammatory cytokines. The signal transduction pathway from cytokine receptors to NF-κB nuclear localization has been elucidated by several laboratories, and has recently been reviewed in detail by Baeuerle *(92)* (*see* **Fig. 1**). When ligand binds to the trimeric TNF receptor, the TNF receptor-associated death domain (TRADD), the receptor-interacting protein (RIP) and the TNF receptor associated factor-2 (TRAF2) are recruited to the receptor's cytoplasmic domain. Similarly, the

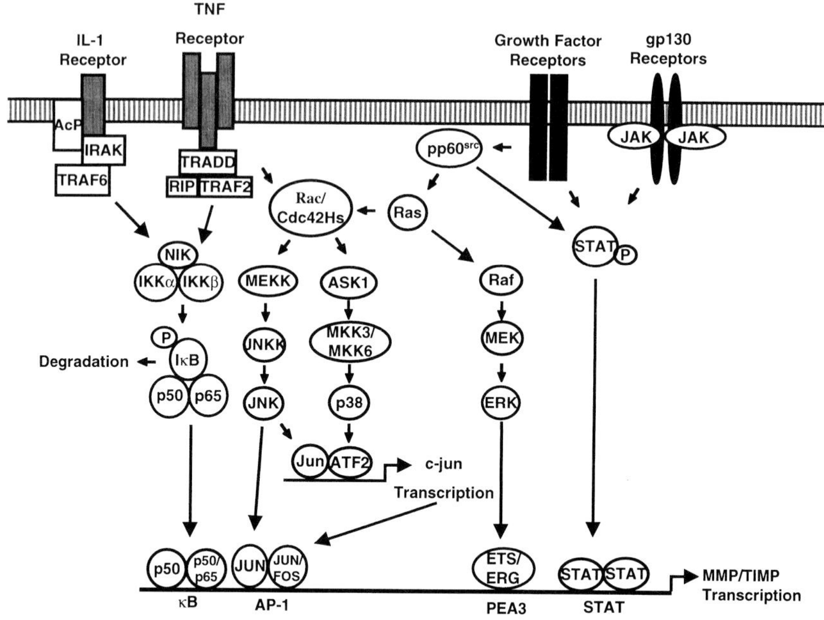

Fig. 1. Signaling pathways that can contribute to MMP and TIMP gene transcription.

IL-1 receptor recruits additional proteins including the accessory protein (AcP), the IL-1 receptor associated kinase (IRAK) and a protein related to TRAF2 known as TRAFF6. Both the TNF and IL-1 receptor complexes then recruit and activate the NF-κB-inducing kinases (NIK), which can activate NF-κB when overexpressed in cells. NIK is thought to activate NF-κB by associating with and activating the alpha and beta isoforms of the I Kappa B Kinase (IKKα, IKKβ). The IKKs phosphorylate IκB, which is the cytosolic inhibitor of NF-κB. Once phosphorylated IκB becomes ubiquitinated, which targets IκB for proteosome-mediated proteolytic degradation, and this event frees the p50 and p65 subunits of NF-κB to translocate to the nucleus and transactivate a multitude of genes, including MMP-1 and MMP-9.

Inflammatory cytokines and growth factors also activate the Mitogen-activated Protein Kinases (MAPK), which are a group of proline-directed, serine/threonine kinases that phosphorylate and activate specific transcription factors in the nucleus *(93,94)*. Three groups of MAPK have been described to date; the Extracellular Stimulus-regulated Kinases (ERK1, p44; ERK2, p42) *(95)*, the c-Jun N-terminal kinases (JNK1, JNK2 and JNK3) *(96)*, and the p38 kinase *(97)*. The three MAPK pathways have analogous intermediate signaling proteins that are responsible for propagation of activation signals *(98,99)* (**Fig. 1**). Small GTP-binding proteins, which are activated by ligand-bound transmembrane receptors, in turn activate serine/threonine kinases known as the MAPK Kinase Kinases (MAPKKK). The MAPKKK phosphorylate and activate the MAPK Kinases, which ultimately activate the MAPK by phosphorylating them on threonine and tyrosine residues. Once active, the MAPK translocate to the nucleus where they phosphorylate and activate the transactivation potential of specific transcription factors required for MMP transcription.

The ERK pathway is activated by growth factors *(100)*, cell-matrix interactions and tumor promoters *(100)*, and requires specific intermediates, including the small GTP-binding protein Ras, the MAPKKK Raf, and the MAPKK known as MEK *(99)*. Although inhibition of the ERKs blocks AP-1-driven transcription *(100)*, these MAPK also phosphorylate ETS family transcription factors which target PEA3 sites in MMP promoters. Specifically, ETS2, which positively regulates MMP-1 and MMP-3 transcription *(35)*, demonstrates enhanced transactivation potential when phosphorylated by ERKs *(101)*. Interestingly, phosphorylation of ETS1 on threonine 38 by ERK2 results in reduced ETS1 activation by growth factors *(102)*, suggesting complex regulation of ETS family members by the ERK pathway.

The JNKs pathway is also activated by growth factors *(103)*, in addition to inflammatory cytokines IL-1 and TNF *(96,104)*. Whereas the small GTP-binding protein Ras is sometimes an intermediate in JNK activation, other related proteins are also involved, including Rac/CdC42Hs *(105)*. The GTP-binding proteins activate the MAPKKK known as MEKK and ASK1, which in turn phosphorylate and activate the MAPKK known as JNKK. Once phosphorylated by JNKK, the JNKs can then initiate MMP transcription by phosphorylating and enhancing the transactivation potential of the AP-1 subunits c-Jun and ATF2 *(96)*.

Like the JNKs, the p38 MAPK is also activated by the cytokines IL-1 and TNF *(106)*, as well as the Gram-negative bacterial component LPS *(97)*. Whereas the GTP-binding proteins Ras and Rac/Cdc42Hs are involved, only the MAPKKK ASK-1 contributes directly to the p38 pathway *(99)*. ASK-1 primarily phosphorylates and activates the MAPKK known as MKK3 and MKK6, which phosphorylate and activate p38. An important divergence of the JNK and p38 pathways is the transcription factors that are targeted. Although both JNK and p38 phosphorylate ATF2, c-Jun is not a substrate of p38 *(99)*. Instead, p38 phosphorylates MAPK-activated Protein Kinase-2 (MAPKAPK-2), which phosphorylates Heat Shock Protein 27. Although a specific inhibitor of p38 blocks MMP-1 and MMP-3 induction *(107,108)*, it is currently not clear how the p38 pathway contributes to MMP or TIMP gene expression. One possibility is that phosphorylation of ATF2 by p38 activates the c-*jun* promoter. Increased synthesis of c-Jun may then contribute to AP-1-driven MMP-1 and MMP-3 transcription.

A distinct protein kinase cascade that regulates MMP and TIMP transcription is the Janus Kinases (JAK) and Signal Transducers and Activators of Transcription (STAT) pathway. The JAKs are a family of tyrosine kinases that associate with the cytoplasmic domains of growth factor receptors and gp130-containing cytokine receptors *(109)*. Upon ligand binding, the JAKs phosphorylate themselves and a family of transcription factors known as the STATs. Once phosphorylated, the STATs dimerize and translocate to the nucleus, where they bind to specific transcriptional response elements. Recently, Korzus and colleagues have demonstrated that STATs, which are activated by Oncostatin M, interact with AP-1-binding proteins to maximally transactivate the human MMP-1 and TIMP-1 promoters in human dermal fibroblasts *(110)*. However, a separate study demonstrated that the STAT-binding site of the murine TIMP-1 promoter did not bind STAT in response to IL-6 or Oncostatin M in HepG2 cells *(53)*. Additional studies may determine if this apparent paradox is due to differences in the cell types examined.

The src family of nonreceptor tyrosine kinases may be important upstream activators of the STAT and MAPK pathways that lead to MMP and TIMP transcription. When cells express the oncogene v-*src*, which is a constitutively active tyrosine kinase, the DNA-binding activity of STAT3 is enhanced *(111)*, as well as the kinase activity of ERK *(112)*. Furthermore, when cells adhere to fibronectin, an event that triggers MMP gene expression *(19)*, the Focal Adhesion Kinase and c-*src* cooperate to activate the ERK pathway *(113)*. Thus, src kinases may be an essential signaling hub for signaling pathways that drive MMP transcription. This proposed role for src kinases is supported by two lines of evidence. First, herbimycin A, which is a potent inhibitor of src family kinases, effectively blocks activation of MMP-1 transcription by multiple stimuli *(24,114–116)*. Second, the oncogenic kinase v-src activates the MMP-1 promoter through PEA3 *(12)* and STAT-binding sites, and requires functional ERK *(22)*.

5. Cell Type-Specific Expression of MMP and TIMP Genes

The mechanisms of gene regulation that have been described are characteristic of the cell types used in individual studies, and by no means represent global pathways. In fact, there is emerging evidence that expression of some MMP and TIMP genes is restricted to specific cell types. Furthermore, the same agonist can have opposite effects on gene expression depending on the cell type examined. These cell-specific responses may allow expression of appropriate ratios of MMP and TIMP proteins during developmental and inflammatory events.

The three interstitial collagenases, MMP-1, MMP-8, and MMP-13 have similar substrate specificities in that they all degrade collagens I, II, and III

(**Table 1**). However, their cellular expression is quite different, suggesting distinct roles in tissue remodeling. MMP-1 is by far the most broadly expressed of these enzymes, and can be induced in fibroblasts, chondrocytes, macrophages, endothelial cells, and keratinocytes *(117)*. MMP-8, also known as neutrophil collagenase is stored in neutrophils as granules that are released upon cellular activation. Whereas expression of MMP-8 is primarily restricted to neutrophils, it has also recently been found in osteoarthritic chondrocytes *(118)*. MMP-13, which is the newest interstitial collagenase to be identified, was first cloned from a human breast carcinoma *(88)*, and has since been shown to be expressed in other invasive carcinomas *(119,120)*, as well as the stromal cells surrounding these types of tumors *(121)*. The expression of MMP-13 by normal cells is currently being characterized and its restricted expression may be important during developmental and homeostatic processes.

In the embryo, MMP-13 is restricted to osteoblasts, hypertrophic chondrocytes and fibroblasts on the inner side of calvarial bone of the skull *(122,123)*. This restricted expression of MMP-13 around cartilage and bone during embryogenesis implicates this MMP in the orchestration of normal skeletal development. MMP-13 is also expressed by adult articular chondrocytes that have been stimulated with inflammatory cytokines and in chondrocytes from patients with osteoarthritis *(2,124,125)*, thus suggesting a role for this enzyme in this disease. Initial studies suggested that MMP-13 was restricted to chondrocytes within the joint, and not expressed by other mesenchymal cells such as synovial fibroblasts *(2,125)*. However, others have documented MMP-13 expression in rheumatoid synovium *(126)* and its presence in synovial fluid *(123)*, suggesting that MMP-13 may also contribute to the progression of rheumatoid arthritis. Cultured synovial fibroblasts also express MMP-13, albeit at lower levels than chondrocytes *(89)*. The constrained expression of MMP-13 in synovial fibroblasts may be a function of the fact that these cells reside in vascularized tissue and are more likely than chondrocytes to be exposed to inflammatory cytokines. Since MMP-13 is ten times more effective than MMP-1 at degrading type II collagen *(124)*, which is a major component of articular cartilage, unrestricted activation of gene expression could have severe consequences for the joint.

Expression of MMP-13 is also restricted among cells of the skin. Dermal fibroblasts express MMP-1, but not MMP-13 when cultured on plastic *(127)*. However, these fibroblasts will express MMP-13 when cultured in three dimensional collagen gels *(127)*. In deep chronic ulcer beds, fibroblasts express MMP-13, whereas MMP-1 is restricted to re-epithelializing keratinocytes that contact type I collagen *(116,127,128)*. Thus, perhaps contact with matrix proteins, which is known to play an important role in MMP-1 gene expression *(116,129)*, is the critical signal for cell type-specific expression of MMP-13.

In addition to differential expression of MMPs during development and wound healing, regulation by various agonists is often cell-type specific. As previously stated, steroid hormones are potent inhibitors of MMP gene expression in fibroblasts. However, although cortisol suppresses MMP-13 expression in rat fibroblasts, this hormone activates MMP-13 in osteoblasts from the same animals *(85)*. Similarly, retinoic acid increases MMP-2 and MMP-13 mRNA in chondrocytes *(118,130,131)*. The retinoic acid effect in chondrocytes may be due to the ability of this steroid to increase c-*fos* and c-*jun* gene expression in these cells *(132)*. This selective increase in MMP gene expression in osteoblasts and chondrocytes may explain why glucocorticoid therapies can promote bone resorption *(133)* and why retinoids can augment the severity of collagen-induced arthritis in the rat *(134)*.

Interestingly, in contrast to fibroblasts, TGF-β activates MMP-1 in normal keratinocytes *(135)* and MMP-13 expression in transformed keratinocytes *(45)*. A possible mechanism for the cell-specific expression MMP-1 is that in keratinocytes, TGF-β induces c-*jun* *(135)*, which is a positive regulator of AP-1 sites *(136)*, although in fibroblasts, TGF-β activates jun-B *(135)*, which is an inhibitor of AP-1-mediated transcription *(136)*. This differential regulation of MMP-1 by TGF-β may be necessary to orchestrate reepithelialization by keratinocytes in response to wounding of the skin.

I have already discussed the fact that IL-1 can induce MMP-1 through increased transcription and stabilization of mRNA *(18)*. However, certain cells preferentially use these mechanisms to increase gene expression. In human skin fibroblasts, IL-1 only slightly activates the MMP-1 promoter, but potently increases mRNA stability. In contrast, a breast carcinoma cell line demonstrated greater promoter responsiveness to IL-1, and no transcript stabilization *(18)*. Thus, the cell lineage not only determines if an agonist will stimulate MMP gene expression, but also defines the level of gene regulation employed.

There are two MMP family members that are expressed only in single-cell lineages, and up-regulation of these enzymes can contribute directly to disease progression. In contrast to most MMPs, which are expressed by stromal cells, matrilysin (MMP-7) is expressed exclusively in cells of epithelial origin *(137)*. Specifically, MMP-7 is expressed by the epithelium of the human *(138)* and rat endometrium *(139)*, and it contributes to tissue remodeling during the menstrual cycle. In addition, the epithelial cells of several exocrine glands, including the mammary, parotid, prostate, pancreas, liver, and lung peribronchial glands constitutively secrete MMP-7 at their luminal surfaces *(140)*. There this MMP is thought to act as a "pipe cleaner" of the exocrine ducts, keeping them clear of matrix proteins *(140)*. MMP-7 expression is increased in tumors of epithelial origin, including carcinomas of the breast *(141)*, colon and stomach *(142)*, head and neck, lung *(143)* as well as aggressive infiltrative carcinomas of the skin *(144)*. Whereas overexpression of MMP-7 in a prostate carcinoma results in a more invasive phenotype *(145)*, anti-sense reduction of MMP-7 expression in colon carcinomas reduced tumorigenicity of these cells *(146)*. These studies demonstrate that cell-specific expression of MMP-7 plays an important role in normal epithelial cell function and can potentiate the growth and spread of epithelial-derived tumors.

The second MMP family member that is exclusively expressed in one type of cell is macrophage metalloelastase (MMP-12) *(147)*. In mice that have been chronically exposed to cigarette smoke, pathological changes occur in the lungs that are characteristic of emphysema in humans, including enlargement of bronchioles, alveolar ducts and alveoli, and the accumulation of alveolar macrophages *(148)*. Since macrophages are involved in the pathogenesis of emphysema, Shapiro and colleagues hypothesized that the macrophage-specific MMP-12 could contribute to the pathophysiology of this disease *(149)*. To test this, the MMP-12 gene was deleted in mice by gene targeting *(150)*, and the MMP-12 deficient mice were challenged with chronic exposure to cigarette smoke *(148)*. In contrast to wild-type mice, MMP-12-deficient mice did not develop the macrophage infiltration and pathological changes typical of this emphysema model. Thus, macrophage-specific expression of MMP-12, possibly through generation of monocyte chemotactic activity *(148)* is required for the progression of emphysema in mice. While an exact mechanism has not yet been elucidated, this cell-specific MMP may someday be a target of therapies for environmentally induced pulmonary disease in humans.

6. Conclusion and Future Prospects

Throughout development and adult homeostasis, disruption of the balance between MMP and TIMP gene expression can contribute to the progression of several diseases, including arthritis, cancer, cardiovascular, and pulmonary disease. Much research in recent years has focused on the development of drugs that inhibit MMP catalytic activity in peripheral tissues *(151,152)*. However, since the broad spectrum MMP inhibitors are the most effective clinically *(153)*, this approach may not be effective for disorders involving the expression of specific MMP family members by specific cells. Alternatively, recent work on the intracellular mechanisms of MMP and TIMP gene expression has demonstrated the complexity of approaching disease intervention at this level. This work has begun to characterize the spatial and temporal expression of MMP and TIMP genes, to define the cellular kinase cascades that propagate extracellular signals to the nucleus, and to identify nuclear factors that are activated by these cascades to initiate transcription. Thus, the prospects of designing therapies that target specific signal transduction pathways and transcription factor activation look promising for the intervention of MMP-mediated diseases. A major hurdle will be to deliver such drugs efficaciously directly to the cell types that are expressing MMPs and/or TIMPs at inappropriate levels. Furthermore, since it has been established that posttranscriptional mechanisms can have a profound effect on the amount of MMP and TIMP gene expression, future therapies will also have to address this level gene regulation.

Acknowledgments

I would like to thank Dr. Constance Brinckerhoff for critical reading of this chapter. I would also like to acknowledge those investigators whose contributions to this field were not included due to limitations of space. My work is supported by a grant from the National Institute of Arthritis and Musculoskeletal and Skin Diseases (K01 AR02024).

References

1. Firestein, G. S., Paine, M. M., and Littman, B. H. (1991) Gene expression (collagenase, tissue inhibitor of metalloproteinases, complement, and HLA-DR) in rheumatoid arthritis and osteoarthritis synovium. Quantitative analysis and effect of intra-articular corticosteroids. *Arthritis Rheum.* **34**, 1094–2105.
2. Reboul, P., Pelletier, J. P., Tardif, G., Cloutier, J. M., and Martel-Pelletier, J. (1996) The new collagenase, collagenase-3, is expressed and synthesized by human chondrocytes but not by synoviocytes. A role in osteoarthritis. *J. Clin. Invest.* **97**, 2011–2019.
3. Wernicke, D., Seyfert, C., Hinzmann, B., and Gromnica-Ihle, E. (1996) Cloning of collagenase 3 from the synovial membrane and its expression in rheumatoid arthritis and osteoarthritis. *J. Rheumatol.,* **23**, 590–595.
4. Clark, I. M., Powell, L. K., Ramsey, S., Hazleman, B. L., and Cawston, T. E. (1993) The measurement of collagenase, tissue inhibitor of metalloproteinases (TIMP), and collagenase-TIMP complex in synovial fluids from patients with osteoarthritis and rheumatoid arthritis. *Arthritis Rheum.* **36**, 372–379.
5. Firestein, G. S., Paine, M. M., and Boyle, D. L. (1994) Mechanisms of methotrexate action in rheumatoid arthritis. Selective decrease in synovial collagenase gene expression. *Arthritis Rheum.* **37**, 193–200.
6. Hembry, R. M., Bagga, M. R., Reynolds, J. J., and Hamblen, D. L. (1995) Immunolocalisation studies on six matrix metalloproteinases and their inhibitors, TIMP-1 and TIMP-2, in synovia from patients with osteo- and rheumatoid arthritis. *Annals of the Rheumatic Diseases.* **54**, 25–32.
7. Coussens, L. M. and Werb, Z. (1996) Matrix metalloproteinases and the development of cancer. *Chemistry & Biology.* **3**, 895–904.
8. Chambers, A. F. and Matrisian, L. M. (1997) Changing views of the role of matrix metalloproteinases in metastasis. *J. Nat. Cancer Institute.* **89**, 1260–1270.
9. Henney, A. M., Wakeley, P. R., Davies, M. J., Foster, K., Hembry, R., Murphy, G., and Humphries, S. (1991) Localization of stromelysin gene expression in

atherosclerotic plaques by in situ hybridization. *Proc. Nat. Acad. Sci. USA* **88**, 8154–8158.

10. Libby, P. (1995) Molecular bases of the acute coronary syndromes. *Circulation.* **91**, 2844–2850.

11. Angel, P., Imagawa, M., Chiu, R., Stein, B., Imbra, R. J., Rahmsdorf, H. J., Jonat, C., Herrlich, P., and Karin, M. (1987) Phorbol ester-inducible genes contain a common cis elecent recognized by a TPA-modulated trans-acting factor. *Cell,* **49**, 729–739.

12. Gutman, A. and Wasylyk, B. (1990) The collagenase gene promoter contains a TPA and oncogene-responsive unit encompassing the PEA3 and AP-1 binding sites. *Embo J.* **9**, 2241–2246.

13. Auble, D. T. and Brinckerhoff, C. E. (1991) The AP-1 sequence is necessary but not sufficient for phorbol induction of collagenase in fibroblasts. *Biochem.* **30**, 4629–4635.

14. Mitchell, P. G. and Cheung, H. S. (1991) Tumor necrosis factor alpha and epidermal growth factor regulation of collagenase and stromelysin in adult porcine articular chondrocytes. *J. Cell Phys.* **149**, 132–140.

15. Mauviel, A., Kahari, V. M., Chen, Y. Q., Kurkinen, M., Evans, C. H., and Uitto, J. (1992) Transcriptional activation of fibroblast stromelysin-1 and collagenase gene expression by a novel lymphokine, leukoregulin, *Trans. Assoc. Am. Phys.* **105**, 100–109.

16. James, T. W., Wagner, R., White, L. A., Zwolak, R. M., and Brinckerhoff, C. E. (1993) Induction of collagenase and stromelysin gene expression by mechanical injury in a vascular smooth muscle-derived cell line. *J. Cell Phys.* **157**, 426–437.

17. Chamberlain, S. H., Hemmer, R. M., and Brinckerhoff, C. E. (1993) Novel phorbol ester response region in the collagenase promoter binds Fos and Jun. *J. Cell Biochem.* **52**, 337–351.

18. Vincenti, M. P., Coon, C. I., Lee, O., and Brinckerhoff, C. E. (1994) Regulation of collagenase gene expression by IL-1 beta requires transcriptional and post-transcriptional mechanisms *Nucleic Acids Res.* **22**, 4818–4827.

19. Tremble, P., Damsky, C. H., and Werb, Z. (1995) Components of the nuclear signaling cascade that regulate collagenase gene expression in response to integrin-derived signals *J. Cell Biol.* **129**, 1707–1720.

20. White, L. A. and Brinckerhoff, C. E. (1995) Two activator protein-1 elements in the matrix metalloproteinase-1 promoter have different effects on transcription and bind Jun D, c-Fos, and Fra-2. *Matrix Biol.* **14**, 715–725.

21. Doyle, G. A. R., Pierce, R. A., and Parks, W. C. (1997) Transcriptional induction of collagenase-1 in differentiated monocyte-like (U937) cells is regulated by AP-1 and an upstream C/EBP-beta site. *J. Biol. Chem.* **272**, 11,840–11,849.

22. Vincenti, M. P., Schroen, D. J., Coon, C. I., and Brinckerhoff, C. E. (1998) V-src activation of the collagenase-1) matrix metalloproteinase-1 promoter through PEA3 and STAT: Requirement of extracellular signal-regulated kinases and inhibition by retinoic acid receptors. *Mol. Carcin.* **21**.

23. Pierce, R. A., Sandefur, S., Doyle, G. A., and Welgus, H. G. (1996) Monocytic cell type-specific transcriptional induction of collagenase. *J. Clin. Invest.* **97**, 1890–1899.
24. Hurley, M. M., Marcello, K., Abreu, C., Brinckerhoff, C. E., Bowik, C. C., and Hibbs, M. S. (1995) Transcriptional regulation of the collagenase gene by basic fibroblast growth factor in osteoblastic MC3T3-E1 cells. *Biochem. Biophys. Res. Com.* **214**, 331–339.
25. Overall, C. M., Wrana, J. L., and Sodek, J. (1991) Transcriptional and post-transcriptional regulation of 72-kDa gelatinase/type IV collagenase by transforming growth factor-beta 1 in human fibroblasts. Comparisons with collagenase and tissue inhibitor of matrix metalloproteinase gene expression. *J. Biol. Chem.* **266**, 14,064–14,071.
26. Frisch, S. M. and Ruley, H. E. (1987) Transcription from the stromelysin promoter is induced by interleukin-1 and repressed by dexamethasone. *J. Biol. Chem.* **262**, 16,300–16,304.
27. Quinones, S., Saus, J., Otani, Y., Harris, E. D., Jr., and Kurkinen, M. (1989) Transcriptional regulation of human stromelysin. *J. Biol. Chem.* **264**, 8339–8344.
28. Sirum, K. L. and Brinckerhoff, C. E. (1989) Cloning of the genes for human stromelysin and stromelysin 2: differential expression in rheumatoid synovial fibroblasts. *Biochem.* **28**, 8691–8698.
29. Wasylyk, C., Gutman, A., Nicholson, R., and Wasylyk, B. (1991) The c-Ets oncoprotein activates the stromelysin promoter through the same elements as several non-nuclear oncoproteins. *Embo. J.* **10**, 1127–1134.
30. Sirum-Connolly, K. and Brinckerhoff, C. E. (1991) Interleukin-1 or phorbol induction of the stromelysin promoter requires an element that cooperates with AP-1. *Nucleic Acids Res.* **19**, 335–341.
31. Buttice, G., Quinones, S., and Kurkinen, M. (1991) The AP-1 site is required for basal expression but is not necessary for TPA-response of the human stromelysin gene. *Nucleic Acids Res.* **19**, 3723–3731.
32. Mauviel, A., Kahari, V. M., Kurkinen, M., Evans, C. H., and Uitto, J. (1992) Leukoregulin, a T-cell derived cytokine, upregulates stromelysin-1 gene expression in human dermal fibroblasts: evidence for the role of AP-1 in transcriptional activation. *J. Cell Biochem.* **50**, 53–61.
33. Buttice, G. and Kurkinen, M. (1993) A polyomavirus enhancer A-binding protein-3 site and Ets-2 protein have a major role in the 12-O-tetradecanoylphorbol-13-acetate response of the human stromelysin gene. *J. Biol. Chem.* **268**, 7196–7204.
34. Quinones, S., Buttice, G., and Kurkinen, M. (1994) Promoter elements in the transcriptional activation of the human stromelysin-1 gene by the inflammatory cytokine, interleukin-1. *Biochem. J.* **302**, 471–477.
35. Buttice, G., Duterque-Coquillaud, M., Basuyaux, J. P., Carrere, S., Kurkinen, M., and Stehelin, D. (1996) Erg, an Ets-family member, differentially regulates human collagenase1 (MMP1) and stromelysin1 (MMP3) gene expression by physically interacting with the Fos/Jun complex. *Oncogene.* **13**, 2297–2306.

36. Kirstein, M., Sanz, L., Quinones, S., Moscat, J., Diaz-Meco, M. T., and Saus, J. (1996) Cross-talk between different enhancer elements during mitogenic induction of the human stromelysin-1 gene. *J. Biol. Chem.* **271**, 18,231–18,236.

37. Gaire, M., Magbanua, Z., McDonnell, S., McNeil, L., Lovett, D. H., and Matrisian, L. M. (1994) Structure and expression of the human gene for the matrix metalloproteinase matrilysin. *J. Biol. Chem.* **269**, 2032–2040.

38. Klein, R. D., Borchers, A. H., Sundareshan, P., Bougelet, C., Berkman, M. R., Nagle, R. B., and Bowden, G. T. (1997) Interleukin-1beta secreted from monocytic cells induces the expression of matrilysin in the prostatic cell line LNCaP. *J. Biol. Chem.* **272**, 14,188–14,192.

39. Sato, H. and Seiki, M. (1993) Regulatory mechanism of 92 kDa type IV collagenase gene expression which is associated with invasiveness of tumor cells. *Oncogene.* **8**, 395–405.

40. He, C. (1996) Molecular mechanism of transcriptional activation of human gelatinase B by proximal promoter. *Cancer Letters.* **106**, 185–191.

41. Yokoo, T. and Kitamura, M. (1996) Dual regulation of IL-1 beta-mediated matrix metalloproteinase-9 expression in mesangial cells by NF-kappa B and AP-1, *Am. J. Physiol.* **270**, F123–F130.

42. Monet-Kuntz, C., Cuvelier, A., Sarafan, N., and Martin, J. P. (1997) Metalloelastase expression in a mouse macrophage cell line—regulation by 4beta-phorbol 12-myristate 13-acetate, lipopolysaccharide and dexamethasone. *Eur. J. Biochem.* **247**, 588–595.

43. Grumbles, R. M., Shao, L., Jeffrey, J. J., and Howell, D. S. (1996) Regulation of rat interstitial collagenase gene expression in growth cartilage and chondrocytes by vitamin D3, interleukin-1 beta, and okadaic acid. *J. Cell Biochem.* **63**, 395–409.

44. Pendas, A. M., Balbín, M., Llano, E., Jimenez, M. G., and López-Otín, C. (1997) Structural analysis and promoter characterization of the human collagenase-3 gene (MMP13). *Genomics.* **40**, 222–233.

45. Johansson, N., Westermarck, J., Leppa, S., Hakkinen, L., Koivisto, L., Lopez-Otin, C., Peltonen, J., Heino, J., and Kahari, V. M. (1997) Collagenase 3 (matrix metalloproteinase 13) gene expression by HaCaT keratinocytes is enhanced by tumor necrosis factor alpha and transforming growth factor beta. *Cell Growth Differ.* **8**, 243–250.

46. Tardif, G., Pelletier, J. P., Dupuis, M., Hambor, J. E., and Martel-Pelletier, J. (1997) Cloning, sequencing and characterization of the 5'-flanking region of the human collagenase-3 gene. *Biochem. J.* **323**, 13–16.

47. Alitalo, R., Partanen, J., Pertovaara, L., Holtta, E., Sistonen, L., Andersson, L., and Alitalo, K. (1990) Increased erythroid potentiating activity/tissue inhibitor of metalloproteinases and jun/fos transcription factor complex characterize tumor promoter-induced megakaryoblastic differentiation of K562 leukemia cells. *Blood* **75**, 1974–1982.

48. Campbell, C. E., Flenniken, A. M., Skup, D., and Williams, B. R. (1991) Identification of a serum- and phorbol ester-responsive element in the murine tissue inhibitor of metalloproteinase gene. *J. Biol. Chem.* **266**, 7199–7206.

49. Colige, A. C., Lambert, C. A., Nusgens, B. V., and Lapiere, C. M. (1992) Effect of cell-cell and cell-matrix interactions on the response of fibroblasts to epidermal growth factor in vitro. Expression of collagen type I, collagenase, stromelysin and tissue inhibitor of metalloproteinases. *Biochem. J.* **285**, 215–221.

50. Edwards, D. R., Rocheleau, H., Sharma, R. R., Wills, A. J., Cowie, A., Hassell, J. A., and Heath, J. K. (1992) Involvement of AP1 and PEA3 binding sites in the regulation of murine tissue inhibitor of metalloproteinases-1 (TIMP-1) transcription. *Biochim. Biophys. Acta.* **1171**, 41–55.

51. Clark, I. M., Rowan, A. D., Edwards, D. R., Bech-Hansen, T., Mann, D. A., Bahr, M. J., and Cawston, T. E. (1997) Transcriptional activity of the human tissue inhibitor of metalloproteinases 1 (TIMP-1) gene in fibroblasts involves elements in the promoter, exon 1 and intron 1. *Biochem. J.* **324**, 611–617.

52. Doyle, G. A., Saarialho-Kere, U. K., and Parks, W. C. (1997) Distinct mechanisms regulate TIMP-1 expression at different stages of phorbol ester-mediated differentiation of U937 cells. *Biochem.* **36**, 2492–2500.

53. Botelho, F. M., Edwards, D. R., and Richards, C. D. (1998) Oncostatin M stimulates c-Fos to bind a transcriptionally responsive AP- 1 element within the tissue inhibitor of metalloproteinase-1 promoter (In Process Citation). *J. Biol. Chem.* **273**, 5211–5218.

54. Cook, T. F., Burke, J. S., Bergman, K. D., Quinn, C. O., Jeffrey, J. J., and Partridge, N. C. (1994) Cloning and regulation of rat tissue inhibitor of metalloproteinases-2 in osteoblastic cells. *Arch. Biochem. Biophys.* **311**, 313–320.

55. Leco, K. J., Khokha, R., Pavloff, N., Hawkes, S. P., and Edwards, D. R. (1994) Tissue inhibitor of metalloproteinases-3 (TIMP-3) is an extracellular matrix-associated protein with a distinctive pattern of expression in mouse cells and tissues. *J. Biol. Chem.* **269**, 9352–9360.

56. Su, S., Dehnade, F., and Zafarullah, M. (1996) Regulation of tissue inhibitor of metalloproteinases-3 gene expression by transforming growth factor-beta and dexamethasone in bovine and human articular chondrocytes. *DNA Cell Biol.* **15**, 1039–1048.

57. Huttenlocher, A., Werb, Z., Tremble, P., Huhtala, P., Rosenberg, L., and Damsky, C. H. (1996) Decorin regulates collagenase gene expression in fibroblasts adhering to vitronectin. *Matrix Biol.* **15**, 239–250.

58. Anglard, P., Melot, T., Guerin, E., Thomas, G., and Basset, P. (1995) Structure and promoter characterization of the human stromelysin-3 gene. *J. Biol. Chem.* **270**, 20.337–20.344.

59. Guerin, E., Ludwig, M. G., Basset, P., and Anglard, P. (1997) Stromelysin-3 induction and interstitial collagenase repression by retinoic acid. Therapeutical implication of receptor-selective retinoids dissociating transactivation and AP-1-mediated transrepression. *J. Biol. Chem.* **272**, 11,088–11,095.

60. Ahmad, A., Marshall, J. F., Basset, P., Anglard, P., and Hart, I. R. (1997) Modulation of human stromelysin 3 promoter activity and gene expression by human breast cancer cells. *Int. J. Cancer* **73**, 290–296.

61. Gutman, A., Wasylyk, C., and Wasylyk, B. (1991) Cell-specific regulation of onco-gene-responsive sequences of the c-fos promoter. *Mol. Cell Biol.* **11**, 5381–5387.

62. Gutman, A. and Wasylyk, B. (1991) Nuclear targets for transcription regulation by oncogenes. *Trends Genet.* **7**, 49–54.

63. Vincenti, M. P., Coon, C. I., and Brinckerhoff, C. E. (1998) NF-kB/p50 acti-vates an element in the distal matrix metalloproteinase-1 promoter in interleukin-1ß-stimulated synovial fibroblasts. *Arthritis Rheum.* **41**, 1987–1994.

64. Ponton, A., Coulombe, B., Steyaert, A., Williams, B. R., and Skup, D. (1992) Basal expression of the gene (TIMP) encoding the murine tissue inhibitor of metalloproteinases is mediated through AP1- and CCAAT-binding factors. *Gene.* **116**, 187–194.

65. Coulombe, B., Ponton, A., Daigneault, L., Williams, B. R., and Skup, D. (1988) Presence of transcription regulatory elements within an intron of the virus-inducible murine TIMP gene. *Mol. Cell Biol.* **8**, 3227–3234.

66. Kikuchi, K., Kadono, T., Furue, M., and Tamaki, K. (1997) Tissue inhibitor of metalloproteinase 1 (TIMP-1) may be an autocrine growth factor in scleroderma fibroblasts. *J. Invest. Dermatol.* **108**, 281–284.

67. Wright, J. K., Clark, I. M., Cawston, T. E., and Hazleman, B. L. (1991) The secretion of the tissue inhibitor of metalloproteinases (TIMP) by human syn-ovial fibroblasts is modulated by all-trans-retinoic acid. *Biochim. Biophys. Acta.* **1133**, 25–30.

68. Auble, D. T. and Brinckerhoff, C. E. (1990) Regulation of Collagenase Gene Expression in Synovial Fibroblasts. *Annals. NY Acad. Sci.* **580**, 355–373.

69. Jonat, C., Stein, B., Ponta, H., Herrlich, P., and Rahmsdorf, H. J. (1992) Positive and negative regulation of collagenase gene expression. *Matrix Supplement* **1**, 145–155.

70. Lotan, R. (1996) Retinoids and their receptors in modulation of differentiation, development, and prevention of head and neck cancers. *Anticancer Res.* **16**, 2415–2419.

71. Lotan, R. (1996) Retinoids in cancer prevention. *FASEB J.* **10**, 1031–1039.

72. Nicholson, R. C., Mader, S., Nagpal, S., Leid, M., Rochette-Egly, C., and Chambon, P. (1990) Negative regulation of the rat stromelysin gene promoter by retinoic acid is mediated by an AP1 binding site. *EMBO J.* **9**, 4443–4454.

73. Pan, L.-Y., Chamberlain, S. H., Auble, D. T., and Brinckerhoff, C. E. (1992) Differential regulation of collagenase gene expression by retinoic acid recep-tors-alpha, beta, and gamma. *Nucleic Acids Res.* **20**, 3105–3111.

74. Pan, L., Eckhoff, C., and Brinckerhoff, C. E. (1995) Suppression of Collagenase Gene Expression by all-trans and 9-cis Retinoic Acid is Ligand Dependent and Requires Both RARs and RXRs. *J. Cell Biochem.* **57**, 575–589.

75. Schroen, D. J. and Brinckerhoff, C. E. (1996) Inhibition of rabbit collagenase (matrix metalloproteinase-1; MMP-1) transcription by retinoid receptors: evi-dence for binding of RARs/RXRs to the -77 AP-1 site through interactions with c-Jun. *J. Cell Phys.* **169**, 320–332.

76. Edwards, D. R., Murphy, G., Reynolds, J. J., Whitham, S. E., Docherty, A. J., Angel, P., and Heath, J. K. (1987) Transforming growth factor beta modulates the expression of collagenase and metalloproteinase inhibitor. *EMBO J.* **6**, 1899–1904.

77. Kerr, L. D., Miller, D. B., and Matrisian, L. M. (1990) TGF-beta 1 inhibition of transin/stromelysin gene expression is mediated through a Fos binding sequence. *Cell* **61**, 267–278.

78. Rydziel, S., Varghese, S., and Canalis, E. (1997) Transforming growth factor beta1 inhibits collagenase 3 expression by transcriptional and post-transcriptional mechanisms in osteoblast cultures (published erratum appears in J. Cell Physiol. 1997 May;171(2):234). *J. Cell Phys.* **170**, 145–152.

79. Rutter, J. L., Benbow, U., Coon, C. I., and Brinckerhoff, C. E. (1997) Cell-type specific regulation of human interstitial collagenase-1 gene expression by interleukin-1 beta (IL-1 beta) in human fibroblasts and BC-8701 breast cancer cells. *J. Cell Biochem.* **66**, 322–336.

80. Hou, X., Johnson, A. C., and Rosner, M. R. (1994) Induction of epidermal growth factor receptor gene transcription by transforming growth factor beta 1: association with loss of protein binding to a negative regulatory element. *Cell Growth & Diff.* **5**, 801–809.

81. Delany, A. M. and Brinckerhoff, C. E. (1992) Post-transcriptional regulation of collagenase and stromelysin gene expression by epidermal growth factor and dexamethasone in cultured human fibroblasts. *J. Cell Bioche.* **50**, 400–410.

82. Tamai, K., Ishikawa, H., Mauviel, A., and Uitto, J. (1995) Interferon-gamma coordinately upregulates matrix metalloprotease (MMP)-1 and MMP-3, but not tissue inhibitor of metalloproteases (TIMP), expression in cultured keratinocytes. *J. Invest. Dermatol.* **104**, 384–390.

83. Shaw, G. and Kamen, R. (1986) A conserved AU sequence from the 3'untranslated region of GM-CSF mRNA mediates selective mRNA degradation. *Cell* **46**, 659–997.

84. Brewer, G. (1991) An A+U-rich element RNA-binding factor regulates c-myc mRNA stability in vitro. *Mol. Cell Biol.* **12**, 2460–2466.

85. Delany, A. M., Jeffrey, J. J., Rydziel, S., and Canalis, E. (1995) Cortisol increases interstitial collagenase expression in osteoblasts by post-transcriptional mechanisms. *J. Biol. Chem.* **270**, 26,607–26,612.

86. Varghese, S., Delany, A. M., Liang, L., Gabbitas, B., Jeffrey, J. J., and Canalis, E. (1996) Transcriptional and posttranscriptional regulation of interstitial collagenase by platelet-derived growth factor BB in bone cell cultures. *Endocrinol.* **137**, 431–437.

87. Rydziel, S., Varghese, S., and Canalis, E. (1997) Transforming growth factor beta1 inhibits collagenase 3 expression by transcriptional and post-transcriptional mechanisms in osteoblast cultures. *J. Cell Phys.* **170**, 145–152.

88. Freije, J. M., Diez-Itza, I., Balbin, M., Sanchez, L. M., Blasco, R., Tolivia, J., and Lopez-Otin, C. (1994) Molecular cloning and expression of collagenase-3, a novel human matrix metalloproteinase produced by breast carcinomas. *J. Biol. Chem.* **269**, 16,766–16,773.

89. Vincenti, M. P., Coon, C. I., Mengshol, A. J., Yocum, S., Mitchell, P., and Brinckerhoff, C. E. (1998) Cloning of the gene for interstitial collagenase-3 (matrix metalloproteinase-13) from rabbit synovial fibroblasts: differential expression with collagenase-1 (matrix metalloproteinase-1). *Biochem J.* **331**, 341–346.

90. Edwards, D. R., Waterhouse, P., Holman, M. L., and Denhardt, D. T. (1986) A growth-responsive gene (16C8) in normal mouse fibroblasts homologous to a human collagenase inhibitor with erythroid-potentiating activity: evidence for inducible and constitutive transcripts. *Nucleic Acids Res.* **14**, 8863–8878.

91. Waterhouse, P., Khokha, R., and Denhardt, D. T. (1990) Modulation of translation by the 5' leader sequence of the mRNA encoding murine tissue inhibitor of metalloproteinases. *J. Biol. Chem.* **265**, 5585–5589.

92. Baeuerle, P. A. (1998) Pro-inflammatory signaling: last pieces in the NF-kappaB puzzle? *Current Biol.* **8**, R19–22.

93. Su, B. and Karin, M. (1996) Mitogen-activated protein kinase cascades and regulation of gene expression. *Current Opin. Immunol.* **8**, 402–411.

94. Robinson, M. J. and Cobb, M. H. (1997) Mitogen-activated protein kinase pathways. *Curr. Opin. Cell Biol.* **9**, 180–186.

95. Cobb, M. H., Hepler, J. E., Cheng, M., and Robbins, D. (1994) The mitogen-activated protein kinases. ERK1 and ERK2. *Seminars Cancer Biol.* **5**, 261–268.

96. Gupta, S., Barrett, T., Whitmarsh, A. J., Cavanagh, J., Sluss, H. K., Derijard, B., and Davis, R. J. (1996) Selective interaction of JNK protein kinase isoforms with transcription factors. *Embo. J.* **15**, 2760–2770.

97. Han, J., Lee, J. D., Bibbs, L., and Ulevitch, R. J. (1994) A MAP kinase targeted by endotoxin and hyperosmolarity in mammalian cells., *Science.* **265**, 808–811.

98. Matsuda, S., Gotoh, Y., and Nishida, E. (1994) Signaling pathways mediated by the mitogen-activated protein (MAP) kinase kinase/MAP kinase cascade. *J. Leukocyte Biol.* **56**, 548–553.

99. Minden, A. and Karin, M. (1997) Regulation and function of the JNK subgroup of MAP kinases. *Biochim. Biophys. Acta.* **1333**, F85–F104.

100. Frost, J. A., Geppert, T. D., Cobb, M. H., and Feramisco, J. R. (1994) A requirement for extracellular signal-regulated kinase (ERK) function in the activation of AP-1 by Ha-Ras, phorbol 12-myristate 13-acetate and serum. *Proc. Natl. Acad. Sci. USA* **91**, 3844–3848.

101. McCarthy, S. A., Chen, D., Yang, B. S., Garcia Ramirez, J. J., Cherwinski, H., Chen, X. R., Klagsbrun, M., Hauser, C. A., Ostrowski, M. C., and McMahon, M. (1997) Rapid phosphorylation of Ets-2 accompanies mitogen-activated protein kinase activation and the induction of heparin-binding epidermal growth factor gene expression by oncogenic Raf-1. *Mol. Cell Biol.* **17**, 2401–2412.

102. Rabault, B., Roussel, M. F., Quang, C. T., and Ghysdael, J. (1996) Phosphorylation of Ets1 regulates the complementation of a CSF-1 receptor impaired in mitogenesis. *Oncogene* **13**, 877–881.

103. Whitmarsh, A. J., Shore, P., Sharrocks, A. D., and Davis, R. J. (1995) Integration of MAP kinase signal transduction pathways at the serum response element. *Science* **269**, 403–407.

104. Sluss, H. K., Barrett, T., Derijard, B., and Davis, R. J. (1994) Signal transduction by tumor necrosis factor mediated by JNK protein kinases. *Mol. Cell Biol.* **14**, 8376–8383.

105. Minden, A., Lin, A., Claret, F. X., Abo, A., and Karin, M. (1995) Selective activation of the JNK signaling cascade and c-Jun transcriptional activity by the small GTPases Rac and Cdc42Hs. *Cell* **81**, 1147–1157.

106. Raingeaud, J., Gupta, S., Rogers, J. S., Dickens, M., Han, J., Ulevitch, R. J., and Davis, R. J. (1995) Pro-inflammatory cytokines and environmental stress cause p38 mitogen-activated protein kinase activation by dual phosphorylation on tyrosine and threonine, *J. Biol. Chem.*, **270**, 7420–7426.

107. Reunanen, N., Westermarck, J., Hakkinen, L., Holmstrom, T. H., Elo, I., Eriksson, J. E., and Kahari, V. M. (1998) Enhancement of fibroblast collagenase (matrix metalloproteinase-1) gene expression by ceramide is mediated by extracellular signal-regulated and stress-activated protein kinase pathways. *J. Biol. Chem.* **273**, 5137–5145.

108. Ridley, S. H., Sarsfield, S. J., Lee, J. C., Bigg, H. F., Cawston, T. E., Taylor, D. J., DeWitt, D. L., and Saklatvala, J. (1997) Actions of IL-1 are selectively controlled by p38 mitogen-activated protein kinase: regulation of prostaglandin H synthase-2, metalloproteinases, and IL-6 at different levels. *J. Immunol.* **158**, 3165–3173.

109. Darnell, J. E., Jr. (1997) STATs and gene regulation, *Science,* **277**, 1630–1635.

110. Korzus, E., Nagase, H., Rydell, R., and Travis, J. (1997) The mitogen-activated protein kinase and JAK-STAT signaling pathways are required for an oncostatin M-responsive element-mediated activation of matrix metalloproteinase 1 gene expression. *J. Biol. Chem.* **272**, 1188–1196.

111. Yu, C. L., Meyer, D. J., Campbell, G. S., Larner, A. C., Carter-Su, C., Schwartz, J., and Jove, R. (1995) Enhanced DNA-binding activity of a Stat3-related protein in cells transformed by the Src oncoprotein. *Science* **269**, 81–83.

112. Troppmair, J., Bruder, J. T., Munoz, H., Lloyd, P. A., Kyriakis, J., Banerjee, P., Avruch, J., and Rapp, U. R. (1994) Mitogen-activated protein kinase/extracellular signal-regulated protein kinase activation by oncogenes, serum, and 12-O-tetradecanoylphorbol-13-acetate requires Raf and is necessary for transformation. *J. Biol. Chem.* **269**, 7030–7035.

113. Schlaepfer, D. D. and Hunter, T. (1997) Focal adhesion kinase overexpression enhances ras-dependent integrin signaling to ERK2/mitogen-activated protein kinase through interactions with and activation of c-Src. *J. Biol. Chem.* **272**, 13,189–91,315.

114. Daphna-Iken, D., and Morrison, A. R. (1995) Interleukin-1 beta induces interstitial collagenase gene expression and protein secretion in renal mesangial cells. *Am. J. Physiol.* **269**, F831–F837.

115. Vincenti, M. P., Coon, C. I., White, L. A., Barchowsky, A., and Brinckerhoff, C. E. (1996) Src-related tyrosine kinases regulate transcriptional activation of the interstitial collagenase gene, MMP-1, in interleukin-1-stimulated synovial fibroblasts. *Arthritis Rheum.* **39**, 574–582.

116. Sudbeck, B. D., Parks, W. C., Welgus, H. G., and Pentland, A. P. (1994) Collagen-stimulated induction of keratinocyte collagenase is mediated via tyrosine kinase and protein kinase C activities. *J. Biol. Chem.* **269**, 30,022–30,029.

117. Vincenti, M. P., White, L. A., Schroen, D. J., Benbow, U., and Brinckerhoff, C. E. (1996) Regulating expression of the gene for matrix metalloproteinase-1 (collagenase): mechanisms that control enzyme activity, transcription, and mRNA stability. *Crit. Rev. Euk. Gene Express.* **6**, 391–411.

118. Shlopov, B. V., Lie, W. R., Mainardi, C. L., Cole, A. A., Chubinskaya, S., and Hasty, K. A. (1997) Osteoarthritic lesions: involvement of three different collagenases. *Arthritis Rheum.* **40**, 2065–2074.

119. Airola, K., Johansson, N., Kariniemi, A. L., Kahari, V. M., and Saarialho-Kere, U. K. (1997) Human collagenase-3 is expressed in malignant squamous epithelium of the skin. *J. Invest. Dermatol.* **109**, 225–231.

120. Johansson, N., Airola, K., Grenman, R., Kariniemi, A. L., Saarialho-Kere, U., and Kahari, V. M. (1997) Expression of collagenase-3 (matrix metalloproteinase-13) in squamous cell carcinomas of the head and neck. *Am. J. Pathol.* **151**, 499–508.

121. Uria, J. A., Stahle-Backdahl, M., Seiki, M., Fueyo, A., and Lopez-Otin, C. (1997) Regulation of collagenase-3 expression in human breast carcinomas is mediated by stromal-epithelial cell interactions. *Cancer Res.* **57**, 4882–4888.

122. Johansson, N., Saarialho-Kere, U., Airola, K., Herva, R., Nissinen, L., Westermarck, J., Vuorio, E., Heino, J., and Kahari, V. M. (1997) Collagenase-3 (MMP-13) is expressed by hypertrophic chondrocytes, periosteal cells, and osteoblasts during human fetal bone development. *Developmental Dynamics* **208**, 387–397.

123. Stahle-Backdahl, M., Sandstedt, B., Bruce, K., Lindahl, A., Jimenez, M. G., Vega, J. A., and López-Otin, C. (1997) Collagenase-3 (MMP-13) is expressed during human fetal ossification and re-expressed in postnatal bone remodeling and in rheumatoid arthritis. *Lab. Invest,* **76**, 717–728.

124. Mitchell, P. G., Magna, H. A., Reeves, L. M., Lopresti-Morrow, L. L., Yocum, S. A., Rosner, P. J., Geoghegan, K. F., and Hambor, J. E. (1996) Cloning, expression, and type II collagenolytic activity of matrix metalloproteinase-13 from human osteoarthritic cartilage. *J. Clin. Invest.* **97**, 761–768.

125. Borden, P., Solymar, D., Sucharczuk, A., Lindman, B., Cannon, P., and Heller, R. A. (1996) Cytokine control of interstitial collagenase and collagenase-3 gene expression in human chondrocytes (published erratum appears in J Biol Chem 1996 Dec 27;271(52):33706). *J. Biol. Chem.* **271**, 23,577–23,581.

126. Lindy, O., Konttinen, Y. T., Sorsa, T., Ding, Y., Santavirta, S., Ceponis, A., and Lopez-Otin, C. (1997) Matrix metalloproteinase 13 (collagenase 3) in human rheumatoid synovium. *Arthritis Rheum.* **40**, 1391–1399.

127. Vaalamo, M., Mattila, L., Johansson, N., Kariniemi, A. L., Karjalainen-Lindsberg, M. L., Kahari, V. M., and Saarialho-Kere, U. (1997) Distinct populations of stromal cells express collagenase-3 (MMP-13) and collagenase-1 (MMP-1) in chronic ulcers but not in normally healing wounds. *J. Invest. Dermatol.* **109**, 96–101.

128. Saarialho-Kere, U. K., Kovacs, S. O., Pentland, A. P., Olerud, J. E., Welgus, H. G., and Parks, W. C. (1993) Cell-matrix interactions modulate interstitial collagenase expression by human keratinocytes actively involved in wound healing. *J. Clin. Invest.* **92**, 2858–2866.

129. Vogel, W., Gish, G. D., Alves, F., and Pawson, T. (1997) The discoidin domain receptor tyrosine kinases are activated by collagen. *Mol. Cell* **1**, 13–23.

130. Nie, D., Ishikawa, Y., Yoshimori, T., Wuthier, R. E., and Wu, L. N. (1998) Retinoic acid treatment elevates matrix metalloproteinase-2 protein and mRNA levels in avian growth plate chondrocyte cultures. *J. Cell Biochem.* **68**, 90–99.

131. Fosang, A. J., Last, K., Knauper, V., Murphy, G., and Neame, P. J. (1996) Degradation of cartilage aggrecan by collagenase-3 (MMP-13). *FEBS Letters* **380**, 17–20.

132. Ballock, R. T., Heydemann, A., Wakefield, L. M., Flanders, K. C., Roberts, A. B., and Sporn, M. B. (1994) Inhibition of the chondrocyte phenotype by retinoic acid involves upregulation of metalloprotease genes independent of TGF-beta. *J. Cell Phys.* **159**, 340–346.

133. Kelly, W. N., Harris, E. D., Ruddy, S., and Sledge, C. B. (1993) Textbook of Rheumatology. W.B. Saunders, Philadelphia.

134. Trentham, D. E. and Brinckerhoff, C. E. (1982) Augmentation of collagen arthritis by synthetic analogues of retinoic acid. *J. Immunol.* **129**, 2668–2672.

135. Mauviel, A., Chung, K. Y., Agarwal, A., Tamai, K., and Uitto, J. (1996) Cell-specific induction of distinct oncogenes of the Jun family is responsible for differential regulation of collagenase gene expression by transforming growth factor-beta in fibroblasts and keratinocytes. *J. Biol. Chem.* **271**, 10,917–10,923.

136. Karin, M. (1995) The Regulation of AP-1 Activity by Mitogen-activated Protein Kinases. *Cell* **270**, 16,483–16,486.

137. Wilson, C. L., Heppner, K. J., Rudolph, L. A., and Matrisian, L. M. (1995) The metalloproteinase matrilysin is preferentially expressed by epithelial cells in a tissue-restricted pattern in the mouse. *Mol. Biol. Cell* **6**, 851–869.

138. Rodgers, W. H., Osteen, K. G., Matrisian, L. M., Navre, M., Giudice, L. C., and Gorstein, F. (1993) Expression and localization of matrilysin, a matrix metalloproteinase, in human endometrium during the reproductive cycle. *Am. J. Ostet. Gynecol.* **168**, 253–260.

139. Woessner, J. F., Jr. (1996) Regulation of matrilysin in the rat uterus. *Biochem. Cell Biol.* **74**, 777–784.

140. Saarialho-Kere, U. K., Crouch, E. C., and Parks, W. C. (1995) Matrix metalloproteinase matrilysin is constitutively expressed in adult human exocrine epithelium. *J. Invest. Dermatol.* **105**, 190–196.

141. Basset, P., Bellocq, J. P., Wolf, C., Stoll, I., Hutin, P., Limacher, J. M., Podhajcer, O. L., Chenard, M. P., Rio, M. C., and Chambon, P. (1990) A novel metalloproteinase gene specifically expressed in stromal cells of breast carcinomas. *Nature* **348**, 699–704.

142. McDonnell, S., Navre, M., Coffey, R. J., Jr., and Matrisian, L. M. (1991) Expression and localization of the matrix metalloproteinase pump-1 (MMP-7) in human gastric and colon carcinomas. *Mol. Carcinogen.* **4**, 527–533.

143. Muller, D., Breathnach, R., Engelmann, A., Millon, R., Bronner, G., Flesch, H., Dumont, P., Eber, M., and Abecassis, J. (1991) Expression of collagenase-related metalloproteinase genes in human lung or head and neck tumours. *Int. J. Cancer* **48**, 550–556.

144. Karelina, T. V., Goldberg, G. I., and Eisen, A. Z. (1994) Matrilysin (PUMP) correlates with dermal invasion during appendageal development and cutaneous neoplasia. *J. Invest. Dermatol.* **103**, 482–487.

145. Powell, W. C., Knox, J. D., Navre, M., Grogan, T. M., Kittelson, J., Nagle, R. B., and Bowden, G. T. (1993) Expression of the metalloproteinase matrilysin in DU-145 cells increases their invasive potential in severe combined immunodeficient mice. *Cancer Res.* **53**, 417–422.

146. Witty, J. P., McDonnell, S., Newell, K. J., Cannon, P., Navre, M., Tressler, R. J., and Matrisian, L. M. (1994) Modulation of matrilysin levels in colon carcinoma cell lines affects tumorigenicity in vivo. *Cancer Res.* **54**, 4805–4812.

147. Shapiro, S. D., Griffin, G. L., Gilbert, D. J., Jenkins, N. A., Copeland, N. G., Welgus, H. G., Senior, R. M., and Ley, T. J. (1992) Molecular cloning, chromosomal localization, and bacterial expression of a murine macrophage metalloelastase. *J. Biol. Chem.* **267**, 4664–4671.

148. Hautamaki, R. D., Kobayashi, D. K., Senior, R. M., and Shapiro, S. D. (1997) Requirement for macrophage elastase for cigarette smoke-induced emphysema in mice. *Science* **277**, 2002–2004.

149. Shapiro, S. D. (1994) Elastolytic metalloproteinases produced by human mononuclear phagocytes. Potential roles in destructive lung disease. *Am. J. Resp. & Crit. Care Med.* **150**, S160–164.

150. Shipley, J. M., Wesselschmidt, R. L., Kobayashi, D. K., Ley, T. J., and Shapiro, S. D. (1996) Metalloelastase is required for macrophage-mediated proteolysis and matrix invasion in mice. *Proc. Nat. Acad. Sci. USA* **93**, 3942–3946.

151. De Clerck, Y. A., Shimada, H., Taylor, S. M., and Langley, K. E. (1994) Matrix metalloproteinases and their inhibitors in tumor progression. *Annals of the New York Academy of Sciences* **732**, 222–232.

152. Vincenti, M. P., Clark, I. M., and Brinckerhoff, C. E. (1994) Using inhibitors of metalloproteinases to treat arthritis. Easier said than done? (see comments). *Arthritis Rheum.* **37**, 1115–1126.

153. Rasmussen, H. S. and McCann, P. P. (1997) Matrix metalloproteinase inhibition as a novel anticancer strategy: a review with special focus on batimastat and marimastat. *Pharmacol. Therapeut.* **75**, 69–75.

154. Gross, J. and Lapiere, C. (1962) Collagenolytic activity in amphibian tissues: a tissue culture assay. *Proc. Natl. Acad. Sci.* **48**, 1014–1022.

155. Gross, R. H., Sheldon, L. A., Fletcher, C. F., and Brinckerhoff, C. E. (1984) Isolation of collagenase cDNA clone and measurement of changing collagenase mRNA levels during induction in rabbit synovial fibroblasts. *Proc. Natl. Acad. Sci. USA* **81**, 1981–1985.

156. Goldberg, G., Wilhelm, S. M., A., K., Bauer, E. A., Grant, G. A., and Eisen, A. Z. (1986) Human fibroblast collagenase: complete primary structure and

homology to an oncogene transforation-induced rat protein. *J. Biol. Chem.* **261**, 5645–5650.

157. Woessner, J. F., Jr. (1991) Matrix metalloproteinases and their inhibitors in connective tissue remodeling. *Faseb. J.* **5**, 2145–2154.

158. Nagase, H. (1997) Activation Mechanisms of Matrix Metalloproteinases. *Biol Chem.* **378**, 151–160.

159. Collier, I. E., Bruns, G. A., Goldberg, G. I., and Gerhard, D. S. (1991) On the structure and chromosome location of the 72- and 92-kDa human type IV collagenase genes. *Genomics.* **9**, 429–434.

160. Matrisian, L. M., Leroy, P., Ruhlmann, C., Gesnel, M. C., and Breathnach, R. (1986) Isolation of the oncogene and epidermal growth factor-induced transin gene: complex control in rat fibroblasts. *Mol. Cell Biol.* **6**, 1679–1686.

161. Sanchez-Lopez, R., Nicholson, R., Gesnel, M. C., Matrisian, L. M., and Breathnach, R. (1988) Structure-function relationships in the collagenase family member transin. *J. Biol. Chem.* **263**, 11,892–11,899.

162. Muller, D., Quantin, B., Gesnel, M. C., Millon-Collard, R., Abecassis, J., and Breathnach, R. (1988) The collagenase gene family in humans consists of at least four members. *Biochem. J.* **253**, 187–192.

163. Quantin, B., Murphy, G., and Breathnach, R. (1989) Pump-1 cDNA codes for a protein with characteristics similar to those of classical collagenase family members. *Biochem.* **28**, 5327–5334.

164. Hasty, K. A., Pourmotabbed, T. F., Goldberg, G. I., Thompson, J. P., Spinella, D. G., Stevens, R. M., and Mainardi, C. L. (1990) Human neutrophil collagenase. A distinct gene product with homology to other matrix metalloproteinases. *J. Biol. Chem.* **265**, 11,421–11,424.

165. Wilhelm, S. M., Collier, I. E., Marmer, B. L., Eisen, A. Z., Grant, G. A., and Goldberg, G. I. (1989) SV40-transformed human lung fibroblasts secrete a 92-kDa type IV collagenase which is identical to that secreted by normal human macrophages (published erratum appears in J. Biol. Chem. 1990 Dec 25;265(36):22570). *J. Biol. Chem.* **264**, 17,213–17,221.

166. Masure, S., Nys, G., Fiten, P., Van Damme, J., and Opdenakker, G. (1993) Mouse gelatinase B. cDNA cloning, regulation of expression and glycosylation in WEHI-3 macrophages and gene organisation. *Eur. J. Biochem.* **218**, 129–141.

167. Breathnach, R., Matrisian, L. M., Gesnel, M. C., Staub, A., and Leroy, P. (1987) Sequences coding for part of oncogene-induced transin are highly conserved in a related rat gene, *Nucleic Acids Res.* **15**, 1139–1151.

168. Quinn, C. O., Scott, D. K., Brinckerhoff, C. E., Matrisian, L. M., Jeffrey, J. J., and Partridge, N. C. (1990) Rat collagenase. Cloning, amino acid sequence comparison, and parathyroid hormone regulation in osteoblastic cells. *J. Biol. Chem.* **265**, 22,342–2237.

169. Pendas, A. M., Matilla, T., Estivill, X., and López-Otín, C. (1995) The human collagenase-3 (CLG3) gene is located on chromosome 11q22.3 clustered to other members of the matrix metalloproteinase gene family. *Genomics* **26**, 615–618.

170. Sato, H., Takino, T., Okada, Y., Cao, J., Shinagawa, A., Yamamoto, E., and Seiki, M. (1994) A matrix metalloproteinase expressed on the surface of invasive tumour cells (see comments). *Nature* **370**, 61–65.

171. Will, H. and Hinzmann, B. (1995) cDNA sequence and mRNA tissue distribution of a novel human matrix metalloproteinase with a potential transmembrane segment. *Eur. J. Biochem.* **231**, 602–608.

172. Takino, T., Sato, H., Shinagawa, A., and Seiki, M. (1995) Identification of the second membrane-type matrix metalloproteinase (MT-MMP–2) gene from a human placenta cDNA library. MT-MMPs form a unique membrane-type subclass in the MMP family. *J. Biol. Chem.* **270**, 23,013–23,020.

173. Puente, X. S., Pendas, A. M., Llano, E., Velasco, G., and López-Otín, C. (1996) Molecular cloning of a novel membrane-type matrix metalloproteinase from a human breast carcinoma. *Cancer Res.* **56**, 944–949.

174. Stolow, M. A., Bauzon, D. D., Li, J., Sedgwick, T., Liang, V. C., Sang, Q. A., and Shi, Y. B. (1996) Identification and characterization of a novel collagenase in Xenopus laevis: possible roles during frog development. *Mol. Biol. Cell* **7**, 1471–1483.

175. Yang, M., Murray, M. T., and Kurkinen, M. (1997) A novel matrix metalloproteinase gene (XMMP) encoding vitronectin-like motifs is transiently expressed in Xenopus laevis early embryo development. *J. Biol. Chem.* **272**, 13,527–13,533.

176. Cossins, J., Dudgeon, T. J., Catlin, G., Gearing, A. J., and Clements, J. M. (1996) Identification of MMP-18, a putative novel human matrix metalloproteinase. *Biochem. Biophys. Res. Com.* **228**, 494–498.

177. Pendas, A. M., Knauper, V., Puente, X. S., Llano, E., Mattei, M. G., Apte, S., Murphy, G., and Lopez-Otin, C. (1997) Identification and characterization of a novel human matrix metalloproteinase with unique structural characteristics, chromosomal location, and tissue distribution. *J. Biol. Chem.* **272**, 4281–4286.

178. Llano, E., Pendas, A. M., Knauper, V., Sorsa, T., Salo, T., Salido, E., Murphy, G., Simmer, J. P., Bartlett, J. D., and Lopez-Otin, C. (1997) Identification and structural and functional characterization of human enamelysin (MMP–20). *Biochem* **36**, 15,101–15,118.

179. Docherty, A. J., Lyons, A., Smith, B. J., Wright, E. M., Stephens, P. E., Harris, T. J., Murphy, G., and Reynolds, J. J. (1985) Sequence of human tissue inhibitor of metalloproteinases and its identity to erythroid-potentiating activity. *Nature* **318**, 66–69.

180. Goldberg, G. I., Strongin, A., Collier, I. E., Genrich, L. T., and Marmer, B. L. (1992) Interaction of 92-kDa type IV collagenase with the tissue inhibitor of metalloproteinases prevents dimerization, complex formation with interstitial collagenase, and activation of the proenzyme with stromelysin. *J. Biol. Chem.* **267**, 4583–4591.

181. Gomez, D. E., Alonso, D. F., Yoshiji, H., and Thorgeirsson, U. P. (1997) Tissue inhibitors of metalloproteinases: structure, regulation and biological functions. *Eur. J. Cell Biol.* **74**, 111–122.

182. Stetler-Stevenson, W. G., Krutzsch, H. C., and Liotta, L. A. (1989) Tissue inhibitor of metalloproteinase (TIMP-2). A new member of the metalloproteinase inhibitor family. *J. Biol. Chem.* **264**, 17,374–17,378.
183. Goldberg, G. I., Marmer, B. L., Grant, G. A., Eisen, A. Z., Wilhelm, S., and He, C. S. (1989) Human 72-kilodalton type IV collagenase forms a complex with a tissue inhibitor of metalloproteases designated TIMP-2. *Proc. Nat. Acad. Sci. USA* **86**, 8207–8211.
184. Zucker, S., Drews, M., Conner, C., Foda, H. D., DeClerck, Y. A., Langley, K. E., Bahou, W. F., Docherty, A. J., and Cao, J. (1998) Tissue inhibitor of metalloproteinase-2 (TIMP-2) binds to the catalytic domain of the cell surface receptor, membrane type 1-matrix metalloproteinase 1 (MT1-MMP). *J. Biol. Chem.* **273**, 1216–1222.
185. Apte, S. S., Mattei, M. G., and Olsen, B. R. (1994) Cloning of the cDNA encoding human tissue inhibitor of metalloproteinases-3 (TIMP-3) and mapping of the TIMP3 gene to chromosome 22. *Genomics* **19**, 86–90.
186. Apte, S. S., Olsen, B. R., and Murphy, G. (1995) The gene structure of tissue inhibitor of metalloproteinases (TIMP)-3 and its inhibitory activities define the distinct TIMP gene family (published erratum appears in J Biol Chem 1996 Feb 2;271(5):2874). *J. Biol. Chem.* **270**, 14,313–14,318.
187. Apte, S. S., Hayashi, K., Seldin, M. F., Mattei, M. G., Hayashi, M., and Olsen, B. R. (1994) Gene encoding a novel murine tissue inhibitor of metalloproteinases (TIMP), TIMP-3, is expressed in developing mouse epithelia, cartilage, and muscle, and is located on mouse chromosome 10. *Developmental Dynamics.* **200**, 177–197.
188. Apte, S. S., Olsen, B. R., and Murphy, G. (1996) The gene structure of tissue inhibitor of metalloproteinases (TIMP)-3 and its inhibitory activities define the distinct TIMP gene family. *J. Biol. Chem.* **271**, 2874.
189. Greene, J., Wang, M., Liu, Y. E., Raymond, L. A., Rosen, C., and Shi, Y. E. (1996) Molecular cloning and characterization of human tissue inhibitor of metalloproteinase 4. *J. Biol. Chem.* **271**, 30,375–30,380.
190. Leco, K. J., Apte, S. S., Taniguchi, G. T., Hawkes, S. P., Khokha, R., Schultz, G. A., and Edwards, D. R. (1997) Murine tissue inhibitor of metalloproteinases-4 (Timp-4): cDNA isolation and expression in adult mouse tissues. *FEBS Letters* **401**, 213–217.

6

Models for Gain-of-Function and Loss-of-Function of MMPs

Transgenic and Gene Targeted Mice

Lisa M. Coussens, Steven D. Shapiro, Paul D. Soloway, and Zena Werb

1. Introduction

The most powerful approach for studying gene function in an intact animal is to regulate the levels of the gene product and thereby see gains-of-function or losses-of-function. The occasional mutation in the genes for the matrix metalloproteinases or their inhibitors, or polymorphism in their promoters that alter transcriptional regulation has been identified in humans and has helped define the function of these proteins. With ever increasing sophistication in producing targeted mutations in mice, there are now available null mutation in most of the known genes for the matrix metalloproteinases and their inhibitors. A number of mouse strains with ectopic expression of normal and mutant proteins have also been made. These transgenic mice are giving us new insights into the processes of development and pathogenesis.

2. Matrix Metalloproteinases

Mice with a transgenic gain-of-function or with a targeted gene knockout provide means to modulate matrix metalloproteinase (MMP) expression specifically in vivo, permitting the determination of MMP function in health and disease. **Table 1** summarizes the mice that have transgenic and gene targeted MMPs and their phenotypes to date. Together these studies have begun to uncover the role of MMPs in a variety of physiologic and pathologic processes.

From: *Methods in Molecular Biology, vol. 151: Matrix Metalloproteinase Protocols*
Edited by: I. Clark © Humana Press Inc., Totowa, NJ

Table 1
Phenotypes of MMP Transgenic and Gene Targeted Mice

Mouse		Phenotype	Refs.
Transgenic overexpression			
MMP-1	**Lung:**	Enlarged airspaces/emphysema	*(58)*
	Skin:	Acanthosis, hyperkeratosis, increased susceptibility to tumorigenesis	*(30)*
MMP-3	**Breast:**	Enhancedmammary gland involution, apoptosis	*(32–34)*
		Precocious alveolar branching morphogenesis	
		Mammary gland tumor formation	
MMP-7	**Breast:**	Increased β-casein expression in virgins	*(26)*
	Testis:	Disorganized testis and disintegration of interstitial tissue	
		Surrounding seminiferous tubules, and infertility	
MMP-9		DNA 5' flanking sequences –522–+19 direct appropriate reporter expression	*(4)*
		Developmental and injury-induced expression of LacZ reporter	
Gene targeted knock-Out			
MMP-2/null		Reduced angiogenesis and tumor progression	*(43)*
MMP-3/null		Contact dermatitis: Impaired onset/T cell proliferation	*(62)*
		Diminished tumor incidence and reduced malignancy	*([a])*
MMP-7/null		Decreased intestinal tumorigenesis	*(24)*
		Impaired tracheal wound repair	*(13)*
MMP-9/null		Impaired primary angiogenesis in bone growth plates	*(1)*
		Resistant to bullous pemphigoid	*(44)*
		Normal neutrophil migration, and lung renal development *(3)*	
		Contact dermatitis: Delayed resolution	*(62)*
		Diminished tumor incidence with enhanced malignant potential	*([a])*
MMP-11/null		Decreased chemical induced mutagenesis	*(40)*
MMP-12/null		Impaired macrophage proteolysis	*(18)*
		Resistant to cigarette smoke-induced emphysema	*(59)*
Related null mutants/transgenics			
c-Fos inducible overexpression		Induced collagenase expression in bone	*(6)*
C'ase-resistant type I collagen		Marked dermal fibrosis	*(6)*
		Impaired post-partum uterine involution	
		MMP-13 N-telopeptide cleavage accounts for bone resorption during embryonic and early life	
uPA/null		Unable to activate pro-MMPs	*(17)*
		On ApoE[-/-] background, protected from atherosclerotic macrophage infiltration and microaneurysm formation	

[a]Coussens, Hanahan, and Werb, unpublished observations.

150

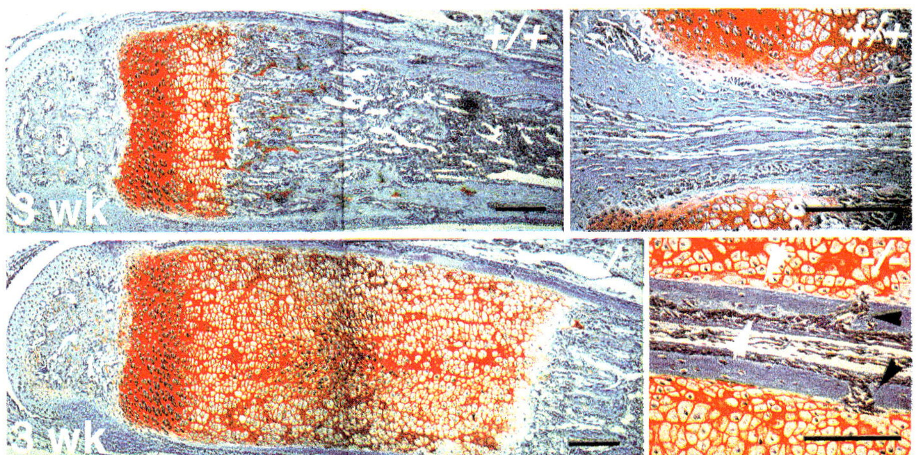

Fig. 1. Histological analysis of the metatarsals of wild type and gelatinase B-null mice. Histological sections of 3-wk-old metatarsals. hc: hypertrophic cartilage, tb: trabecular bone. Magnified view on the right lower panel shows cortical bone surrounding the hypertrophic cartilage in gelatinase B-null mice. Black arrows point to capillaries invading from the cortical bone. Wild-type (+/+), knockout mice (–/–).

2.1. Physiological Processes

MMPs, which are usually undetectable in adult mammals under normal circumstances, are prominently expressed during reproduction. From the maternal side, MMPs are associated with menstruation, ovulation, uterine implantation, parturition, and postpartum uterine and mammary gland involution. From the offspring's perspective, MMPs are believed to be required for trophoblast implantation, embryonic growth, and tissue morphogenesis. Yet, none of the individual MMP mutant mice generated to date has an embryonic lethal phenotype. Mice deficient in gelatinase B (MMP-9/null) demonstrate morphologic abnormalities at the site of implantation, but these defects are not lethal. Moreover, all mothers null for MMPs are capable of delivering and nurturing healthy pups.

2.1.1. Embryonic Development

2.1.1.1. GELATINASE B (MMP-9)

The major defect in MMP-9/null mice is delayed long bone growth and development *(1)*. Long bones develop from mesenchymal condensations where cartilage cells differentiate and deposit a cartilage matrix. Blood vessels invade and degrade the cartilage matrix, cartilage cells undergo apoptosis followed by proliferation of osteoblasts and endochondral ossification converting the tissue into mature bone. In MMP-9/null mice, there is delayed vascular invasion in skeletal growth plates, resulting in an excessively wide zone of hypertrophic cartilage and delayed ossification (*see* **Fig. 1**). MMP-9 is required to initiate primary angiogenesis in the cartilage growth plate, probably through generation of an angiogenic signal (or less likely degradation of an angiogenesis inhibitor). The mechanism of this phenotype may not involve degradation of a structural or adhesive matrix protein.

Although this phenotype is marked during growth, adult mice show only a 10% shortening in the long bones. This is not meant to diminish the importance of this finding but, rather it emphasizes that until careful analyses of all MMP/ null mice are performed conclusions regarding the role of individual MMPs in growth and development are premature. The overall minimal phenotypes observed to date may be due to overlapping functions of MMPs expressed in the same organs, safeguarding the host from untoward consequences of individual MMP mutations. Generation of double and multiple MMP/null mice may be required to unmask full MMP functions. Alternatively, MMPs may not be needed for grossly normal development and growth in the mouse.

Of note, MMP-9 is present in the mesenchyme of embryonic kidneys, and branching morphogenesis of ureteric buds is specifically blocked in metanephric organ culture by antisera to MMP-9 but not anti-MMP-2 IgG *(2)*. However, no abnormalities are found upon analysis of neonatal and adult MMP-9/ null mice by light microscopy or immunofluorescence for basement membrane proteins. Renal function in adult mice is also normal *(3)*. Although the discrepancies between these studies may result from differences in study design; in vivo versus in vitro, or antibodies overestimate whereas knockouts underestimate abnormalities, it is clear that the MMP-9/null mice do show developmental delay in embryonic kidney morphogenesis (B. Lelongt, P. Ronco and Z. Werb, unpublished observations).

Transgenic technology may also be used to determine promoter elements required for tissue and developmental-specific expression in vivo. DNA sequences –522 to +19 (containing NF-κB, AP-2, and Sp1 sites) in the 5' flanking region of the rabbit MMP-9 promoter are necessary and sufficient to direct appropriate developmental and injury-induced expression of the lacZ reporter in transgenic mice *(4)*. Prominent localization of LacZ in developing bones and neural tissue largely parallel endogenous gene expression. β-galactosidase activity is also induced in an in vivo acute full-thickness skin wound model. MMP-9 promoter activity is localized to basal epidermal keratinocytes at the migrating front. Transgene expression in vivo differs in comparison to migrating epithelial cells in culture. Much of the DNA between –522 to +19 is not required in vitro, but is essential in vivo. In culture, promoter activity is never silent, expression is uniform throughout the epithelium. This study highlights the limitations of extrapolating findings in cell culture to complex organisms.

2.1.1.2. COLLAGENASES

The collagen content of different tissues during development and in the adult is tightly regulated by synthesis and degradation. The profile of mouse collagenases appears to be quite different than in human. Collagenase 3 (MMP-13) is the dominant mouse collagenase, however, recently expression of mouse collagenase-2 (MMP-8) has been reported *(5)*. MMP-8 expression is detected during the late stages of mouse embryogenesis, coinciding with the appearance of hematopoietic cells. In addition, MMP-8 is highly expressed in postpartum uterus (d 1–5). To date, collagenase 1 (MMP-1) has not been detected in rodents. However, overexpression of human collagenase 1 (MMP-1) in transgenic mice results in pathologic phenotypes in the lung and skin (*see* **Subheading 2.2.3.** and **Table 1**).

Gene targeting of mouse collagenases is not yet complete, but over-expression of a transgene encoding a collagenase-resistant type I collagen results in late embryonic lethality *(6)*. Replacement of native collagen with this mutant form by homologous recombination was not lethal, but did result in impaired postpartum uterine involution with persistent collagenous bundles in the uterine wall *(6)*. As the mutant mice aged they also developed dermal fibrosis, similar to scleroderma. Of note, the mutation in both the transgene and gene targeted mutant is at the Gly75-Ile76 collagenase cleavage site of the $\alpha 1$(I) chain. However, MMP-13 also cleaves collagen in the nonhelical N-telopeptide domain. Thus, degradation of collagen is possible in these studies, perhaps leading to underestimation of the requirement of collagenases during development.

2.1.2. Mammary Gland Development

MMPs are highly expressed during mammary gland development and involution *(7)*. Over-expression of stromelysin-1/MMP-3 produces precocious development in virgin mice and unscheduled involution during pregnancy *(8–10)*. These phenotypes are reversed in animals over-expressing TIMP-1 driven by the ß actin promoter *(9)*. The MMP nulls obtained to date have very mild mammary gland phenotypes, suggesting that several MMPs may contribute to the proteinase-dependent MMP functions. The MMP-3/null mice exhibit decreased branching morphogenesis during postpubertal development and accelerated adipocyte differentiation during involution (C. M. Alexander, M. Sciabica, and Z. Werb, unpublished observations).

2.1.3. Wound Repair

In epithelial cell migration during normal wound healing, Parks and coworkers hypothesize that interaction of keratinocyte integrins $\alpha 2\beta 1$ with native type I collagen in a provisional wound matrix induces collagenase-1 (MMP-1) expression *(11)*. By cleaving collagen, the initial high affinity contact is loosened, releasing the cell, which then migrates to "grab" high affinity $\alpha2\beta1$ bonds of undigested collagen ahead in the open wound. Indeed, keratinocytes can migrate on native collagen but not on a collagenase-resistant collagen matrix *(12)*. Whether collagenases are required for normal wound healing of the skin in mice awaits further study and gene targeting.

In the lung, in contrast to skin, matrilysin (MMP-7) is the sole MMP expressed by airway epithelial cells *(13)*. MMP-7 expression increases in migrating epithelial cells in wounded human and mouse tracheas. In wounded human tissue, hydroxamates inhibit epithelial cell migration >80%. In mice, reepithelialization of wounded trachea is entirely absent in MMP-7/null mutant mice. These studies demonstrate the requirement for MMPs in normal wound repair and highlight tissue specific differences in MMP involvement.

2.2. Pathological Processes

2.2.1. Cardiovascular Disease

Atherosclerosis is a chronic inflammatory process whereby plaques are formed in the intimal layer of the vessel wall as a result of accumulation of extracellular matrix (ECM), smooth muscle cells, and lipid-laden macrophages. In coronary arteries plaques may become unstable and rupture, triggering intravascular thrombosis leading to myocardial infarction. The atherosclerotic vessel wall may also dilate due to destruction of the media, leading to aneurysm formation and rupture of the weakened vessel wall. Recently, plasminogen activators and several MMPs have been detected in association with human atherosclerotic arteries *(14)* and abdominal aortic aneurysms *(15)*.

Mice with a targeted disruption of the apolipoprotein E gene (ApoE/null) have a delayed clearance of lipoproteins. When fed a "Western diet," serum cholesterol levels reach 1,400–2,000 mg/dL, and fatty streaks progressing to fibrous plaques develop at branch points of major vessels. The formation of this lesion is associated with macrophage recruitment causing disruption of the medial elastic lamina and microaneurysm formation. Complex lesions with plaque rupture and hemorrhage have yet to be observed for any model of atherosclerosis in the mouse *(16)*.

To investigate the role of plasminogen activators [tissue-type (t-PA) and urokinase-type (u-PA)], Carmeleit and colleagues crossed ApoE/null mice with u-PA/null and t-PA/null mice. ApoE/null X u-PA/null mice, but not ApoE/null or ApoE/null X t-PA/null are protected from macrophage-mediated destruction of medial elastic lamina and microaneurysm formation *(17)*. Macrophages line up along elastic lamina, but they do not penetrate or disrupt these matrix structures. Because the ability of macrophages to degrade and migrate through elastin is most likely due to macrophage elastase (MMP-12) *(18)*, rather than nonelastolytic plasmin, these findings suggest that plasmin activates MMP proenzymes. Indeed, in the absence of u-PA, macrophages are unable to convert macrophage pro-MMPs (MMP-3, MMP-9, MMP-12, and MMP-13) into their active forms in a reconstituted system. Generation of ApoE/null X MMP/null mice is currently in progress to determine directly the role of MMPs in development of these vascular lesions.

Interestingly, a human polymorphism in the stromelysin-1 (MMP-3) promoter, which diminishes stromelysin-1 expression, is associated with enhanced progression of atherosclerosis *(19)*. Together, these and other studies suggest that MMPs initially maintain patency of the atherosclerotic vascular lumen at the risk of subsequent plaque rupture.

2.2.2. Cancer

An emerging view of epithelial tumorigenesis includes the realization that tumors evolve in a multi-step manner and are composed not only of neoplastic epithelial cells, but also of inflammatory cells (lymphocytes, macrophages, mast cells, neutrophils), and cells of mesenchymal origin (fibroblasts, smooth muscle, and endothelial cells). Malignant progression of neoplastic cells is largely driven by intrinsic factors, e.g., expression of oncogenes and/or loss of expression of tumor suppressor genes. Such changes typically occur early in the evolution of a neoplastic mass and are involved in resultant phenotypic changes leading to acquisition of malignant traits, namely, the ability to invade into and grow in ectopic tissue environments. Whereas cancer research has historically focused on such intrinsic factors, it has become increasingly clear that the microenvironment surrounding the neoplastic mass is an integral partner in the neoplastic process. Consequently, extrinsic factors, such as extracellular proteinases, e.g., MMPs, whose expression in neoplasms is largely of stromal cell origin (**Table 2**), are now believed to regulate key events other than tumor cell invasion.

Utilizing the power of mouse genetics to generate gain of function/overexpression transgenic mice and loss of function/under-expression gene-knockout mice, we are now getting our first glimpses into how significant the contribution by MMPs and the stromal microenvironment is regarding *de novo* malignancy. To date, there are only a handful of studies utilizing transgenic and gene deficient MMP mice in the context of neoplastic progression (**Table 1**). These studies have begun to demonstrate causal roles for MMPs in such pathological processes, as well as unmasking exciting and unanticipated biological functions for MMPs, well beyond matrix destruction.

Table 2
MMP Expression In Human Epithelial Tumors

	Breast	Colorectal	Head and neck	Skin	
				BCC	SCC
Collagenases					
Collagenase-1	F	C, F	F	F	C, F
Collagenase-2	nd	I, M	nr	nr	nr
Collagenase-3	E	nr	E, F	E	nr
Collagenase-4	nr	nr	nr	nr	nr
Gelatinases					
Gelatinase A	F	F	C, F	F	F
Gelatinase B	C, I, M	I, M	I	E, I	E, F, I
Stromelysins 1 and 2					
Stromelysin-1	F	C, F	C, F	F	nr
Stromelysin-2	nd	nd	F	nr	nr
Membrane-type MMPs					
MT1-MMP	F	F	C, F	F	F
MT2-MMP	nr	nr	nr	nr	nr
MT3-MMP	nr	nr	nr	nr	nr
MT4-MMP	nr	nr	nr	nr	nr
Other MMPs					
Matrilysin	E	E	nd	E	nd
Stromelysin-3	F	F	F	F	F
Metalloelastase	M	M	M	M	M
Enamelysin	nr	nr	nr	nr	nr
MMP-19	nr	nr	nr	nr	nr

[a]Basal Cell Carcinoma, BCC; Squamous Cell Carcinoma, SCC; Epithelial, E; Capillary-associated, C; Fibroblast, F; Inflammatory cell, I; Macrophage, M; not detected, nd; not reported, nr.
[b]Data from references *28, 29, 56, 108*.

2.2.2.1. MATRILYSIN (MMP-7)

Whereas most MMPs are expressed primarily by cells of mesenchymal and/ or monocytic origin, matrilysin (MMP-7) mRNA is readily detected in glandular epithelia, e.g., breast and colon, in normal tissue with increased expression observed in adenomas, carcinomas, and adenocarcinomas *(20)* of humans *(21)* and mice *(22)*. Ectopic expression of matrilysin cDNA in colon carcinoma cell lines increases tumorgenicity although not affecting metastasis in nude mice *(23)*. Matrilysin-deficient mice *(24)*, although harboring no apparent phenotype, when crossed with mice carrying the APC*min* mutation, predisposed to intestinal adenomas, resulted in an approx 60% reduction in mean tumor multiplicity and average tumor diameter *(24)*. Surprisingly, matrilysin immunolocalizes to the apical rather than basal surface of premalignant cells, suggesting that it may promote tumor development by degrading molecules other than those of the extracellular matrix *(24)*. Matrilysin is also located apically in normal epithelium, again suggesting the existence of apical non-matrix substrates *(21,25)*. Mice that overexpress a matrilysin transgene driven by the mouse mammary tumor virus (MMTV) promoter exhibit precocious mammary gland differentiation, premalignant mammary gland lesions, and male infertility *(25,26)*. In addition, mammary tumorigenesis in transgenic mice that express an MMTV-driven *neu* oncogene was accelerated when these MMTV-*neu* mice were crossed with the matrilysin overexpressing MMTV-matrilysin transgenic mice, again suggesting a role for matrilysin in promoting early tumor formation *(26)*.

2.2.2.2. COLLAGENASES

Four MMP collagenases have been identified: collagenase-1 (MMP-1), collagenase-2 (MMP-8), collagenase-3 (MMP-13), and Xenopus collagenase-4 (MMP-18). Collagenases -1, -2, and -3 are detected in humans, and collagenase-2 *(27)* and -3 are found in rodents *(28)*. Due to their unique ability to readily cleave the triple-helical domain of native fibrillar collagens, e.g., collagens I, II, and III, these enzymes have historically been thought of as key matrix destabilizers facilitating invasion of diverse cell types *(28)*. In human malignancies, expression of collagenase-1 is restricted to reactive stromal cells, and in colorectal cancer is associated with poor prognosis *(29)*. In contrast, collagenase-2 expression appears restricted to inflammatory cells, and collagenase-3 mRNA detectable in both tumor epithelia and reactive fibroblasts (**Table 2**). Thus, whereas these genes share considerable homology, their regulation is not coordinate. Additional insight into collagenase function has come from transgenic expression of human collagenase-1 in mouse skin *(30)*. These mice develop hyperproliferative skin lesions that are increasingly sensitive to malignant conversion induced by chemical carcinogens *(30)*. Interestingly expression of this gene product, while normally not observed in skin epithelia, does not result in altered keratinocyte differentiation (as does over-expression of matrilysin in the mammary gland) or in frank invasion of keratinocytes into underlying stroma prior to treatment with carcinogens. This implies that some targets for collagenase-1, when cleaved, can release proliferation and/or cell cycle restraints imposed by the local microenvironment. Furthermore, this data supports a role for this MMP as an "initiator," inducing proliferation and rendering cells more sensitive to carcinogen-induced promotion.

2.2.2.3. STROMELYSINS

Stromelysin-1 (MMP-3) is exclusively expressed by reactive stromal cells during neoplastic progression (**Table 2**). MMP-3/null mice exhibit hypomorphic virgin and pregnant mammary glands, altered involution, and delayed closure of excisional wounds (M. Sciabica, M. Sternlicht, and Z. Werb, unpublished observations). Furthermore, when crossed with transgenic mice at risk for squamous cell carcinomas *(31)*, loss of MMP-3 results in a diminution of both tumor incidence and malignant grade, without apparent consequence to earlier neoplastic stages prior to malignant conversion (Coussens, Hanahan, and Werb, unpublished observations). In contrast, mice that express an auto-activating stromelysin-1 transgene targeted to mammary epithelium display precocious virgin glandular development, unscheduled apoptosis during mid-pregnancy, formation of a reactive stroma, and the development of premalignant and malignant mammary gland lesions *(32–36)*. Moreover, stromelysin-1 can trigger epithelial-to-mesenchymal phenotypic conversion of cultured mammary epithelial cells and induces otherwise noninvasive and nontumorigenic cells to form highly infiltrative tumors in immunocompromised mice *(35,37)*. These data indicate that in addition to its traditional proinvasive activity *(37)*, stromelysin-1 can initiate the development of early premalignant lesions in the breast and can foster late phenotypic changes associated with more aggressive tumor behavior.

Stromelysin-3 (MMP-11) was initially identified from a subtracted breast cancer cDNA library and found to be exclusively expressed by stromal cells surrounding invasive tumors *(38)*. It represents an independent prognostic factor for disease-free survival in certain breast cancers, and even though it is normally expressed by adjacent stromal cells of mesenchymal origin rather than the malignant epithelial cells themselves, it is apparently induced in certain advanced cancers when the epithelial tumor cells undergo epithelial-to-mesenchymal conversion *(39)*. Moreover, MMP-11/null mice develop fewer and smaller chemically-induced tumors than wild-type mice, and whereas wild-type fibroblasts foster the tumorgenicity of human MCF7 breast cancer cells in nude mice, stromelysin-3-deficient fibroblasts do not *(40)*. Further evidence indicates that this apparent tumor promoting activity of stromelysin-3 derives from its paracrine ability to release and/or activate growth factors that are sequestered in the extracellular matrix *(40)*.

2.2.2.4. GELATINASES

The gelatinases, gelatinase A (MMP-2), and gelatinase B (MMP-9) have long been thought primarily responsible for basement membrane degradation facilitating malignant cell invasion into adjacent tissue compartments. Recent experiments clearly support an expanded role for these MMPs.

Gelatinase A expression during epithelial tumorigenesis is limited to stromal fibroblasts (**Table 2**). Gelatinase A associates with plasma membranes via its C-terminal domain to $\alpha V\beta 3$ integrins on melanoma cells and angiogenic blood vessels; this interaction enhances tumor growth *(41)*. Autolytic processing of gelatinase A with release of the C-terminal domain competes with cell surface binding of the enzyme; this inhibits tumor growth by limiting angiogenesis *(42)*. Consistent with these results, MMP-2/null host mice exhibit impaired primary tumor growth and decreased experimental metastases of B16-BL6 melanoma and Lewis Lung carcinoma (LLC) cells due primarily to diminished rate of neovascularization *(43)*. This is the first study to date, providing direct evidence that host-derived gelatinase A plays a specific role in angiogenesis regulating tumor progression in vivo.

Gelatinase B (MMP-9), on the other hand, is quite unique in that it is almost exclusively expressed by reactive fibroblasts and inflammatory cells in and around carcinomas (**Table 2**). The exception lies in that keratinocytes on the edge of hyperproliferative epithelia adjacent to wounds or ulcers and within hair follicles, also express gelatinase B mRNA (Coussens, Hanahan, and Werb, unpublished observations). The significance of inflammatory cell expression in epidermal-dermal interaction has been shown by the observation that gelatinase B/null mice are resistant to bullous pemphigoid *(44)*. Interestingly, gelatinase B has never been detected in screens looking to identify malignancy-associated genes. Why is this? During squamous cell carcinoma (SCC) development in mice expressing human papilloma virus 16 (HPV16) in basal keratinocytes *(31)*, gelatinase B mRNA and encoded enzyme are most abundant during the onset and development of the neovasculature (Coussens, et al., manuscript submitted) suggesting a role in angiogenesis. However, when HPV16 transgenic mice are bred into the gelatinase B/null background, epithelial proliferation, angiogenesis, tumor incidence, and overall malignancy of escaping tumors is severely compromised (Coussens, Hanahan, and Werb, unpublished observations). These results suggest that growth-promoting signals generated in the absence of gelatinase B are not sufficient to induce extensive early epithelial hyperproliferation, a provocative result in light of the fact that these cells remain oncogene-positive. Correspondingly, generation of a reactive stroma with intense inflammatory cell infiltration and dramatic neovascularization characteristic of dysplasias in HPV16 mice is dampened and delayed in HPV16-gelatinase B/null. Whether this delay and diminution of characteristic stromal reorganization is due to lack of gelatinase B within the stroma directly, or diminished signals originating from the epidermis is unclear. In vitro proliferation experiments suggests that gelatinase B is not a direct mitogen for either dermal fibroblasts or capillary endothelial cell; however, gelatinase B can induce otherwise non-angiogenic tissue to initiate angiogenic activity suggesting that stromal targets for gelatinase B may in turn act as either mitogens and/or morphogens.

Taken together, these data argue that gelatinase B is a critical component of epithelial tumor development, a somewhat surprising result as gelatinase B is largely synthesized by reactive inflammatory cells. Moreover, that the small number of tumors escaping in the HPV16 gelatinase B/null mice are more malignant and metastatic compared to controls suggests that analysis of HPV gelatinase B/null tumors will provide insight into genetic elements involved in elaborating more aggressive disease. Whether this altered malignancy reflects a suppressive role for gelatinase B in the tumor environment remains to be determined.

2.2.2.5. MACROPHAGE METALLOELASTASE (MMP-12)

Although MMPs commonly facilitate neoplastic progression, proteolytic cleavage products may inhibit angiogenesis, limiting tumor progression. This was first apparent with the isolation of angiostatin from the urine of mice with Lewis lung cell (LLC) carcinomas *(45)*. Angiostatin, a plasminogen cleavage product containing kringle regions 1–4, inhibits endothelial cell proliferation and is believed to be responsible for maintaining LLC lung metastases in a dormant state.

Generation of angiostatin in primary LLC tumors correlated with the presence of macrophages and macrophage elastase *(*MMP-12*) (46)*. The importance of MMP-12 in limiting lung metastasis growth in the LLC model has been confirmed by use of mice deficient in macrophage elastase (MMP-12/null) by gene targeting (Grisolano and Shapiro, unpublished observations). However, preliminary studies suggest that local expression of MMP-12 in macrophages surrounding the secondary lung metastases limit growth, in part through generation of angiostatin. This effect may also be related to MMP-2 processing or other mechanisms.

Several other MMPs are also capable of generating angiostatin and other antiangiogenic fragments of plasminogen *(47)*. The kringle 5 domain by itself appears to have the greatest capacity to inhibit endothelial cell proliferation *(48)*. Serine proteinases, including plasmin, in association with an extracellular reductase that reduces disulfide bonds in plasmin, also triggers generation of angiostatin *(49,50)*. In addition to angiostatin, other proteolytic fragments, most prominently endostatin, a 20-kDa C-terminal fragment of type XVIII collagen, effectively inhibit tumor angiogenesis *(51)*. In fact, treatment of several tumors in mice with endostatin results in prolonged tumor dormancy *(52)*

2.2.2.6. SUMMARY

Collectively, this body of work suggests that proteinases may benefit the host or the tumor depending upon spatial expression, proteolytic capacity, and binding affinity for matrix versus neoplastic cells. Nevertheless, nonspecific MMP inhibitors, such as active site zinc chelating hydroxamates, are effective in inhibiting growth of several primary tumors and metastases in animal models *(53)* including LLC *(54)*. The ability of these compounds to inhibit growth of neoplasms suggests that tumors use MMPs more effectively than the host; when "all weapons are off the street," the host has the advantage. Development of more specific MMP inhibitors combined with anti-angiogenic agents, such as angiostatin, endostatin, or $\alpha v\beta 3$, might prove optimal in clinical treatment of particular tumors. Hence, MMPs are now believed to regulate critical events other than tumor cell invasion, e.g., liberation of bioactive fragments of extracellular matrix, releasing sequestered growth factors and angiogenic mediators, and in the shedding of cell surface receptors and ligands effecting bioavailability of chemokines and cytokines *(55–57)*, all of which represent extrinsic events effecting neoplastic proliferation and/or growth and expansion. Thus, the behavior of neoplastic cells can no longer be understood without taking into account their active and reciprocal dialogue with the dynamic microenvironment in which they live.

2.2.3. Pulmonary Emphysema

Pulmonary emphysema is defined as enlargement of terminal airspaces, accompanied by destruction of their walls. Emphysema is the major component of the morbidity and mortality of chronic obstructive pulmonary disease (COPD) which is currently the fourth leading cause of death in the US and one of the causes whose incidence is increasing. COPD may grow to epidemic proportions given the large increase in smoking worldwide.

Typically, cigarette smoke induces a chronic inflammatory response with accumulation of macrophages, and to a much lesser extent neutrophils, in respiratory bronchioles and alveolar space. Release of inflammatory cell proteinases, in excess of inhibitors, coupled with abnormal matrix repair lead to emphysema. Inherited deficiency of $\alpha 1$-antitrypsin, the primary inhibitor of neutrophil elastase a serine proteinase, predisposes individuals to early onset emphysema, and intrapulmonary instillation of elastolytic enzymes such as neutrophil elastase in experimental animals causes lung destruction characteristic of emphysema. Together, these findings have led to the hypothesis that emphysema results from proteolytic injury, presumably by neutrophil elastase, directed against elastin.

This hypothesis was challenged when D'Armiento and colleagues *(58)* found that a human collagenase-1 (MMP-1) transgene driven by the haptoglobin reporter unexpectedly resulted in lung-specific expression in several independent founder lines. These mice develop enlarged airspaces characteristic of emphysema. This was the first demonstration that an MMP could directly cause emphysema. Since MMP-1 is inactive against mature elastin, this result suggested that collagen degradation was sufficient to cause emphysema. However, it is not certain whether the alveolar pathology in these animals is due to destruction of collagen in mature lung tissue, or, more likely, whether expression of the transgene during growth and development interfered with normal elastic fiber assembly, perhaps through destruction of the elastic fiber microfibrillar scaffold. Inducible MMP-1 overexpression would be required to differentiate abnormal alveolar development from destruction of mature alveolar spaces which defines emphysema.

Fortunately, like humans, mice chronically exposed to cigarette smoke develop a macrophage predominant inflammatory infiltrate in the lungs, followed by airspace enlargement *(59)*. Exposure of wild-type MMP-12 (+/+) mice to long-term cigarette smoke exposure leads to inflammatory cell recruitment followed by alveolar space enlargement similar to the pathologic defect in humans. However, mice deficient in macrophage elastase (MMP-12/null) are protected from development of emphysema despite heavy long-term smoke-exposure (*see* **Fig. 2**). Surprisingly, MMP-12/null mice also fail to recruit monocytes into their lungs in response to cigarette smoke *(59)*. Because MMP-12 and most other MMPs are only expressed upon differentiation of monocytes to macrophages, it appears unlikely that monocytes require MMP-12 for transvascular migration. More likely, cigarette smoke induces resident macrophages, which are present in lungs of MMP-12/null mice, to produce MMP-12 that in turn cleaves elastin thereby generating fragments chemotactic for monocytes (Paige and Shapiro, unpublished observations). This positive feedback loop would perpetuate macrophage accumulation and lung destruction. The concept that proteolytically generated elastin fragments mediate monocyte chemotaxis is not original. Independent studies by Senior and Mecham *(60)* and Hunninghake and colleagues *(61)* from the early 1980s demonstrated that elastase-generated elastin fragments were chemotactic for monocytes and fibroblasts. Gene targeting is merely reinforcing this as a major in vivo mechanism of macrophage accumulation in a chronic inflammatory condition.

Whether human emphysema is also dependent on this single MMP is of course uncertain. At the very least, this study demonstrates a critical role of macrophages in the development of emphysema and unmasks a proteinase-dependent mechanism of inflammatory cell recruitment that may have broader biological implications.

2.2.4. Bullous Pemphigoid

Bullous pemphigoid is an autoimmune subepidermal blistering disease characterized by deposition of autoantibodies at the basement membrane zone. In an experimental model of bullous pemphigoid in mice, the blistering is mediated by antibodies directed against the hemidesmosomal protein BP180

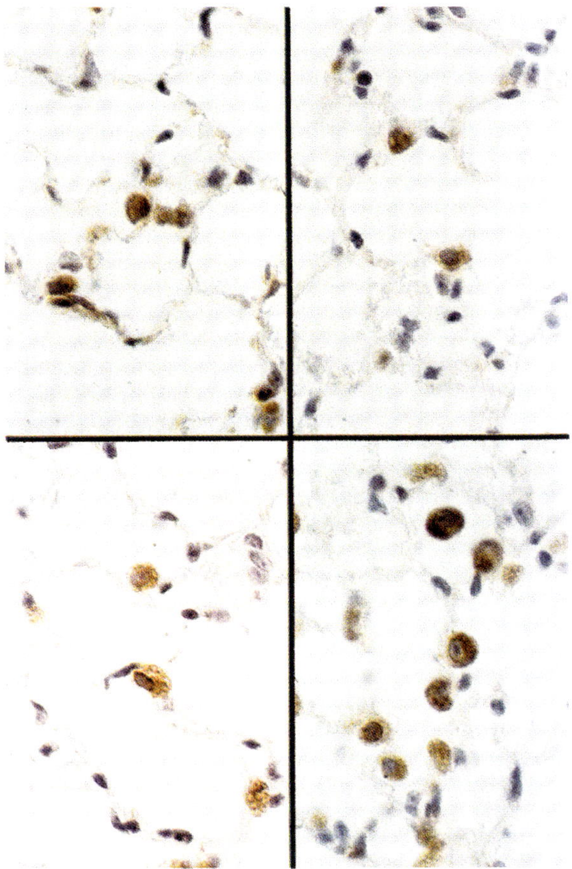

Fig. 2. Macrophage elastase deficient mice fail to accumulate macrophages in response to cigarette smoke. Wild-type (MME+/+) and macrophage elastase-deficient (MME/null) mice were exposed to cigarette smoke for six months. Smoke-exposed (MME/null, *upper right*; MME+/+, *lower right*) mice and age-matched controls (MME/null, *upper left*; MME+/+, *lower left*) were then sacrificed, lungs inflated, fixed, and mid-saggital sections stained for Mac-3. Note, cigarette smoke exposure resulted in fourfold increase in number of alveolar and interstitial macrophages in wild-type mice. However, in MME/null mice, while there were equal number of constitutive macrophages in the lungs compared to wild-type mice, there was no significant additional recruitment in response to cigarette smoke exposure *(59)*.

(collagen XVII), and depends on complement activation and neutrophil infiltration. In contrast to wild-type littermates, MMP-9/null mice are resistant to the blistering effects in this model, despite deposition of autoantibodies and neutrophil recruitment equivalent to wild-type *(44)*. Whether MMP-9 directly causes blistering or augments neutrophil elastase activity by degrading α1-antitrypsin is being investigated.

2.2.5. Contact Hypersensitivity

Contact hypersensitivity (CHS) is a classic experimental model for investigation of antigen specific T cell mediated immune response. Topical application of 2,4-dinitrobenzenesulfonic acid (DNBS) to previously sensitized wild-type mice resulted in dermal infiltration of lymphocytes and monocytes, dilation and increased permeability of cutaneous blood vessels, and dermal edema. In contrast, MMP-3/null mutant mice showed a markedly impaired CHS response to DNBS, although they responded normally to a nonspecific irritant, phenol *(62)*. Intradermal injection of MMP-3 just prior to DNBS restored CHS response to DNBS. Lymphocytes from lymph nodes of MMP-3/null mice did not proliferate in response to DNBS (but responded identically as wild-type to concanavalin A). MMP-9/null mutant mice, unlike MMP-3/null mice, exhibited a normal onset of CHS at day 1, but unlike wild-type mice, inflammation persisted at 7 d *(62)*. These results suggest that MMP-3 is required for initiation and MMP-9 for resolution of inflammation in CHS.

3. Tissue Inhibitors of Metalloproteinases

3.1. TIMP Transgenic Mice

Mice deficient for TIMPs and transgenic mice carrying portions of the TIMP genes have been generated and used to study the in vivo functions of TIMPs, their patterns of expression and to identify sequences important for their regulation (**Table 3**).

3.1.1. TIMP-1

The first TIMP transgenics expressed the *Escherichia coli* β-galactosidase gene under control of a 5' region of the mouse TIMP-1 gene containing -2158 to -58 bp upstream of the initiator ATG and were used to follow patterns of TIMP-1 promoter activity *(63–65)*. One of five founders was subjected to detailed analysis for β-galactosidase expression. β-galactosidase staining patterns agreed well with TIMP-1 expression as assayed by RNA *in situ* hybridization in zones of ossification and extracellular matrix deposition, particularly around the mandible, clavicle, and cervical ve22rtebrae, in the day 13.5 to 17.5 developing embryo *(66)*. However, staining for β-galactosidase is also found in sites were TIMP-1 mRNA is not detected, for example in hair follicles, primarily in Henle's layer of the inner root sheath during the mid anagen (growing) phase of the hair cycle *(64)*.

3.1.2. TIMP-3

A set of six transgenic lines designed to probe the expression pattern of TIMP-3 during development was also generated using human TIMP-3 sequences from –940 to +277 relative to the transcriptional start site fused to

Table 3
Mice with Genetically Regulated TIMP Levels

TIMP TRANSGENICS

Transgene	Promoter	Phenotype	Refs.
LacZ	**Mouse Timp-1**	Blue histochemical staining of tissues expressing TIMP-1	*(63–65)*
Human TIMP-1	**Human β-Actin**	Reduced decidual remodeling and growth Reduced apoptosis of mammary epithelium when an autoactivating MMP-3 transgene is present	*(68,77)*
Mouse Timp-1	**Mouse MHC Class I H2**	Reduced formation of hepatocellular carcinomas Reduced intradermal growth and spontaneous metastasis of T-cell lymphoma	*(81, 83,84)*
Mouse Timp-1 antisense	**Mouse MHC Class I H2**	Increased formation of hepatocellular carcinomas. Enhanced intradermal growth and spontaneous and experimental metastasis of T-cell lymphoma	*(81,83)*
Mouse Timp-1	**Mouse Metallothionein-1**	Reduced experimental metastasis to the brain of fibrosarcoma cells.	*(84)*
Human TIMP-1	**Rat LFABP**	Increased formation of *Min*-induced colon tumors. Founder-dependent effect.	*(85)*
LacZ	**Human TIMP-3**		*(67)*

Timp Knockouts

Locus	Allele	Phenotype	Ref.
Timp-1	**Null**	Complement-dependent hyper–resistance to corneal infection with *P. aeruginosa*	*(101,102)*
Timp-2	**Null**	Impaired proMMP-2 activation	[a]
Timp-2	**Truncated peptide**	Impaired proMMP-2 activation	[b]

[a]Wang et al., in preparation.
[b]Caterina et al., in preparation.

β-galactosidase *(67)*. The spatial pattern of β-galactosidase staining shows some variation among the different founders. However, staining in placenta and fetal and early postnatal vertebral discs, developing ear and choroid plexus agrees with results from *in situ* analysis of RNA and immunohistochemical staining for protein, although there are differences in the timing of expression *(68–72)*. TIMP-3 expression detected by direct methods in heart and skeletal muscle of mice *(69)* and in lungs and heart of rat *(73)* is not detected in the transgenics. The transgenics reveal β-galactosidase staining in tissues in which TIMP-3 has not been detected by direct methods, including developing bones, the choriod layer of the eye, and in webs between developing digits.

It is not clear if the transgenic approaches accurately reflect expression of the endogenous mouse TIMP loci. Targeted mutagenesis of TIMP loci by knock in of histochemical markers may prove useful for future studies of the temporal and spatial expression patterns of TIMPs. Alternatively, cell lineages that express TIMPs at any stage of development can be permanently marked by knocking in the cre recombinase at TIMP loci, and mating those strains of mice with one containing a cre–activatable histochemical marker *(74)*. Application of this method to the TIMP-1 locus may reveal staining in all embryonic and adult tissues, given that TIMP-1 mRNA is detected in the two cell embryo by RT–PCR *(75)*. In those studies, it is not possible to distinguish maternal transcripts from those made by the embryo.

3.2. TIMPs in Development

The most extensive use of TIMP transgenic mice has been to elucidate the in vivo functions of TIMP-1 and the consequences of its altered expression. Transgenic mice with modified TIMP expression reported to date have been made with either the TIMP-1 gene or antisense expression constructs of TIMP-1. Most studies have focused on the role of TIMP-1 in normal developmental processes or in tumorigenesis.

Transgenic overexpression of human TIMP-1 by the human β-actin promoter delays matrix degradation in the mammary gland and the involution process after weaning *(76)*. The delay in matrix turnover is presumably due to a reduction in net MMP activity caused by increased TIMP-1 levels. These studies of mammary development have been extended using progeny of mice carrying the same TIMP-1 transgene crossed with mice transgenic for an autoactivating MMP-3 allele. These progeny have a range of MMP-3 activity ranging from low (TIMP-1 transgene only) to higher (wild-type mice and mice with both the TIMP-1 and MMP-3 transgenes) to highest (autoactivating allele of MMP-3 only). Under conditions of increased MMP-3 activity, the mammary alveolar epithelial cells of mice undergo unscheduled apoptosis during late pregnancy. When the MMP-3 transgene is balanced by the TIMP-1 overexpressing allele, premature apoptosis is suppressed *(77)*. The MMP-3 transgene-dependent apoptosis correlates with the accumulation of breakdown products in the mammary gland of the basement membrane protein nidogen/entactin. These breakdown products are not seen in wild-type, TIMP-1 transgenic, or TIMP-1/MMP-3 double transgenic mice at the same time point in pregnancy. These results along with others *(78)* provide in vivo correlates of a model based on in vitro experiments that describes integrin-ECM cross linking as essential for preventing apoptosis of epithelial cells *(79,80)*. In yet other studies with the TIMP-1 transgenic mice, early post-implantation development was shown to be altered. The cross sectional area of embryonic day 6.5 decidua and their overall length were reduced in TIMP-1 transgenic mice *(68)*.

3.3. TIMPs in Cancer

Independent transgenic lines carrying either a *Timp-1* sense or antisense expression cassette driven by the murine MHC class I H2 promoter were used to investigate the influences of increased or diminished TIMP-1 synthesis on hepatocellular carcinoma formation induced by an SV40 T antigen transgene *(81)*. Mice hemizygous for the *Timp-1* sense or antisense expression cassettes were bred with mice hemizygous for an SV40 T antigen transgene driven by the liver-specific C-reactive protein *(82)*. Tumor growth in T antigen positive progeny who also carried the *Timp-1* expression cassettes was compared to growth in mice that were transgenic for only T antigen. Mice with increased TIMP-1 levels due to presence of the *Timp-1* sense transgene showed less tumor development when compared to nontransgenic littermates. Furthermore, mice with reduced TIMP-1 levels due to the *Timp-1* antisense transgene showed more rapid tumor development than their nontransgenic littermates. Although a direct comparison of *Timp-1* sense and *Timp-1* antisense mice revealed no differences in tumor development, this may be due to other genetic differences that existed between the two *Timp-1* transgenic lines that arose as a consequence of developing the transgenic mice in outbred backgrounds.

These same transgenic mice were challenged intradermally with the a *LacZ*-tagged variant of the T-cell lymphoma line L.CI.5s *(83)*. Primary tumor growth was reduced and time to death was increased in the *TIMP-1* overexpressing line. In contrast, primary tumor growth was accelerated and in the time to death was hastened in the *Timp-1* underexpressing line. Qualitative differences were observed in the liver metastases seen in the two lines of mice with more diffuse infiltrates observed in the *Timp-1* underexpressing line. When mice were given an intravenous challenge of the same cells, tumor burden was unchanged in the *Timp-1* overexpressing line but increased in the *Timp-1* underexpressing line. Because little TIMP-1 is made by the liver, it is not clear how a further reduction in expression affects growth of tumors there. However, TIMP-1 can be turned on in liver during growth of some tumors and TIMP-1 made and secreted by other tissues may affect growth of tumors in the liver.

Overexpression of *TIMP-1* in the brain by the mouse metallothionein-1 promoter could confer a two to four reduction in brain metastasis of a *LacZ*-tagged FS/L10 fibosarcoma cells in experimental metastasis assays *(84)*. The magnitude of the change depended upon the transgenic founder used.

Although the mechanisms by which host TIMP-1 synthesis inhibited tumor growth and metastasis in the above studies are unknown, many other studies have shown that TIMPs can inhibit tumor cell invasion and angiogenesis.

Although the results are consistent with the long held view that, in general, TIMPs can inhibit tumor growth and metastasis, exceptions to the rule have been found that call the paradigm into question. A pair of transgenic mice were developed that carry a human *TIMP-1* transgene driven by the rat liver fatty acid binding protein. Human and/or mouse *TIMP-1* were activated in the colons of transgenic animals. These were interbred with mice carrying the *Min* mutation in the *Apc* gene that led to the formation of multiple intestinal tumors. The frequency of *Min*-induced tumor formation was either unaffected or mildly stimulated by TIMP-1 depending upon the founder line contributing the *TIMP-1* transgene *(85)*.

An additional exception to the observations that TIMP-1 is an inhibitor of tumor growth and metastasis was apparent in TIMP-1/null mice. *Timp-1* was mutated by gene targeting in mouse embryonic stem (ES) cells using two separate targeting vectors. Mice deficient for TIMP-1 were generated with replacement vector-targeted ES cells and co-isogenic pairs of mutant or wild-type tumorigenic cell lines were generated using insertion vector-targeted ES cells. The mice and cells were used in experimental metastasis assays to evaluate the effects of TIMP-1 loss by a tumor and tumor host on metastasis *(86)*. The results showed that the *Timp-1* genotype of the tumor host had no effect on metastasis to the lung of any of four cell lines tested, however, the *Timp-1* genotype of the tumor did. Two pairs of cell lines behaved in the predicted manner; the TIMP-1/null member of the pair was more invasive than the wild-type member of the pair. However, for a third pair, the TIMP-1/null line was less invasive than the wild-type line. When the *timp-1* mutation was crossed into the *Min* background, loss of TIMP-1 had no effect on frequency of appearance or histological appearance of intestinal tumors (Soloway, unpublished results).

Collectively, these results reveal that the influence of host TIMP-1 synthesis on metastasis is highly varied. Up-regulation of host synthesis of TIMP-1 could inhibit tumor development in the SV40 T antigen-induced liver tumor model, and in a founder-dependent manner promote tumor development in the *Min*-induced colon tumor model. Induction of TIMP-1, however, had no effect on tumor invasion of liver in the T-cell lymphoma model of experimental metastasis, but did inhibit experimental metastasis to the brain of fibrosarcoma cells. Down regulation of TIMP-1 could allow enhanced tumor development of T antigen-induced liver tumors and of L.CI.5s T-cell lymphomas but had no effect on experimental metastasis to the lung or of *Min*-induced intestinal tumors.

The complex actions of host TIMP-1 expression on tumor development and metastasis may relate to the fact that TIMP-1 has been shown to have many, sometimes opposing activities. TIMP-1 has been shown to inhibit invasion of basement membranes by tumorigenic cells in vitro *(87)*, apoptosis of certain cell lines (Stetler-Stevenson, unpublished results), angiogenesis *(88)* and the release of the growth factor, IGF2, from its inactivating IGF2 binding proteins *(89,90)*. TIMPs have also been shown to be growth promoting for a variety of cell types *(91–96)*. The importance of TIMP-1 levels for growth of a given tumor will depend upon the relative sensitivity of the tumor to the growth enhancing versus the growth inhibiting properties of TIMP-1. Local concentrations of TIMP-1 are likely to be very important in this regard. Furthermore, a given tumor that is inhibited by high concentrations of TIMP-1 may not respond significantly to loss of TIMP-1.

An additional complication is due to the likelihood that TIMPs are coordinately regulated. TIMP-1 loss could affect the accumulation of mRNA transcribed from other *Timp* genes, the greatest being a six fold elevation of Timp-3 transcript accumulation in the heart (Soloway, unpublished) and *(97)*. Changes in production of other TIMPs that result from an alteration in TIMP-1 levels may affect activities specific for those TIMPs that may in turn have different effects on tumor development. For example, TIMP-2 inhibits bFGF-induced endothelial cell proliferation which may suppress angiogenesis *(98)*. If under conditions of TIMP-1 up-regulation, TIMP-2 levels drop, endothelial cell proliferation and angiogenesis may be stimulated. In summary, the variable effects of host TIMP-1 on tumor development for a given tumor model may be due to the fact that some of the TIMP-1 activities may or may not be important or rate limiting for development or metastasis of that particular tumor type.

3.4. TIMPs in Steroidigonesis

TIMP-1/null mice have also been used in a number of assays to evaluate the role of TIMP-1 in processes other than tumorigenesis. There have been reports that TIMP-1 is a stimulator of steroid synthesis by sertoli cells in vitro *(99)*. Using TIMP-1/null mice, Nothnick and colleagues compared serum estradiol-17 β and progesterone content to wild-type animals and found no differences *(100)*. However, failure to detect changes in steroidogenesis in vivo in the absence of TIMP-1 does not necessarily indicate that TIMP-1 does stimulate steroid synthesis, only that loss of TIMP-1 does not reduce it *(97)*. The role of TIMP-1 in steriodogenesis may be more directly addressed by study of TIMP-1 overexpressing animals.

3.5. TIMPs in Host Defense

A surprising phenotype of TIMP-1/null mice is their hyper-resistance to corneal infection by *Pseudomonas aeruginosa*. Infections become established equally well in mutant and wild-type mice, but coincident with the maximal inflammatory response 24 h after infection, the bacterial burden in mutant mice drops to a level 500-fold lower than in wild-type mice. The hyper-resistance to infection is dependent upon intact complement and hematopoietic systems

(101,102). The increased resistance of mutant mice to infection is associated with increased serum complement activity, increased permeability of the vasculature to plasma proteins and increased kinetics of leukocyte margination to the vascular endothelium in response to proinflammatory factors. Each of these phenotypes may have a common mechanism and collectively enhance complement-dependent immune responses in TIMP-1/null mice. In other experiments studying inflammation in the mutant animals, inflammatory responses were elevated (K. Gijbels et al. in preparation). Together, these results suggest that TIMP-1 limits inflammation in vivo.

3.6. TIMPs in Angiogenesis

The role of TIMP-1 in retinal neovascularization has been studied using a VEGF transgene-dependent model of neovascularization *(103)*. *VEGF* transgenic mice were crossed with both *TIMP-1* transgenic *(77)* and TIMP-1/null animals. Retinal neovascularization was assayed by counting lesion number and area *(104)*. Given previous studies describing TIMP-1 as an inhibitor of neovascularization *(88)* it was predicted that neovascularization would be increased in TIMP-1/null animals and reduced in the *TIMP-1* transgenics. Surprisingly, the opposite trend was observed. In *VEGF/TIMP-1* double transgenic animals, the total area of the neovascular lesions was greater than in *VEGF* single transgenic mice. In addition, neovascular lesions were smaller in TIMP-1/null mice, although the total lesion size was unchanged by loss of TIMP-1. It is not clear if views regarding the influences of TIMP-1 on neovascularization need to be reexamined, or if in these experiments there was a compensating down-regulation of TIMP-2 in the *TIMP-1* transgenics and up-regulation in the knockouts. An alteration in TIMP-2 levels, given TIMP-2's inhibitory effect on endothelial cell proliferation *(98)*, may be at the root of this apparent paradox.

3.7. TIMP-2 Deficient Mice

More recently, TIMP-2/nullmice were developed by two independent groups. A line made by Henning Birkedal-Hansen's group gives rise to a truncated peptide with an internal deletion that has extremely weak MMP inhibitory activity (Caterina et al., in preparation). A second allele made by Paul Soloway's group is a null mutation (Wang et al., in preparation). Cells and tissues from both strains of mice fail to activate proMMP-2, consistent with predictions based on biochemical studies into the mechanisms of activation of proMMP-2 *(105,106)*. Studies reveal clear functional differences in vivo between TIMPs-1 and -2. TIMP-1/null mice have normal activation of proMMP-2 whereas responses to infection are normal in TIMP-2/null mice *(107)*.

4. Notes

1. The gene for MT1-MMP (MMP-14) has recently been knocked out *(108)*. MT-1-MMP deficiency causes craniofacial dysmorphism, arthritis, osteopenia, dwarfism, and fibrosis of soft tissues.

Acknowledgments

We wish to thank all past and present members of our various laboratories for scientific contributions not mentioned here. We also acknowledge our collaborators and colleagues for communicating data prior to publication. This work has been supported by grants from the National Institutes of Health.

References

1. Vu, T. H., Shipley, J. M., Bergers, G., Berger, J. E., Helms, J. A., Hanahan, D., Shapiro, S. D., Senior, R. M., and Werb, Z. (1998) MMP-9/gelatinase B is a key regulator of growth plate angiogenesis and apoptosis of hypertrophic chondrocytes. *Cell* **93**, 411–422.
2. Lelongt, B., Trugnan, G., Murphy, G., and Ronco, P. M. (1997) Matrix metalloproteinases MMP2 and MMP9 are produced in early stages of kidney morphogenesis but only MMP9 is required for renal organogenesis in vitro. *J. Cell. Biol.* **136**, 13,663–13,673.
3. Minor, J. H., Betsuyaku, T., Shipley, J. M., and Senior, R. M. (1997) Renal function is normal in gelatinase B deficient mice. *Mol. Biol. Cell* **8**, 403a.
4. Mohan, R., Rinehart, W. B., Bargagna-Mohan, P., and Finis, M. E. (1998) Gelatinase B/lacZ transgenic mice, a model for mapping gelatinase B expression during developmental and injury-related tissue remodeling. *J. Biol. Chem.* **273**, 25,903–25,914.
5. Balbin, M., Fueyo, A., Knauper, V., Pendas, A. M., Lopez, J. M., Jimenez, M. G., Murphy, and G., López-Otín, C. (1998) Collagenase 2 (MMP-8) expression in murine tissue-remodeling processes. Analysis of its potential role in postpartum involution of the uterus. *J. Biol. Chem.* **273**, 23,959–23,968.
6. Liu, X., Wu, H., Byrne, M., Jeffrey, J., Krane, and S., Jaenisch, R. (1995) A targeted mutation at the known collagenase cleavage site in mouse type I collagen impairs tissue remodeling. *J. Cell. Biol.* **130**, 227–237.
7. Lund, L. R., J. Romer, J., Thomasset, N., Solberg, H., Pyke, C., Bissell, M. J., Dano, K., and Werb, Z. (1996) Two distinct phases of apoptosis in mammary gland involution: Proteinase-independent and -dependent pathways. *Development* **122**, 181–193.
8. Sympson, C. J., Talhouk, R. S., Alexander, C. M., Chin, J. R., Clift, S. M., Bissell M. J., and Werb, Z. (1994) Targeted expression of stromelysin-1 in mammary gland provides evidence for a role of proteinases in branching morphogenesis and the requirement for an intact basement membrane for tissue-specific gene expression. *J. Cell Biol.* **125**, 681–693. [Published correction appears in *J. Cell Biol.* **132,** 753 (1996)].

9. Alexander, C. M., Howard, E. W., Bissell, M. J., and Werb, Z. (1996). Rescue of mammary epithelial cell apoptosis and entactin degradation by a TIMP-1 transGene *J. Cell Biol.* **135**, 1669–1677.

10. Thomasset, N., Lochter, A., Sympson, C. J., Lund, L. R., Williams D. R., Behrendtsen, O. , Werb, Z., and Bissell, M. J. (1998) Expression of autoactivated stromelysin-1 in mammary glands of transgenic mice leads to a reactive stroma during early Development. *Am. J. Path.* **153**, 457–467.

11. Sudbeck, B. D., Pilcher, B. K., Welgus, H. G., and Parks, W. C. (1997) Induction and repression of collagenase-1 by keratinocytes is controlled by distinct components of different extracellular matrix compartments. *J. Biol. Chem.* **272**, 22,103–22,110.

12. Pilcher, B. K., Dumin, J. A., Sudbeck, B. D., Krane, S. M., Welgus, H. G., and Parks, W. C. (1997) The activity of collagenase-1 is required for keratinocyte migration on a type I collagen matrix. *J. Cell. Biol.* **137**, 1445–1457.

13. Dunsmore, S. E., Saarialh-Kere, U. K., Roby, J. D., Wilson, C. L., Matrisian, L. M., Welgus, H. G., and Parks, W. C. (1998) Matrilysin expression and function in airway epithelium. *J. Clin. Invest.* **102**, 1321–1331.

14. Libby, P. (1995) Molecular bases of the acute coronary syndromes. *Circulation* **91**, 2844–2850.

15. Thompson, R. W., Mertens, R. A., Liao, S., Holmes, D. R., Mecham, R. P., Welgus, H. G., and Parks, W. C. (1995) Production and localization of 92-kD gelatinase in abdominal aortic aneurysms: an elastolytic metalloproteinase expressed by aneurysm-infiltrating macrophages. *J. Clin. Invest.* **96**, 318–326.

16. Breslow, J. (1996) Mouse models of atherosclerosis. *Science* **272**, 685–688.

17. Carmeleit, P., Moons, L., Lijnen, R., Crawley, J., Tipping, P., Drew, A., Eeckhout, Y., Shapiro, S. D., Lupu, F., and Collen, D. (1997) Plasmin predisposes to atherosclerotic aneurysm formation by activation of matrix metalloproteinases. *Nature Genetics* **17**, 439–444.

18. Shipley, J. M., Wesselschmidt, R. L., Kobayashi, D. K., Ley, T. J., and Shapiro, S. D. (1996) Metalloelastase is required for macrophage-mediated proteolysis and matrix invasion in mice. *Proc. Natl. Acad. Sci.* **93**, 3942–3946.

19. Ye, S., Eriksson, P., Hamsten, A., Kurkinen, M., Humphries, S. E., and Henney, A. M. (1996) Progression of coronary atherosclerosis is associated with a common genetic variant of the human stromelysin-1 promoter which results in reduced gene expression. *J. Biol. Chem.* **271**, 13,055–13,060.

20. Wilson, C. L. and Matrisian, L. M. (1996) Matrilysin: An epithelial matrix metalloproteinase with potentially novel functions. *Int. J. Biochem. Cell. Biol.* **28**, 123–136.

21. Saarialho-Kere, U. K., Crouch, E. C., and Parks, W. C. (1995) Matrix metalloproteinase matrilysin is constitutively expressed in adult human exocrine epithelium. *J. Invest. Dermatol.* **105**, 190–196.

22. Wilson, C. L., Heppner, K. J., Rudolph, L. A., and Matrisian, L. M. (1995) The metalloproteinase matrilysin is preferentially expressed by epithelial cells in a tissue-restricted pattern in the mouse. *Mol. Biol. Cell* **6**, 851–869.

23. Witty, J. P., McDonnell, S., Newell, K. J., Cannon, P., Navre, M., Tressler, R. J., and Matrisian, L. M. (1994) Modulation of matrilysin levels in colon carcinoma cell lines affects tumorgenicity in vivo. *Cancer Res.* **54**, 4805–4812.

24. Wilson, C. L., Heppner, K. J., Labosky, P. A., Hogan, B. L., and Matrisian, L. M. (1997) Intestinal tumorigenesis is suppressed in mice lacking the metalloproteinase matrilysin. *Proc. Nat. Acad. Sci.* **94**, 1402–1407.

25. Rudolph-Owen, L. A. and Matrisian, L. (1998) Matrix metalloproteinases in remodeling of the normal and neoplastic mammary gland. *J. Mammary Gland Neoplasia* **3**, 177–189.

26. Rudolph-Owen, L. A., Cannon, P., and Matrisian, L. (1998) Overexpression of the matrix metalloproteinase matrilysin results in premature mammary gland differentiation and male infertility. *Mol. Biol. Cell.* **9**, 421–435.

27. Lawson, N. D., Khana-Gupta, A., and Berliner, N. (1998) Isolation and characterization of the cDNA for mouse neutrophil collagenase: Demonstration of shared negative regulatory pathways for neutrophil secondary granule protein gene expression. *Blood* **91**, 2517–2524.

28. Coussens, L. M. and Werb, Z. (1996) Matrix metalloproteinases and the development of cancer. *Chem. and Biol.* **3**, 895–904.

29. Murray, G. I., Duncan, M. E., O'Neil, P., Melvin, W. T., and Fothergill, J. E. (1996) Matrix metalloproteinase-1 is associated with poor prognosis in colorectal cancer. *Nature Med.* **2**, 461–461.

30. D'Armiento, J., DiColandrea, T., Dalal, S. S., Okada, Y., Huang, M. T., Conney, A. H., and Chada, K. (1995) Collagenase expression in transgenic mouse skin causes hyperkeratosis and acanthosis and increases susceptibility to tumorigenesis. *Mol. Cell. Biol.* **15**, 5732–5739.

31. Coussens, L. M., Raymond, W. W., Bergers, G., et al. (1999) Inflammatory mast cells up-regulate angiogenesis during squamous epthelial carcinogenesis. *Genes Dev.* **13**, 1382–1397

32. Alexander, C. M., Hansell, E. J., Behrendtsen, O., Flannery, M. L., Kishnani, N. S., Hawkes, S. P., and Werb, Z. (1996) Expression and function of matrix metalloproteinases and their inhibitors at the maternal-embryonic boundary during mouse embryo implantation. *Development* **122**, 1723–1736.

33. Sympson, C. J., Talhouk, R. S., Alexander, C. M., Chin, J. R., Clift, S. M., and Bissell, M. J. (1994) Targeted expression of stromelysin-1 in mammary gland provides for a role of proteinases in branching morphogenesis and the requirement for an intact basement membrane for tissue-specific gene expression. *J. Cell. Biol.* **125**, 681–693.

34. Lochter, A., Srebrow, A., Sympson, C. J., Terracio, N., Werb, Z., and Bissell, M. J. (1997) Misregulation of stromelysin-1 expression in mouse mammary tumor cells accompanies acquisition of stromelysin-1-dependent invasive properties. *J. Biol. Chem.* **272**, 5007–5015

35. Sternlicht, M. D., Lochter, A., Bissell, M. J., and Werb, Z. (1997) Ectopic expression of an autoactivating form of stromelysin-1 promotes mammary tumor formation in transgenic mice and in mice injected with mammary epi-

thelial cells containing an induciible transgene *Breast Cancer Res. Treat.* **46**, 28.

36. Sympson, C. J., Talhouk, R. S., Bissell, M. J., and Werb, Z. (1995) The role of metalloproteinases and their inhibitors in regulating mammary epithelial morphology and function in vivo. *Persp. Drug Discovery Design 2*, 401–411.

37. Lochter, A., Srebrow, A., Sympson, C. J. Terracio, N., Werb, Z., and Bissell, M. J. (1997) Misregulation of stromelysin-1 expression in mouse mammary tumor cells accompanies acquisition of stromelysin-1-dependent invasive properties. *J. Biol. Chem.* **272**, 5007–5015.

38. Basset, P., Bellocq, J. P., Wolf, C., Stoll, I., Hutin, P., Limacher, J. M., Podhajcer, O. L., Chenard, M. P., Rio, M. C., and Chambon, P. (1990) A novel metalloproteinase gene specifically expressed in stromal cells of breast carcinomas. *Nature* **348**, 699–704.

39. Ahmad, A., Hanby, A., Dublin, E., Poulsom, R., Smith, P., Barnes, D., Rubens, R., Anglard, P., and and Hart, I. (1998) Stromelysin 3: an independent prognostic factor for relapse-free survival in node-positive breast cancer and demonstration of novel breast carcinoma cell expression. *Am. J. Path.* **152**, 721–728.

40. Masson, R., Lefebvre, O., Noel, A., El Fahime, M., Chenared, M.-P., Wendling, C., Kebers, F., LeMeur, M., Dierich, A., Foidart, J-M., Basset, P., and Rio, M.-C. (1998) In vivo evidence that the stromelysin-3 metalloproteinase contributes in a paracrine manner to epithelial cell malignancy. *J. Cell. Biol.* **140**, 1535–1541.

41. Brooks, P. C., Stromblad, S., Sanders, L. C., von Schalscha, T. L., Aimes, R. T., Stetler-Stevenson, W. G., Quigley, J. P., and Cheresh, D. A. (1996) Localization of matrix metalloproteinase MMP-2 to the surface of invasive cells by interaction with integrin aVb3. *Cell* **85**, 683–693.

42. Brooks, P. C., Silletti, S., von Schalscha, T. L., Friedlander, M., and Cheresh, D. A. (1998) Disruption of angiogenesis by PEX, a noncatalytic metalloproteinase fragment with integrin binding activity. *Cell* **92**, 391–400.

43. Itoh, T., Tanioka, M., Yoshida, H., Yoshioka, T., Nishimoto, H., and Itohara, S. (1998) Reduced angiogenesis and tumor progression in gelatinase A-deficient mice. *Cancer Res.* **58**, 1048–1051.

44. Liu, Z., Shipley, J. M., Vu, T. H., Zhou, X., Diaz, L. A., Werb, Z., and Senior, R. M. (1998) Gelatinase B-deficient mice are resistant to experimental bullous pemphigoid. *J. Exp. Med.* **188**, 475–482.

45. O'Reilly, M. S., Holmgren, L., Shing, Y., Chen, C., Rosenthal, R. A., Moses, M., Lane, W. S., Cao, Y., Sage, E. H., and Folkman, J. (1994) A novel angiogenesis inhibitor which mediates the suppression of metastasis by a Lewis Lung carcinoma. *Cell* **79**, 315–328.

46. Dong, Z., Kumar, R., Yang, X., and Fidler, I. J. (1997) Macrophage-derived metalloelastase is responsible for the generation of angiostatin in Lewis lung carcinoma. *Cell* **88**, 801–810.

47. Patterson, B. C. and Sang, Q. A. (1997) Angiostatin-converting enzyme activities of human matrilyisin (MMP-7) and gelatinase B/type IV collagenase (MMP-9). *J. Biol. Chem.* **272**, 28,823–28,825.

48. Cao, Y., Chen, A., An, S. A., Ji, R. W., Davidson, D., and Llinas, M. (1997) Kringle 5 of plasminogen is a novel inhibitor of endothelial cell growth. *J. Biol. Chem.* **272**, 22,924–22,928.

49. Stathakis, P., Fitzgerald, M., Matthias, L. J., Chesterman, C. N., and Hogg, P. J. (1997) Generation of angiostatin by reduction and proteolysis of plasmin: catalysis by a plasmin reductase secreted by cultured cells. *J. Biol. Chem.* **272**, 20,641–20,645.

50. Gately, S., Twardowski, S. P., Stack, M. S., Patrick, M., Boggio, L., Cundiff, D. L., Schnaper, H. W., Madison, L., Volpert, O., Bouck, N., Enghild, J., Kwaan, H. C., and Soff, G. (1996) Human prostate carcinoma cells express enzymatic activity that converts human plasminogen to the angiogenesisis inhibitor, angiostatin. *Cancer Res.* **56**, 4887–4890.

51. O'Reilly, M., Boehm, T., Shing, Y., Fukai, N., Vasios, G., Lane, W. S., Flynn, E., Birkhead, J. R., Olsen, B. R., and Folkman, J. (1997) Endostatin: An endogenous inhibitor of angiogenesis and tumor growth. *Cell.* **88,** 277–285.

52. Boehm, T., Folkman, J., Browder, T., and O'Reilly, M. S. (1997) Antiangiogenic therapy of experimental cancer does not induce acquired drug resistance. *Nature* **390**, 404–407.

53. Brown, P. D. (1997) Matrix metalloproteinase inhibitors in the treatment of cancer. *Med. Oncol.* **14**, 1–10.

54. Anderson, I. C., Shipp, M. A., Docherty, A. P., and Teicher, B.A. (1996) Combination therapy including a gelatinase inhibitor and cytotoxic agent reduces local invasion and metastasis of murine Lewis lung carcinoma. *Cancer Res.* **56**, 710–715.

55. Werb Z. (1997) ECM and cell surface proteolysis: Regulating cellular ecology. *Cell* **91**, 439–442.

56. Wernert, N. (1997) The multiple roles of tumor stroma. *Virchows Arch.* **430**, 433–443.

57. Lukashev, M. E. and Werb, Z. (1998) ECM signalling: orchestrating cell behaviour and misbehaviour. *Trends in Cell Biol.* **8**, 437–441.

58. D'Armiento, J., Dalal, S. S., Okada, Y., Berg, R. A., and Chada, K. (1992) Collagenase expression in the lungs of transgenic mice causes pulmonary emphysema. *Cell* **71**, 955–961.

59. Hautamaki, R. D., Kobayashi, D. K., Senior, R. M., and Shapiro, S. D. (1997) Macrophage elastase is required for cigarette smoke-induced emphysema in mice. *Science* **277**, 2002–2004.

60. Senior, R. M., Griffin, G. L., and Mecham, R. P. (1980) Chemotactic activity of elastin-derived peptides. *J. Clin. Invest.* **66**, 859–862.

61. Hunninghake, G. W., Davidson, J. M., Rennard, S., Szapiel, S., Gadek, J. E., and Crystal, R. G. (1981) Elastin fragments attract macrophage precursors to diseased sites in pulmonary emphysema. *Science* **212**, 925–927.

62. Wang, M., Qin, X., Mudgett, J. S., Ferguson, T. A., Senior, R. M., and Welgus, H. G. (1998) Matrix metalloproteinase deficiencies affect contact hypersensitivity: stromelysin-1 deficiency prevents the response and gelatinase B deficiency prolongs the response (Submitted for publication).

63. Flenniken, A. M. and Williams, B. R. (1990) Developmental expression of the endogenous TIMP gene and a TIMP-lacZ fusion gene in transgenic mice. *Genes Dev.* **4**, 1094–106.

64. Kawabe, T. T., Rea, T. J. , Flenniken, A. M., Williams, B. R., Groppi, V. E., and Buhl A. E. (1991) Localization of TIMP in cycling mouse hair. *Development* **111**, 877–879.

65. Flenniken, A. M., Campbell, C. E., and Williams, B. R. (1992) Regulation of TIMP gene expression in cell culture and during mouse embryogenesis. *Matrix Suppl.* **1**, 275–280.

66. Nomura, S., Hogan, B. L., Wills, A. J., Heath, J. K., and Edwards D. R., (1989) Developmental expression of tissue inhibitor of metalloproteinase (TIMP) RNA. *Development* **105**, 575–583.

67. Zeng, Y., Rosborough, R. C., Li, Y., Gupta, A. R., and Bennett, J. (1998) Temporal and spatial regulation of gene expression mediated by the promoter for the human tissue inhibitor of metalloproteinases-3 (TIMP- 3)-encoding Gene *Dev. Dyn.* **211**, 228–237.

68. Alexander, C. M., Hansell, E. J., Behrendtsen, O., Flannery, M. L., Kishnani, N. S., Hawkes, S. P., and Werb, Z. (1996) Expression and function of matrix metalloproteinases and their inhibitors at the maternal-embryonic boundary during mouse embryo implantation. *Development* **122**, 1723–36.

69. Apte, S. S., Hayashi, K. Seldin, M. F., Mattei, M. G., Hayashi, M., and Olsen, B. R. (1994) Gene Encoding a Novel Murine Tissue Inhibitor Of Metalloproteinases (Timp), Timp-3, Is Expressed In Developing Mouse Epithelia, Cartilage, and Muscle, and Is Located On Mouse Chromosome 10. *Dev. Dynamics* **200**, 177–197.

70. Hurskainen, T., Hoyhtya, M., Tuuttila, A., Oikarinen, A., and Autio-Harmainen, H. (1996) mRNA expressions of TIMP-1, -2, and -3 and 92-KD type IV collagenase in early human placenta and decidual membrane as studied by in situ hybridization. *J. Histochem. CytoChem.* **44**, 1379–1388.

71. Leco, K. J., Khokha, R., Pavloff, N., Hawkes, S. P., and Edwards, D. R. (1994) Tissue Inhibitor Of Metalloproteinases-3 (Timp-3) Is an Extracellular Matrix-Associated Protein With a Distinctive Pattern Of Expression In Mouse Cells and Tissues. *J. Biol. Chem.* **269**, 9352–9360.

72. Silbiger, S. M., Jacobsen, V. L., Cupples, R. L., and Koski, R. A. (1994) Cloning Of Cdnas Encoding Human Timp-3, a Novel Member Of the Tissue Inhibitor Of Metalloproteinase Family. *Gene* **141**, 293–297.

73. Wu, I. and Moses, M. A. (1996) Cloning and expression of the cDNA encoding rat tissue inhibitor of metalloproteinase 3 (TIMP-3). *Gene* **168**, 243–246.

74. Zinyk, D. L., Mercer, E. H., Harris, E., Anderson, D. J., and Joyner, A. L. (1998) Fate mapping of the mouse midbrain-hindbrain constriction using a site-specific recombination system. *Curr. Biol.* **8**, 665–668.

75. Brenner, C. A., Adler, R. R., Rappolee, D. A., Pedersen, R. A., and Werb, Z. (1989) Genes for extracellular-matrix-degrading metalloproteinases and their inhibitor, TIMP, are expressed during early mammalian development. *Genes Dev.* **3**, 848–859.

76. Werb, Z., Ashkenas, J., MacAuley, A., and Wiesen, J. F. (1996) Extracellular matrix remodeling as a regulator of stromal-epithelial interactions during mammary gland development, involution and carcinogenesis. *Braz. J. Med. Biol. Res.* **29**, 1087–1097.

77. Alexander, C. M., Howard, E. W., Bissell, M. J., and Werb, Z. (1996) Rescue of mammary epithelial cell apoptosis and entactin degradation by a tissue inhibitor of metalloproteinases-1 transGene *J. Cell Biol.* **135**, 1669–1677.

78. Boudreau, N., Sympson, C. J., Werb, Z., and Bissell, M. J. (1995) Suppression of ICE and apoptosis in mammary epithelial cells by extracellular matrix. *Science* **267**, 891–893.

79. Frisch, S. M. and Francis, H. (1994) Disruption of Epithelial Cell-Matrix Interactions Induces Apoptosis. *J. Cell Biol.* **124**, 619–626.

80. Cardone, M. H., Salvesen, G. S., Widmann, C., Johnson, G., and Frisch, S. M. (1997) The Regulation Of Anoikis - Mekk-1 Activation Requires Cleavage By Caspases. *Cell* **90**, 315–323.

81. Martin, D. C., Ruther, U., Sanchez-Sweatman, O. H., Orr, F. W., and Khokha, R. (1996) Inhibition of SV40 T antigen-induced hepatocellular carcinoma in TIMP-1 transgenic mice. *OncoGene* **13**, 569–576.

82. Ruther, U., Woodroofe, C., Fattori, E., and Ciliberto, G. (1993) Inducible formation of liver tumors in transgenic mice. *OncoGene* **8**, 87–93.

83. Kruger, A., Fata, J. E., and Khokha, R. (1997) Altered tumor growth and metastasis of a T-cell lymphoma in Timp-1 transgenic mice. *Blood* **90**, 1993–2000.

84. Kruger, A., Sanchez-Sweatman, O. H., Martin, D. C., Fata, J. E., Ho, A. T., Orr, F. W., Ruther, U., and Khokha, R. (1998) Host TIMP-1 overexpression confers resistance to experimental brain metastasis of a fibrosarcoma cell line. *OncoGene* **16**, 2419–2423.

85. Goss, K. J., Brown, P. D., and Matrisian, L. M. (1998) Differing effects of endogenous and synthetic inhibitors of metalloproteinases on intestinal tumorigenesis. *Int. J. Cancer* **78**, 629–635.

86. Soloway, P. D., Alexander, C. M., Werb, Z., and Jaenisch, R. (1996) Targeted mutagenesis of *Timp*-1 reveals that lung tumor invasion is influenced by *Timp*-1 genotype of the tumor but not by that of the host. *OncoGene* **13**, 2307–2314.

87. Mignatti, P., Robbins, E., and Rifkin, D. B. (1986) Tumor invasion through the human amniotic membrane: requirement for a proteinase cascade. *Cell* **47**, 487–498.

88. Moses, M. A. and Langer, R. (1991) A metalloproteinase inhibitor as an inhibitor of neovascularization. *J. Cell BioChem.* **47**, 230–235.

89. Fowlkes, J. L., Enghild, J. J., Suzuki, K., and Nagase, H. (1994) Matrix metalloproteinases degrade insulin-like growth factor-binding protein-3 in dermal fibroblast cultures. *J. Biol. Chem.* **269**, 25,742–25,746.

90. Thrailkill, K. M., Quarles, L. D., Nagase, H., Suzuki, K., Serra, D. M., and Fowlkes, J. L. (1995) Characterization of insulin-like growth factor-binding protein 5-degrading proteases produced throughout murine osteoblast differentiation. *Endocrinology* **136**, 3527–3533.

91. Gasson, J. C., Golde, D. W., Kaufman, S. E., Westbrook, C. A., Hewick, R. M., Kaufman, R. J., Wong, G. G., Temple, P. A., Leary, A. C., and Brown, E. L. (1985) Molecular characterization and expression of the gene encoding human erythroid-potentiating activity. *Nature* **315**, 768–771.

92. Hayakawa, T., Yamashita, K., Tanzawa, K., Uchijima, E., and Iwata, K. (1992) Growth-promoting activity of tissue inhibitor of metalloproteinases-1 (TIMP-1) for a wide range of cells. A possible new growth factor in serum. *Febs Lett.* **298**, 29–32.

93. Stetler-Stevenson, W. G., Bersch, N., and Golde, D. W. (1992) Tissue inhibitor of metalloproteinase-2 (TIMP-2) has erythroid-potentiating activity. *Febs Lett.* **296,** 231–234.

94. Matsumoto, H., Ishibashi, Y., Ohtaki, T., Hasegawa, Y., Koyama, C., and Inoue, K. (1993) Newly established murine pituitary folliculo-stellate-like cell line (TtT/GF) secretes potent pituitary glandular cell survival factors, one of which corresponds to metalloproteinase inhibitor. *Biochem. Biophys. Res. Comm.* **194**, 909–915.

95. Satoh, T., Kobayashi, K., Yamashita, S., Kikuchi, M., Sendai, Y., and Hoshi, H. (1994) Tissue inhibitor of metalloproteinases (TIMP-1) produced by granulosa and oviduct cells enhances in vitro development of bovine embryo. *Biol. Repro.*, **50**, 835–844.

96. Nemeth, J. A. and Goolsby, C. L. (1993) TIMP-2, a growth-stimulatory protein from SV40-transformed human fibroblasts. *Exp. Cell Res.* **207**, 376–382.

97. Nothnick, W. B., Soloway, P. D., and Curry, T. E. (1998) Pattern of mRNA *Timp-1* Gene *Biol. Repro.* **59**, 364–370.

98. Murphy, A. N., Unsworth, E. J., and Stetler-Stevenson, W. G. (1993) Tissue inhibitor of metalloproteinases-2 inhibits bFGF-induced human microvascular endothelial cell proliferation. *J. Cell Physiol.* **157**, 351–358.

99. Boujrad, N., Ogwuegbu, S. O., Garnier, M., Lee, C. H., Martin, B. M., and Papadopoulos, V. (1995) Identification of a stimulator of steroid hormone synthesis isolated from testis. *Science* **268**, 1609–1612.

100. Nothnick, W. B., Soloway, P., and Curry, T. E. (1997) Assessment of the role of tissue inhibitor of metalloproteinase-1 (Timp-1) during the periovulatory period in female mice lacking a functional Timp-1 Gene. *Biol. Repro.* **56**, 1181–1188.

101. Yoon, B. J., Osiewicz, K., Weaver, K., Potter, W., Johnston, B., Preston, M. J., Jaenisch, R., Pier, G. B., Kubes, P., Dougherty, T., and Soloway, P. D. TIMP-1-regulates innate immune responses to infection (Manuscript submitted).

102. Osiewicz, K., McGarry, M., and Soloway, P. D. Hyper-resistance to infection in TIMP-1-deficient mice is neutrophil-dependent but not immune cell autonomous. *Ann. NY Acad. Sci.* (In press).

103. Okamoto, N., Tobe, T., Hackett, S. F., Ozaki, H., Vinores, M. A., LaRochelle W., Zack, D. J., and Campochiaro, P. A. (1997) Transgenic mice with increased

expression of vascular endothelial growth factor in the retina: a new model of intraretinal and subretinal neovascularization. *Am. J. Pathol.* **151**, 281–291.

104. Tobe, T., Yamada, H., Yamada, E., Okamoto, N., Zack, D. J., Werb, Z., Soloway, P. D., and Campochiaro, P. A. Increase in the Ratio of Tissue Inhibitor of Metalloproteinases-1 to Metalloproteinases Promotes Vascular Endothelial Growth Factor-Induced Neovascularization in the Retina (Manucscript Submitted).

105. Strongin, A. Y., Collier, I., Bannikov, G., Marmer, B. L., Grant, G. A., and Goldberg, G. I. (1995) Mechanism of cell surface activation of 72-kDa type IV collagenase. Isolation of the activated form of the membrane metalloprotease. *J. Biol. Chem.* **270**, 5331–5338.

106. Will, H., Atkinson, S. J., Butler, G. S., Smith, B., and Murphy, G. (1996) The soluble catalytic domain of membrane type 1 matrix metalloproteinase cleaves the propeptide of progelatinase A and initiates autoproteolytic activation. Regulation by TIMP-2 and TIMP-3. *J. Biol. Chem.* **271**, 17,119–17,123.

107. Wang, Z. and Soloway, P. D. TIMP-1 and TIMP-2 perform different functions in vivo. *Ann. NY Acad. Sci.* In press.

108. Holmbeck, K., Bianco, P., Caterina, J., et al. (1999) MT1-MMP-deficient mice develop dwarfism, oseopenia, arthritis, and connective tissue disease due to inadequate collagen turnover. *Cell* **99**, 81–92.

II

EXPRESSION AND PURIFICATION OF MMPs AND TIMPs

7

Expression of MMPs and TIMPs in Mammalian Cells

Karen M.-L. Yeow, Blaine W. Phillips, P. Paul Beaudry,
Kevin J. Leco, Gillian Murphy, and Dylan R. Edwards

1. Introduction

Expression of recombinant matrix metalloproteinases (MMPs) and tissue inhibitors of metalloproteinases (TIMPs) in mammalian cells is an important step in their functional characterization. Also, transient transfection analysis of promoter constructs driving CAT or luciferase reporter genes is a mainstay of gene regulation studies. One of the critical advantages of mammalian expression over bacterial systems for production of functional proteins is that problems of refolding, which are particularly significant for the MMPs and TIMPs, are generally not encountered. Transient expression in COS cells is very rapid and can be useful in many situations, for instance to demonstrate activity or to compare the characteristics of mutant proteins. It can also generate sufficient quantities of protein for biochemical and cell biological studies, but most often bulk production will require stable expression. In this chapter we will discuss our experiences with transient expression in COS-1 and C3H 10T1/2 mouse fibroblasts, and describe two systems that we have used for constitutive stable expression, namely BHK (Baby Hamster Kidney) and the NS0 myeloma cell line. This chapter complements the information in Chapter 11 by Butler, d'Ortho, and Atkinson who have described stable, tetracycline-regulated expression of MT1-MMP in CHOL cells. Also, the reader should note that many expression vector systems are now available commercially, and we have not attempted here to provide comparisons.

The transfection protocol that we have used for most transient and stable expression studies is the Chen and Okayama procedure *(1)*. Stable expression in

From: *Methods in Molecular Biology, vol. 151: Matrix Metalloproteinase Protocols*
Edited by: I. Clark © Humana Press Inc., Totowa, NJ

BHK TK21- cells involves the use of the pNUT vector, the key feature of which is a mutated dihydrofolate reductase (DHFR) gene driven by the SV40 early promoter *(2)*. This encodes an altered DHFR protein that has a 270-fold lower affinity for the competitive inhibitor methotrexate compared to the wild-type protein *(3)*, allowing rapid selection of stably transfected cells in high concentrations of methotrexate and, in theory, avoiding the need for DHFR-deficient recipient cells. Transfectants carrying high copy numbers of the inserted vector are selected by placing the cells in a high concentration of methotrexate (0.5 m*M*) immediately after recovery from the transfection procedure. Expression of the gene of interest is under the control of the metallothionein-1 promoter. After the selection process, cells can be passaged and maintained in normal media, except that every fourth passage, the selection pressure has to be reapplied in order to maintain high levels of expression.

Stable expression of MMPs and TIMPs in mouse myeloma cells *(4,5)* has proved a useful large scale production system since these cells make negligible amounts of these proteins and can, in some instances, generate up to 30 mg protein per litre (e.g., TIMP-1). Transfection of expression vectors into these cells is most efficiently carried out by electroporation. Expression of the target protein may be effected using commercial or custom-made hCMV driven vectors with antibiotic selection cassettes *(6)* but the most efficient vector, combining both selection and amplification modes, employs the lack of glutamine synthetase production by these cells (*see* **Note 1**) *(7)*.

2. Materials

2.1. Transient Transfection

1. COS-1 cells seeded at 1×10^6 cells per 10cm tissue culture dish; C3H 10T1/2 at 5×10^5 cells/dish.
2. pXMT2 expression vector (*see* **Fig. 1**)
3. 2X BES buffered saline (BBS): 50 m*M* BES, 280 m*M* NaCl, 1.5 m*M* Na$_2$HPO$_4$; pH to exactly 6.96 with HCl.
4. 1 *M* CaCl$_2$ solution (sterile).
5. TE buffer: 10 m*M* Tris-HCl, 1 m*M* EDTA, pH 8.0; sterile.
6. DNA: 1–5 μg/μL in TE or H$_2$O.
7. Tissue culture medium: 5–10% Fetal Calf Serum (FCS) in Dulbecco's Modified Eagle's Medium : Ham's F12 (DMEM-F12), 0.5% antibiotics, for routine maintenance of cells.
8. Tissue culture flasks (75 cm^2).
9. 5 mL Falcon snap cap tubes.
10. Tissue culture incubator set at 3% CO$_2$/ 97% air atmosphere, another at 5% CO$_2$/95% air atmosphere.

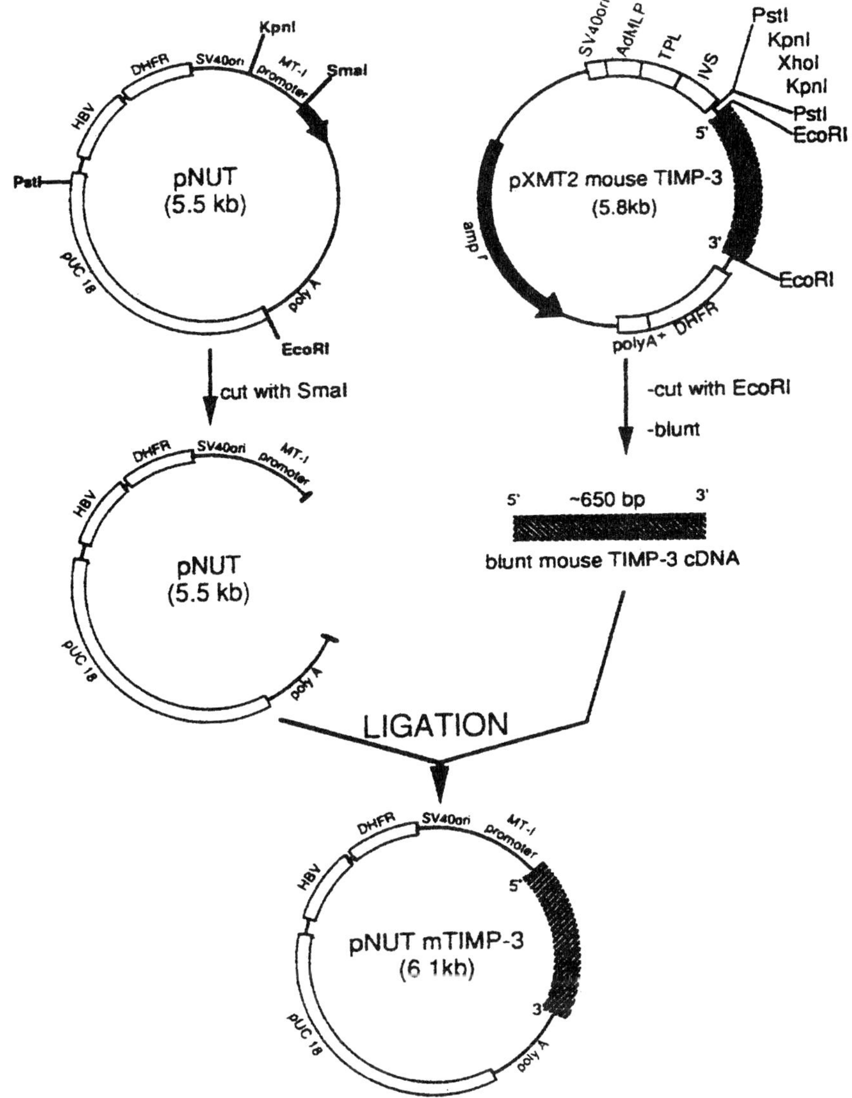

Fig. 1. Maps of pXMT2 and pNUT and construction of a pNUT-mouse TIMP-3 expression vector.

2.2. Stable Transfection in BHK Cells

1. BHK TK21 cells seeded at 5×10^5 cells per 10 cm tissue culture dish.
2. pNUT expression vector (*see* **Fig. 1**).
3. Methotrexate (sodium injection solution) 25 mg/mL.

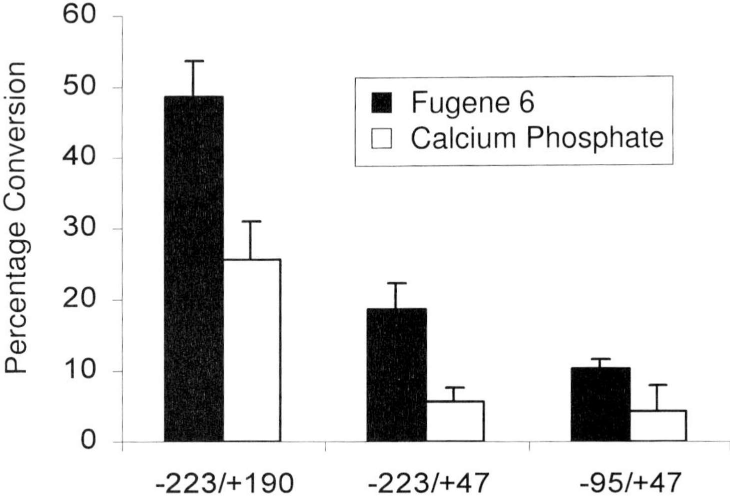

Fig. 2. Transient transfection of TIMP-1-CAT reporter plasmids into mouse C3H10T1/2 cells using the Chen and Okayama and Fugene 6 procedures. The results represent triplicate transfections with the indicated plasmids.

3. Methods

3.1 Calcium Phosphate/DNA Precipitation Transfection

The transfection method that we have found to be most reliable and versatile is a standard Chen and Okayama protocol (*1*). Since there are some critical differences between this procedure and a related protocol described in Chapter 11 by Butler et al., the version that we use will be described in detail. This method works well for introducing supercoiled plasmid DNA into all the adherent cells that we have used to date. However, we have recently found that a commercial transfection system that uses cationic lipids (Fugene 6; Roche Bioscience) may be superior for fibroblastic cells (*see* **Fig. 2**). We have not provided details of the Fugene protocol, which follows the manufacturer's instructions precisely.

The following is our standard transfection procedure:

1. The day before transfecting, plate cells out at a density of 5×10^5 or 1×10^6 cells (depending on cell type) per 10cm petri dish. Each plate of cells should contain 10 mL DMEM:F12 with 5% or 10% (vol/vol) FCS (depending on the cell type, *see* **Note 2**), 0.5% (vol/vol) antibiotics in DMEM-F12. The cells should be evenly spread out all over the plate (*see* **Note 3**). Leave two extra plates of cells for untransfected controls.
2. Aliquot 1 mL 2X BBS solution into Falcon tubes. Allow one tube of BBS per tube of DNA mixture.

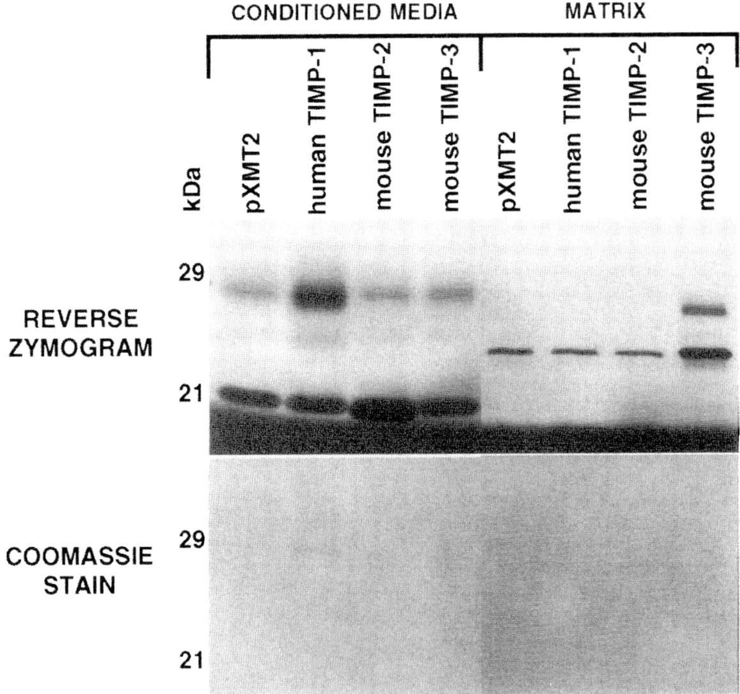

Fig. 3. Transient expression of TIMPs in COS-1 cells. Samples of conditioned media were concentrated 10-fold using Centricon filters, then 10 µL samples were analyzed by reverse zymography (*upper panel*) or SDS-PAGE with Coomassie stain (*lower panel*).

3. Make the DNA mix: use 10 µg of DNA per plate of cells to be transfected. Typically, we transfect two plates of cells per construct, so a total of 20 µg DNA is used. Add the appropriate amount of DNA to enough TE buffer to produce a final volume of 900 µL. Mix carefully, avoiding formation of droplets on the sides of the tube which may alter the final DNA concentration. Then add 100 µL 1 *M* CaCl₂ solution to the TE-DNA mixture, and gently shake the tube until the solutions are completely mixed. Set aside.

4. Slowly add the DNA mixture dropwise to a tube of 2X BBS while vortexing vigorously.

5. Leave the tube undisturbed for exactly 20 min after the DNA mix has been added to the BBS (*see* **Note 4**).

6. Add 1 mL of the DNA/BBS mixture (10 µg DNA) to each of the two plates of cells to be transfected with this particular construct. Make sure that the tube is mixed before adding the DNA. Swirl the plates gently to ensure even mixing of the DNA in the media.

7. Leave the cells at 3% CO_2 overnight (*see* **Note 5**).

8. The following day, change to fresh serum-containing medium and leave the cells to recover at 5% CO_2 for 24 h. All subsequent cell culture is done in a regular 5% CO_2 atmosphere.

9. After the recovery period, the medium is changed for fresh medium, the composition of which depends on the nature of the experiment. For expression of recombinant proteins in COS-1 cells, change the medium to serum-free DMEM:F12 and collect conditioned media either 24 or 48 h later. Some expressed proteins such as TIMP-3 are deposited in the ECM, which is collected by washing the cell monolayer in PBS and then detaching the cells with 5 m*M* EDTA, 5 m EGTA *(8)*. The ECM proteins remaining on the dish are removed by scraping in small volume of SDS-PAGE sample buffer (without DTT) for analysis by SDS-PAGE and reverse zymography or zymography (*see* Chapter 23 by Hawkes, Li, and Taniguchi). **Figure 3** shows results of reverse zymography for a variety of mouse and human TIMP genes transiently expressed in COS-1 cells.

10. For transient transfection analysis of reporter constructs, after the recovery period, change the medium to serum-free DMEM:F12 for 6–24 h, then add stimuli (serum, growth factors, etc) as appropriate for a further 18–24 h. Wash the cells with PBS twice, then harvest by scraping the cells from the dish in 0.25 *M* Tris-HCl, pH 7.8, and lyse with 3 cycles of freeze-thaw. Alternatively, yields can be improved by using a commercially available reporter lysis buffer (Promega, Madison, WI). After centrifugation to remove cell debris, protein concentrations in the lysates are measured by Bradford assay, and equal amounts of protein extracts are used for CAT assays.

3.2 Stable Expression in BHK Cells

1. We use the same Chen and Okayama procedure with transfection with supercoiled pNUT expression vectors for generation of stably transfected BHK cells. Construction of a typical pNUTmouseTIMP-3 expression vector is shown in **Fig. 1**.

2. Follow the protocol in **Subheading 3.1.** up to **step 9**, then after the recovery period change the medium to 5% FCS, 0.5% antibiotics, and 5 m*M* methotrexate in DMEM-F12. The untransfected cells should also be subjected to selection, in order to see how long it takes for untransfected cells to die off.

3. Change the selection media every day for the first week or so, until more of the cells begin to die off. Then the media can be changed every 2 d, until foci of transfectants appear on the plates. This may take 2–3 wk. The untransfected cells typically die off slowly, between 10–16 d (*see* **Note 6**).

4. Once the foci of cells have formed, keep changing the media until they have expanded somewhat. Then, they can be isolated with cloning rings or trypsinized and pooled onto a fresh plate. Spreading the cells out more evenly allows faster growth.

5. The cells are typically maintained in methotrexate for the first 3 passages. After this time, they can be maintained in normal media lacking methotrexate. However, the cells should be passaged in methotrexate every 4th passage to maintain

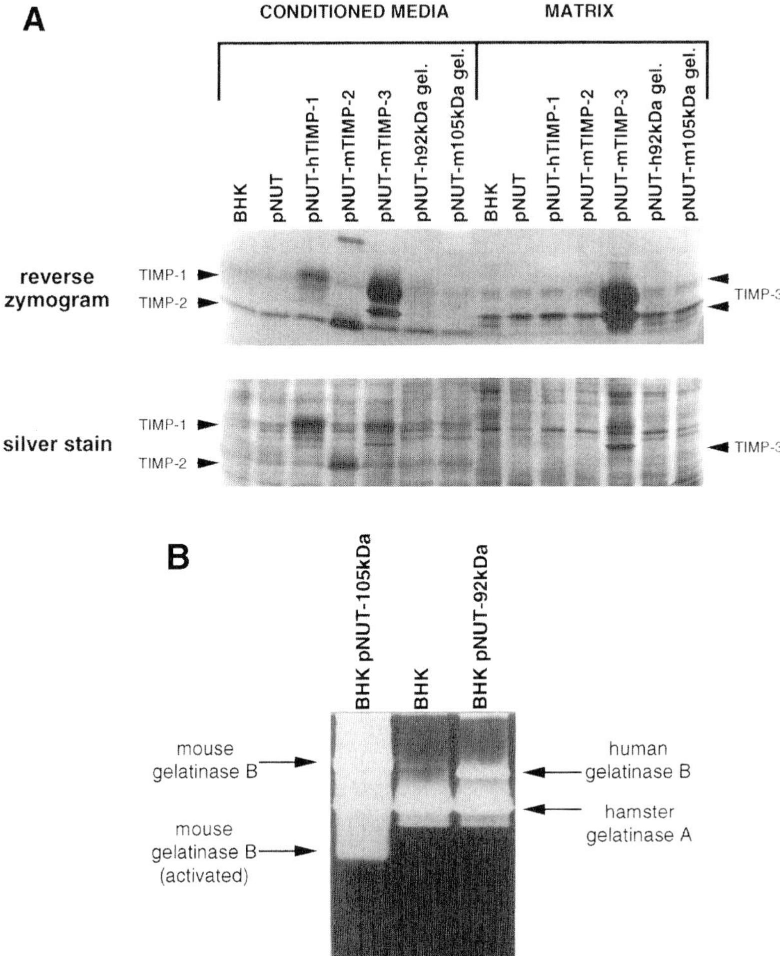

Fig. 4. Stable expression of TIMPs and MMPs in BHK cells. Samples of unconcentrated conditioned media (20 μL) or ECM (5 μL) were analyzed by (**A**) reverse zymography or SDS-PAGE with silver stain; (**B**) zymography.

selection pressure. Stocks of cells should be frozen after at least one passage in normal media, preferably early on after the selection process has been completed. The BHK cells grow very rapidly, so we recommend using 1/20 or 1/50 split ratios every 3–4 d.

6. Conditioned media or ECM samples can be harvested from cells growing in methotrexate-free media, and tested by zymography/reverse zymography, for the expression of MMPs or TIMPs, as described in **step 9** in **Subheading 3.1. Figure 4** shows zymography and reverse zymography analysis of medium and ECM

from BHK cells stably expressing a variety of TIMPs (**Fig. 4A**) and mouse and human MMP-9 (**Fig. 4B**) compared to control empty pNUT vector or untransfected BHK cells (*see* **Note 7**).

4. Notes

1. NSØ mouse myeloma cells *(9)* are obtainable from the European Collection of Cell Cultures (ECACC, cat. no. 85110503). Transfection with vectors of the pEE series that incorporate an hCMV promoter expression cassette with an SV40 promoter driven glutamine synthetase cDNA allows selection of cells that can grow in a glutamine free medium *(7)*. Amplification of expression levels is possible by titrated use of the glutamine synthetase inhibitor methionine sulphoximine.
2. For most cell types use DMEM:F12 with 10% (vol/vol) FCS. BHK cells need only 5% FCS.
3. Before transfecting, ensure that the cells are in the best possible condition. Use cells that are subconfluent and growing prior to plating for transfection. Do not use cells that have become confluent. On the day of transfection, make sure they look healthy and are evenly spread out on the plate.
4. Add the DNA to the BBS slowly, i.e., one drop per second, to ensure that the DNA precipitate is as fine as possible. The DNA should appear as a fine cloudy mist in the tube. Ensure that the DNA is well mixed in the media.
5. We find this to be the most important aspect of the transfection procedure.
6. Selection will take approx 3 wk. It takes the cells 2–3 wk to die off, and for stable transfectants to appear. You will see many floaters and dead cells for the first week or so, but be patient. It is important to change the medium 24–48 h for the first 2 wk, to ensure that there is a constant selection pressure. Once the stable transfectants are established, they will grow fairly rapidly.
7. Yields of expressed protein from the pNUT/BHK system can be variable, as shown by the much greater level of expression of mouse MMP-9 compared to human MMP-9 in **Fig. 4**. We estimate that expression of TIMP-1, -2 and -3 in the BHK system is in the range 1-10 mg/litre in conditioned medium. All data in **Figs. 3** and **4** were obtained with pooled stable clones, so it is likely that expression could be pushed higher by picking individual clones.

References

1. Chen, C. and Okayama, H. (1987) High-efficiency transformation of mammalian cells by plasmid DNA. *Mol. Cell. Biol.* **7**, 2745–2752.
2. Palmiter, R. D., Behringer, R. R., Quaife, C. J., Maxwell, F., Maxwell, I. H., and Brinster, R. L. (1987) Cell lineage ablation in transgenic mice by cell-specific expression of a toxin gene. *Cell* **50**, 435–43.
3. Funk, W. D., MacGillivray, R. T., Mason, A. B., Brown, S. A., and Woodworth, R. C. (1990) Expression of the amino-terminal half-molecule of human serum transferrin in cultured cells and characterization of the recombinant protein. *Biochemistry* **29**, 1654–1660.

4. Knäuper, V., Docherty, A. J. P., Smith, B., Tschesche, H., and Murphy, G. (1997). Analysis of the contribution of the hinge region of human neutrophil collagenase (HNC, MMP-8) to stability and collagenolytic activity by alanine scanning mutagenesis. *FEBS Letts.* **405**, 60–64.

5. Knäuper, V., López-Otin, C., Smith, B., and Murphy, G. (1996). Purification and characterisation of human collagenase-3. *J. Biol. Chem.* **271**, 1544–1550.

6. Stephens, P. and Cockett, M. I. (1989) The construction of a highly efficient and versatile set of mammalian expression vectors. *Nucl. Acids Res.* **17**, 7110.

7. Bebbington, C. R., Renner, G., Thomson, S., King, D., Abrams, D., and Yarranton, G. T. (1992) High-level expression of a recombinant antibody from myeloma cells using a glutamine synthetase gene as an amplifiable selectable marker. *Bio/Technology* **10**, 169–195.

8. Leco, K. J., Khokha, R., Pavloff, N., Hawkes, S. P., and Edwards, D. R. (1994). Tissue inhibitor of metalloproteinases-3 (TIMP-3) is an extracellular matrix-associated protein with a distinctive pattern of expression in mouse cells and tissues. *J. Biol. Chem.* **269**, 9352–9360.

9. Galfre, G. and Milstein, C. (1981) Preparation of monoclonal antibodies: strategies and procedures. *Meths. Enzymol.* **73(B)**, 3–46.

8

Expression of Recombinant Matrix Metalloproteinases in *Escherichia coli*

L. Jack Windsor and Darin L. Steele

1. Introduction

With the advent of recombinant DNA technology, numerous systems have been utilized for the overexpression of proteins. Recombinant protein expression in *Escherichia coli (E. coli)* typically provides large quantities of the protein of interest in a relatively short period of time. The expression can result in the accumulation of the recombinant protein to levels approaching 30–50% of the total *E. coli* protein. Expression of matrix metalloproteinases (MMPs) in *E. coli* has proven to be very useful in generating proteins for structural and functional studies, including X-ray crystallography. Numerous MMPs, including altered forms, have been successfully expressed in and purified from *E. coli*. These include forms of stromelysin-1, stromelysin-2, stromelysin-3, matrilysin, elastase, collagenase-1, collagenase-3, neutrophil collagenase, and membrane type-1 MMP *(1–20)*. The expression of a truncated form of human stromelysin-1 (SL-1) will be used to illustrate the methods utilized for the expression of a matrix metalloproteinase in *E. coli* and for its extraction, refolding, and purification.

2. Materials

1. pGEMEX-1 expression system (Promega; Madison, WI).
2. Matrix Metalloproteinase cDNA.
3. *E. coli* strains CJ236, DH5α, and BL21 (DE3).
4. Helper phage VCS-M13 (Stratagene; La-Jolla, CA).
5. Oligonucleotide primers.
6. Restriction enzymes, T7 DNA polymerase, and T4 DNA ligase.
7. Agarose and DNA sequencing gel equipment.

From: *Methods in Molecular Biology, vol. 151: Matrix Metalloproteinase Protocols*
Edited by: I. Clark © Humana Press Inc., Totowa, NJ

8. DYT (double yeast tryptone) media.
9. Ampicillin.
10. IPTG (isopropyl-β-D-thio-galactopyranoside).
11. French Press.
12. PMSF (phenylmethylsulfonyl fluoride).
13. Standard (or wash) buffer: 50 mM Tris-HCl, pH 7.5, 0.2 M NaCl, 5 mM CaCl$_2$, and 1 μM ZnCl$_2$ at 4°C.
14. Extraction buffer: 6 M urea in 50 mM Tris-HCl, pH 7.5, 0.2 M NaCl, 5 mM CaCl$_2$, and 1 μM ZnCl$_2$ at 4°C.
15. SDS-PAGE (sodium dodecyl sulfate-polyacrylamide gel electrophoresis) equipment.
16. Whatman filter paper #1.
17. Chromatography equipment.
18. Sephacryl S-200 HR (Pharmacia; Piscataway, NJ).
19. Anti-MMP antibody.
20. CNBr-activated Sepharose 4B (Pharmacia).
21. APMA (aminophenyl-mercuric acetate).
22. Mca peptide (Bachem; King of Prussia, PA).
23. Casein.

3. Methods

The methods described below outline (1) the construction of the expression plasmid, (2) the induction of protein expression, (3) the extraction of the protein from *E. coli*, (4) the refolding and purification of the protein, and (5) the characterization of the protein.

3.1. Expression Plasmid

The construction of the expression plasmid for a truncated form of stromelysin-1 (SL-1) is described in **Subheading 3.1.1.–3.1.5.** This includes (a) the description of the expression vector, (b) the description of the SL-1 cDNA, (c) cloning, (d) the deletion of the coding regions for the T7 gene 10 protein in the pGEMEX-1 vector and for the signal peptide of SL-1, and (e) the deletion of the coding region for the hemopexin-like domain of SL-1.

3.1.1. pGEMEX-1 Expression Vector

The pGEMEX-1 (**Fig. 1A**) expression system (Promega) is based on the T7 expression system developed by Studier *(21)*. Sequences inserted into the multiple cloning site can be expressed as T7 gene 10 fusion proteins that are transcriptionally initiated from the pGEMEX-1 T7 promoter when transformed into *E. coli* DE3 cells, which contain an IPTG (isopropyl-β-D-thio-galactopyranoside) inducible T7 RNA polymerase gene. Expression of the recombinant protein alone can be accomplished by removing the coding region for the T7 gene 10 protein. The pGEMEX-1 expression vector contains a ph-age

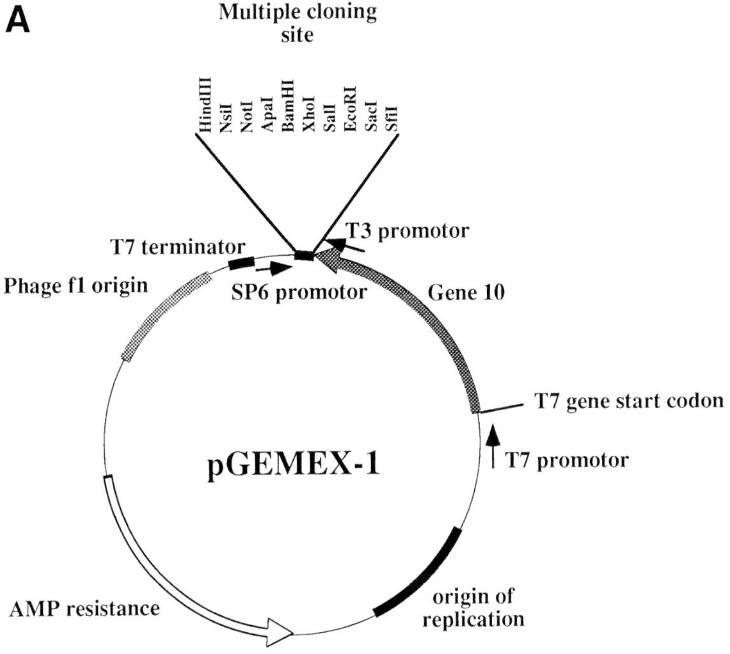

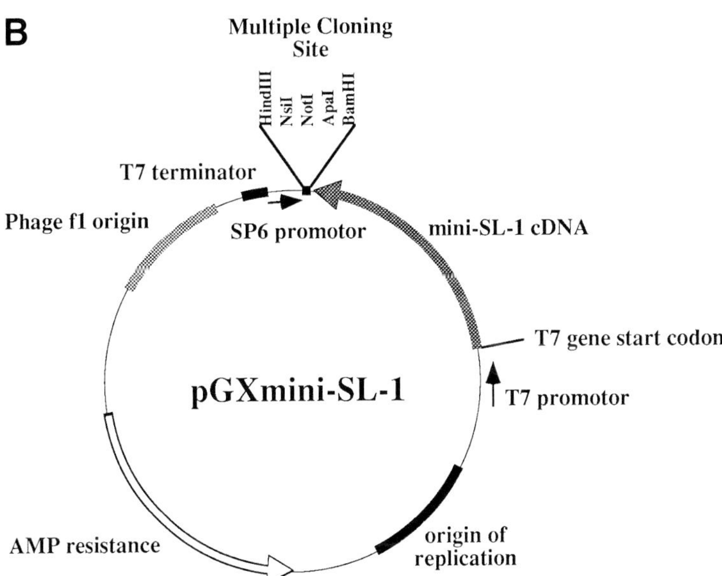

Fig. 1 (**A**) Schematic drawing of pGEMEX-1 expression plasmid adapted from Promega (Madison, WI). (**B**) Schematic drawing of expression plasmid pGXmini-SL-1 for the expression of truncated SL-1.

f1 origin of replication that provides a mechanism for the production of single-stranded DNA, which can then be used in site-directed mutagenesis in order to construct expression plasmids and/or mutants in the recombinant protein. The pGEMEX-1 expression vector also contains an ampicillin resistance gene for selectivity. These and other characteristics make pGEMEX-1 a suitable expression system to use for the expression of truncated SL-1 and other proteins (*see* **Note 1**).

3.1.2. cDNA

A plasmid (pBS'SL) containing the cDNA for human SL-1 was kindly donated by Dr. Goldberg *(22)*. It contained the initiation methionine codon followed by 1431 nucleotides coding for the 477 amino acid preprostromelysin-1 protein along with a portion of the 3' untranslated region. The coding region for SL-1 was removed from pBS'SL by digestion with EcoRI and BamHI by standard molecular biology techniques *(23)* (*see* **Note 2**).

3.1.3. Cloning

DNA manipulations were performed by standard recombinant DNA methods *(23)* to construct the expression plasmid and are not described here in detail due to space limitations. The expression plasmid for truncated SL-1 (mini-SL-1) was constructed by adding BamHI linkers to the EcoRI-BamHI fragment from pBS'SL. After digestion with BamHI and ligation into the BamHI restriction site of the pGEMEX-1 vector, the DNA was transformed into *E. coli* DH5α cells by standard methods *(23)*. The *E. coli* DH5α cells were then plated on DYT (double yeast and tryptone) plates containing ampicillin (50–100 µg/mL) and incubated overnight at 37°C. Single colonies were selected and grown overnight in DYT with ampicillin. The plasmid DNA was then isolated *(23)* and checked for the presence of the insert and for the correct orientation using restriction enzyme digestions and DNA sequencing *(24,25)*.

3.1.4. Removal of the Coding Regions for the SL-1 Signal Peptide and for the T7 Gene 10 Protein

The first 18 amino acids of human SL-1 constitute the signal peptide that is removed upon secretion into the extracellular environment of eukaryotic cells and is not necessary for protein expression in prokaryotic cells such as *E. coli*. The signal peptide sequence of the SL-1 insert and the coding sequence for the T7 gene 10 protein in pGEMEX-1 were removed by site-directed mutagenesis *(26)* so that only the SL-1 protein would be produced. Single-stranded DNA was produced from *E. coli* CJ236 cells (previously transformed with pGEMEX-1 containing the cDNA for SL-1) infected with helper phage VCS-M13 (Stratagene, La-Jolla, CA) as described by the manufacturer. Site-directed mutagenesis was performed using this single-stranded DNA, the oligonucleotide primer 5' GGAGATATACAT**ATG**TATCCGCTGGATGGAGCTGC 3', T7 DNA polymerase, and T4 DNA ligase. This oligonucleotide primer was also specifically designed to place the initiation codon ATG in frame with the coding region of the SL-1 cDNA. This expression plasmid, pGSL-1 *(9)*, containing the full-length cDNA for SL-1 minus the signal peptide was verified using restriction enzyme digestions and DNA sequencing.

3.1.5. Deletion of the Coding Region
for the Hemopexin-Like Domain (Carboxyl Terminus) of SL-1

Single-stranded DNA was produced from *E. coli* CJ236 cells transformed with the pGSL-1 plasmid using helper phage VCS-M13 as described by the manufacturer (Stratagene). The pGSL-1 single-stranded DNA and the primer 5' TATGGACCTCCCCCGGATCCC**TGATAG**AGAGATATGTAGAAG 3' were used to remove the coding sequence for the hemopexin-like domain of SL-1 by introducing the stop codons TGA and TAG after the codon for amino acid 253 of SL-1. This resulted in expression plasmid pGXmini-SL-1 (**Fig. 1B**), which contained the coding region of SL-1 minus the signal peptide and the hemopexin-like domain (mini-SL-1). Before attempting protein expression, the expression construct was verified by restriction enzyme digestions and DNA sequencing.

3.2. Protein Induction

The next steps in this process involve the transformation of *E. coli* with the expression plasmid for mini-SL-1 followed by the induction with IPTG to initiate protein expression (*see* **Note 3**).

3.2.1. E. coli Transformation and Plasmid Selection

1. Transform DH5α *E. coli* cells with plasmid pGXmini-SL-1 DNA using standard molecular biology methods *(23)*.
2. Plate cells on DYT plates containing ampicillin and incubate overnight at 37°C.
3. Select single colonies and grow overnight at 37°C in DYT media containing ampicillin. Take care selecting the single colonies from the plates to avoid cross contamination from the other colonies.
4. Freeze samples of the overnight cultures at –70°C in DYT media containing 25% glycerol.
5. Purify the plasmid DNA from the remaining overnight cultures by standard methods *(23)*.

3.2.2. Induction

1. Transform *E. coli* BL21 (DE3) cells with the pGXmini-SL-1 plasmid DNA using standard methods *(23)* and grow overnight at 37°C in DYT media with ampicillin.

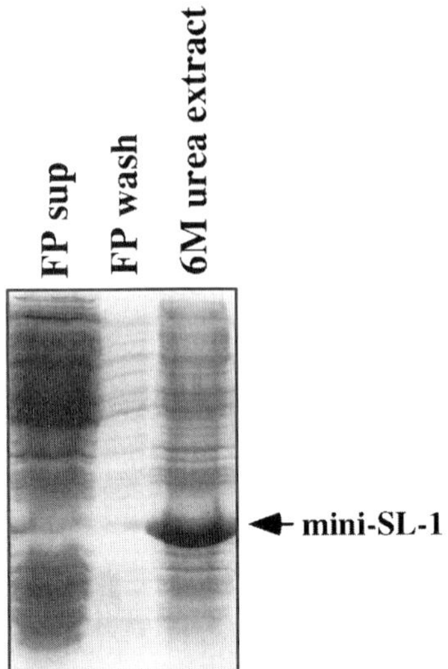

Fig. 2. Extraction of mini-SL-1 from *E. coli*. The supernatant following passage through a French press (FP sup), the wash supernatant of the pellet after passage through the French press (FP wash), and the *6 M* urea extract of the pellet containing the inclusion bodies (*6 M* urea extract) were resolved in a 10% SDS-PAGE gel and stained with Coomassie blue.

2. Freeze samples of the overnight cultures at –70°C in DYT media containing 25% glycerol.
3. Inoculate individual liters of DYT containing ampicillin with aliquots (2–4 mL) of the overnight cultures and allow to grow to a density of 10^8 cells/mL.
4. Induce the cells with IPTG (1 m*M*) for 2–3 h. It should be noted that the induction time can be extended to increase expression, however for mini-SL-1 this usually results in more heterogeneity in the protein.

3.2.3. Harvest

1. After 2–3 h of induction, harvest the cells by centrifugation (30,000*g*; 30 min) and resuspend the cell pellet in standard buffer.
2. After resuspension of the pellet, the cells can either be directly extracted or frozen at –20°C for as long as a month before extraction without significant breakdown of the mini-SL-1 protein.

3.3. Protein Extraction

The steps described in **Subheadings 3.3.1.** and **3.3.2.** outline the procedure for extraction of the protein from the *E. coli* cells in a manner that yields significant quantities of native-like protein after refolding (*see* **Note 4**).

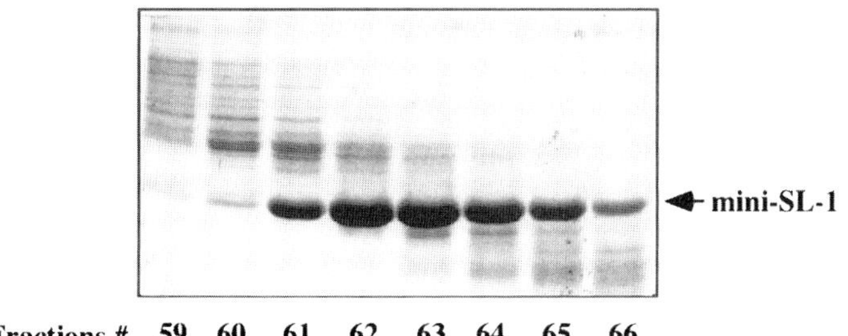

Fig. 3. Gel filtration column. Fractions collected from a Sephacryl S-200 HR column were resolved in a 10% SDS-PAGE gel and stained with Coomassie blue.

3.3.1. Cell Disruption

1. Add PMSF (phenylmethylsulfonyl fluoride) to a final concentration of 1 mM to freshly resuspended cells or fully thawed frozen cells.
2. Pass the cells through a French press at 10,000–15,000 psi, then repeat to ensure complete cell lysis.

3.3.2. Isolation of Inclusion Bodies and Extraction

1. Centrifuge the cell lysate (30,000g; 20 min) and retain both supernatant (to check for presence of the mini-SL-1 protein) and pellet (*see* **Fig. 2**).
2. Wash the pellet (which contains the protein laden inclusion bodies) with standard buffer by vigorously pipeting to a homogeneous suspension, then re-pellet by centrifugation as in **step 1**.
3. Solubilize the pellet from **step 2** in extraction buffer and extract at 4°C on an orbital mixer for 2–3 h or overnight. Centrifuge as in **step 1** and retain the supernatant (extract).
4. The extract can be frozen at –20°C for up to 2 wk before purification.

3.4. Protein Refolding and Purification

Described below are the steps that can be utilized in the separation of the mini-SL-1 protein from the *E. coli* proteins and its progressive purification to a homogeneous sample (*see* **Note 5**).

3.4.1. Gel Filtration

1. Apply the urea extract onto a gel filtration column (2.5 cm × 88 cm) containing Sephacryl S-200 HR (Pharmacia) equilibrated with extraction buffer and elute with the same buffer.
2. Identify fractions of interest by Coomassie blue staining of a SDS-PAGE gel (*see* **Fig. 3**) and verify by a Western blot probed with anti-SL-1 antibody IID4 or equivalent *(27)*.

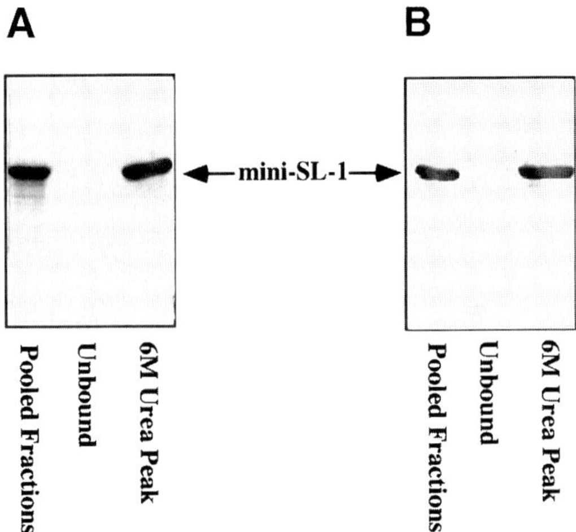

Fig. 4. Affinity purification of fractions from the gel filtration column. Selected fractions from the Sephacryl S-200 HR column were pooled, diluted and/or dialyzed to remove the urea, and passed over the anti-SL-1 antibody IID4 affinity column. The bound material was eluted with 6 *M* urea (peak) and resolved in 12.5% SDS-PAGE gels. Companion gels were stained with Coomassie blue (**A**) or transferred to nitrocellulose for Western blot analysis using 5 µg/mL of antihuman-SL-1 monoclonal antibody IID4 (**B**) as described in **ref.** *(27)*.

3.4.2. Refolding

1. Pool the fractions of interest and dilute in the extraction buffer to a protein concentration less than 1mg/mL. This dilution reduces steric hindrance by providing more space for refolding to occur, thus allowing for more efficient protein refolding and decreased protein precipitation.
2. Dialyze at 4°C against at least a 100-fold volume of standard buffer to remove the urea.
3. Alternatively, dilute the pooled fractions more than 100-fold with standard buffer to lower the urea and protein concentrations and incubate at 4°C overnight.
4. After refolding, concentrate the sample and store at –20°C until use. If the sample is to be further purified, it can be frozen at –20°C for a few days before proceeding to the next step.

3.4.3. Affinity Chromatography after Gel Filtration

The purity of the samples from the gel filtration varies, but greater than 80% pure protein can be obtained from this step alone (**Fig. 4A**, pooled fractions). A small amount of contaminating sub-28,000 kDa proteins can be seen in the dialyzed pooled fractions from the gel filtration column (**Fig. 4A**, pooled fractions). To remove these proteins, the samples can be further purified by affin-

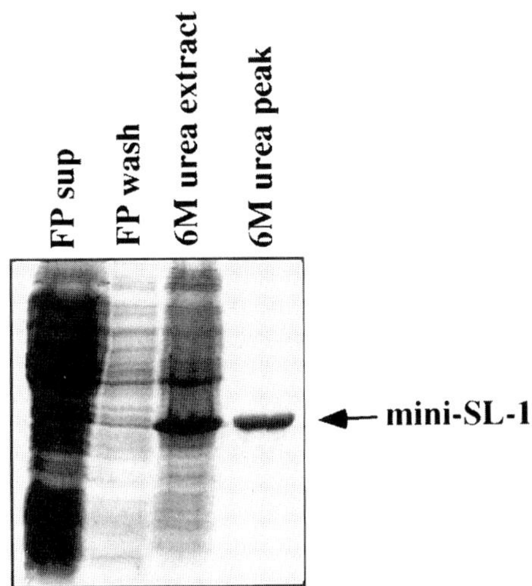

Fig. 5. Affinity purification of 6 *M* urea extract. The 6 *M* urea extract was diluted more than 100-fold before passage over the anti-SL-1 antibody IID4 affinity column without prior passage over the gel filtration column. The supernatant following passage through a French press (FP sup), the wash supernatant of the pellet after passage through the French press (FP wash), the 6 *M* urea extract of the pellet containing the inclusion bodies (6 *M* urea extract), and the peak eluted from the anti-SL-1 antibody IID4 affinity column with 6 *M* urea (6 *M* urea peak) were resolved in a 12.5% SDS-PAGE gel and stained with Coomassie blue.

ity chromatography using anti-SL-1 monoclonal antibody IID4 or equivalent, coupled to CNBr-activated Sepharose 4B *(27)*.

1. Apply the diluted or dialyzed sample from gel filtration above to the anti-SL-1 monoclonal antibody Sepharose affinity column pre-equilibrated with standard buffer.
2. Wash the column with 3–5 column volumes of standard buffer or until the absorbance at 280nm returns to that of the buffer.
3. Elute the bound material with extraction buffer (this results in mini-SL-1 which is greater than 98% pure, **Fig. 4B**)
4. Dialyze the sample against standard buffer to remove the urea. An estimated 10–20 mgs per liter of mini-SL-1 can be obtained in this manner by growing individual liters of *E. coli* at 37°C. It is possible to increase the yield to 20–40 mgs per liter by optimizing the growth conditions in a fermentor.

3.4.4. Affinity Chromatography without Gel Filtration

Affinity chromatography can also be performed using the 6 *M* urea extract of the inclusion body pellet without prior passage over the gel filtration column.

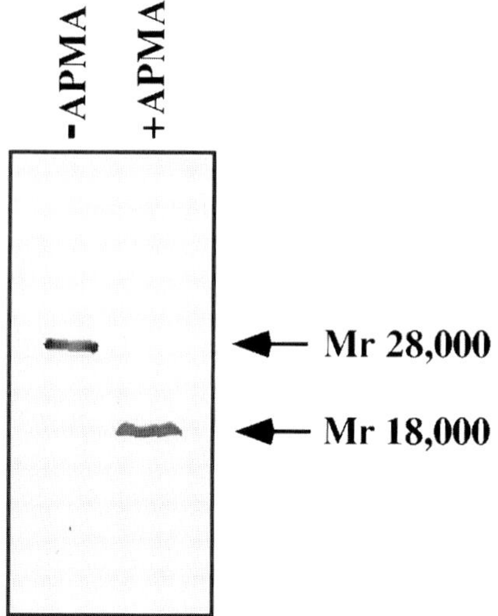

Fig. 6. APMA-induced activation. Samples of the affinity purified mini-SL-1 were incubated with and without 1 m*M* APMA at 37°C for 12 h. The samples were resolved in an 12.5% SDS-PAGE gel and transferred to nitrocellulose for Western blot staining with anti-SL-1 monoclonal antibody IID4 (5 µg/mL) as previously described *(27)*.

1. Dilute the urea extract dropwise into greater than 100 vol of standard buffer (to minimize protein precipitation).
2. Filter the diluted extract through #1 Whatman filter paper
3. Apply to the affinity column, wash, and elute as above.

This method also yields greater than 98% pure mini-SL-1 (*see* **Fig. 5**, 6 *M* urea peak). However, this method results in a greater loss of the protein due to precipitation during the dilution step when compared to the samples passed over the gel filtration column first.

3.5. Characterization

The purified mini-SL-1 can be assayed for catalytic competency by assessing its ability to undergo APMA (aminophenyl-mercuric acetate)-induced activation, its ability to cleave casein in a zymogram, and its ability to cleave a synthetic peptide.

3.5.1. APMA-Induced Activation

1. Incubate samples of mini-SL-1 in the presence and absence of 1 m*M* APMA for 12 h at 37°C.
2. Resolve the samples in a 12.5 % SDS-PAGE gel.

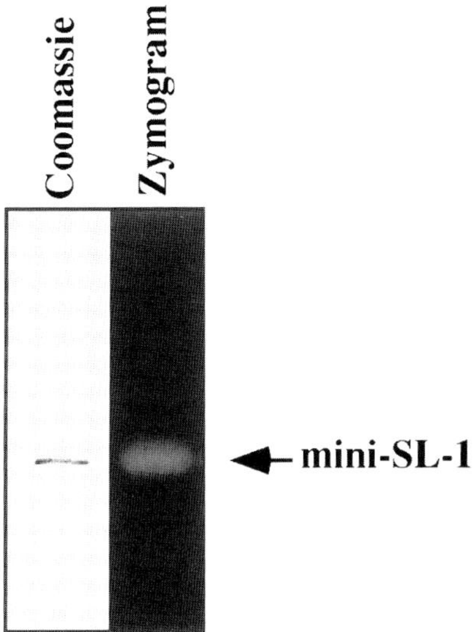

Fig. 7. Casein zymography of mini-SL-1. A sample of the affinity purified mini-SL-1 was resolved in a 10% SDS-PAGE gel and stained with Coomassie blue. A companion sample was resolved in a 10% SDS-PAGE gel copolymerized with 1 mg/mL of casein (zymogram) and incubated at 37°C for 16 h and stained with Coomassie blue to visualize the lytic band.

3. Transfer to nitrocellulose and analyze by Western blot utilizing anti-SL-1 monoclonal antibody IID4 or equivalent *(27)*.

The APMA induces the mini-SL-1 protein (Mr 28,000) to autocatalytically convert (*see* **Fig. 6**) to the lower molecular weight active form (Mr 18,000). This conversion demonstrates that the mini-SL-1 is catalytically competent and able to process its own propeptide during APMA-induced self-activation.

3.5.2. Casein Zymography

1. Resolve a sample of the mini-SL-1 in a 10% SDS-PAGE gel co-polymerized with 1 mg/mL casein.
2. After electrophoresis, wash the gel vigorously in three changes of 1% Triton X-100 and incubate in standard buffer for 16 h at 37°C *(9)*.
3. Stain with Coomassie blue to visualise bands of lysis.

The correctly folded mini-SL-1 is capable of cleaving the casein incorporated in the gel as visualized by the zone of clearing (*see* **Fig. 7**, zymogram).

3.5.3. Cleavage of the Mca-Peptide

1. Incubate the coumarinyl peptide derivative Mca-P-L-G-L-Dpa-A-R (2.5 μM) *(28)* with 3 nM mini-SL-1 (activated with APMA) in standard buffer at 23°C.

2. Monitor emerging fluorescence to calculate k_{cat}/K_m. A k_{cat}/K_m of 33.3 $\mu M^{-1}h^{-1}$ is comparable to that previously reported for mini-SL-1 and indicates that the protein is correctly refolded *(9)*.

4. Notes

1. The pGEMEX-1 expression vector has been utilized very effectively in producing recombinant MMPs *(3,9,10)*. Other expression vectors have also proven to be very effective in producing recombinant MMPs in *E. coli*. Some of these include (1) pET plasmids *(4–6,11,15–17)*, (2) pRSET plasmids *(1,12,14)*, and (3) pBluescript plasmids *(8,30)*. All of these are based on the T7 expression system and take advantage of the same strategies for protein expression. The selection of an appropriate expression vector depends on the objectives set forth in the investigation.
2. With the rapid advancement in molecular biology techniques, multiple methods can be utilized in the expression of proteins and the construction of expression plasmids. Methods such as cloning by PCR (polymerase chain reaction) have been very effective in the constructon and reconstruction of the cDNAs, as well as the expression plasmids for the MMPs.
3. Several modifications to the described methods can be performed to optimize protein expression. Fermentation facilities, which optimize the growth conditions for the bacteria, can be used to produce larger volumes of the *E. coli*. This not only results in more protein overall, but more protein per liter of *E. coli*. A problem that has been observed is that the expression plasmid in *E. coli* BL21 (DE3) cells can be unstable. Therefore, instead of expressing the protein from a frozen stock of transformed *E. coli* BL21 (DE3) cells, transformation of the *E. coli* BL21 (DE3) cells may have to be performed just prior to scaling up for expression. Also, it has been observed that the expression of more protein can be obtained if the overnight culture used to inoculate the liters is in log phase. Another modification includes changing the temperature at which the cells are grown and induced to 30°C instead of 37°C to optimize protein expression and to minimize breakdown of the protein.
4. Alternative methods to the use of the French press for *E. coli* cell lysis include treatment with lysozyme, freeze-thawing, and sonication or some combination of these. The utilization of a protease inhibitor (for example PMSF) or a cocktail of protease inhibitors is contingent on the amount of breakdown of the recombinant MMP that occurs throughout the extraction process. This chapter describes the use of urea to extract the inclusion bodies, but 4–6 *M* guanidine hydrochloride can also be used. For full-length CL-1 extracted with 6 *M* guanidine hydrochloride, a stepwise dilution to approx 2 *M* guanidine hydrochloride with 20% glycerol, 2.5 m*M* glutathione (reduced and oxidized), and 5 m*M* CaCl$_2$ followed by buffer exchange to remove denaturant and thiol reagents was determined to yield the best results *(7)*. The refolding steps, especially for the full-length MMPs, have been the greatest challenge in attempting to maximize yield of native-like

protein. Determination of which extraction and refolding methods to be utilized appears to vary from protein to protein.

5. Most purification schemes of MMPs from *E. coli* inclusion bodies involve extraction with a denaturant followed by a gel filtration step to separate the recombinant protein from the *E. coli* proteins. In this report, an antibody affinity column has been utilized to purify the protein further. Since an affinity purification step may not always be available, other purification steps have been utilized that include an anion and/or cation exchange column (Q-Sepharose or S-Sepharose, respectively) (*11*).

Acknowledgments

The authors thank Dr. G. I. Goldberg for providing the cDNA for human stromelysin-1. We also thank Drs. Susan B. LeBlanc and Grazyna Galazka for construction of plasmids pGSL-1 and pGXmini-SL-1. This work was supported by USPHS grants DE08228 (to S. M. Michalek), DE10631 (to J. A. Engler), DE/CA 11910 (to J. A. Engler), and AR44701 (to L. J. Windsor).

References

1. Rosenfeld, S. A., Ross, O. H., Corman, J. I., Pratta, M. A., Blessington, D. L., Fesser, W. S., and Freimark, B. D. (1994) Production of human matrix metalloproteinase 3 (stromelysin) in *Escherichia coli*. *Gene* **139**, 281–286.
2. Salowe, S. P., Marcy, A. I., Cuca, G. C., Smith, C. K., Kopka, I. E., Hagmann, W. K., and Hermes, J. D. (1992) Characterization of zinc-binding sites in human stromelysin-1: stoichiometry of the catalytic domain and identification of a cysteine ligand in the proenzyme. *Biochemistry* **31**, 4535–4540.
3. Ye, Q.-Z., Johnson, L. L., Hupe, D. J., and Baragi, V. (1992) Purification and characterization of the human stromelysin catalytic domain expressed in *Escherichia coli*. *Biochemistry* **31**, 11,231–11,235.
4. Gronski, Jr., T. J., Martin, R. L., Kobayashi, D. K., Walsh, B. C., Holman, M. C., Huber, M., Van Wart, H. E., and Shapiro, S. D. (1997) Hydrolysis of a broad spectrum of extracellular matrix proteins by human macrophage elastase. *J. Biol. Chem.* **272**, 121,89–12,194.
5. Ho, T. F., Qoronfleh, M. W., Wahl, R. C., Pulvino, T. A., Vavra, K. J., Falvo, J., Banks, T. M., Brake, P. G., and Ciccarelli, R. B. (1994) Gene expression, purification and characterization of recombinant human neutrophil collagenase. *Gene* **146**, 297–301.
6. Lemaitre, V., Jungbluth, A., and Eeckhout, Y. (1997) The recombinant catalytic domain of mouse collagenase-3 depolymerizes type I collagen by cleaving its aminotelopeptides. *Biochem. Biophys. Res. Commun.* **230**, 202–205.
7. Zhang, Y. and Gray, R. D. (1996) Characterization of folded, intermediate, and unfolded states of recombinant human interstitial collagenase. *J. Bio. Chem.* **271**, 8015–8021.

8. Windsor, L. J., Birkedal-Hansen, H., Birkedal-Hansen, B., and Engler, J. A. (1991) An internal cysteine plays a role in the maintenance of the latency of human fibroblast collagenase. *Biochemistry* **30**, 641– 647.

9. Windsor, L. J., Steele, D. L., LeBlanc, S. B., and Taylor, K. B. (1997) Catalytic domain comparisons of human fibroblast-type collagenase, stromelysin-1, and matrilysin. *Biochim. Biophys. Acta.* **1334**, 261–272.

10. Ye, Q.-Z., Johnson, L. L., and Baragi, V. (1992) Gene synthesis and expression in *E. coli* for pump, a human matrix metalloproteinase. *Biochem. Biophys. Res. Commun.* **186**, 143–149.

11. Marcy, A. I., Eiberger, L. L., Harrison, R., Chan, H. K., Hutchinson, N. I., Hagmann, W. K., Cameron, P. M., Boulton, D. A., and Hermes, J. D. (1991) Human fibroblast stromelysin catalytic domain: expression, purification, and characterization of a C-terminally truncated form. *Biochem.* **30**, 6476–6483.

12. Pendas, A. M., Knauper, V., Puente, X. S., Llano, E., Mattei, M.-G., Apte, S., Murphy, G., and López-Otin, C. (1997) Identification and characterization of a novel human matrix metalloproteinase with unique structural characteristics, chromosomal location, and tissue distribution. *J. Biol. Chem.* **272**, 4281–4286.

13. Kinoshita, T., Sato, H., Takino, T., Itoh, M., Akizawa, T., and Seiki, M. (1996) Processing of a precursor of 72-kilodalton type IV collagenase/gelatinase A by a recombinant membrane-type I matrix metalloproteinase. *Cancer Res.* **56**, 2535–2538.

14. Freimark, B. D., Feeser, W. S., and Rosenfeld, S. A. (1994) Multiple sites of the propeptide region of human stromelysin-1 are required for maintaining a latent form of the enzyme. *J. Biol. Chem.* **269**, 26,982–26,987.

15. Murphy, G., Segain, J.-P., O'Shea, M., Cockett, M., Ioannou, C., Lefebre, O., Chambon, P., and Basset, P. (1993) The 28-kDa N-terminal domain of mouse stromelysin-3 has the general properties of a weak metalloproteinase. *J. Biol. Chem.* **268**, 15,435–15,441.

16. Shapiro, S. D., Griffin, G. L., Gilbert, D. J., Jenkins, N. A., Copeland, N. G., Welgus, H. G., Senior, R. M., and Ley, T. (1992) Molecular cloning, chromosomal localization, and bacterial expression of a murine macrophage metalloelastase. *J. Biol. Chem.* **267**, 4664–4671.

17. Lichte, A., Kolkenbrock, H., and Tschesche, H. (1996) The recombinant catalytic domain of membrane-type matrix metalloproteinase-1 (MTI-MMP) induces activation of progelatinase A and progelatinase A complexed with TIMP-2. *FEBS Lett.* **397**, 277–282.

18. Sato, H., Kinoshita, T., Takino, T., Nakayama, K., and Seiki, M. (1996) Activation of a recombinant membrane type 1-matrix metalloproteinase (MT1-MMP) by furin and its interaction with tissue inhibitor of metalloproteinases (TIMP-2). *FEBS Lett.* **393**, 101–104.

19. Itoh, M., Masuda, K., Ito, Y., Akizawa, T., Yoshioka, M., Imai, K., Okada, Y., Sato, H., and Seiki, M. (1996) Purification and refolding of recombinant human proMMP-7 (pro-matrilysin) expressed in *Escherichia coli* and its characterization. *J. Biochem.* **119**, 667–673.

20. Welch, A. R., Holman, C. M., Browner, M. F., Gehring, M. R., Kan, C.-C. and Van Wart, H. E. (1995) Purification of human matrilysin produced in *Escherichia coli* and characterization using a new optimized fluorogenic peptide substrate. *Archives of Biochemistry and Biophysics* **324**, 59–64.

21. Studier, F. W. and Moffatt, B. A. (1986) Use of Bacteriophage T7 RNA Polymerase to Direct Selective High-level Expression of Cloned Genes. *J. Mol. Biol.* **189**, 113–130.

22. Wilhelm, S. M., Collier, I. E., Kronberger, A., Eisen, A. Z., Marmer, B. L., Grant, G. A., Bauer, E. A., and Goldberg, G. I. (1987) Human skin fibroblast stromelysin: Structure, glycosylation, substrate specificity, and differential expression in normal and tumorigenic cells. *Proc. Natl. Acad. Sci. USA* **84**, 6725–6729.

23. Sambrook, J., Fritsch, E. F., and Maniatis, T. (1989) *Molecular Cloning, A Laboratory Manual*, Second ed. Cold Spring Harbor, New York: Cold Spring Harbor Laboratory Press.

24. Sanger, F., Nicklen, S., and Coulson, A. R. (1977) DNA sequencing with chain-terminating inhibitors. *Proc. Natl. Acad. Sci. USA* **74**, 5463–5467.

25. Biggin, M. D. , Gibson, T. J., and Hong, G. F. (1983) Buffer gradient gels and ^{35}S label as an aid to rapid DNA sequence determination. *Proc. Natl. Acad. Sci. USA* **80**, 3963–3965.

26. Zoller, M. J. and Smith, M. (1983) Oligonucleotide-directed mutagenesis of DNA fragments cloned into M13 vectors. *Methods Enzymol.* **100**, 468–500.

27. Windsor, L. J., Grenett, H., Birkedal-Hansen, B., Bodden, M. K., Engler, J. A., and Birkedal-Hansen, H. (1993) Cell-type-specific regulation of SL-1 and SL-2 genes. Induction of SL-2, but not SL-1, in human keratinocytes in response to cytokines and phorbolesters. *J. Biol. Chem.* **266**, 13,064–13,069.

28. Knight, C. G., Willenbrock, F., and Murphy, G. (1992) A novel coumarin-labelled peptide for sensitive continuous assays of the matrix metalloproteinases. *FEBS Lett.* **296**, 263–266.

29. Lovejoy, B., Cleasby, A., Hassell, A. M., Longley, K., Luther, M. A., Weigl, D., McGeehan, G., McElroy, A. B., Drewry, D., Lambert, M. H., and Jordan, S. R. (1994) Structure of the catalytic domain of fibroblast collagenase complexed with an inhibitor. *Science* **263**, 375–377.

30. Windsor, L. J., Bodden, M. K., Birkedal-Hansen, B., Engler, J. A., and Birkedal-Hansen, H. (1994) Mutational analysis of residues in and around the active site of human fibroblast-type collagenase. *J. Biol. Chem.* **269**, 26,201–26,207.

9

Expression of Human Collagenase I (MMP-1) and TIMP-1 in a Baculovirus-Based Expression System

Rüdiger Vallon and Peter Angel

1. Introduction

Resident cells of tissues are capable of secreting an array of structurally related zinc endopeptidases known as matrix metalloproteinases (MMPs). They initiate the degradation of the surrounding macromolecules of the extracellular matrix (ECM), mostly proteoglycans and specific types of collagen. ECM degradation presumably contributes to the initial phase of tissue remodeling inherent to the physiological processes of morphogenesis, angiogenesis, involution of the uterus, bone resorption, inflammation, and wound healing. The aberrant activity of MMPs has been found to be involved in a variety of pathological processes, e.g., rheumatoid joint destruction, corneal ulceration, metastasis of tumor cells, and various genetic diseases *(1,2)*.

The various members of the MMP family contain several distinct domains that are highly conserved. Despite this high degree of similarity, the MMPs identified so far, differ in substrate specificity and expression in response to extracellular stimuli suggesting that each individual member of the MMP family has a distinct function in the physiological and pathological processes listed above *(3,4)*. Furthermore, the activity of these enzymes is tightly regulated on several levels: regulation of transcription, activation of the latent pro-enzyme and interaction with specific inhibitors of MMPs (TIMP-1, TIMP-2, TIMP-3, TIMP-4) *(2,5)*. Studies in cell culture systems have revealed multiple mechanisms that positively or negatively interfere with the transcription of MMPs. Transcription is enhanced by carcinogens, cytokines and tumor promoters and repressed by steroid hormones and by the E1A protein of adenovirus *(7–9)*.

From: *Methods in Molecular Biology, vol. 151: Matrix Metalloproteinase Protocols*
Edited by: I. Clark © Humana Press Inc., Totowa, NJ

The well documented evidence for the funcional involvement of MMPs in the diseases described above makes them attractive targets for therapeutical intervention. One possiblity would be to interfere with the expression of specific MMP proteins. The much more promising approach, however, may be the possibility to shift the equilibrium between MMPs and TIMPs toward the inhibitors resulting in efficient interference with MMP activity. This can be achieved by using recombinant TIMPs or specific small molecular weight inhibitors mimicking TIMP action *(10)*. To use this type of approach and to develop highly efficient inhibitors for clinical therapy, knowledge of the structural requirements of MMPs and the inhibitors as well as the modes of interactions is absolutely required.

In general, to answer this type of question in vitro studies using recombinant proteins are the method of choice. Several systems have been established to express recombinant MMPs, including bacteria, yeast and insect and mammalian cells *(11–18)*. However, although isolates of specific MMP and TIMP proteins expressed exogenously in mammalian cells may be contaminated by endogenous MMP and TIMP protein, the expression of MMP-1 and TIMP-1 in bacteria results in the formation of nonsoluble proteins; therefore, denaturation and renaturation steps are required which bear the risk of obtaining incorrectly folded proteins. We and others have experienced the Baculovirus-based expression system in SF-9 insect cells for expression of MMP and TIMP proteins in eukaryotes *(19–24)*. Major advantages of the Baculovirus-based expression system are the high yield of soluble proteins of interest. In this chapter, with examples from expression of recombinant MMP-1 and TIMP-1 proteins, we will describe the method in detail to enable interested scientists in the field to use the Baculovirus expression system for the synthesis of other MMPs and TIMPs as soluble and bioactive proteins.

2. Materials

1. Spodoptera frugiperda (SF-9) cells (American Type Culture Collection CRL1711).
2. TC 100 medium (Gibco-BRL Life Technologies).
3. Grace's medium (Gibco-BRL Life Technologies).
4. Insect-XPRESS medium (Gibco-BRL Life Technologies).
5. Fetal bovine serum (FBS, Gibco-BRL Life Technologies).
6. TC 100 medium containing 10% FBS.
7. TC 100 medium containing 5% FBS.
8. Grace's medium containing 10% FBS.
9. Cell scraper (Costar).
10. Linearized BaculoGold^{TM} Baculovirus DNA (Pharmingen).
11. PLV 1392/1393 Transfer Vector (Pharmingen).
12. DOTAP transfection reagent (Roche Biochemicals).
13. Transfection buffer: 20 mM HEPES, pH 7.4, 150 mM NaCl.

Table 1
Growth and Infection Conditons for SF-9 Cells

	Cell density	Minimum volume of inoculum	Final incubation volume
96 well plate	2.0×10^4/well	10 µL	100 µL
24 well plate	6.0×10^5/well	200 µL	500 µL
25 cm² flask	3.0×10^6/well	1 mL	5 mL
175 cm²flask	2.0×10^7	4 mL	20 mL
spinner	1.5–2.0/mL	based on MOI	based on size

14. Lysis buffer: 0.5 *M* NaOH, 1.5 *M* NaCl.
15. MMP buffer: 50 m*M* Tris-HCl, pH 7.5, 10 m*M* $CaCl_2$, 1 µ*M* $ZnCl_2$.

3. Methods

3.1. Cloning of Baculovirus Expression Vectors Encoding MMP-1 and TIMP-1 Proteins

1. Excise the cDNA containing human wild-type MMP-1 *(25)* from the plasmid pBAT human MMP-1 (*26*; Vallon, R. and Angel, P., unpublished) using restriction enzymes SmaI/NotI and subclone this into the SmaI/NotI digested transfer vector PLV-1393 (Pharmingen).
2. Generate the mutant form of MMP-1 missing the C-terminal hemopexin like domain (Δ248–450) by PCR using RSV-MMP-1 (an RSV-LTR based expression vector *(27)* containing MMP-1 coding sequences) as template and oligonucleotides 5'-AATGAATTCCTCGAGCTAAGGATTTTGGGAACG-3'(nucleotides 852–856 of human MMP-1) and 5'-GAATTCCGCATTGCAGAGAT-3' (RSV-LTR). The cDNA of human TIMP-1 *(28)* can be amplified from a human cDNA library *(29)* using the oligonucleotides 5' HT (5'-GGGGATCCAAGCT TACCATGGCCCCCTTTGAGCCCCT-3') and 3' HT (5'-CCCTCGAGAATT CAGGCTATCTGGGACCGC 3').
3. Digest the resulting PCR product with BamHI/EcoRI and subclone this into the BamHI/EcoRI digested transfer vector PLV-1393. The correct sequences of PLV-transfer constructs can be confirmed by sequencing.

3.2. Culturing of Insect Cells

1. SF-9 cells are grown either in TC-100 medium containing 10% FBS or in serum free medium (Insect-XPRESS) at 27°C in flasks, Petri dishes or spinner culture bottles. No specific adjustment of CO_2 conditions is required for culturing SF-9 cells.
2. Twice a week confluent cells are either diluted (cells in suspension) or removed from the flasks or Petri dishes (cells growing as monolayers) using cell scrapers

(instead of trypsin), diluted and seeded at the appropriate cell densities (for example, *see* **Table 1**).

3.3. Transfection of Insect Cells

1. Remove the medium from a 175 cm² flask containing logarithmically growing SF-9 cells and replace it with 5 mL fresh TC-100 medium (10% FBS).
2. Remove the SF-9 cells from the flask using a cell scraper. Determine the percentage of viable cells using trypan blue staining (cells which incorporate trypan blue are considered to be nonviable).
3. For each transfection, seed 1.5×10^6 viable SF-9 cells in a 25 cm² flask in TC-100 medium containing 10% FBS and incubate for 40 min at 27°C in order to let the cells adhere.
4. In parallel, prepare the transfection mix containing 3 µg of the MMP-1 or TIMP-1 expression vector plasmid (*see* **Note 1**) and 0.5 µg linearized BaculoGold™ Baculovirus DNA. Pipet the viral DNA with a pipet where the tip is cut off in order to minimize shearing forces on the DNA. Adjust the DNA mix to a final volume of 100 µL using transfection buffer (20 nm*M* HEPES pH 7.4, 150 m*M* NaCl).
5. In a separate tube, mix 30 µL of DOTAP and 70 µL of transfection buffer. Ten minutes before the transfection, add the DOTAP mix to the DNA mix.
6. Remove the medium from the cells and replace it with 1 mL of Grace's medium. Add 1mL of Grace's medium to the transfection mix and add this dropwise to the cells using a 1 mL plastic pipet.
7. After 4h incubation at 27°C, add 2 mL of Grace's medium containing 10% FBS. Incubate the cells for a further 4 d at 27°C.
8. Harvest the supernatant containing the recombinant Baculovirus and use this for determination of the virus titer by end-point dilution.

3.4. Determination of the Virus Titer by End-Point Dilution

1. For the determination of the virus titer, make 10-fold dilutions of the virus in the range of $10^{-1} - 10^{-8}$ in a volume of 1000 µL .
2. Seed 2×10^4 SF-9 cells in 100 µL of TC-100 medium containing 5%FBS in each well of a 96 well plate. After 30 min add 50 µL of the appropriate virus dilution to all wells row by row. Incubate infected cells for 5 d at 27°C.
3. Transfer supernatants to a new plate and keep at 4°C. Lyse cells using 200 µL of lysis buffer (0.5–*M* NaOH, 1.5–*M* NaCl) per well for 20 min.
4. Transfer DNA to Hybond N+ membrane using a dot blot apparatus. Dry the membrane and crosslink the DNA to the membrane with UV light, followed by Southern blot hybridization at standard high stringency conditions (*see* **Note 2**) and subsequent exposure to phosphoimager plates (*see* **Fig. 1**).
5. Calculation of the 50% end-point is performed as described *(30)* by interpolation from the cumulative frequencies of positive and negative responses that are closest to the 50% rate. The dilutions which are used for the calculation range from the highest yielding only positive responses to that one giving only negative results. In our example (*see* **Fig 1** and **Table 2**) the two dilutions of interest are 10^{-2} (all

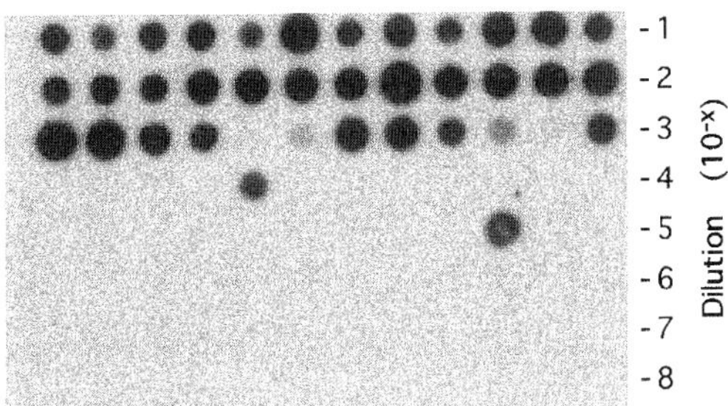

Fig. 1. End-point dilution of recombinant Baculovirus containing WT-MMP-1

Table 2
Calculation of the 50% End-Point

Dilution	10^{-2}	10^{-3}	10^{-4}	10^{-5}	10^{-6}
Positive rate	12/12	10/12	1/12	1/12	0/12
Positive number	12	10	1	1	0
Negative number	0	2	11	11	12
Positive total	24	12	2	1	0
Negative total	0	2	13	24	36
Positive rate total	24/24	12/14	2/15	1/24	0/36
% Positive	100%	85.7	13.3	4.1	0

positive) to 10^{-6} (all negative). The proportionate distance (PD) between these two dilutions is inferred by linear interpolation according to the formula:

PD = ([%next above 50%] – 50%) / ([next above 50%] – [% next below 50%]).

In our example PD = 0.49 = (85.7–50)/(85.7–13.3).

For the 50% end-point the following formula is used: Log lower dilution (the dilution next above 50) = –3.0 minus the proportionate distance (0.49 in the example) This results in Sum (50% end-point) = –3.49.

This results in a TCID**50** = $10^{-3.49}$ = 1/(3.08 × 10^3). The reciprocal of this result is the titer in number of infective units per volume of inoculum. Since 50 mL of virus were used per well the final titer is: 3.08 × 10^3/0.05 = 6.1 × 10^4 TCID$_{50}$/µL. This is converted to PFU/ml by the following formula: TCID$_{50}$/mL /0.69 = PFU/mL. In our case 6.1 × 10^4/0.69 = 8.8 × 10^4/mL.

6. The supernatants which were infectious at the highest dilution are used for further expansion of the virus, followed by selection of a virus stock with a higher virus titer and finally for the expression of recombinant proteins. Alternatively, expression of recombinant proteins can be checked immediately.

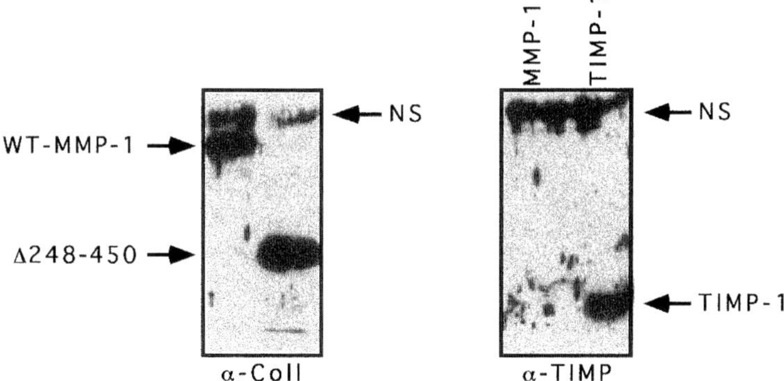

Fig. 2. Expresion of recombinant TIMP-1 and wild-type and mutant MMP-1 proteins. Culture medium of insect cells infected with recombinant Baculovirus containing Δ248-450 MMP-1 (*left panel*) WT-MMP-1 (*both panels*), and TIMP-1 (*right panel*, indicated on the top) sequences was analysed by western blot using a MMP-1 (*left panel*) or TIMP-1 (*right panel*)-specific antiserum.

3.5. Expression of Recombinant Proteins

1. To check for the expression of recombinant TIMP-1 and wild type or mutant MMP-1 proteins, seed 2×10^5 SF-9 cells/well in 24-well plates in TC-100 containing 5% FBS and incubate for 30 min at 27°C.
2. In parallel, prepare dilutions from the supernatants which were infectious at the highest dilution in the end-point dilution experiment of 50 µL, 10 µL, and 1 µL supernatant in a final volume of 200 µL of TC100 containing 5% FBS.
3. After 30 min exchange the culture medium with the virus dilutions and incubate the cells for 1 h at 27°C.
4. Adjust the volume of each well to a final volume of 1 mL by the addition of TC-100 containing 5%FBS and incubate the cells for 4 d at 27°C.
5. Transfer the supernatant containing extracellular virus and secreted proteins to a sterile centrifuge tube and centrifuge for 15 min at 1000*g*. Keep the supernatant as a stock at 4°C for future inoculum.
6. The cell pellets as well as 10 µL of the supernatant are used for analysis of expression of the recombinant proteins by Western blot.
7. The conditions for the further expansion of the virus and the large scale expression of recombinant MMP-1 and TIMP-1 proteins, such as cell density and the minimal volume of inoculum, are given in **Table 1** (if a specific MOI is not absolutely required).

For this type of large scale production of recombinant proteins we have used SF-9 cells which have been adapted to grow in serum free medium (Insect-XPRESS, Serva) and in spinner culture bottles (up to 500 mL). Western-blot analysis of conditioned medium of such cultures is shown in **Fig 2**. In general we

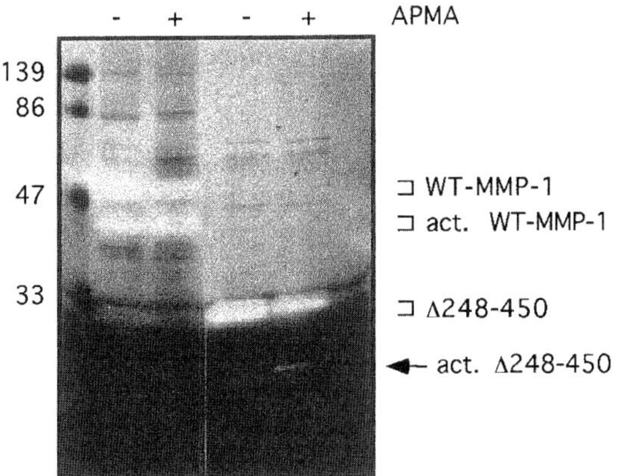

Fig. 3. Zymogram of recombinant latent and activated (by APMA; indicated on the top) WT-MMP-1 and Δ248-450 MMP-1 proteins.

obtained up to 50 µg recombinant protein/mL medium (*see* **Notes 3** and **4**). Others have reported even higher yields of recombinant proteins *(20,22,24)*.

3.6. Recombinant WT or Mutant MMP-1 and TIMP-1 are Biologically Active

To confirm that the recombinant proteins expressed in the Baculovirus system were correctly folded, as indicated by characteristic biological activities, we tested MMP-1 and TIMP-1 for enzymatic and inhibitory activities, respectively.

1. Harvest supernatants of SF-9 cells infected with recombinant Baculovirus containing latent forms of WT-MMP-1 or Δ248-450-MMP-1 after 5 d of infection.
2. Incubate supernatants for 90 min at 37°C in the presence or absence of APMA (final concentration 1 m*M*) in order to cleave the pro-domain by autolytic activity *(31)*. Alternatively, activation of latent forms of MMP-1 proteins can be achieved by addition of trypsin (10 µg/mL, then inactivate trypsin with 50 µg/mL soybean trypsin inhibitor).
3. Load samples (10 µL) on a 12% SDS polyacrylamide gel in which β-Casein (1 mg/mL) is co-polymerized. Separate proteins under nonreducing conditions.
4. After electrophoresis, incubate the gel twice for 15 min each time in 1% Triton X-100 and then at 37°C in 50 m*M* Tris pH 7.5, 10 m*M* $CaCl_2$, 1 m*M* $ZnCl_2$ for 4–24 h.
5. Stain the gel with Coomassie Blue. After destaining, proteolytic activity can be detected as clear zones in the otherwise blue-stained gel (*see* **Fig. 3**).
6. In order to detect the inhibitory activity of TIMP-1 we used the collagenase assay in solution. [14]C label type I collagen as described (*see* Cawston et al. Chapter 22 and **ref. 32**). Mix supernatants containing latent or activated (by trypsin) MMP-1 proteins with the same volume of supernatants of mock-infected cells, or from cells which were infected with a recombinant TIMP-1 Baculovirus.

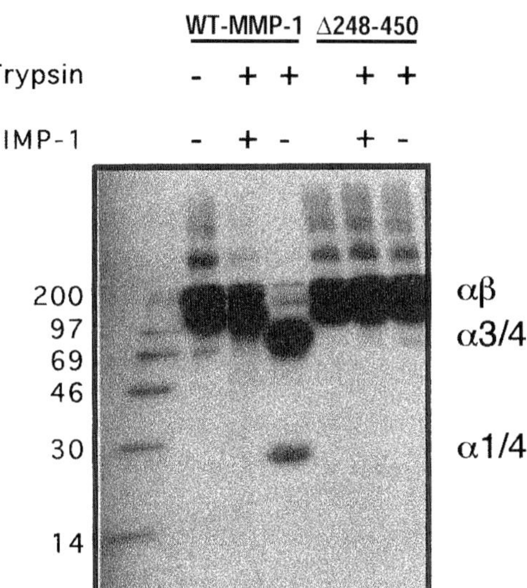

Fig. 4. Collagenase-assay in solution measuring collagenolytic activity of latent and activated (by trypsin, indicated on the top) WT-MMP-1 and Δ248–450 MMP-1 proteins in the presence or absence of recombinant TIMP-1.

7. Incubate the mixtures with 100 μg of labeled collagen at 25°C for 24 h and seperate on SDS-PAGE (*see* **Fig. 4**). Activated, but not latent WT-MMP-1 is capable of cleaving native type I collagen (compare lanes 2 and 4) whereas the mutant form Δ248-450-MMP-1 is not. In the presence of recombinant TIMP-1 cleavage of type I collagen is totally abolished (compare lanes 3 and 4) confirming biological activity of MMP-1 and TIMP-1 proteins.

Using the Baculovirus expression system and additional mutants of MMP-1 we were able to identify the catalytic domain of activated MMP-1 to be absolutely required for its interaction with TIMP-1 *(23)*.

4. Notes

In the present paper we have described in detail the use of the Baculovirus expression system for generation of recombinant interstitial collagenase (MMP-1) and TIMP-1 proteins in SF-9 cells. In order to successfully use this system for expression of recombinant proteins we would like to stress additional important points.

1. SF-9 cells are incubated for four days with the transfection supernatant in the absence of antibiotics. In order to avoid contamination of the transfected cells during this time period it is highly recommended to use plasmid preparations of

    highest quality for the transfection experiment, which has been precipitated with ethanol, washed with 70% ethanol and finally resuspended in sterile water.

2. As an alternative to using methods with radioactively labeled probes for the determination of the end-point dilution one could use a PCR-based screen using specific primers for your gene of interest. However, if PCR is the method of choice great care should be taken to avoid even minor traces of cross contamination of cells and DNAs otherwise these would result in false positive PCR signals.

3. To gain an additional rapid and efficient step for purification of recombinant proteins fusion of a tag, such as the FLAG- or the HA-tag, either to the N- or C-terminal end of the protein may be considered. However, these additional sequences may affect the expression, stability, enzymatic activity, or other biological features of your protein of interest.

4. For each protein to be expressed in SF-9 cells using Baculovirus, the optimal time point of harvesting the supernatant and/or the infected cells must be determined empirically.

The expression of recombinant MMP-1 and TIMP-1 proteins using the Baculovirus expression system has major advantages compared to other systems, such as bacterial expression or expression in mammalian cells. First, SF-9 insect cells are devoid of any detectable collagenolytic activity, which makes this system highly specific for the study of exogenously expressed MMP-1. Second, in contrast to bacterially expressed MMP-1 proteins Baculovirus-based recombinant proteins are efficiently expressed in a soluble form and are secreted into the culture medium. Therefore, a denaturation/renaturation protocol, which is required to solubilize bacterially expressed MMP-1 proteins but bears the risk of generating noncorrectly folded proteins *(13)* can be omitted. Although we have not performed structural analysis of the recombinant expressed proteins using CD or NMR spectroscopy, there is good evidence that these proteins are correctly folded. First, MMP-1 proteins containing an intact catalytic center can be autoactivated by treatment with APMA. Second, exposure of recombinant MMP-1 proteins to trypsin results in all cases to stable products representing the activated enzyme *(23)*. The lack of complete degradation of recombinant MMP-1 proteins has been considered as a criterion for appropriate folding of MMP-1 *(13)*. The wild-type MMP-1 is able to digest native collagen type I. Addition of supernatants from infected SF-9 cells containing rTIMP-1 inhibits the cleavage of native type I collagen in contrast to supernatants of mock infected cells *(23,* data not shown), which also confirms correct folding of rTIMP-1. These results demonstrate that the expression of recombinant MMP-1 and TIMP-1 in SF-9 cells using the Baculovirus expression system leads to biological active proteins. These facts, taken together with the amount of recombinant protein which can be obtained (routinely 3–200 µg/mL conditioned medium; *19,22–24)* clearly demonstrates the efficacy of the

Baculovirus expression system for the generation of recombinant MMP-1 and TIMP-1 proteins. Moreover, the use of the Baculovirus expression system for the generation of recombinant MMP-1 and TIMP-1 proteins provides a powerful tool to study the protein-protein interaction of MMP-1 and TIMP-1. This study will lead to a better understanding of the mechanism underlying the TIMP-1/MMP-1 interaction and will be extremely useful for the generation of specific inhibitors of MMP-1. These inhibitors will provide new ways of therapeutic intervention of diseases such as rheumatoid arthritis and tumor metastasis.

Acknowledgments

We thank Dr. Niamh Keon for critical reading of the manuscript and Sibylle Teurich for technical assistence. Work was supported by grants from the Deutsche Forschungsgemeinschaft (An 182/6-2 and He 581/8-2).

References

1. Matrisian, L. M. (1992) The matrix degrading metalloproteinases. *Bioessays* **14**, 455–463.
2. Birkedal-Hansen, H., Moore, W. G. I., Bodden, M. K., Windsor, L. J.,Birkedal-Hansen, B., DeCarlo, A., and Engler, J. A. (1993) Matrix metalloproteinase. A review. *Crit. Rev. Oral. Biol. Med.* **4**, 197–250.
3. Dioszegi, M., Cannon, P., and van Wart, H. (1995) Verterbrate collagenases. *Methods Enzymol.* **248**, 413–431.
4. Sang, Q. A. and Douglas, D. A. (1996) Computational sequence analysis of matrix metalloproteinases. *J. Protein Chem.* **15,** 137–160.
5. Murphy, G. and Willenbrock, F. (1995) Tissue inhibitors of metalloendopeptidases. *Methods Enzymol.* **248**, 496–510.
6. Greene, J., Wang, M., Raymond, L. A., Liu, Y. E., Rosen, C., and Shi, Y. E. (1996) TIMP-4, a novel human tissue inhibitor of metalloproteinase, was identified and cloned. *J. Biol. Chem.* **271**, 30,375–30,380.
7. Angel, P. and Karin, M. (1991) The role of jun, fos and the AP-1 complex in cell proliferation and transformation. *Biochem. Biophys. Acta.* **1072**, 129–157.
8. Vincenti, M. P., White, L. A., Schrön, D. J., Benbow, U., and Brinckerhoff, C. E. (1996) Regulating expression of the gene for matrix metalloproteinase-1 (collagenase): mechanisms that control enzyme activity, transcription, and mRNA stability. *Crit. Rev. Eukaryot. Gene Expr.* **6 (4)**, 391–411.
9. Borden, P. and Heller, R. A. (1997) Transcriptional control of matrix metalloproteinases and the tissue inhibitors of matrix metalloproteinases. *Crit. Rev. Eukaryot. Gene Expr.* **7, (1–2)**, 159–178.
10. Docherty, A.J., O'Connell, J., Crabbe, T., Angal, S., and Murphy, G. (1992) The matrix metalloproteinases and their natural inhibitors: prospects for treating degenerative tissue diseases. *Trends Biotechnol.* **10 (6)**, 200-207.

11. O'Hare, M. C., Clarke, N. J., and Cawston, T. E. (1992) Production in Escherichia coli of porcaine type I collagenase as a fusion protein with β-galactosidase. *Gene* **111**, 245–248.

12. Housley, T. J., Baumann, A. B., Braun, I. D., Davis, G., Seperack, P. K., and Wilhelm, S. M. (1993) Recombinant Chinese hamster ovary cell matrix metalloprotease-3 (MMP-3, stromelysin-1). Role of calcium in promatrix metalloprotease-3 (pro-MMP-3, prostromelysin-1) activation and thermostability of the low mass catalytic domain of MMP-3. *J. Biol. Chem.* **268** (**6**), 4481–4487.

13. Windsor, L. J., Bodden, M. K., Birkedal-Hansen, B., and Engler, J. A., and Birkedal-Hansen, H. (1994) Mutational analysis of residues in and around the active site of human fibroblast-type collagenase. *J. Biol. Chem.* **269**, 26,201–26,207.

14. Freimark, B. D., Feeser, W. S., and Rosenfeld, S. A. (1994) Multiple sites of the propeptide region of human stromelysin-1 are required for maintaining a latent form of the enzyme. *J. Biol. Chem.* **269** (**43**), 26,982–26,987.

15. Rosenfeld, S. A., Ross, O. H., Hillman, M. C., Corman, J. I., and Dowling, R. L. (1996) Production and purification of human fibroblast collagenase (MMP-1) expressed in the methylotrophic yeast Pichia pastoris. *Protein Expr. Purif.* **7**, (**4**), 423–430.

16. Zhang, Y. and Gray, R. D. (1996) Characterization of folded, intermediate, and unfolded states of recombinant human interstitial collagenase. *J. Biol. Chem.* **271**, (**14**), 8015–80211.

17. Knäuper, V., Cowell, S., Smith, B., López-Otin, C., O'Shea, M., Morris, H., Zardi, L., and Murphy, G. (1997) The role of C-terminal domain of human collagenase-3 (MMP-13) in the activation of procollagenase-3 substrate specificity and tissue inhibitor of metalloproteinse interaction. *J. Biol. Chem.* **272** (**12**), 7608–7616.

18. Lemaitre, V., Jungbluth, A., and Eeckhout, Y. (1997) The recombinant catalytic domain of mouse collagenase-3 depolymerizes type I collagen by cleavingits aminotelopeptides. *Biochem. Biophys. Res. Commun.* **230** (**1**), 202–205.

19. Gomez, D. E., Lindsay, C. K., Cottam, D. W., Nason, A. M., and Thorgeirsson, U. P. (1994) Expression and characterization of human tissue inhibitor of metalloproteinases 1 in a baculovirus-insect cell system. *Biochem. Biophys. Res. Commun.* **203** (**1**), 237–243.

20. Cocuzzi, E. T., Walther, S. E., and Denhardt, D. T. (1994) Expression and secretion of active mouse TIMP-1 using a baculovirus expression vector. *Inflammation* **18** (**1**), 35–43.

21. Bergmann, U., Tuuttila, A., Stetler-Stevenson, W. G., and Tryggvason, K. (1995) Autolytic activation of recombinant human 72 kilodalton type IV collagenase. *Biochemistry* **34** (**9**), 2819–2825.

22. Kurschat, P., Graeve, L., Erren, A., Gatsios, P., Rose-John, S., Roeb, E., Tschesche, H., Koj, A., and Heinrich, P. C. (1995) Expression of a biologically active murine tissue inhibitor of metalloproteinases-1 (TIMP-1) in baculovirus-infected insect cells. Purification and tissue distribution in the rat. *Eur. J. Biochem.* **234** (**2**) 485–491.

23. Vallon, R., Müller, R., Moosmayer, D., Gerlach, E., and Angel, P. (1997) The catalytic domain of activated collagenase I (MMP-1) is absolutely required for interaction with its specific inhibitor, tissue inhibitor of metalloproteinases-1 (TIMP-1). *Eur. J. Biochem.* **244 (1)**, 81–88.

24. Liu, Y. E., Wang, M., Greene, J., Su, J., Ullrich, S., Li, H., Sheng, S., Alexander, P., Sang, Q. A., and Shi, Y. E. (1997) Preparation and characterization of recombinant tissue inhibitor of metalloproteinase 4 (TIMP-4). *J. Biol. Chem.* **272**, **(33)** 20,479–20,483.

25. Whitham, S. E., Murphy, G., Angel, P., Rahmsdorf, H. J., Smith, B. J., Lyons, A., Harris, T. J. Reynolds, J. J., Herrlich, P., and Docherty, A. (1986)Comparison of human stromelysin and collagenase by cloning and sequence analysis. *Biochem. J.* **240**, 913–916.

26. Annweiler, A., Hipskind, R. A., and Wirth, T. (1991) A strategy for efficient in vitro translation using the rabbit β-globin leader sequence. *Nucleic Acids Res.* **19**, 3750.

27. Angel, P. Hattori, K., Smeal, T., and Karin, M. (1988a) The Jun proto-oncogene is positively autoregulated by its product, Jun/AP1. *Cell* **55**, 875–885.

28. Docherty, A. J. P., Lyons, A., Smith, B. J., Wright, E. M., Stephens, P. E., Harris, T. J. R., Murphy, G., and Reynolds, J. J. (1985) Sequence of human tissue inhibitor of metalloproteinases and its identity to erythroid-potentiating activity. *Nature* **318**, 66–69.

29. Angel, P., Allegretto, E. A., Okino, S., Hattori, K., Boyle, W. J., Hunter, T., and Karin, M. (1988) Oncogene Jun encodes a sequence specific trans-activator similar to AP1. *Nature* **332**, 166–171.

30. Reed, L. J. and Muench, H. (1938). A simple method of estimating fifty percent endpoints. *Amer. J. Hygiene* **27**, 493–497.

31. Springman, E. B., Angleton, E. L., Birkedal-Hansen, H., and van Wart, H. E. (1990) Multiple modes of activation of latent human fibroblast collagenase. Evidence for the role of a Cys-73 active-size zinc complex in latency and a "cysteine switch" mechanism for activation. *Proc. Natl. Acad. Sci. USA* **87**, 364–368.

32. Cawston, T. E. and Murphy, G. (1981) Mammalian collagenase. *Methods Enzymol.* **80**, 711–722.

10

Expression of Recombinant Matrix Metalloproteinases in Yeast

Glenn A. Doyle

1. Introduction

1.1. Practicality of Pichia pastoris System

Deciding what system to use for recombinant matrix metalloproteinase (MMP) expression is often a matter of practicality (i.e., simplicity, cost effectiveness, time considerations). Of course, the final judgment of practicality is determined by the production of sufficient quantities of functional MMPs. A system that combines the benefits of bacterial and cultured cell systems, without their disadvantages, would be of great use.

Yeast offer an attractive alternative to working in other eukaryotic expression systems or in bacteria *(1)*. Like bacteria, yeast are easy to manipulate, relatively inexpensive to grow, and require a limited amount of equipment for small to mid-scale protein production. Large-scale recombinant protein production is also possible using a fermenter. In contrast to mammalian cells, yeast do not express endogenous matrix metalloproteinases (MMPs) or tissue inhibitors of matrix metalloproteinases (TIMPs). Yet, as eukaryotes, they have similar pathways as mammalian cells for protein folding, posttranslational modification and secretion. Expression of recombinant proteins in *Saccharomyces cerevisiae* is often hampered by low yields and can result in undesirable protein modification, such as hyperglycosylation *(1–3)*. Therefore, a method for recombinant protein production in the methylotrophic yeast *Pichia pastoris* was developed as an alternative to using bacteria, *S. cerevisiae* or other eukaryotic systems *(4,5)*.

From: *Methods in Molecular Biology, vol. 151: Matrix Metalloproteinase Protocols*
Edited by: I. Clark © Humana Press Inc., Totowa, NJ

Table 1
Features of Available Expression Vectors

Vector	*E. coli*	*P. pastoris*	Targeting	Other features
pHILS1	Amp	HIS4	S; PHO1	
pPIC9	Amp	HIS4	S; AMF	
pPIC9K	Amp	HIS4, G418	S; AMF	Multiple copy selection
pPICZa(A,B,C)	Zeo	Zeo	S; AMF	myc and 6xHis carboxy tags
pHILD2	Amp	HIS4	I	
pAO815	Amp	HIS4	I	In vitro multi copy system
pPIC3.5	Amp	HIS4	I	
pPIC3.5K	Amp	HIS4, G418	I	Multiple copy selection
pPICZ(A,B,C)	Zeo	Zeo	I	myc and 6xHis carboxy tags

The selectable marker(s) for growth in bacteria and yeast are shown. Targeting of the expression product, secreted (S) or intracellular (I), is indicated. Other features, such as the ability to create (pAO815) or to select for ("K" series) multiple copies of integrated genes, or to tag the expressed protein ("Z" series) are also indicated.

Production of recombinant MMPs in the *Pichia* system has been achieved with various degrees of success *(6–10)*. Yields of approx 2–3 mg/L and 5–10 mg/L of proMMP-1 and proMMP-9, respectively, have been reported *(7,8)*. Perhaps the most reproducible success has been achieved for proMMP-2 and proMMP-9 *(6,8–10)*. It is worth noting that, at least for proMMP-1 and proMMP-9, the biochemical properties of the recombinantly expressed full-length proteins closely resemble those of proteins purified from mammalian sources *(7,8)* (Doyle, G. A., unpublished observation). Of course, as with many recombinant systems, there are pitfalls (*see* **Notes** Section). I leave it to the reader to decide whether trying the *Pichia* system is worthwhile for their purposes.

Kits for recombinant protein production in *Pichia* are commercially available (Invitrogen). These kits include extremely detailed methods manuals. Therefore, I focus this chapter on the methods I and others have used specifically for MMP production.

1.2. Characteristics of Pichia pastoris

P. pastoris shares many of the characteristics that make *S. cerevisiae* so useful for basic research: (1) few equipment requirements, (2) relatively low cost, (3) rapid and scalable growth at 30°C, and, perhaps most important, (4) easy use and genetic manipulation. Indeed, many protocols developed for *S. cerevisiae* are applicable to working with *P. pastoris* *(11,12)*. Yet, certain characteristics of *P. pastoris* make it more desirable for production of recom-

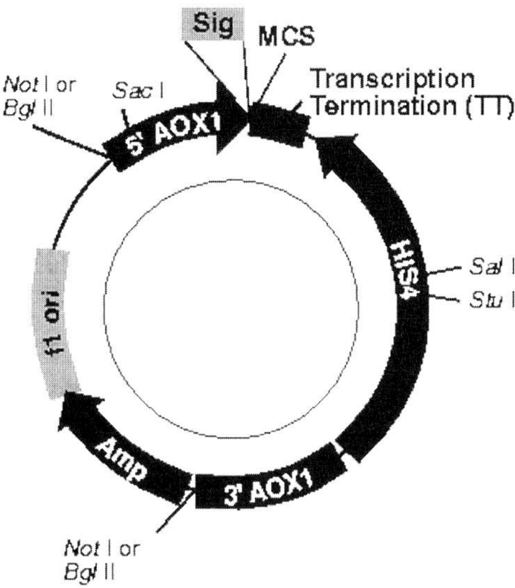

Fig. 1. Schematic representation of *Pichia* system vectors. In general, vectors contain a multiple cloning site (MCS), AOX1 sequences for homologous recombination, and selectable genes for growth in bacteria (Ampicillin, Zeocin) and/or yeast (HIS4, Zeocin, Kanamycin). A signal peptide (Sig) for secretion of the expressed protein is also found in some vectors (*see* **Subheading 1.3.** and **Table 1** for more details).

binant mammalian proteins. Two characteristics in particular enhance the production of high levels of concentrated recombinant protein: (1) its metabolic pathway for methanol utilization; and (2) its capacity to grow at extremely high densities *(1,2)*. These features allow extremely strong and specific induction of recombinant protein when *Pichia* is grown with methanol as its carbon source (*see* **Subheading 1.3.**). In addition, glycosylation of secreted recombinant proteins by *P. pastoris* is more similar to mammalian N-linked glycosylation than *S. cerevisiae*, with little, if any, hyperglycosylation or O-linked glycosylation *(2,3)*. Perhaps one disadvantage of working with *P. pastoris* is that it is not as extensively studied as *S. cerevisiae*. Thus, one has limited options when difficulties are encountered using this system (*see* **Note 8**).

1.3. Overview of the Pichia pastoris *System*

To efficiently use methanol as its sole carbon source, *Pichia pastoris* requires the enzyme alcohol oxidase 1 (Aox1) *(13)*. Growth of *Pichia* on methanol leads to powerful transcriptional induction of the *AOX1* gene *(14)*. Indeed, *AOX1* expression levels can reach 5% of total yeast mRNA and 30% of total yeast protein. Another alcohol oxidase gene, *AOX2*, is also induced by methanol, but to a much lesser extent than *AOX1* *(13,14)*. Accordingly, the *Pichia*

Table 2
Available *Pichia* Strains[a]

Pichia strain	Parental genotype	Parental phenotype
X33	wild-type	His$^+$ Mut$^+$
GS115	*his4*	His$^-$ Mut$^+$
KM71	*his4 arg4 aox1::ARG4*	His$^-$ Mut^s
SMD1168	*his4 pep4*	His$^-$ Mut$^+$ Pep$^-$
SMD1165	*his4 prb1*	His$^-$ Mut$^+$ Prb$^-$
SMD1163	*his4 pep4 prb1*	His$^-$ Mut$^+$ Pep$^-$Prb$^-$

[a]*his4* = histidinol dehydrogenase, *aox1* = alcohol oxidase 1, *arg4* = argininosuccinate lyase, *pep4* = protease A, *prb1* = protease B.

system employs heterologous expression vectors containing the 5' flanking region of the *AOX1* gene to drive high level expression of recombinant (fusion) proteins upon methanol induction (**Fig. 1**) *(15)*. **Table 1** gives a list of available vectors and their salient features (*see* **Note 1** and **4**).

There are currently six strains of *P. pastoris* readily available for use in recombinant protein expression projects (**Table 2**). Parental strains having the wild-type genes for both *AOX1* and *AOX2* (X33, GS115, and SMD strains) are considered <u>M</u>ethanol <u>u</u>tilization positive (Mut$^+$). When the *AOX1* gene is disrupted (e.g., in strain KM71), *AOX2* gene activity allows *Pichia* to grow slowly on methanol *(14)*. This slow growth phenotype is refered to as <u>M</u>ethanol <u>u</u>tilization <u>s</u>low (Mut^s). Most *Pichia* strains carry a mutation in the *HIS4* gene (*his4*) that prevents them from growing on media lacking histidine. Therefore, most *Pichia* expression vectors carry the *HIS4* gene, which can complement the *his4* gene of the *Pichia* strains *(4)*. When *Pichia* strains (except X33) are transformed with these expression vectors, the yeast are initially screened for His$^+$ clones by growing them on plates lacking histidine (*see* **Subheading 3.2.**). Depending on the yeast strain, targeting strategy, and vector used, one must employ slightly different screening procedures following the His$^+$ phenotype screen (*see* **Notes 2–4**).

Yeast are transformed (**Subheading 3.2.**) with linearized vector DNA which incorporates into the yeast genome by homologous recombination events. The use of different restriction enzymes to linearize the DNA allows one to target recombination to either the *AOX1* or *his4* locus of the parental yeast strain. In general, three kinds of recombination events occur: (1) gene insertion at the *AOX1* locus; (2) gene insertion at the *his4* locus; and (3) gene replacement at the *AOX1* locus. Gene insertion strategies can result in multiple integration events with a higher frequency than gene replacement strategies (*see* **Notes 2–4** for other considerations). Therefore, the subsequent discussion of methods focuses on gene insertion strategies targeted to the *AOX1* locus.

The *Pichia* system offers the choice of either intracellular protein expression or secretion of an expressed fusion protein *(1,2)*. Secretion is a convenient first step in recombinant protein purification and is useful and effective for expression of the MMPs. Secretion of recombinant proteins is usually accomplished by fusing either the *S. cerevisiae* alpha-mating factor (AMF) or the *P. pastoris* acid phosphatase (PHO1) signal peptide in-frame with the coding sequence of the desired cDNA *(1,2)*. An alternative approach is to subclone a cDNA that includes the native pre-domain secretion signal into one of the "intracellular" expression vectors *(6)*. The methods for creation and screening of clones are independent of the secretion signal used (*see* **Subheadings 3.2. and 3.3.**).

2. Materials

2.1. Solutions, Media and Plates for Growth and Screening of Pichia.

2.1.1. Stock Solutions

1. 13.4% YNB: 134 g Yeast Nitrogen Base (with $[NH]_2SO_4$, without amino acids) in 1L water, filter sterilize, store at 4°C.
2. 0.02% Biotin: 20 mg biotin in 100 mL water, filter sterilize, store at 4°C.
3. 0.4% Histidine: 400 mg L-histidine in 100 mL water. If necessary, dissolve at no greater than 50°C. Filter sterilize, store at 4°C.
4. 20% Dextrose: 200 g D-glucose in 1L water. Autoclave 15 min or filter sterilize, store at room temperature.
5. 5% Methanol: 5 mL methanol in 95 mL water. Filter sterilize, store at 4°C.
6. 10% Glycerol: 100 mL glycerol in 900 mL water. Autoclave or filter sterilize, store at room temperature.
7. 0.5% Amino Acids: 500 mg each of L-glutamic acid, L-methionine, L-lysine, L-leucine, L-isoleucine in 100 mL water. Filter sterilize, store at 4°C.
8. 1 *M* potassium phosphate buffer, pH 6.0: Mix 132 mL 1M K_2HPO_4 with 868 mL 1 *M* KH_2PO_4. Confirm pH is 6.0, adjust with KOH if necessary. Autoclave, store at room temperature. Also, see **Notes 9–11** for other pH considerations.

2.1.2. Liquid Media and Plates

1. YPD (Yeast Extract Peptone Dextrose) (1L): mix 10 g yeast extract and 20 g peptone in 900 mL water. For plates, add 20 g agar/L. Autoclave 20 minutes, set solution at 60°C to prevent overcooling. Add 100 mL 20% dextrose, store liquid at room temperature, plates at 4°C.
2. RDB and RDHB (Regeneration Dextrose [± Histidine] Base) (1L): dissolve 186 g sorbitol in 700 mL water. For plates, add 20 g/L agar. Autoclave 20 minutes, set solution at 60°C to prevent overcooling. Add 300 mL prewarmed (45°C) stock

solutions mixture of: 100 mL 20% dextrose, 100 mL 13.4% YNB, 2 mL 0.02% biotin, 10 mL 0.5% amino acids, 88 mL sterile water (Note: for RDHB, mixture should contain 78 mL sterile water and 10 mL 0.4% histidine). Store at 4°C.

3. RD and RDH (<u>R</u>egeneration <u>D</u>extrose ± <u>H</u>istidine) Top Agar (1L): same as for RDB and RDHB, but use 10 g/L agar. Use as molten solution at 45°C.

4. MGY (<u>M</u>inimal <u>G</u>lycerol) (1L): aseptically mix 800 mL sterile water, 100 mL 13.4% <u>Y</u>NB, 2 mL 0.02% biotin, and 100 mL 10% glycerol. For MGYH add 10 mL 0.4% histidine. Store at 4°C.

5. MD (<u>M</u>inimal <u>D</u>extrose) (1L): autoclave 800 mL water 20 min. For plates, add 15 g/L agar before autoclaving. Cool to 60°C, then add: 100 mL 13.4% YNB, 2 mL 0.02% biotin, 100 mL 20% dextrose. For MDH, add 10 mL 0.4% histidine. Store media or plates at 4°C.

6. MM (<u>M</u>inimal <u>M</u>ethanol) (1L): same as for MD, but add 100 mL 5% methanol instead of dextrose. For MMH, add 10 mL 0.4% histidine. Store media and plates at 4°C.

7. BMG (<u>B</u>uffered <u>M</u>inimal <u>G</u>lycerol) (1L): autoclave 700 mL water 20 min, cool to room temperature. Add 100 mL 1 *M* potassium phosphate buffer, pH 6.0, 100 mL 13.4% YNB, 2 mL 0.02% biotin, 100 mL 10% glycerol. Store at 4°C.

8. BMM (<u>B</u>uffered <u>M</u>inimal <u>M</u>ethanol) (1L): same as for BMG, but add 100 mL 5% methanol instead of glycerol. Store at 4°C.

9. BMGY (<u>B</u>uffered <u>G</u>lycerol-complex Medium) (1L): autoclave 10 g yeast extract and 20 g peptone in 700 mL water, 20 min. Cool to room temperature. Add 100 mL 1 *M* potassium phosphate buffer, pH 6.0, 100 mL 13.4% YNB, 2 mL 0.02% biotin, 100 mL 10% glycerol. Store at 4°C.

10. BMMY (<u>B</u>uffered <u>M</u>ethanol-complex Medium) (1L): same as BMGY, but add 100 mL 5% methanol instead of glycerol. Store at 4°C.

2.2. Generation of proMMP Expression Constructs

1. proMMP cDNA template for high fidelity PCR.
2. pPIC-9 (or other) vector DNA (Invitrogen).
3. Oligomer primers for PCR containing appropriate restriction enzyme sites.
4. Deep Vent™ DNA Polymerase and 10X Deep Vent™ buffer (New England Biolabs).
5. dNTPs: Prepare equimolar stock solution at 10 m*M* each in water.
6. Restriction Enzymes, T4 DNA Ligase and 10X Buffers (New England Biolabs).
7. Dideoxy Sequencing kit (ABI Prism or other).
8. Qiaex extraction kit (Qiagen).

2.3. Transformation

2.3.1. Spheroplasting

Most reagents provided in kits from Invitrogen.

1. GS115 (or other) *Pichia* strain (Invitrogen).
2. Linearized DNA (5–20 µg) in 5–10 µL TE: 10 m*M* Tris-HCl, 1 m*M* EDTA, pH 8.0.

3. YPD medium and plates (warmed to room temperature).
4. RDB and RDHB plates (warmed to room temperature).
5. 5% SDS in water (20 mL, prepare fresh).
6. RD molten agar, 100 mL (prepare fresh, keep at 45°C).
7. 1 *M* Sorbitol.
8. SE: 1 *M* sorbitol, 25 m*M* EDTA, pH 8.0.
9. SCE: 1 *M* sorbitol, 1 m*M* EDTA, 10 m*M* sodium citrate buffer, pH 5.8.
10. CaS: 1 *M* sorbitol, 10 m*M* Tris-HCl, pH 7.5, 10 m*M* $CaCl_2$.
11. 1 *M* DTT (analytical grade) in water.
12. 3 mg/mL Zymolase slurry in water (Invitrogen).
13. 40% (w/w) PEG 3350 (Reagent grade) in water.
14. CaT: 20 m*M* Tris-HCl, pH 7.5, 20 m*M* $CaCl_2$.
15. SOS: 1*M* sorbitol, 0.3X YPD, 10 m*M* $CaCl_2$.
16. SED: 19 mL of SE and 1 mL of 1 *M* DTT (make fresh).
17. PEG/CaT: 1:1 mix of 40% PEG and CaT (make fresh).
18. Two water baths: one at 30°C and one at 45°C.
19. UV-visible spectrophotometer.

2.3.2. Electroporation

1. GS115 (or other) *Pichia* strain (Invitrogen).
2. YPD medium.
3. MD plates (warmed to room temperature).
4. RDB plates (warmed to room temperature).
5. Sterile water (ice-cold).
6. 1 *M* Sorbitol (ice-cold).
7. Linearized DNA (5–20 μg) in 5–10 μL TE: 10 m*M* Tris-HCl, 1 m*M* EDTA, pH 8.0.
8. Electroporator (Bio-Rad GenePulser) with 0.2 cm cuvets.

2.4. Screen for Mut Phenotype and Immunoblot Analyses for Secreted Protein Expression Levels

2.4.1. Colony-Based Mut Phenotype Screen and Semi-Quantitative Immunoblot Screen for Rapid Identification of Secreting Expressors

1. Primary antibody against MMP of interest.
2. HRP-conjugated secondary antibody (Amersham).
3. Nitrocellulose Disks (Schleicher and Schuell).
4. Whatman 3MM chromatography paper.
5. MD plates.
6. MM plates.
7. Tris-buffered saline (TBS): 20 m*M* Tris-HCl, pH 7.6, 137 m*M* NaCl.
8. TBST: TBS with 0.1% Tween 20.
9. Blotto: 5% milk in TBST.
10. ECL kit (Amersham or Pierce Chemical).
11. Sterile Toothpicks.

2.4.2. Quantitative SDS-PAGE-Based Immunoblot Screening for High Expressors of Secreted Protein

1. BMGY and BMMY media.
2. BMG and BMM media.
3. MG and MM media.
4. Appropriate percentage SDS-PAG (8% for proMMP-9) minigel.
5. Nitrocellulose, Whatman 3MM paper and Electroblot Transfer Apparatus (Bio-Rad).
6. TBS, TBST, and Blotto (*see* **Subheading 2.4.1.**, **steps 7–9**).
7. Primary antibody to MMP of interest.
8. HRP-conjugated secondary antibody (Amersham).
9. ECL kits (Amersham or Pierce).

2.5. Purification of Secreted proMMP-9

2.5.1. Optimization of Conditions for Production

1. Optimal media for production (determined in **Subheading 3.3.3.**).
2. Protease Inhibitors: phenylmethylsulfonylfluoride (PMSF), and EDTA.
3. Casamino acids.

2.5.2. Purification of Full-Length proMMP-9 by Gelatin Affinity Column

1. Gelatin-Sepharose (Pharmacia or Sigma).
2. 10X column loading buffer: 50 mM Tris-HCl, pH7.5, 1.5 M NaCl, 25 mM CaCl$_2$, 50 mM EDTA.
3. Column wash buffer: 10 mM Tris-HCl, pH 7.5, 1M NaCl, 5 mM CaCl$_2$, 1 mM EDTA.
4. Column elution buffer: 10 mM Tris-HCl, pH 7.5, 1 M NaCl, 5 mM CaCl$_2$, ($\pm$ 10% dimethyl sulfoxide).

3. Methods

3.1. Generation of Expression Constructs

Standard PCR and subcloning techniques *(12,16,17)* are used to create heterologous expression constructs containing the proMMP coding sequences in-frame with either the AMF (pPIC vectors) or the PHO1 (pHIL-S1) signal peptide sequences (*see* **Note 1** for considerations). Alternatively, the full-length preproMMP cDNA is subcloned into an intracellular expression vector (e.g., pHIL-D2). If PCR is used, it is recommended that the sequence be checked for mutations by sequencing *(18)*.

1. Once the heterologous gene is constructed, linearize the DNA by digestion with *Bgl* II (or other) restriction enzyme (*see* **Note 2** for considerations).
2. Purify the linearized DNA by Qiaex extraction or similar method. Elute DNA from beads in 5–10 µL of TE for use in transformation (*see* **Subheading 3.2.1.** or **3.2.2.**). (Organic extraction and ethanol precipitation can also be used to purify linearized DNA, but transformation efficiency is often decreased.)

3.2. Generation of Pichia Clones

3.2.1. Transformation by Spheroplasting

Spheroplasting is the most efficient means of transforming *Pichia (4)*. It involves the digestion of the yeast's outer cell wall with Zymolase, a β-1, 3-glucanase. The resulting cell is a transformation competent spheroplast. Unlike electroporation, spheroplasting requires regeneration of the cell wall before additional screening can be done. Other methods for yeast transformation can be used *(11,19)*, but spheroplasting and electroporation are more efficient *(4,20)*.

3.2.1.1. PREPARATION OF PICHIA FOR SPHEROPLASTING

1. Streak GS115 (or other) strain onto a YPD plate and incubate at 30°C for 2 d.
2. Inoculate 10 mL of YPD (in 50 mL conical tube) with a single colony of GS115. Grow overnight at 30°C, shaking at 250–300 rpm.
3. Set up three 500 mL flasks, each with 200 mL YPD. Inoculate the flasks with 5, 10 or 20 µL of cells from the overnight culture. Grow overnight at 30°C, shaking at 250–300 rpm. Plan to have transformation solutions and RDB and RDBH plates at room temperature.
4. Next day, check OD_{600} of each of the three overnight cultures. Harvest the cells from the culture that is at an OD_{600} between 0.2 and 0.3 by centifuging at 1500g for 5 min at room temperature. Decant the supernatant.
5. Wash the cells in 20 mL sterile water. Resuspend the cells by swirling the tube. Transfer cells to a sterile 50 mL conical tube.
6. Pellet the cells by centrifuging as in **step 4**. Decant the supernatant.
7. Wash cell pellet by resuspending in 20 mL fresh SED. Pellet cells by centrifuging as in **step 4**. Decant the supernatant.
8. Wash cells in 20 mL of 1 *M* sorbitol. Pellet cells by centrifuging as in **step 4**. Decant the supernatant.
9. Resuspend cells by swirling in 20 mL SCE buffer. Divide cells evenly into two 50 mL conical tubes. Label these tube "A" and tube "B."

3.2.1.2. PREPARATION OF SPHEROPLASTS

1. Set UV-visible spectrophotometer at 800 nm. Blank with a mixture of 800 µL 5% SDS and 200 µL SCE.
2. Label seventeen 1.5-mL microfuge tubes: 0, 2, 4, 5, 6, 7, 8, 9, 10, 15, 20, 25, 30, 35, 40, 45, 50. These numbers represent time (minutes) after Zymolase addition (**step 4**). Add 800 µL 5% SDS to each.
3. Remove 200 µL of cells from tube A (*see* **Subheading 3.2.1.1.**, **step 9**) and add them to 800 µL 5% SDS in tube labeled "0." Set aside on ice.
4. To remainder of tube "A", add 7.5 µL of Zymolase slurry. Gently mix cells by inversion and incubate at 30°C.

5. At the times indicated by the numbered 1.5-mL microfuge tubes, remove 200 µL of cells from the 50 mL conical tube and add to the appropriate tube. Mix and set aside on ice.
6. Read the OD_{800} for all samples.
7. Determine the percent of spheroplasting at each time point using the equation:
   % Spheroplasting = $100 - [(OD_{800}$ at time t / OD_{800} at time 0$) \times 100]$
8. Determine the time resulting in 70% spheroplasting. Add 7.5 µL Zymolase slurry to tube B (**Subheading 3.2.1.1.**, **step 9**) and incubate for the time needed to obtain 70% spheroplasting.
9. Harvest spheroplasts by centrifugation at 750*g* for 10 min at room temperature. Decant and discard the supernatant.
10. Wash the spheroplasts by gently dispersing (tap the tube) them in 10 mL 1 *M* sorbitol. Collect spheroplasts by centrifuging as in **step 9**. Decant the supernatant.
11. Wash spheroplasts by gently dispersing them in 10 mL CaS. Collect spheroplasts as in **step 9** and decant the supernatant.
12. Gently resuspend spheroplasts in 0.6 mL CaS. Use spheroplasts immediately for transformation (enough for 6 transformations).

3.2.1.3. SPHEROPLAST TRANSFORMATION

1. Prepare fresh PEG/CaT solution (1 mL per transformation).
2. Place 100 µL of spheroplasts into 15 mL snap-cap tube.
3. Add 10 µg of linearized DNA and incubate at room temperature for 10 min.
4. Add 1 mL PEG/CaT to suspension of DNA and spheroplasts, mix gently by tapping tube. Incubate at room temperature for 10 min.
5. Recover spheroplasts by centrifuging at 750*g* for 10 min at room temperature. Carefully aspirate PEG/CaT supernatant. Drain any excess by inverting the tube.
6. Resuspend pellet in 150 µL SOS medium. Incubate at room temperature for 20 min.
7. Add 850 µL of 1 *M* sorbitol.

3.2.1.4. PLATING TRANSFORMED CELLS FOR REGENERATION AND HIS$^+$ SCREEN

1. For each transformation, have 3 tubes containing 10 mL of molten RD agar at 45°C.
2. Mix 100, 200, and 300 µL of each transformation solution (**Subheading 3.2.1.3.**, **step 7**) with 10 mL molten RD agar. Quickly cap tube, mix by inversion and pour onto RDB bottom plates.
3. As a control for His$^+$ selection, 100 µL of untransformed spheroplasts (no DNA added) should be mixed with 900 µL of 1 *M* sorbitol. 100 µL of this mixture is then added to 10 mL of molten RD agar and plated on a RDB bottom plate.
4. For cell viability control, 100 µL of untransformed spheroplasts should be mixed with 900 µL of 1 *M* sorbitol. 100 µL of this mixture is then added to 10 mL of molten RDH agar and plated on a RDHB bottom plate.
5. Allow top agar to harden. Invert plates and incubate at 30°C for 4–6 d.

3.2.2. Transformation of Pichia by Electroporation **(20)**

Although spheroplasting results in higher transformation efficiency and probably a greater frequency of multicopy integrants than electroporation *(20)*, the method is somewhat labor intensive. Additionally, regeneration is required before further screening can be done. Thus, electroporation has been used as an alternative method to rapidly create and screen *Pichia* clones *(20)*. When using the appropriate vectors, this method also allows rapid primary screening for high copy integrants without the need for regeneration (*see* **Note 4**).

3.2.2.1. PREPARATION OF *PICHIA* FOR ELECTROPORATION

1. Streak GS115 (or other) strain onto a YPD plate and incubate at 30°C for 2 d.
2. Inoculate 5 mL of YPD (in 50 mL conical tube) with single colony of GS115. Grow overnight at 30°C, shaking at 250–300 rpm.
3. In a 2 L flask, inoculate 500 mL of fresh YPD medium with 100–500 µL of overnight culture. Grow overnight at 30°C until $OD_{600} = 1.3 - 1.5$.
4. Collect cells by centrifuging at 1500*g* for 5 min at 4°C. Decant supernatant and resuspend cells in 500 mL of ice-cold sterile water.
5. Collect cells as in **step 4**. Resuspend cells in 250 mL of ice-cold sterile water.
6. Collect cells as in **step 4**. Resuspend cells in 20 mL of ice-cold 1 *M* sorbitol.
7. Collect cells as in **step 4**. Resuspend cells in 1 mL of ice-cold 1 *M* sorbitol. Use cells immediately for best transformation efficiency.

3.2.2.2. ELECTROPORATION AND PLATING *PICHIA* FOR HIS⁺ SCREEN (ADAPTED FROM **20**)

1. Set electroporator (Bio-Rad Gene Pulser) to 1500 V, 25 µF and 400 ohms. Bring MD plates to room temperature.
2. Mix 80 µL of cells (**Subheading 3.2.2.1.**, **step 7**) with 1–3 µg of linearized DNA (in TE). Transfer mixture to an ice-cold 0.2 cm electroporation cuvette. Avoid creating bubbles.
3. Incubate on ice for 5 min.
4. Pulse the cells and immediately add 1 mL of ice-cold 1 *M* sorbitol to the cuvet. Transfer cells to a 1.5 mL-Eppendorf tube.
5. Plate cells by spreading 200–600 µL onto MD plates.
6. Invert plates and incubate at 30°C for 2–3 d.

3.3. Screening Pichia Clones for Mut Phenotype and Secreting Expressors **(21)**

3.3.1. Screening for Mut Phenotype

1. Bring MD and MM plates to room temperature.
2. With sterile toothpick, pick His⁺ colony from RD top agar (*see* **Subheading 3.2.1.4.**) or from MD plate (*see* **Subheading 3.2.2.2.**). Using a grid pattern, touch

the colony sequentially onto the MM plate and then onto the MD plate. Using a fresh toothpick each time, repeat until enough (~100) colonies have been picked and spotted (2 plates, 52 colonies per plate).
3. Incubate plates at 30°C for 2 d. Add methanol (100 μL) to top of inverted MM plate every 24 h.
4. After 2 d, score the colonies for either Mut⁺ or Mutˢ phenotype. The Mut⁺ colonies should be roughly the same size on both plates, while the Mutˢ colonies should be large on the MD plate and small on the MM plate.

3.3.2. Colony-Based Immunoblot for Identifying Secreting Expressors

If a good polyclonal antibody is not available, proceed to **Subheading 3.3.3.**

1. After scoring, place the MD plates at 4°C for later use.
2. With sterile forceps, carefully place a nitrocellulose disk over the colonies on the MM plate. Poke three holes asymmetrically through the nitrocellulose with a sterile 18-gauge syringe needle. Be sure to note the relative positions of the colonies to the holes. Place three pieces of 3MM Whatman paper on top of the nitrocellulose, followed by a stack of several paper towels. Place about 50 g on top of the pile to weigh it down, (identification of high expressers may be aided by not using weight).
3. Incubate at 30°C for 1–3 h.
4. Remove the stack and lift the nitrocellulose from the plate.
5. Remove the yeast from the nitrocellulose filter by washing twice in ~25 mLs TBS.
6. Block the filter in Blotto for at least 30 min at room temperature.
7. Incubate the filter with primary antibody in Blotto for 45 min at room temperature with shaking.
8. Wash the filter at room temperature three times for 15 min each with TBST.
9. Incubate the filter with HRP-conjugated secondary antibody in Blotto for 20 min at room temperature with shaking.
10. Wash the filter at room temperature three times for 10 min each with TBST.
11. Develop the blot with ECL reagents and expose to film.

3.3.3. SDS-PAGE-Based Immunoblot Screen for Secreting Expressors

Once candidate clones have been identified, a standard immunoblot assay is used to determine more accurately the relative amounts of secreted protein and to assess the production of full-length product. Additionally, an initial media optimization can be done at this step. If a good polyclonal antibody is not available, choose several clones (six to ten) for growth and induction. Production of secreted full-length protein can be assessed by Coomassie or silver-staining the SDS-PAGE gel. Alternatively, zymography can be used for detecting active MMPs (*see* **Notes 6** and **7** for considerations).

1. For each candidate clone, one colony on the MD plate (**Subheading 3.3.1.**) is used to inoculate 10 mL (in 50 mL conical tube) of MGY, BMG or BMGY (*see* **Note 5**).

2. Grow cultures at 30°C 250–300 rpm until OD_{600} = 2 – 6 (overnight).
3. Centifuge at 1500g for 5 min at room temperature to harvest cells.
4. Decant the supernatant and resuspend cell pellet in 2 mL of MM, BMM or BMMY (*see* **Note 5**).
5. Grow at 30°C 250–300 rpm adding methanol to 0.5% every 24 h. Remove 200 µL sample from the culture and transfer to microfuge tube. Repeat every 24 h for 4–6 d (*see* **Note 5**).
6. Pellet cells in microfuge at 3000g and transfer supernatant to fresh tube. Freeze samples at –80°C or process immediately for SDS-PAGE.
7. For SDS-PAGE, add 20 µL of media sample to 20 µL of 2X Laemmli buffer *(22)*. Boil for 5 min and load 20–30 µL on 10% SDS-PAG minigel. Coomassie or silver stain gel or proceed with immunoblot (*see* **Note 6**).
8. Transfer gel to nitrocellulose. Rinse membrane in TBS.
9. Proceed with immunoblot as in **Subheading 3.3.2.**, **steps 6–11**.

3.4. Scaled-Up Production and Purification of MMP-9

Once the highest expresser has been determined, production can be scaled up. If proteolysis is a problem, various parameters can be changed to try to enhance stability (*see* **Notes 5, 8, 10,** and **11**). Such optimizations should be done on a small scale first (**Subheading 3.3.3.**).

3.4.1. Large-Scale Growth and Induction

1. Streak clone onto YPD plate and incubate at 30°C for 2 d.
2. From a single colony, inoculate 25 mL of optimal glycerol medium (BMG, pH 5.0 for proMMP-9, or determined in **Subheading 3.3.3.**) in a 250 mL baffled flask. Grow at 30°C 250–300 rpm until OD_{600} = 2 – 6 (overnight).
3. Use the 25 mL culture to inoculate 1L of optimal glycerol medium in a 4L baffled flask. Grow at 30°C 250–300 rpm until OD_{600} = 2 – 6.
4. Centrifuge cells in sterile bottles at 3000g for 5 min at room temperature. Decant the supernatant and resuspend the pellet:
   a. For Mut⁺ clones, to an OD_{600} = 1.0 (2–6 L) in optimal methanol medium (BMM, pH 5.0 for proMMP-9). Divide between several 4L baffled flasks.
   b. For Mut^s clones, in 100–200 mL of optimal methanol medium. Transfer cells to 1L baffled flask.
5. Induce cells at 30°C for optimal time as determined in **Subheading 3.3.3.** (For MMP-9, 24 h in BMM, pH 5.0). Cells can be pelleted and resuspended in fresh BMM for sequential inductions (*see* **Notes 5** and **11**). For longer inductions, add methanol to 0.5% every 24 h to account for evaporation.
6. Pellet cells by centrifuging at 3000g for 5 min at 4°C. Transfer and save supernatant. Process conditioned medium immediately for MMP purification or rapid-freeze medium and store at –80°C until needed.

3.4.2. Purification of MMP-9 **(23)**

For each clone, the level of expression may vary. Therefore, the size of the gelatin-Sepharose column needed should be determined empirically based on the quantity of secreted protein in the conditioned medium.

1. Add 1/10 vol of 10X column load buffer to conditioned medium. Add PMSF to a final concentration of 100 m*M* (*see* **Note 7**). Mix.
2. Load conditioned medium onto gelatin-Sepharose affinity column.
3. Wash column with 20 bed volumes of 1X column load buffer.
4. Wash column with 20 bed volumes of column wash buffer.
4. Elute bound proMMP-9 from column with a shallow (0–10%) DMSO gradient in column elution buffer. Collect small fractions to enhance yield of purified full-length proMMP-9 (*see* **Note 11**).
5. Chase any remaining proMMP-9 off the column with two bed volumes of column elution buffer containing 10% DMSO.
6. Assess fractions for full-length proMMP-9 by Coomassie or silver staining SDS-PAG minigel, immunoblot and/or zymography (*see* **Notes 6** and **7**).
7. Pool appropriate fractions and dialyze to completion against desired buffer.

4. Notes

1. In general, the alpha-mating factor (AMF) signal has been used most successfully for secretion of recombinant proteins in the *Pichia* system *(2,24)*. The AMF signal peptide works for secretion of full-length proMMP-9 and proMMP-1 *(7,8)* (Doyle, G. A., unpublished observation) and the acid phosphatase (PHO1) signal functions well for expression of carboxy-truncated proMMP-9 *(9)*. Notably, Sreekrishna et al. find that the native signal peptides of preproMMP-1, preproMMP-2, preproMMP-3 and preproMMP-9, as well as that of TIMP-1, are functional in the *Pichia* system *(6)*. When fusing sequences to the AMF signal, it is important not to disrupt the Kex2 cleavage site (Glu-Lys-Arg-Xxx). Efficient cleavage (between Arg-Xxx) is somewhat dependent on the amino acid in the P_1' site *(26)*. In the native AMF signal peptide sequence, a Glu-Ala repeat follows the Arg residue. Thus, at the very least, one should place a Glu residue in the P_1' site. But, it is recommended that the diamino repeat (Glu-Ala-Glu-Ala) be included. The diamino repeat is removed from the amino terminus of the secreted fusion protein by the *STE13* gene product *(25)*. When using the PHO1 signal, it is necessary to keep or recreate the Ala-Arg-Glu sequence for efficient cleavage *(1,24)*.
2. As mentioned in **Subheading 1.3.**, recombination can be targeted to either the *AOX1* or *his4* locus by using different restriction enzymes to linearize the DNA construct. The example given in **Subheading 3.1.**, **step 1**, (digestion with *Bgl* II) would target recombination to the *AOX1* locus. Insertions at either the 5' or 3' region of the *AOX1* locus would occur at a higher frequency, but gene replacement is also possible. If the cDNA to be expressed contains a *Bgl* II site, a number of other sites can be used to linearize the construct (*see* **Fig. 1** and Invitrogen manual). A unique *Sac* I site, located in the *AOX1* 5' flanking region, can be used

to specifically target insertions to the 5′ region of the *Pichia AOX1* locus. Alternatively, the construct can be linearized with either *Sal* I or *Stu* I, at unique sites located in the *HIS4* gene, to target insertions to the *Pichia his4* locus. It should be noted that the pHIL-D2 vector has *Not* I restriction sites flanking the *AOX1* sequences rather than the *Bgl* II sites found in the other vectors (**Fig. 1**).

3. Because gene replacement requires a double crossover event, it occurs at lower frequency than gene insertion, which requires only a single crossover event. When insertion strategies are used, transformed clones should have the same Mut phenotype as the parental strain. For example, KM71 clones will be Mut^s, while GS115 or SMD clones will be Mut$^+$. If an *AOX1* gene replacement strategy is pursued, GS115 and SMD clones would be screened for a His$^+$Mut^s phenotype. A replacement strategy can be used with the KM71 strain, but is not preferred. An added advantage to using an insertion strategy is that, although less frequent than single insertions, multiple insertions do occur at a rate of 1–10%. Clones containing multiple integration events often produce very high levels of recombinant protein *(21)*. Therefore, strategies to create *Pichia* transformants with insertion events are recommended (*see* **Note 4**).

4. The pAO815 vector allows multimers of the gene of interest to be synthesized in vitro before transformation of the yeast. Alternatively, the pPIC3.5K and pPIC9K vectors, carrying the kanamycin resistance gene, allow His$^+$ transformants to be screened for high copy integration clones by selection on YPD plates containing increasing concentrations (0.25 to 2 mg/mL) of G418 *(20)*. The pPICZ and pPICZα vectors, carrying the zeocin resistance gene, are an exception to screening for His$^+$ clones. When using these vectors, screening for positive transformants is done by plating transformed yeast on zeocin containing plates. Similar to the pPIC3.5K and pPIC9K vectors, one can select for high copy integrants by growing cells on YPD plates containing increasing concentrations of zeocin. The zeocin vectors are specifically designed to target multiple insertion events to the 5′ flanking region of the *AOX1* locus.

5. The optimal conditions for growth and induction will vary depending on the protein being expressed and will need to be refined before large-scale production is tried. The pH of the induction medium can influence protein stability, particularly in strains that are not protease deficient. Therefore, initially screening clones in one of the buffered medium sets (BMG/BMM) under standard pH (6.0) conditions is recommended. Once the highest producer(s) are identified, different induction conditions can be compared. During the initial screen, the optimal induction time for production of full-length protein can also be determined. If proteolysis is a problem, shorter induction times or different pH conditions may increase yields of full-length protein. Additionally, including Casamino acids, yeast extract, peptone, and/or protease inhibitors in the induction medium may enhance yields of full-length protein. Please *see* **Notes 8–11** for discussion of MMP considerations.

6. The level of expression for each *Pichia* clone will vary due to differences in the number of integration events and to the behavior of the particular protein being

expressed. At times, Coomassie or silver staining may not be sensitive enough to detect the amount of expressed protein in medium samples. Using the sensitive ECL-based immunoblotting technique allows the detection of small quantities of full-length and truncated protein. But, immunoblotting relies on a good polyclonal antibody and gives no information regarding the functional activity of the expressed protein. Zymography should allow detection of small amounts of full-length proMMPs, as well as some degraded or autoactivated MMP product present in the sample. However, truncated forms of a MMP that are no longer active against the substrate will not be detected. Thus, it is recommended that, if possible, both immunoblotting and zymography be used to assess production and activity of expressed proMMPs. Once purified, proMMPs can be used in native activation and activity assays.

7. Endogenous gelatinolytic activity has been observed by gelatin zymography using conditioned complex medium from negative control *Pichia (8)* and from *Pichia* cell lysates (Doyle, G. A., unpublished observation). Apparently, the use of minimal (BMM) rather than complex (BMMY) medium for induction alleviates the former problem *(8)*. Two gelatinolytic activities are seen in whole cell lysates, one migrating at ~60 kDa and the other at ~82 kDa. Each of these bands is PMSF, but not EDTA inhibitable. These PMSF inhibitable proteases do not bind gelatin-Sepharose (Doyle, G. A., unpublished observation). As long as there is not appreciable cell death during induction, intracellular proteases should not be problematic. However, as a precaution, during purification, I put PMSF in proMMP-9 conditioned medium before loading the sample on the gelatin-Sepharose column. I also recommend comparing gelatin zymograms done in the presence and absence of PMSF when *Pichia* strains, clones and production conditions are initially analyzed for use.

8. In general, changing yeast strains, medium conditions, or adding protease competitors or inhibitors to the induction medium can alleviate problems with protein degradation. For proMMPs, many of these approaches have proven unsuccessful *(6,8)* (Doyle, G. A., unpublished observation). However, all options have not been exhaustively studied.

9. Masure et al. have noted high molecular weight complexes of proMMP-9 being formed. It is unclear why the complexes form, but it does not appear to be due to glycosylation (Opdenakker, G., personal communication). Presumably, these complexes are due to inter-molecular disulfide bonds. A similar phenomenon is observed when the human MMP-9 fibronectin type II repeats are expressed in GS115 (Doyle, G. A., unpublished observation). In the latter case, high molecular weight complexes were formed in medium at pH 6.0, but were eliminated when induction medium was buffered to pH 5.0. It is not known if adjusting the pH of the induction media will eliminate the high molecular weight complexes observed for murine proMMP-9 in SMD1168 *(8)*.

10. The most prevalent problem with MMP expression in *Pichia* is degradation of the full-length proMMP; either by autocatalysis or by secreted or released (upon cell death) *Pichia* proteases. Apparent autocatalytic activation of proMMP-1 has

hampered the ability to produce and purify high yields of full-length proMMP-1 *(6,7)*. In addition, autocatalysis and yeast cell toxicity drastically limit success with production of proMMP-3 in *Pichia (6)*. Autocatalytic activation seems to be less of a problem for production and purification of full-length proMMP-2 and proMMP-9 *(6,8,11)*. However, degradation does occur (*see* **Notes 8** and **11**).

Various proMMPs have been expressed in either GS115, KM71 or SMD1168 strains with similar results *(6–8)* (Doyle, G. A., unpublished observation). Thus, autocatalysis appears to be independent of the yeast strain used *(6)*. To my knowledge, the SMD1165 *(his4, prb1)* and SMD1163 *(his4, pep4, prb1)* protease deficient strains have not been tried for proMMP expression. Although it is unlikely that using these strains will solve the problem of autocatalysis, trying them might prove advantageous to the problem of *Pichia*-mediated proteolysis of certain MMPs (*see* **Note 11**).

Adjusting the pH of or adding 1% Casamino acids to the induction medium does not appear to help with the problem of autocatalytic activation *(6)*. Thus far, the use of only two protease inhibitors in the induction media has been reported to help stabilize the proMMPs against autocatalysis: TIMP-1 and EDTA *(6)*. For obvious reasons, inclusion of TIMP-1 (or other TIMPs) in the production system is not prefered. Addition of 5 m*M* EDTA to the induction medium improved proMMP stability to some extent, but resulted in the accumulation of a 50 kDa protein, likely exo-β-1,3-glucanase, in the culture medium *(6)*. It is unclear whether this protein would interfere with production and purification of full-length proMMPs. Presumably, its activity might alter and weaken the cell wall of *Pichia*, resulting in unwanted cell lysis.

11. In GS115, human proMMP-9 or its isolated tandem fibronectin type II repeats are degraded when induced in either buffered (pH 6.0) minimal or complex medium (Doyle, G. A., unpublished observation). Since both proMMP-9 and the fibronectin type II repeats are cleaved, this degradation is not autocatalytic and is likely due to a *Pichia* protease. The cleavage occurs within the fibronection type II repeats and results in the inactivation of the proMMP-9 product as determined by gelatin zymography (Doyle, G A., unpublished observation). It is not clear if the same cleavage is occuring during production of murine proMMP-9 in SMD1168. The use of the double protease deficient strain SMD1163 may eliminate this particular proteolytic event. Fortunately, full-length human proMMP-9 produced in GS115 or other strains can be purified away from the degradation products by gelatin affinity chromatography with a shallow DMSO gradient during the elution step.

    The addition of inhibitors such as 1% casamino acids to the induction medium does not prevent proteolysis of proMMP-9 expressed in GS115 or SMD1168 *(6–8)* (Doyle, G. A., unpublished observation). Although changing the pH of the induction medium does not seem to decrease autocatalysis *(6)*, it can help stabilize proMMPs against certain *Pichia*-mediated proteolysis. Buffering the induction medium to pH 5.0 or 6.0 diminishes proMMP-9 degradation *(8)* (Doyle, G. A., unpublished observation). The very low pH (3.0) obtained during extended induction periods in unbuffered medium results in greater murine proMMP-9 degradation *(8)*. If some of the observed degradation is in fact due to the release of intracellular proteases caused by *Pichia* cell death, the addition of more general protease inhibitors (PMSF and so on) to the induction medium might be helpful. To my knowledge, this has not been tried. In any case, one must determine if addition of such inhibitors would interfere with the ability of *Pichia* to produce and/or secrete the induced protein.

Adjusting the time of the induction period can help to increase the yield of purified full-length protein. Improved yields of full-length murine proMMP-9, and perhaps other proMMPs, can be accomplished by using sequential 24 h inductions in fresh medium (buffered or unbuffered) rather than extended induction periods *(8)*.

Acknowledgments

I gratefully acknowledge David Miles of Invitrogen for allowing the reproduction of some of their manual protocols and vector maps. I would also like to thank Koti Sreekrishna and Ghislain Opdenakker for helpful communications, and Linda Robinson for helpful comments on the chapter.

References

1. Buckholz, R. G. and Gleeson, M. A. G. (1991) Yeast systems for the commercial production of heterologous protein. *Bio/Technology* **9**, 1067–1072.
2. Cregg, J. M., Vedvick, T. S., and Raschke, W. C. (1993) Recent advances in the expression of foreign genes in *Pichia pastoris. Bio/Technology* **11**, 905–910.
3. Grinna, L. S. and Tschoop, J. F. (1989) Size distribution and general structural features of N-linked oligosaccharides from the methylotrophic yeast *Pichia pastoris. Yeast* **5**, 107–115.
4. Cregg, J. M., Barringer, K. J., and Hessler, A. Y. (1985) *Pichia pastoris* as a host system for transformations. *Mol. Cell. Biol.* **5**, 3376–3385.
5. Cregg, J. M., Tschopp, J. F., Stillman, C., Siegel, R., Akong, M., Craig, W. S., Buckholz, R. G., Madden, K. R., Kellaris, P. A., Davis, G. R., Smiley, B. L., Cruze, J., Torregrossa, R., Velicelebi, G., and Thill, G.P. (1987) High-level expression and efficient assembly of hepatitis B surface antigen in the methylotrophic yeast *Pichia pastoris. Bio/Technology* **5**, 479–485.
6. Sreekrishna, K., Brankamp, R. G., Kropp, K. E., Blankenship, D. T., Tsay, J.-T., Smith, P. L., Wierschke, J. D., Subramaniam, A., and Birkenberger, L. A. (1997) Strategies for optimal synthesis and secretion of heterologous proteins in the methylotrophic yeast. *Pichia pastoris. Gene* **190**, 55–62.
7. Rosenfeld, S. A., Ross, O. H., Hillman, M. C., Corman, J. I., and Dowling, R. L. (1996) Production and purification of human fibroblast collagenase (MMP-1) expressed in the methylotrophic yeast. *Pichia pastoris. Protein Expression and Purification* **7**, 423–430.
8. Masure, S., Paemen, L., Van Aelst, I., Fiten, P., Proost, P., Billiau, A., Van Damme, J., and Opdenakker, G. (1997) Production and characterization of recom-

binant active mouse gelatinase B from eukaryotic cells and *in vivo* effects after intravenous administration. *Eur. J. Biochem.* **24**, 21–30.

9. Shipley, J. M., Doyle, G. A., Fliszar, C. J., Ye, Q.-Z., Johnson, L. L., Shapiro, S. D., Welgus, H. G., and Senior, R. M. (1996) The structural basis for the elastolytic activity of the 92-kDa and 72-kDa gelatinases: Role of the fibronectin type II-like repeats. *J. Biol. Chem.* **271**, 4335–4341.

10. Ro, N., Smith, M. M., Narve, M., and Das, G. (1997) Human gelatinase B (MMP-9) fragment expression using pre-pro mating factor alpha secretion signal, in *Current Topics in Gene Expression Systems,* Invitrogen, Inc., San Diego, *1956*, 56.

11. Guthrie, C. and Fink, G. R. (1991) Guide to yeast genetics and molecular biology, in *Methods in Enzymology*, (J. N. Abelson and M. I. Simon, eds.), San Diego, CA, Academic Press.

12. Ausubel, F. M., Brent, R., Kingston, R. E., Moore, D. D., Seidman, J. G., Smith, J. A., and Struhl, K. (1994) *Current Protocols in Molecular Biology,* Greene Publishing Associates and Wiley-Interscience, New York.

13. Cregg, J. M., Madden, K. R., Barringer, K. J., Thill, G., and Stillman, C. A. (1989) Functional characterization of the two alcohol oxidase genes from the yeast, *Pichia pastoris. Mol. Cell. Biol.* **9**, 1316–1323.

14. Ellis, S. B., Brust, P. F., Koutz, P. J., Waters, A. F., Harpold, M. M., and Gingeras, T. R. (1985) Isolation of alcohol oxidase and two other methanol regulated genes from the yeast, *Pichia pastoris. Mol. Cell. Biol.* **5**, 1111–1121.

15. Tschopp, J. F., Brust, P. F., Cregg, J. M., Stillman, C., and Gingeras, T. R. (1987) Expression of the *lacZ* gene from two methanol regulated promoters in *Pichia pastoris. Nucleic Acids Res.* **15**, 3859–3876.

16. Sambrook, J., Fritsch, E. F., and Maniatis, T. (1989) Molecular cloning: A laboratory manual, Second edition (Plainview, New York: Cold Spring Harbor Laboratory Press).

17. Pierce, J. C. (1997) In-Frame cloning of synthetic genes using PCR inserts, in *Methods in Molecular Biology*, (B. A. White, ed.), Totowa, New Jersey, Humana Press.

18. Sanger, F., Nicklen, S., and Coulson, A. R., (1977) DNA sequencing with chain-terminating inhibitors. *Proc. Natl. Acad. Sci. USA* **74**, 5463–5467.

19. Gietz, R. D., Schiestl, R. H., Willems, A. R., and Woods, R. A. (1995) Studies on the transformation of intact yeast cells by the LiAc/SS-DNA/PEG procedure. *Yeast* **11**, 355–360.

20. Scorer, C. A., Clare, J. J., McCombie, W. R., Romanos, M. A., and Sreekrishna, K. (1994) Rapid selection using G418 of high copy number transformants of *Pichia pastoris* for high-level foreign gene expression. *Bio/Technology* **12**, 181–184.

21. Wung, J. L. and Gascoigne, N. R. (1996) Antibody screening for secreted proteins expressed in *Pichia pastoris. Biotechniques* **21**, 808, 810, 812.

22. Laemmli, U. K., (1970) Cleavage of structural proteins during the assembly of the head of bacteriophage T4. *Nature* **227**, 680–685.

23. Wilhelm, S. M., Collier, I. E., Marmer, B. L., Eisen, A. Z., Grant, G. A., and Goldberg, G. I. (1989) SV40-transformed human lung fibroblasts secrete a 92-kDa

type IV collagenase which is identical to that secreted by normal human macrophages [published erratum appears in *J. Biol. Chem.* **265,** 22,570 (1990)]. *J. Biol. Chem.* **264**, 17,213–17,221.

24. Laroche, Y., Storme, V., De Meutter, J., Messens, J., and Lauwereys, M. (1994) High-level secretion and very efficient isotopic labeling of tick anticoagulant peptide (TAP) expressed in the methylotrophic yeas., *Pichia pastoris. Bio/Technology* **12**, 1119–1124.

25. Brake, A. J., Merryweather, J. P., Coit, D. G., Heberlein, U. A., Masiarz, F. R, Mullenbach, G. T., Urdea, M. S., Valenzuela, P., Barr, P. J. (1984) Alpha-factor-directed synthesis and secretion of mature foreign proteins in *Saccharomyces cerevisiae. Proc. Natl. Acad. Sci. USA* **81**, 4642–4646.

11

Expression of Recombinant Membrane-Type MMPs

Michael J. Butler, Marie-Pia d'Ortho, and Susan J. Atkinson

1. Introduction

Since the discovery of the membrane-type matrix metalloproteinases (MT-MMPs) which are characterized by the existence of a C-terminal membrane-spanning domain followed by a short cytoplasmic "tail" *(1)* there has been much interest in the production of recombinant protein to facilitate both structural and functional analyses. The proteolytic activity of the membrane-type-1 MMP (MT1-MMP) has been studied using recombinant material produced from expression in *E. coli (2,3)*. These data have been confirmed and extended by reports describing expression of MT1-MMP in mammalian cells *(4–7)*. This chapter will, therefore, focus on methods to produce recombinant MT1-MMP in both types of expression systems and the issues governing the choice of which system to use will be discussed. Overall, we have used the glutathione-S-transferase (GST)-fusion technology to produce material for injection into sheep to raise polyclonal antibodies. For kinetic analysis of catalytic activity we have used *E. coli* expression, both periplasmic, and expression in inclusion bodies (followed by solubilization and refolding). We have found the latter approach to be most reliable and, therefore, describe this method in detail. To study the location and function of the membrane-associated form we have used transient expression in Chinese hamster ovary (CHO) cells. High level expression in stably transfected cells has been more difficult to achieve and this chapter describes the use of an inducible mammalian cell system which gives reasonably high levels of expression of MT1-MMP.

From: *Methods in Molecular Biology, vol. 151: Matrix Metalloproteinase Protocols*
Edited by: I. Clark © Humana Press Inc., Totowa, NJ

1.1. E. coli *Expression Systems for MT-MMP Production*

1.1.1. N-Terminal (6 His-Tagged) Fusion Proteins

We have found the plasmid vector pRSET to be useful for the expression of pro-MT-MMPs. This vector provides a 6 His tag at the amino terminus of the recombinant protein and is followed by a sequence derived from the gene 10 protein of T7 phage. This is highly expressed in *E. coli*, therefore eliminating potential problems due to codon usage or mRNA secondary structure at the beginning of the translation process. Other useful features of this vector are an antibody epitope located after the T7 gene 10 sequence, followed by an enterokinase cleavage site. The latter facilitates removal of most of the N-terminal fusion peptide. The most convenient, reproducible results are obtained using essentially the protocols described by Studier et al. *(8)* with the relatively minor modifications described in the methods section. Overall the cDNA is amplified to modify the 5' and 3' termini to facilitate "in frame" subcloning. The DNA is cloned and analyzed in *E. coli* DH5α (in which no expression of the protein occurs) followed by transformation of the recombinant plasmid DNA into the BL21 host for induction of protein production.

1.1.2. GST-Fusion Proteins

The GST expression system uses the plasmid expression vector pGEX. It has been constructed to direct the synthesis of foreign polypeptides in *E. coli* as fusions with the C-terminus of Sj-26, a 26 kD glutathione S-transferase (GST) encoded by the parasitic helminth *Schisostoma japonicum (9)*.

The GST-fusion system offers several advantages:

1. The expression of the recombinant protein is under the control of the tac-promoter, which allows induction of expression by isopropylthiogalactoside (IPTG).
2. The vectors have been engineered so that the GST carrier can be cleaved from the fusion protein by digestion with specific proteases, thrombin, or blood coagulation factor Xa (for expression of MT1-MMP, we used the vector pGEX-2T where cleavage between carrier protein and MT1-MMP can be performed by thrombin).
3. Fusion-protein can be purified by affinity chromatography on immobilized glutathione.
4. Theoretically, the system provides soluble fusion protein which can be directly purified from crude bacterial lysates under nondenaturing conditions. However, in this particular case, the MT1-MMP-GST fusion protein was recovered in inclusion bodies which required solubilization. We used the procedure described here to obtain ΔMT1-MMP, a C-terminally truncated form of MT1-MMP, lacking the transmembrane domain (residues 511–582).

1.2. Transient Expression in CHO Cells

Transient expression of MT-MMPs provides a relatively quick and simple method for screening newly identified members of the MT-MMP family. For example, transient transfection of MT1-MMP in CHO cells enabled us to demonstrate its ability to activate progelatinase A (proMMP-2) *(5,7)*. Immunocytochemical studies of the transfected cells revealed its location on the cell surface and interactions with TIMP2 and proMMP-2 *(10)*. Additionally, transient transfection of CHO cells on a rhodamine-labeled gelatin substrate and the use of confocal microscopy allowed us to examine the ability of MT1-MMP to degrade the extracellular matrix *(11)*.

A rapidly increasing number of techniques are developing for the transient expression of proteins in mammalian cells. These include electroporation, calcium phosphate precipitation, the use of a variety of cationic lipid reagents, e.g., Lipofectamine, Fugene and so on and the DEAE-dextran technique. Several different cell types have been employed but those most commonly used for MT1-MMP are the cell line, CHO-K1 and two kidney epithelial cell lines derived from African green monkeys, Cos-1, and Cos-7. Described below is the protocol for the successful transient transfection of MT1-MMP into a CH-K1 cell line, CHOL761h *(12)*, using the calcium phosphate precipitation technique essentially as described by Chen and Okayama *(13)*. Three major procedures are involved; culture and subculture of the cells, the transfection procedure itself and finally the detection of the expressed MT1-MMP.

Transient transfection is performed by incubation of freshly subcultured cells at 37°C, 5% CO_2 with preformed calcium phosphate-cDNA precipitates. This is followed by a brief glycerol shock to aid entry of the DNA into the cells. MT1-MMP expression is detected approx 16–24 h after transfection, either by assessing the activation of proMMP-2 by gelatin zymography, by immunolocalization using indirect immunofluorescence or by Western blotting of the cell lysates.

1.3. Inducible Expression in Mammalian Cells Using the Tet-Off System

Initially we attempted to use the NS0 mouse myeloma cell expression system using the strong constitutive human cytomegalovirus (hCMV) promoter which has been very successful for MMPs *(14,15)*. However, we were unable to select high producing clonal cell lines until we cotransfected with TIMP-2 cDNA (unpublished data). Thus, it appears that overexpression of MT1-MMP is relatively toxic to the host cell and this is consistent with reports documenting the secretion of MT1-MMP:TIMP-2 complexes from other stably transfected mammalian cell lines in which the endogenous TIMP-2 production affords protection to the host cell allowing significant (but relatively low) levels of MT1-MMP to be produced.

To circumvent this toxicity problem we have used the Tet-off system to produce full-length membrane associated MT1-MMP. In *E. coli*, the genes of the tetracycline-resistant operon of the Tn*10* transposon are negatively regulated by the Tet repressor (TetR), which blocks transcription by binding to the tet operator sequences (*tetO*) in the absence of Tc *(16)*. These two elements—the TetR protein; and the *tetO* regulatory sequence—provide the basis for the Tet-Off system. The Tet-off system has several advantages over other regulated gene expression systems that have been used in mammalian cells. Thus, pleiotropic effects due to the stimulation of other transcriptional promoters/enhancers observed for heavy metals, steroid hormone or heat shock induced expression systems are avoided. Specific repression of the target genes is maximized using doxycycline which is more potent than tetracycline and can, therefore, be used at a lower concentration to maintain the stably integrated gene in an inactive state. One of the practical difficulties of this type of inducible system is the need to incorporate the control element into the particular cell type of interest. Pilot experiments can be carried out to establish the ability of a host cell to express a specific protein by transiently cotransfecting both the pTRE response element plasmid (carrying the target gene) and the plasmid (pTet-off) containing the control element. Another consideration in choosing a host cell is the ability of the cell to carry out the appropriate posttranslational modifications required for activity of the protein.

We obtained a derivative of the HT1080 human fibrosarcoma cell line, HTC-75, into which the pTet-off control element had been stably transfected *(17)*. This allowed us to test a combination of this host cell and control element for MT1-MMP production. We selected stable cell lines carrying the MT1-MMP cDNA on a pTRE plasmid and screened for inducible expression of MT1-MMP using gelatin zymography to test for activation of the secreted progelatinase A endogenously produced by the cells. Relatively high level expression of the MT1-MMP was confirmed by Western blot analysis of isolated cell membranes from the clonal cell line and by direct immunolocalization of the cells using anti-MT1-MMP antibodies.

2. Materials

2.1. E. coli expression of MT1-MMP

1. *E. coli DH5α* (Gibco-BRL) for DNA cloning and analysis; store at –70°C.
2. *E.coli* BL21(DE3)pLysS (Stratagene) for pRSET expression of protein. (*See* **Note 1**); store at –70°C.
3. 2YT medium: 16 g/L tryptone, 10 g/L yeast extract, 5 g/L NaCl, pH7.4.
4. Ampicillin: 100 mg/mL (1000 × stock soln) in water; store in aliquots at –70°C.

5. Chloramphenicol: 33 mg/mL (1000 × stock soln) in methanol; store in aliquots at –20°C.
6. Buffer (a): 6 *M* urea, 50 m*M* Tris-HCl, pH 7.5, for washing Ni-agarose columns.
7. Buffer (b): 8 *M* urea, 50 m*M* Tris-HCl, pH 7.5, for solubilization of inclusion bodies.
8. Buffer (c): 6 *M* urea, 50 m*M* Tris-HCl, pH 7.5, 0.25 *M* imidazole, for elution of 6 His-tagged recombinant protein. Sterilize buffers (a) – (c) either by autoclaving or filter sterilization and store at 4°C for not more than a few days.

2.2. Transient Expression in CHO Cells

2.2.1. Cell Culture for Transient Transfection

1. Cell line CHOL761h (*see* **Note 2**); store in liquid N_2.
2. Dulbecco's modified Eagles medium (DMEM), (Gibco UK), store at 4°C.
3. Fetal calf serum (FCS), (Globe Pharm). Heat inactivated 56°C, 1 h. Store at –20°C.
4. Penicillin/streptomycin: final concentrations, 100 I U/mL penicillin, 50 µg/mL streptomycin (Sigma).
5. Glutamine: final concentration, 2 m*M* (Gibco, UK).
6. Growth medium: DMEM, 10% FCS with added glutamine (2 m*M*), penicillin (100 I U/mL), streptomycin (50 µg/mL).
7. Trypsin – EDTA: 1% trypsin, 2% EDTA in PBS, (Gibco UK), store –20°C.
8. 24-well culture plates, (Nunc).
9. 8 Chamber Labtek slides (ICN Biomedicals).

2.2.2. Transfection

1. Plasmid cDNAs: MT1-MMP in pEE14.tPA vector, pEE14.tPA vector only (*see* **Note 3**). Plasmids for transfection should be caesium chloride purified or equivalent. Store at –20°C.
2. 2X Hepes buffered saline (2X HBS): 8.18*g* NaCl, 5.59*g* HEPES, 0.2*g* Na_2HPO_4(anhydrous); dissolve in 400 mL distilled H_2O, adjust pH to 7.1 using NaOH, make up volume to 500 mL. Filter sterilize and store at 4°C.
3. 2 *M* $CaCl_2$. Filter sterilize and store at 4°C.
4. 15% (v/v) glycerol in DMEM. Filter sterilize and store at 4°C.
5. Filter sterilized distilled H_2O.
6. 10X sodium butyrate: 100 m*M* sodium butyrate in DMEM. Filter sterilize and store at 4°C.

2.3. MT1-MMP Detection

2.3.1. proMMP-2 Activation: Gelatin Zymography

1. Recombinant human proMMP-2 expressed and purified from e.g., mouse myeloma cell line NS0 *(14)*, *see* Chapter 14. Store at –70°C, pH 6.0 (*see* **Note 4**).
2. SDS-8% polyacrylamide gel with 0.5 mg/mL gelatin incorporated.
3. Calf skin type I collagen: 2 mg/mL calf skin type I collagen (Sigma C9791) in H_2O, 0.02% sodium azide; store at 4°C. Heat denature for 20 min at 60°C just before use (*see* **Note 5**).

4. 2.5% Triton X-100, (Sigma).
5. Tris-CaCl$_2$ buffer (TCAB): 100 mM Tris-HCl, pH 7.9, 30 mM CaCl$_2$, 0.02% NaN$_3$, 0.05% Brij35.
6. Coomassie brilliant blue stain: 0.25% (w/v) Coomassie brilliant blue in 50% (v/v) methanol, 10% (v/v) acetic acid.

2.3.2. Immunolocalization

1. 4% Formaldehyde: Freshly prepared from paraformaldehyde dissolved in PBS pH 7.4. Keeps for approximately one week at room temperature.
2. Primary antibody: Sheep polyclonal antibody against human MT1-MMP *(11)*. Store at –20°C.
3. Secondary antibody: Pig antibody against sheep Fab', conjugated with fluorescein isothiocyanate *(19)*. Store at 4°C.
4. Normal sheep IgG. Store at –20°C.
5. Normal pig serum. Store at –20°C.
6. Citifluor mounting medium, (Agar Scientific, Stansted, Essex, UK).
7. Coverslips 22 × 50 mm.

2.3.3. Western Blots

1. Cell lysis buffer *(20)*: 50 mM Tris-HCl, pH 8.0, 0.15 M NaCl, 10 mM EDTA, 1% (v/v) Triton-X-100, 0.02% (w/v) NaN$_3$, with protease inhibitors, pepstatin A (1 mg/mL) phenylmethylsulfonyl fluoride (PMSF, 100 μM), trans-epoxysuccinyl-1-leucylamido (4-guanidino)-butane (E64, 1 μg/mL) added just before use.
2. Semi-dry transfer buffer: 25 mM Tris base, 192 mM glycine, 20% (v/v) methanol, 0.004% (w/v) sodium dodecyl sulfate (SDS), pH 8.3 at room temperature (no need to pH).
3. Nitrocellulose membrane for enhanced chemiluminescence (ECL) (Amersham International, UK).
4. Ponceau S solution: 1% (w/v) Ponceau S stain in 5% (v/v) acetic acid.
5. Wash buffer: 10 mm Tris-HCl pH 7.4, 150 mM NaCl, 0.5% Tween 20.
6. Blocking buffer: 5% (w/v) milk powder in wash buffer.
7. ECL detection reagents (Pierce and Warriner, UK).

2.4. Cell Culture for Inducible Expression of MT1-MMP

1. HTC-75 cell line or equivalent *(17)*.
2. pTRE plasmid DNA (Clontech Laboratories Inc.).
3. pSV2neo plasmid.
4. Geneticin (G418) (Sigma).
5. TE: 10 mM Tris-HCl, pH 8.0, 1 mM EDTA.
6. DMEM and FCS as for transient transfection (*see* **Subheading 2.2.**)
7. Tet-approved fetal bovine serum (FBS) (Clontech Laboratories) (*see* **Note 6**).
8. Doxycycline HCl, (Sigma).
9. Bio-Rad Gene Pulsar or equivalent for electroporation.

3. Methods

3.1. *E. coli Expression of MT1-MMP*

3.1.1. N-terminal (6 His-Tagged) Fusion Proteins

1. Design appropriate oligonucleotides *(11)* to amplify the desired portion of MT1-MMP cDNA *(see* **Subheading 3.1.2.**).
2. Amplify the cDNA fragment using polymerase chain reaction (PCR) and either clone directly into the appropriately digested pRSET vector (to introduce the cDNA to direct the production of a translational fusion protein) or into a T-vector using *E. coli* DH5α host cells *(see* **Note 7**).
   If desired a small scale expression test can be carried out as follows:
3. Transform the plasmid DNA to be tested into *E. coli* BL21(DE3)pLysS *(see* **Note 8**).
4. Pick a single colony into 10 mL of 2YT medium containing ampicillin (or carbenicillin) and chloramphenicol. Incubate at 37°C until A_{600} is about 0.8 (0.6–1.0).
5. Add IPTG to 0.8 m*M* and incubate at 30°C for 2–3 h *(see* **Note 9**).
6. Remove 1 mL of culture and centrifuge for 1 min in an Eppendorf centrifuge.
7. Resuspend the cell pellet in 100 µL of $1 \times$ SDS sample buffer and analyze 2–5 µL by SDS PAGE *(see* **Note 10**). Protein can be visualized by Coomassie blue staining and the identity of the recombinant protein confirmed initially using the pRSET anti-Xpress antibody after Western blotting.
   A larger scale expression can then be carried out:
8. Start with a fresh transformed colony *(see* **Note 11**) and inoculate into 5 mL 2YT medium containing ampicillin (or carbenicillin) and chloramphenicol. Incubate at 37°C until slightly turbid.
9. Harvest the cells by centrifugation for 1 min in Eppendorf microcentrifuge.
10. Wash the cells once in fresh 2YT medium containing antibiotics.
11. Resuspend the cells in approx 1mL of the same medium and inoculate into a prewarmed flask of medium (e.g., 100 mL in a 1 liter conical flask). Incubate at 37°C until A_{600} is approx 0.8 (0.6–1.0).
12. Add IPTG to 0.8 m*M* and incubate at 30°C for 3–5 h *(see* **Note 9**).
13. Harvest the cells by centrifugation and resuspend the cell pellets in approximately one-tenth of the culture volume. Freeze the cells at –20°C overnight.
14. The next day, thaw the cell paste and sonicate (freeze-thawing causes significant cell lysis and helps to reduce the required sonication time).
15. Collect the recombinant protein as insoluble inclusion body material by centrifugation and wash extensively (to remove cellular proteins) in low ionic strength buffer (e.g., 50 m*M* Tris-HCl, pH 7.5).
16. After washing four to six times, solubilise the protein in 8 *M* urea, 50 m*M* Tris-HCl, pH 7.5 by rotary shaking for one hour at room temperature.
17. Purify the recombinant N-terminal His-tagged fusion protein by applying to a nickel affinity chromatography (e.g. Ni-NTA agarose, Qiagen Ltd.) in 6 *M* urea, 50 m*M* Tris-HCl, pH 7.5.
18. Elute the protein with imidazole at concentrations determined empirically according to the tightness of binding of the protein to the resin *(see* **Note 12**).

The purified, solubilised, recombinant MT-MMP protein is then refolded (*see* **Note 13**).

19. Dialyse the solubilised protein against 6 *M* urea, 50 m*M* Tris-HCl pH 7.5 to remove the imidazole and then rapidly dilute (by dropwise addition) to a protein concentration of approx 50–100 µg/mL (estimated from $A_{280} = 1$, representing a 1 mg/mL solution) in 50 m*M* Tris-HCl, pH 7.5, 100 m*M* NaCl, 1 *M* urea.

20. Incubate the mixture at 4°C overnight and concentrate (e.g., by ultrafiltration) during the following day. Continue further dilution/reconcentration to gradually reduce the urea concentration (*see* **Note 14**).

Using these simple protocols we have been able to produce active material to characterize the catalytic activities of the catalytic domain of the MT-MMPs. Prodomain protein is not usually fully inactive after the refolding indicating some incorrectly folded material but significantly higher activity (measured using fluorigenic substrates [*21*]) is observed after activation to remove the pro-protein sequence.

3.1.2 GST-Fusion Proteins

1. Generate GST-ΔMT1-MMP by performing PCR with MT1-MMP cDNA (*1*) using the primers: 5'-CGTAGATCTGCGCTCGCCTCCCTCGGC-3' and 5'-AAAAGATCTCATGGG CAGCCCATCCAGTC-3'.

2. Subclone the amplified DNA in pIJ2925 (*22*) at the *Bgl*II site and determine the nucleotide sequence from the 5' end to the downstream *Sma*I site of MT1-MMP, and from the 3' end to the upstream *Bam*HI site, using e.g., the Sequenase kit version 1.0 (United States Biochemical).

3. Replace the amplified DNA by the original cDNA sequence between the *Sma*I and *Bam*HI sites.

4. Digest with *Bgl*II and subclone the DNA fragment into the pGEX-2T plasmid at the *Bam*H1 site (using *E. coli* DH5α as host), to encode a GST-MT1-MMP fusion protein.

5. Grow cultures of the transformed strains overnight at 37°C in 10 mL 2YT medium containing 150 µg/mL ampicillin.

6. Dilute the culture 1:10 into fresh prewarmed 2YT containing 150 µg/mL ampicillin; distribute into flasks of appropriate volume and grow at 37°C until an A_{600} of 0.6 is reached.

7. Add IPTG to a final concentration of 0.1 m*M*, and continue the incubation for 4 h at 37°C.

8. Expressed fusion protein can be purified on immobilized glutathione and refolded as above (**Subheading 3.1.1.**)

3.2. Transient Expression in CHO Cells

3.2.1. Culture and Subculture of CHOL761h Cells
for Transient Transfection

1. Revive the cells from liquid N_2 and grow to confluence in growth medium.

2. Day 1: wash cells twice in Hank's Balanced Salt Solution. Add 4mLs trypsin-EDTA to one confluent 75 cm^2 flask of cells and incubate 5–10 min at 37°C until the cells lift off. Add 4 mLs growth medium and spin down the cells for 5 min at 1000g at room temperature. Resuspend the cell pellet in growth medium and divide between three more 75 cm^2 flasks. Incubate overnight at 37°C, 5% CO_2.

3. Day 2: trypsinize cells from all three flasks as before. Resuspend cell pellet in growth medium and count the cells in a haemocytometer. Seed the cells at 5×10^5 cells per well for 24-well plates or at 1.25×10^5 cells per well for Labtek slides (*see* **Note 15**). Incubate 37°C, 5% CO_2.

4. Day 3: remove media and replace with fresh growth medium. Incubate 37°C, 5% CO_2 for three hours. The cells are now ready for transfection (*see* **Note 16**).

3.2.2. Transfection

1. Preform the cDNA-calcium phosphate precipitates. In an appropriately sized tube, mix cDNA (10 µg per well for 24-well plates, 2.5 µg per well for Labtek slides) with 2 M $CaCl_2$ and sterile water so that the final concentration of $CaCl_2$ is 250 mM in a total volume of 80 µL per well (24-well plates) or 20 µL per well (Labtek slides).

2. Add an equal vol per well (i.e., 80 µL or 20 µL) of 2X HBS to the DNA/$CaCl_2$ mix and gently agitate the mixture by taking up and down in a 1 mL pipet tip until a fine precipitate is formed (*see* **Note 17**).

3. Add the DNA mixture carefully, dropwise on to the cells without removing the growth medium. 160 µL per well for 24-well plates, 40 µL per well for Labtek slides.

4. Gently return the cells to the incubator and leave undisturbed at 37°C, 5% CO_2 for 3–4 h.

5. Remove all the media and discard. Add enough 15% glycerol in DMEM to cover the cells (250 µL per well for 24-well plates, 120 µL for Labtek slides). Leave for one minute, then aspirate the glycerol solution thoroughly.

6. Wash the cells briefly with DMEM, then add 1 mL per well of growth medium with 10 mM sodium butyrate added (*see* **Note 18**). Incubate for 24 h at 37°C, 5% CO_2. The cells are now ready for testing for MT1-MMP expression.

3.3. MT1-MMP Detection

3.3.1. ProMMP-2 Activation Detected by Gelatin Zymography

1. Remove growth medium and wash the cells carefully with DMEM. Incubate 16–24 h in 300 µL per well (24-well plates) DMEM containing 4nM recombinant proMMP–2.

2. Harvest the supernatants and centrifuge briefly to bring down the loose cells. Samples may be stored at –20°C at this stage.

3. Prepare SDS–8% polyacrylamide gels with 0.5 mg/mL gelatin incorporated.

4. Prepare 12 µL samples of cell supernatants with Laemmli sample buffer.

5. Electrophorese under non-reducing conditions at 4°C.

6. Remove SDS by incubating gels in 2.5% Triton X-100 for 2 × 15 min.

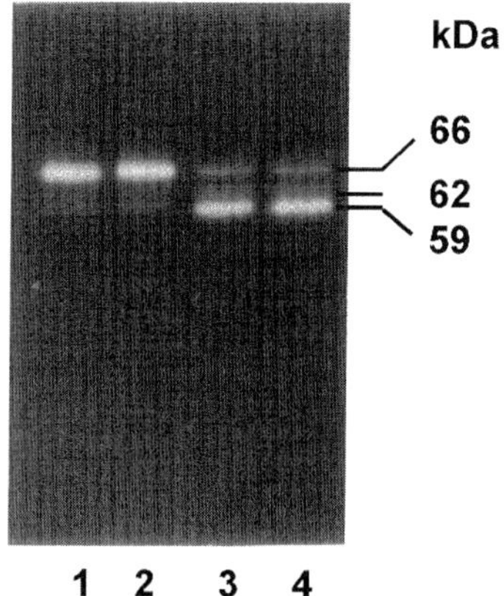

Fig. 1. CHO cells transiently transfected with MT1-MMP process pro MMP-2 to the fully active 59 kDa form. 4 n*M* proMMP-2 was incubated at 37°C for 24 h with CHO cells, 24 h posttransfection. Cell supernatants were analyzed by gelatin zymography. Lanes 1, 2, supernatants from CHO cells transfected with control vector; lanes 3, 4, supernatants from CHO cells transfected with MT1-MMP cDNA.

7. Wash briefly in water.
8. Incubate gel in TCAB buffer at 37°C for 1 1/2 h.
9. Stain with Coomassie brilliant blue stain for 20 min.
10. Destain in 30% (v/v) methanol, 1% (v/v) acetic acid using several changes over 1–2 h.

Gelatinase degradation of gelatin in zymograms is revealed as white zones of lysis on a blue background of residual, Coomassie blue-stained gelatin. SDS activates proMMP-2 to allow gelatin degradation without cleavage of the propeptide, thus MMP-2 in cell supernatants can be detected in the pro, intermediate and fully active forms at 66, 62, and 59 kDa respectively (molecular mass as assessed by non reducing gelatin zymography). Cells that have been successfully transfected with MT1-MMP show partial or complete activation of proMMP-2 compared with vector controls where MMP-2 remains in the proform (*see* **Fig. 1**).

3.3.2. Immunocytochemistry

1. 24 h after transfection wash cells in Labtek slides briefly with PBS.
2. Aspirate and fix for 5 min using 4% formaldehyde.
3. Wash 3× in PBS.
4. Incubate for 1 h at room temperature with the primary antibody, sheep anti-(human MT1-MMP) polyclonal antibody (50 µg/mL). Include 5% normal pig serum should to prevent non-specific binding.

5. Wash carefully 3× in PBS.
6. Incubate the cells for a further 60 min with a FITC conjugated pig Fab' fragment against sheep IgG.
7. Wash carefully 5× in PBS.
8. Mount the slides and view using a fluorescence microscope.
9. MT1-MMP in transiently expressing CHO cells can be seen as a bright ring of green fluorescence in approx 10% of the population. Nontransfected cells, or those with the empty vector and controls incubated with nonspecific IgG show no such fluorescence.

3.3.3. Preparation of Cell Lysates and Western Blot

1. 24 h posttransfection, briefly wash the cells (grown in 24-well plates) twice with ice-cold PBS.
2. Add 25 µL per well of lysis buffer and incubate the cells on ice for 10 min.
3. Remove lyzed cells from each well using a cell scraper and transfer to Eppendorf tubes.
4. Homogenise briefly by passing through a 25 gauge syringe needle.
5. Incubate on ice for a further 30 min.
6. Spin at 10,000*g* for one minute.
7. Transfer supernatants to fresh tubes.
8. Store at –80°C or proceed directly to protein assay and Western blot.
9. Determine the total protein concentration of the lysates using the bicinchoninic acid assay (Sigma).
10. Load samples (25 µg protein per lane) on to SDS-10% polyacrylamide gels and separate lysate proteins by electrophoresis.
11. Transfer proteins to nitrocellulose membranes over one hour using a semi-dry electroblotter.
12. Stain the membrane with Ponceau S solution to check the transfer of proteins.
13. Wash out the Ponceau S stain with wash buffer for approx 10 min.
14. Incubate the blot at 37°C for 1 h in blocking buffer in order to block nonspecific binding sites.
15. Incubate overnight at 4°C in 5 µg/mL affinity purified sheep polyclonal antibody against MT1-MMP in 1:1 blocking buffer/wash buffer.
16. Wash off the primary antibody (*see* **Note 19**) using five changes of wash buffer over 30 min.
17. Incubate for a further 30 min at 37°C in blocking buffer.
18. Incubate for one hour at room temperature in a 1:10,000 dilution of donkey anti-sheep IgG conjugated with horseradish peroxidase.
19. Remove the second antibody and wash with 5 changes of wash buffer over 30 min.
20. Develop using ECL reagents according to the manufacturer's instructions.

MT1-MMP is detected in cell lysates in three forms; 63 kDa (corresponding to the proform), 60 kDa (the fully active form) and 45 kDa (a further degradation product correlating with MMP-2 activation) *(23)*

3.4. Inducible Expression of MT1-MMP Using the Tet-Off System

3.4.1. Production of Stable Cell Lines Inducibly Expressing MT-MMP

1. Subclone the MT1-MMP cDNA into pTRE.
2. Identify transformed colonies of *E.coli* DH5α containing the cDNA subcloned in the appropriate orientation by restriction enzyme analysis and purify plasmid DNA by caesium chloride density gradient centrifugaton or equivalent.
3. Grow HTC-75 cells to approx 80% confluency in growth medium.
4. Digest 40 µg of the MT1-MMP pTRE DNA with ScaI to linearize the DNA in order to promote double crossover recombination events. Combine with 2 µg of pSV2neo. Precipitate the DNA with ethanol and resuspend in 50 µL of TE.
5. Trypsinize the cells and resuspend in complete medium at 1×10^7 cells/mL.
6. In a 0.4 cm cuvet, mix 0.4 mL of cells with the DNA and electroporate at 950 µF and 0.22 kV/cm (t = 20–30 ms).
7. Perform a control plasmid (linearized pTRE vector) transfection to produce vector-only derivative cell lines that will act as a control for the effects of both transfection and expression of the geneticin antibiotic selection.
8. Stand the cells at room temperature for 10 min, then plate the cells at the dilutions below and allow to recover at 37°C in complete medium containing doxycycline. Maintain the doxycycline selection in the medium (5 ng/mL) to ensure repression of the transfected gene during establishment of the stable cell time. A convenient method of plating the cells at this stage is to use 2×24-well plates. Dilute the cells to 24 mL with complete medium containing doxycycline and plate 1 mL aliquots to each of 12 wells. The remaining 12 mL is further diluted to 24 mL with the same medium and 12 mL again plated in 1 mL aliquots. This process is repeated twice more and the cells plated and incubated at 37°C (*see* **Note 20**).
9. After approx 40 h the medium is removed and replaced with fresh complete medium containing geneticin at 1mg/mL (again maintaining doxycycline at 5ng/mL) (*see* **Note 21**).
10. Every 4–5 d change the medium to remove the large numbers of dead (floating) cells.
11. After 2–3 wk, rapidly growing clones of cells appear. Remove them by trypsinization and seed into fresh 24-well plates.
12. As the cells grow, expand them into a 25 cm² flask in the same medium. When confluent, harvest the cells and make stocks in liquid nitrogen. When a convenient number of clones are available test for inducible expression of MT1-MMP.

3.4.2. Testing Inducibility of MT-MMP Expression

Clonally-derived cells resistant to geneticin are easily tested as follows:

1. Seed cells at 10^5 cells/well into duplicate 24-well plates and grow at 37°C in growth medium in the presence or absence of doxycycline but containing 250 µg/mL of geneticin. At this stage it is very important to use tetracycline-free FCS (*see* **Subheading 2.4.** and **Note 6**).

2. When the cells are approaching confluence (after 1–2 d depending on their growth rate) remove the medium and replace with serum-free medium with or without doxycycline.
3. After a further 24 h incubation remove the serum-free medium and test activation of endogenous proMMP-2 by gelatin zymography as described for transiently transfected cells (*see* **Subheading 3.3.1.**). Inducible cell lines show activated MMP-2 in the absence of doxycycline and proMMP-2 in the presence of the drug. We generally test about 20–30 clonal cell lines and obtain 2–3 inducible clones (*see* **Note 22**).

4. Notes

1. The BL21 strains are deficient in the *lon* protease and the *ompT* outer membrane protease, hence, the recombinant proteins are usually more stable, particularly if expressed within insoluble inclusion bodies.
2. CHOL 76lh: this Chinese hamster ovary cell line was derived from the CHO-K1 cell line by transfection with an adenovirus E1A expression vector, pLmE1A.gpt. This induces transactivation of the CMV-MIE promoter in transient transfections, thus increasing expression of the transfected MT1-MMP *(12)*.
3. pEE14.tPA is a eukaryotic expression vector containing the strong constitutive human CMV transcriptional promoter sequence. The cDNA encoding the gene of interest is inserted immediately after the promoter using the *Eco*RI site. Other restriction enzyme sites which can be used in this vector are *Hin*dIII and *Bcl*I. *(24)* Other available cDNA expression vectors can also be used .
4. Recombinant proMMP-2 is liable to partially activate upon storage at –20°C at pH above 6. To retain in the proform, the enzyme should be aliquoted into small volumes e.g., 50 µL, and stored in a sodium cacodylate buffer pH 6.0 at –70°C. Working aliquots should be stored at –20°C and may be thawed and frozen 3–4 times.
5. Cool the gelatin on ice before mixing with gel components
6. In order to maintain maximal induction capability of the MT1-MMP gene cells are grown for expression using Tet approved FBS from Clontech Laboratories (#8630 1). This serum is supplied pretested for the absence of tetracyclines which are often observed in serum preparations due to the widespread use of these drugs in animal feeds.
7. The oligonucleotides may be designed such that the amplified DNA can be cloned using a T-vector. In this case individual transformed colonies must be cultivated for plasmid DNA isolation. The nucleotide sequence from a few colonies is then determined to ensure the fidelity of the PCR reactions; there are usually some errors. These can be minimized by replacement of as much DNA by "parental" cDNA using any available restriction enzyme sites (*see* for example, **Subheading 3.1.2.**). The amplified DNA can then be subcloned into the pRSET vector and used directly for large scale protein production. An alternative approach (which we favor) is to design the oligonucleotides to contain an extra four nucleotides in addition to those encoding the restriction enzyme sites. This satisfies the require-

ment (different for various enzymes) of the restriction enzymes involved and allows the amplified DNA to be digested and cloned directly into the appropriately digested (i.e., two different enzymes producing noncomplementary termini) pRSET DNA. The ligated DNA is transformed into *E. coli* DH5α competent cells and plasmid DNA analyzed by restriction enzyme digestion. Recombinant plasmids which produce the expected sizes of DNA fragments are further transformed into *E. coli* BL21(DE3)pLysS competent cells. Transformed colonies can then be tested in small scale cultures for the production of the expected protein. Plasmid DNA producing such colonies can then be used for nucleotide sequence analysis. This eliminates time wasted in sequencing plasmids in which small deletions or translational frame shifts have occurred.

8. Transformation is best performed as late as possible during the day so that after overnight incubation the colonies are still quite small by the following morning.

9. 37°C can be used but we have never seen any improvement in yield compared to 30°C. The time can also be varied but in our experience overnight incubation does not increase yield and can lead to degradation of the protein.

10. Percentage of acrylamide can be varied to optimize protein analysis dependent on the size of the expected recombinant protein.

11. We have not found it necessary to go through the rigorous replating single colony isolation steps (on agar containing both antibiotics) described by Studier et al. *(8)*. The "compromise" method described here reduces the time required to carry out an induction such that cells can be harvested on the day after transformation rather than three days later. However, we have found severely reduced yields if overnight cultures are used (particularly if stored at 4°C for any period of time) to inoculate the flasks for expression. We prefer, therefore, always to use freshly transformed colonies.

12. Various elution regimes are possible but we find step gradients to be convenient and easily optimized. It is usually convenient to start elution at 50 m*M* imidazole and use 50 m*M* concentration steps.

13. There are many examples of refolding protocols in the literature and it is probably fair to say that most of them work, although not very well. We have not fully optimized the refolding of the recombinant MT-MMPs which we have produced, however, for the purposes of initial characterization of the enzyme activity the simple protocols we have used have been adequate.

14. Alternative protocols use sequential dialyses to reduce the urea levels, but we have found generally high losses due to precipitation in such procedures.

15. Cells are plated at a very high density for maximum positive transfections

16. For successful transfection, cells should be in a healthy, exponential growth phase and this is best achieved by following the double subculture procedure outlined.

17. A transparent tube, e.g., a Bijou, should be used for the preparation of cDNA-calcium phosphate precipitates so that the fine precipitate required is easily observed.

18. Sodium butyrate induces modification of the genomic environment by the acetylation of nuclear proteins. This results in an enhanced level of expression of MT1-MMP.

19. The sheep polyclonal antibody to MT1-MMP may be used several times (approx 6) for Western blotting, providing that it is frozen at −20°C between use.

20. This range of dilutions produces a useful range of cell densities. If the cells are too dense, then the antibiotic selection cannot work because it only affects growing cells. If the cells are too sparse, not enough cells survive the drug selection and too few drug-resistant clones are produced.
21. It is crucial to maintain the doxycycline throughout the selection otherwise no clones are obtained, hence, the level of MT1-MMP expression is obviously toxic to the cells.
22. Some clones were observed to be constitutively expressing activated MMP-2, presumably due to integration of the MT1-MMP cDNA near a very strong transcriptional enhancer which overcame the doxycycline repression

Acknowledgements

HTC-75 cell line and pSV2neo kindly provided by Dr. B. van Steensel, Rockefeller University, New York, USA; MT1-MMP cDNA kindly provided by Professor M. Seiki, University of Tokyo, Japan.

References

1. Sato, H., Takino, T., Okada, Y., Cao, J., Shinagawa, A., Yamamoto, E., and Seiki, M. (1994) A matrix metalloproteinase expressed on the surface of invasive tumor cells. *Nature* (London) **370**, 61–65.
2. Kinoshita, T., Sato, H., Takino, T., Itoh, M., Akizawa, T., and Seiki, M. (1996) Processing of a precursor of 72-kilodalton type IV collagenase/gelatinase A by a recombinant membrane-type 1 matrix metalloproteinase. *Cancer Research* **56**, 2535–2538.
3. Will, H., Atkinson, S. J., Butler, G. S., Smith, B., and Murphy, G. (1996) The soluble catalytic domain of membrane-type 1 matrix metalloproteinase cleaves the propeptide of progelatinase A and initiates autoproteolytic activation. Regulation by TIMP-2 and 3. *J. Biol Chem.* **2711**, 17,119–17,123.
4. Cao, J., Sato, H., Takino, T., and Seiki, M. (1995) The C-terminal region of membrane type matrix metalloproteinase 1 is a functional domain required for progelatinase A activation. *J. Biol. Chem.* **270**, 801–805.
5. Atkinson, S. J., Crabbe, T., Cowell, S., Ward, R. V., Butler, M. J., Sato, H., Seiki, M., Reynolds, J. J., and Murphy, G. (1995) Intermolecular autolytic cleavage can contribute to the activation of progelatinase A by cell membranes. *J. Biol. Chem.* **270**, 30,479–30,485.
6. Imai, K., Ohuchi, E., Aoki, T., Nomura, H., Fujii, Y., Sato, H., Seiki, M., and Okada, Y. (1996) Membrane-type matrix metalloproteinase 1 is a gelatinolytic enzyme and is secreted in a complex with the Tissue Inhibitor of Metalloproteinases 2. *Cancer Research* **56**, 2707–2710.
7. Pei, D. and Weiss, S. J. (1996) Transmembrane-deletion mutants of the membrane-type matrix metalloproteinase-1 process progelatinase A and express intrinsic matrix-degrading activity. *J. Biol. Chem.* **271**, 9135–9140.
8. Studièr, F. W., Rosenberg, A. H., Dunn, J. J., and Dubendorff, J. W. (1990) Use of T7 RNA polymerase to direct the expression of cloned genes. *Methods in Enzymol.* **185**, 61–89.

9. Smith, D. B. and Johnson, K. S. (1988) Single-step purification of polypeptides expressed in *E. coli* as fusions with glutathione S-transferase. *Gene* **67**, 31–40.

10. Butler, G. S., Butler, M. J., Atkinson, S. J., Will, H., Tamura, T., van Westrum, S. S., Crabbe, T., Clements, J., d'Ortho, M-P., and Murphy, G. (1998) The TIMP2 membrane type 1 metalloproteinase "receptor" regulates the concentration and efficient activation of progelatinase α. A kinetic study. *J. Biol. Chem.* **273**, 871–880.

11. d'Ortho, M-P., Stanton, H., Butler, M. J., Atkinson, S. J., Murphy, G., and Hembry, R. M. (1998) MT1-MMP on the cell surface causes focal degradation of gelatin films. *FEBS Lett.* **421**, 159–164.

12. Cockett, M. I., Bebbington, C. R., and Yarranton, G. T.(1991) The use of engineered E1A genes to transactivate the hCMV-MIE promoter in permanent CHO cell lines. *Nucleic Acids Res.* **19**, 319–325.

13. Chen C. and Okayama, H.(1987) High efficiency transformation of mammalian cells by plasmid DNA. *Mol. Cell. Biol.* **7**, 2745–2752.

14. Murphy, G., Allan, J. A., Willenbrock, F., Cockett, M. I., O'Connell, J. P., and Docherty, A. J. P. (1992) The role of C-terminal domain in collagenase and stromelysin specificity. *J. Biol. Chem.* **267**, 9612–9618.

15. Murphy, G., Willenbrock, F., Ward, R. V., Cockett, M. I., Eaton, D., and Docherty, A. J. P. (1992). The C-terminal domain of 72 kDa gelatinase A is not required for catalysis, but is essential for membrane activation and modulates interactions with tissue inhibitors of metalloproteinases. *Biochem. J.* **283**, 637–641.

16. Hillen, W. and Wissman, A. (1989) In: *Protein-Nucleic Acid Interactions, Topics in Molecular and Structural Biology.* (Saenger, W. and Heinemann, U., eds.), Vol. 10, pp. 143–162, MacMillan, London.

17. Van Steensel, B. and de Lange, T. (1997) Control of telomere length by the human telomeric protein TRF1. *Nature* **385**, 740–743.

18. Murphy, G., Houbrechts, A., Cockett, M. I., Williamson, R. A., O'Shea, M., and Docherty, A. J. P. (1991) The N-terminal domain of tissue inhibitor of metalloproteinases retains metalloproteinase inhibitory activity. *Biochemistry* **30**, 8097–8102.

19. Hembry, R. M., Murphy, G., and Reynolds, J. J. (1985) Immunolocalization of tissue inhibitor of metalloproteinases (TIMP) in human cells. Characterization and use of a specific antiserum. *J. Cell Sci.* **73**, 105–119.

20. Lehti, K. Lohi, J. Valtanen, H., and Keski-Oja, J. (1998) Proteolytic processing of membrane-type-1 matrix metalloproteinase is associated with gelatinase A activation at the cell surface. *Biochem. J.* **334**, 345–353.

21. Knight, C. G., Willenbrock, F., and Murphy, G. (1992). A novel coumarin -labeled peptide for sensitive continuous assays of matrix metalloproteinases. *FEBS Lett.* **296**, 263–266.

22. Janssen, G. R. and Bibb, M. J. (1993) Derivatives of pUC18 that have BglII sites flanking a modified multiple cloning site and that retain the ability to identify recombinant clones by visual screening of *Escherichia coli* colonies. *Gene* **124**, 133–134.

23. Stanton, H., Gavrilovic, J., Atkinson, S. J., d'Ortho, M-P., Yamada, K. M., Zardi, L., and Murphy, G. (1998) The activation of proMMP-2 (gelatinase A) by HT 1080 fibrosarcoma cells is promoted by culture on a fibronectin substrate and is concomitant with an increase in processing of MT1-MMP (MMP-14) to a 45 kDa form. *J. Cell Sci.* **111**, 2789–2798.
24. Bebbington, C. R. and Hentschel, C. G. (1987) In: *DNA Cloning* (Glover, D. M. and Hames, B. D., eds.), pp. 163–188, IRL Press, Oxford.

12

Refolding of TIMP-2
from *Escherichia coli* Inclusion Bodies

Richard A Williamson

1. Introduction

E. coli is a convenient host in which to express recombinant proteins. The technology is available to most laboratories as it is relatively inexpensive and does not require extensive expertise. The major drawback of *E. coli* as an expression host is the inability of the organism to carry out many posttranslational modifications, including glycosylation and disulphide bond formation. High-level intracellular expression of many mammalian proteins in *E. coli* results in the formation of large insoluble aggregates, known as inclusion bodies *(1)*. These dense bodies consist predominantly of the misfolded recombinant product, together with components of the transcription/translation machinery (i.e., RNA polymerase, ribosomal RNA, and plasmid DNA). The TIMPs are invariably insoluble when expressed in *E. coli*, their folding requirement for the formation of 6 disulphide bonds being incompatible with the reducing environment of the *E. coli* cell. Fortunately, active, correctly folded recombinant protein can often be recovered from insoluble inclusion bodies by a process of solubilization and in vitro refolding *(2–3)*. Indeed, inclusion body formation has the advantage that the recombinant product often accumulates to high levels in the cell (up to 30% of total cell protein) and allows easy isolation of that protein, a relatively pure and stable form (inclusion bodies are typically >50% recombinant protein).

The refolding protocol presented here was devised for the active N-terminal domain of TIMP-2 (N-TIMP-2) *(4)*. The development of a refolding protocol is often a lengthy empirical task where each parameter at the in vitro refolding stage (e.g., redox potential, residual denaturant concentration, protein concen-

From: *Methods in Molecular Biology, vol. 151: Matrix Metalloproteinase Protocols*
Edited by: I. Clark © Humana Press Inc., Totowa, NJ

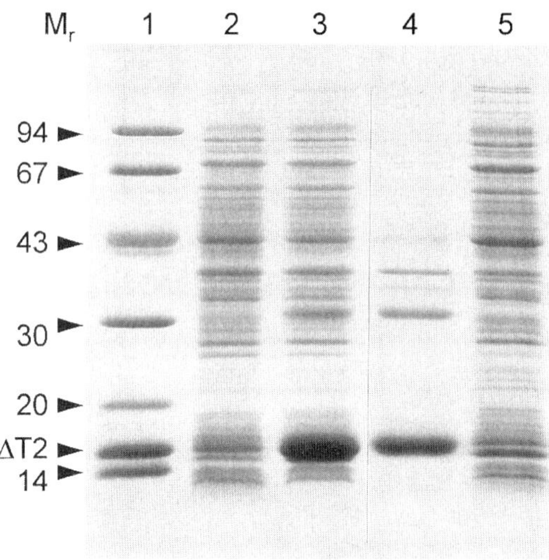

Fig. 1. Expression of pET23d-N-TIMP-2 in E. coli strain BL21 (DE3). Lane 1, molecular weight marker proteins including N-TIMP-2 (ΔT2). Lane 2, *E. coli* before induction with IPTG. Lane 3, *E. coli* after induction with IPTG for 3h. Lane 4, lysis pellet from induced cells (unwashed inclusion bodies). Lane 5, lysis supernatant from induced cells. Samples were run reduced on a 10–20% SDS-PAGE gradient gel.

tration, temperature, pH, time, and so on) must be optimized to achieve maximal yields. For N-TIMP-2, the refolding procedure was developed by following protein refolding in real time using intrinsic fluorescence of the unique tryptophan residue *(5)*. This approach enabled rapid screening of the refolding parameters, and is a generally applicable method if a source of native material is available to characterize the fluorescence properties of the fully folded state.

The N-TIMP-2 refolding procedure can be broken down into several steps.

1. Cell lysis and inclusion body isolation (*see* **Fig. 1**).
2. Inclusion body washing (adapted from Marston et al. *[6]*).
3. N-TIMP-2 reduction and solubilization in 5 *M* guanidinium chloride (GdmCl).
4. In vitro refolding by dilution into refolding buffer (pH 8.75) containing a redox couple of reduced and oxidized glutathione.
5. Protein concentration by ultrafiltration and purification of the refolded form. The procedure can usually be completed in 6 d starting from *E. coli* cell pellets. The refolding protocol has proved to be very robust in our laboratory with over 10 preparations successfully completed. Typical yields of N-TIMP-2 after refolding and purification are 30% of that present in the washed inclusion bodies.

Although the refolding procedure presented here was devised for N-TIMP-2, subsequent work to develop a protocol for full-length TIMP-2 found maximal recoveries under very similar refolding conditions, except for the protein concentration which should be reduced to 10 µg/mL (*see* **Note 4**). Final yields for refolded full-length TIMP-2 are, however, significantly lower than those for the N-terminal domain (typically 16%). To achieve higher yields from the same quantity of inclusion bodies it is possible to recycle the misfolded forms back through the protocol (*see* **Note 11**). This has proved very successful with full-length TIMP-2 where 50% of the starting material was collected as misfolded forms and reprocessed with a refolding and purification yield of 18%.

The procedure starts from *E. coli* cell pellets containing N-TIMP-2 (or full-length TIMP-2) inclusion bodies. In our work, both these molecules are expressed from pET23 vectors (Novagen) in *E. coli* strain BL21 (DE3), where the insert is under the transcriptional control of the T7 promoter. The proteins are expressed as the mature sequence with the addition of an N-terminal methionine residue for initiation of translation. This methionine residue is not present in the final refolded product, however, as it is removed during expression *(5)*. Typical expression yields in Luria-Bertani (LB) medium are 40–60 mg/L for N-TIMP-2 and 20–30 mg/L for the full-length molecule. Minimal medium has also been used extensively in our laboratory for the production of uniformly ^{15}N and ^{13}C labeled TIMP-2 proteins for heteronuclear nuclear magnetic resonance (NMR) studies *(7,8)*. Expression yields of 25–35 mg/L for the N-terminal domain and 12–15 mg/L for full-length TIMP-2 have been obtained in Spizizen's minimal medium using ^{13}C-labeled glucose and ^{15}N-labeled ammonium sulphate as sole carbon and nitrogen sources. The refolding procedure presented here should not be dependent on the *E. coli* expression system used if the protein is produced in a similar form. The effect of fusion partners or expression tags (e.g., glutathione S-transferase or the His-tag) on TIMP-2 refolding has not been evaluated.

2. MATERIALS

2.1. Inclusion Body Recovery and Refolding

Except were explicitly stated, all buffers and reagents should be stored at 4°C and are stable for up to 3 mo. Those buffers and reagents containing free thiol groups (i.e., dithiothreitol (DTT) and reduced glutathione) should only be prepared as required.

1. Lysis buffer: 50 m*M* Tris-HCl, pH 8.0, 2 m*M* EDTA.
2. Lysozyme: 10 mg/mL in lysis buffer. Store at –20°C.
3. Triton X-100: 5% (v/v) in lysis buffer. Store at 4°C.
4. Wash buffer: 50 m*M* Tris-HCl, pH 8.0, 10 m*M* EDTA, 0.5% Triton X-100.
5. Solubilization buffer: 50 m*M* Tris-HCl, pH 8.75, 5 *M* GdmCl.
6. DTT:0.5 *M* in solubilization buffer (prepare immediately before use).

7. Refolding buffer: 50 mM Tris-HCl, pH 8.75, 0.45 M GdmCl, 0.78 mM reduced glutathione, 0.44 mM oxidized glutathione (prepare immediately before use).
8. Dialysis buffer: 50 mM NaH$_2$PO$_4$-NaOH, pH 6.0.

2.2. Purification of Refolded N-TIMP-2

1. Ion exchange buffer A: 50 mM NaH$_2$PO$_4$-NaOH, pH 6.0. (Use within 7 d).
2. Ion exchange buffer B: 50 mM NaH$_2$PO$_4$-NaOH pH 6.0, 0.5 M NaCl. (Use within 7 d).
3. Gel filtration buffer: 25 mM NaH$_2$PO$_4$-NaOH pH 6.0, 100 mM NaCl, 0.03% (w/v) NaN$_3$.

3. Methods

3.1. Inclusion Body Recovery and Refolding

The procedure given below is for cell pellets obtained from 400 mL cultures (final A$_{600}$ of 1.2–1.5). After **step 2**, lysates may be pooled and the procedure scaled-up accordingly.

1. Add 10 mL of lysis buffer to the *E. coli* cell pellet and thoroughly resuspend by pumping in and out of a 10 mL pipet. Transfer the *E. coli* suspension to a 50 mL bench-centrifuge tube and lyse the cells by adding 100 µL of lysozyme and 200 µL of Triton X-100 stock solutions. Mix thoroughly after each addition and incubate at room temperature for 30 min. The solution becomes highly viscous as DNA is released. Cool the lysate on ice.
2. Sonicate the lysis mixture using an appropriate ultrasound generator and probe. Prevent the lysis mixture from warming by placing the tube in an ice-cooled water bath. The viscosity of the mixture should return close to that of water as the DNA is sheared by the effect of the sonication and the mixture becomes whiter in color as the cellular material is solubilized (*see* **Note 1**).
3. Sediment the released N-TIMP-2 inclusion bodies by centrifugation at 12000g for 15 min at 4°C. Decant off the supernatant.
4. Break up the pellet in 2.5 mL of inclusion body wash buffer and transfer to a Teflon or ground-glass homogenizer where it can be thoroughly resuspended (*see* **Note 2**). Recover the inclusion body fraction by centrifugation at 12000g for 10 min at 4°C.
5. Repeat wash **step 4**.
6. Repeat wash **step 4** using 2.5 mL of water in place of the wash buffer.
7. Estimate the total amount of N-TIMP-2 in the inclusion preparation by comparing the intensity of Coomassie blue-stained bands on reducing SDS-PAGE. Use either N-TIMP-2 standards of known amount for comparison, or a different protein of similar molecular weight. Inclusion body samples for SDS-PAGE are conveniently taken after resuspension of the pellet in **step 6**.
8. Add solubilization buffer to the inclusion body pellet to give an N-TIMP-2 concentration of 2 mg/mL. The pellet becomes 'glassy' on contact with the 5 M GdmCl solution and solubilization is best achieved using the Teflon or ground-

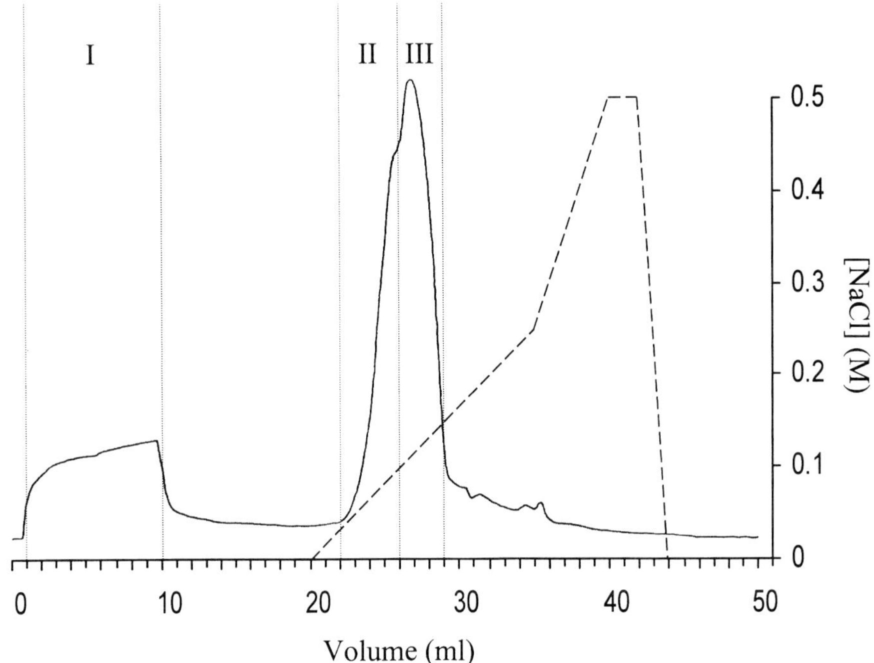

Fig. 2. A_{280} profile from Resource-S chromatograpy of N-TIMP-2. The flow-through fraction (**I**) and the shoulder on the major peak (**II**) contain misfolded forms of N-TIMP-2. Fractions 7, 8 and 9 (**III**) are combined to give the refolded N-TIMP-2 pool. The NaCl gradient used to develop the column is shown by the dashed line and right-hand axis.

    glass homogenizer (*see* **Note 3**). Add DTT from the 0.5 M stock to 11 mM, mix thoroughly and leave at room temperature for 30 min to ensure complete reduction of the recombinant N-TIMP-2.

9. Remove any non-solubilized material by centrifugation at 12000g for 5 min at 4°C.
10. Dilute the solubilized inclusion bodies into refolding buffer (1 in 100) to give a final N-TIMP-2 concentration of 20 µg/mL. Add the solubilized inclusion bodies slowly from a pipet (10 mL over 2–3 min) whilst stirring the refolding buffer to ensure thorough mixing and continue to stir at room temperature for 2 h. Store the refolding mixture at 4°C overnight without stirring.
11. Concentrate the refolding mixture to 1/50–1/100 of its original volume by ultrafiltration using a 3 or 5 kDa molecular weight cut-off membrane (*see* **Note 6**). Either a stirred cell or pumped tangential-flow system may be used.
12. Dialyse the concentrated refolding mixture against 2.5 L of dialysis buffer for 16 h at 4°C. During dialysis a precipitate is formed which consists primarily of misfolded N-TIMP-2.
13. Transfer the contents of the dialysis bag to a centrifuge tube and incubate at 35°C for 45 min. This incubation step causes further precipitation of misfolded

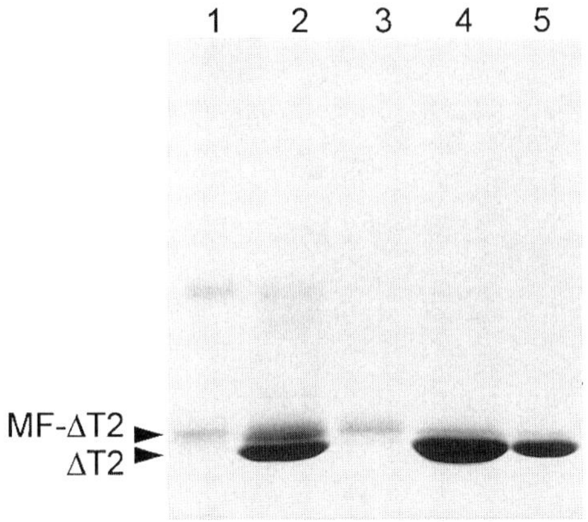

Fig. 3. Non-reducing SDS-PAGE analysis of fractions taken during the N-TIMP-2 refolding and purification procedure. Lane 1, the protein precipitate formed during dialysis against phosphate buffer and the 35°C heat-treatment step. Lane 2, soluble refolded N-TIMP-2 prior to loading onto Resource-S. Lane 3, flow-through fraction containing misfolded forms of N-TIMP-2. Lane 4, the refolded N-TIMP-2 pool collected from the Resource-S column. Lane 5, the refolded N-TIMP-2 pool collected from the Sephacryl S-100 column. Samples were run on a 10–20% SDS-PAGE gradient gel. All loadings were normalized to the same proportion of the total, except for lane 4, where 4-fold more material was loaded to show the contaminants at 30 kDa which were subsequently removed by gel filtration. Misfolded forms of N-TIMP-2 (labeled MF-ΔT2) are incorrectly disulphide bonded and run above the correctly folded material (labeled ΔT2).

N-TIMP-2 (*see* **Note 7**). Remove the precipitate by centrifugation at 1500*g* for 15 min at room temperature and then filter the supernatant through a 0.2 µm filter to ensure complete removal.

3.2. Purification of Refolded N-TIMP-2

The following method is routinely used to purify refolded N-TIMP-2 from misfolded forms and contaminating *E. coli* proteins. This protocol generates N-TIMP-2 judged to be better than 95% pure as determined by image analysis of Coomassie blue-stained SDS-PAGE gels. The Resource-S column is run on Pharmacia's FPLC system (*see* **Note 9**), and the Sephacryl S-100 column is packed following Pharmacia's guidelines. Full-length TIMP-2 behaves very similarly to N-TIMP-2 and can be purified in an identical manner. Non-reducing SDS-PAGE analysis of samples taken during the purification of N-TIMP-2 are shown in **Fig. 3**.

1. Load 10 mL aliquots of the refolded N-TIMP-2 preparation onto a 1 mL Resource-S column (Pharmacia) equilibrated at room temperature in ion

exchange buffer A (*see* **Note 8**). After loading, wash the column with 10 mL (10 column volumes) of buffer A and then elute with a linear gradient of 0–50% buffer B over 15 mL at a flow rate of 1 mL/min. Refolded N-TIMP-2 elutes in fractions 7–9 (1 mL fractions collected from the start of the gradient elution). A misfolded form of N-TIMP-2 runs as a shoulder on the leading edge of the N-TIMP-2 peak and fractions across the N-TIMP-2 peak should be analysed on reducing/non-reducing SDS-PAGE in order to omit this material from the N-TIMP-2 pool (*see* **Note 5**). A typical trace for Resource-S purification of refolded N-TIMP-2 is shown in **Fig. 2**.

2. Load 5–10 mL aliquots of the N-TIMP-2 pool onto a Sephacryl S-100 column ($5.3 \text{ cm}^2 \times 38$ cm; 180 mL bed volume) equilibrated in gel-filtration buffer at 4°C. Elute the column with gel filtration buffer at a flow rate of 1.4 mL/min. N-TIMP-2 elutes at 120–140 mL from the start of elution (*see* **Note 10**).

4. Notes

1. As an alternative to sonication, the viscosity of the lysis mixture can be reduced by treatment with DNase I. To use this method, add $MgCl_2$ to 10 mM and DNase I to 20 µg/mL. Mix well and incubate at room temperature for a further 30 min (or until the viscosity is reduced). Even if DNase I is used, it is still advisable to sonicate the lysis mixture to ensure complete cell breakage and so prevent unnecessary contamination of the inclusion body pellet with *E. coli* proteins.

2. Improved washing of the N-TIMP-2 inclusion body pellet can be achieved by sonication of the resuspended inclusion bodies in wash buffer in **step 4**. This additional step also helps to shear any remaining DNA. N-TIMP-2 inclusion bodies are typically 60–70% pure at the end of the washing stage (**step 6**).

3. If difficulties are experienced solubilizing the inclusion bodies in solubilization buffer (5 M GdmCl), then an alternative strategy is to first resuspend the pellet in 1/6 of the required final volume with water, and then to make up to the final volume with solubilization buffer at 1.2 × concentration (i.e., 60 mM Tris-HCl, 6 M GdmCl, pH 8.75). The inclusion body suspension in water dissolves rapidly in the solubilization buffer, whereas the inclusion body pellet takes on a "glassy" nature and can be difficult to solubilize quickly.

4. Full-length TIMP-2 should be refolded using the same conditions as N-TIMP-2, except the protein concentration should be reduced to 10 µg/mL (solubilize inclusion bodies at 1 mg/mL in solubilization buffer and then dilute 1:100 into refolding buffer as for N-TIMP-2). Refolding yields for both molecules drop significantly if higher protein concentrations are used at the refolding stage.

5. Full-length and N-TIMP-2 refolding is monitored by comparing mobility on SDS-PAGE gels when samples are run with and without reduction. The size of the mobility difference on disulphide bond cleavage is approx 3 kD. Nonreduced SDS-PAGE is an effective method of determining the degree of contamination of the refolded TIMP-2 with misfolded forms. The misfolded material is not correctly disulphide bonded and migrates as a smear above the native molecule (*see* **Fig. 3**).

6. A 10 kDa molecular weight cut-off filter can be used for N-TIMP-2 (14.1 kDa) without suffering substantial losses of material. A 10 kDa molecular weight cut-off filter is recommended for full-length TIMP-2 (21.8 kDa) as flow-rates are approximately twice that of a 3 kDa membrane.
7. The 35°C incubation step is essential to prevent problems with protein precipitation when loading the sample on the Resource-S column at room temperature.
8. 50 m*M* MES buffer (pH 6.0) my be used as an alternative to the phosphate buffer for the ion exchange chromatography. The advantage of MES is that this buffer is compatible with Ca^{2+} ions if the TIMP-2 is to be assayed for MMP inhibitory activity (the MMPs require Ca^{2+} for stability). A wide range of buffers can be used for the subsequent gel filtration step (**step 2**).
9. S-Sepharose FF and mono-S (Pharmacia) can both be used as alternative cation exchange matrices for N-TIMP-2 purification. On both matrices N-TIMP-2 was found to elute slightly later in the NaCl gradient.
10. Gel filtration chromatography is important for the removal of several protein species around 30 kDa. Only one of these bands is seen on reducing SDS-PAGE suggesting that the others are misfolded disulphide-linked N-TIMP-2 dimers. The band seen on reducing SDS-PAGE is thought to be β-lactamase as it is also up-regulated on IPTG induction.
11. The folding efficiencies for the N-terminal domain and full-length TIMP-2 are typically 30% and 15–18%, respectively. A large proportion of the misfolded full-length TIMP-2 precipitates during dialysis into phosphate buffer and the subsequent 35°C incubation step (**steps 12** and **13**). This material, together with the flow-through fraction and leading edge fractions from the ion exchange separation, are easily collected and can be recycled back through the refolding protocol starting at **step 7**. The precipitate can be solubilized directly in 1.2X solubilization buffer (*see* **Note 3**); the ion exchange fractions are concentrated by ultrafiltration and then added to the solubilized pellet such that the final GdmCl concentration is 5 *M*. We have used this recycling approach to obtain good yields of $^{15}N/^{13}C$-labeled TIMP-2 where the high cost of the isotopes necessitate that every effort is made to obtain maximal recoveries. The yields from the reprocessing of TIMP-2 are typically 50–60% of the initial inclusion body refolding.
12. Estimates of the A_{280} extinction coefficients for the N-terminal domain and full-length TIMP-2 were calculated from their tryptophan, tyrosine and cysteine content as described by Gill et al. *(9)*. These values in terms of A_{280} absorbance for a 1 mg/mL solution (0.88 and 1.49 for N-TIMP-2 and full-length TIMP-2, respectively) were used to determine the concentration of refolded TIMP-2 at the end of the purification procedure.
13. The results from biochemical and conformational analysis of refolded N-TIMP-2 can be found in **ref. 5**).

Acknowledgments

The author would like to thank Dessy Natalia, John Sergent, Michael Rae, Mark Carr, and Robert Freedman for their contributions to this work. The DNA for full-length TIMP-2 was kindly provided by Alan Bradbury at the National Institute of Medical Research, Mill Hill, London, U.K. RAW is supported by the Arthritis Research Campaign.

References

1. Thatcher, D. R. and Hitchcock, A. (1994) Protein folding in biotechnology. In *Mechanisms of Protein Folding* (Pain, R. H., ed.) pp. 229-261, Frontiers in Molecular Biology Series, IRL Press, Oxford, UK.
2. Fischer, B., Sumner, I., and Goodenough, P. (1993) Isolation, renaturation and formation of disulphide bonds of eukaryotic proteins expressed in *Escherichia coli* as inclusion bodies. *Biotechnol. Bioengineering* **41**, 3–13.
3. Wetzel, R. (1992) Principles of protein stability. Part 2 - enhanced folding and stabilization of proteins by suppression of aggregation in vitro and in vivo. In *Protein Engineering - A Practical Approach* (Rees, A. R., Sternberg, M. J. E., and Wetzel, R., eds.) pp. 191–219, IRL Press, Oxford, UK.
4. Murphy, G., Houbrechts, A., Cockett, M. I., Williamson, R. A., O'Shea, M., and Docherty, A. J. P. (1991) The N-terminal domain of tissue inhibitor of metalloproteinases retains metalloproteinase inhibitory activity. *Biochem.* **30**, 8097–8102
5. Williamson, R. A., Natalia, D., Gee, C. K., Murphy, G., Carr, M. D., and Freedman, R. B. (1996). Chemically and conformationally authentic active domain of human tissue inhibitor of metalloproteinases-2 refolded from bacterial inclusion bodies. *Eur. J. Biochem.* **241**, 476–483
6. Marston, F. A. O., Lowe, P. A., Doel, M. T., Schoemaker, J. M., White, S., and Angal, S. (1984) Purification of calf prochymosin (prorennin) synthesized in *Escherichia coli. Bio/Technol.* **2**, 800–807
7. Williamson, R. A., Carr, M. D., Frenkiel, T. A., Feeney, J., and Freedman, R. B. (1997). Mapping the binding site for matrix metalloproteinase on the N-terminal domain of the tissue inhibitor of metalloproteinases-2 by NMR chemical shift perturbation. *Biochemistry* **36**, 13,882–13,889
8. Muskett, F. W., Frenkiel, T. A., Feeney, J., Freedman, R. B., Carr, M. D., and Williamson, R. A. (1998). High resolution structure of the N-terminal domain of tissue inhibitor of metalloproteinases-2 and characterisation of its interaction site with matrix metalloproteinase-3. *J. Biol. Chem.* **273**, 21,736–21,743
9. Gill, S. C. and Von Hippel, P. H. (1989) Calculation of molar extinction coefficients from amino acid sequence data. *Anal Biochem* **182**, 319–326.

13

Expression and Refolding of Full-Length Human TIMP-1

Deborah Davis

1. Introduction

Structural analysis of any protein, whether by X-ray crystallography or NMR, requires a reliable source of large amounts of good quality protein. Most proteins are not sufficiently abundant in their natural state to be used as the primary source, and so it is essential for recombinant protein to be expressed in an appropriate system. In many cases the preferred expression system is *E. coli* for ease of handling, generally high yields and, where the protein is required for NMR, relatively inexpensive isotopic labeling. Although an *E. coli* expression has the potential to produce high yields of a protein, it can be found in an insoluble form in inclusion bodies. There are a number of ways in which this problem can be avoided, for example (1) careful choice of growth conditions can reduce the rate of protein production allowing the protein to remain in soluble form; (2) alternatively there are several expression systems that use secretion tags, either to the medium or the periplasmic space where conditions may maintain the protein of interest in solution *(1)*; or (3) another method utilises a low induction of a gene expressing bacteriocin release protein (BRP) which produces holes in the cell wall, expression of this gene at high levels will cause cell death, but at lower levels will induce a leakiness in the cells, allowing protein release into the media *(2)*.

One of the key difficulties with the expression of TIMP-1 (and indeed, all other TIMPs) is the requirement to form all six disulphide bonds correctly *(3,4)*. Since the cytoplasm of *E. coli* is a reducing environment, it is highly unlikely that correct formation of disulphide bonds will occur successfully. Certain bacterial cell lines have been developed that are negative for thioredoxin reduc-

From: *Methods in Molecular Biology, vol. 151: Matrix Metalloproteinase Protocols*
Edited by: I. Clark © Humana Press Inc., Totowa, NJ

tase, which makes the environment less reducing, allowing expression of soluble forms of disulphide bonded proteins *(5)*.

Several laboratories have attempted to express recombinant TIMPs, either as the full-length protein *(6)* or as a truncated form containing only three disulphide bonds. Of these, the most successful appears to be the production of truncated TIMP-2 *(7)* which is also described in this volume (*see* Chapter 12). When applied to full-length human TIMP-1, none of these conditions produce sufficient quantities of correctly folded protein. Consequently a new refolding protocol was necessary to refold the protein from inclusion bodies.

There are few clear rules for refolding proteins, and consequently the optimum conditions often have to be determined empirically, although taking into account the desired final condition for the protein. There are many factors that can influence the outcome of a protein refolding reaction: e.g., protein concentration, whether or not the sample is prepurified, pH, temperature, initial denaturant, method of denaturant removal and the presence of a wide range of stabilizing and solubilizing agents. The effects of some of these conditions are discussed in two reviews *(8,9)* and the references therein.

For human TIMP-1 inserted in the pET11a vector, expression in HM174 cells gave yields of up to 70 mg/L in rich media. All of this protein was found in inclusion bodies. The process of cell lysis removes a large number of contaminating proteins and no further purification appeared necessary for the refolding in this instance. A key feature that proved critical for the success of the preparation is the use of 6 *M* guanidinium chloride (in preference to urea) as the denaturant. It is also essential to have the correct redox potential in the buffer, 10:1 ratio of reduced:oxidized glutathione is used here. If no redox agents are present, the yield is much reduced.

The extremely slow mode of denaturant removal where the buffer is pumped into the denatured protein solution is critical to the refolding. This allows the protein to spend a significant length of time at intermediate denaturant concentrations which may allow for "prefolding" processes to occur as in the renaturation of collagenase-3 *(10)*. This method is approx 50% effective by yield. A mixture of cellular proteins and TIMP-1 can be found as a urea soluble pellet after the refolding and have presumably aggregated during the renaturation process. When refolding any protein it is essential to ascertain that you have a correct conformation by the use of appropriate activity assays and spectroscopic analysis.

2. Materials

2.1. Expression of Recombinant TIMP-1

1. Glycerol stocks of HM174/λDE3 cells containing the pET11a-TIMP-1 vector (vector and cells from Novagen).

2. 2 L flat bottomed conical flasks (sterile).
3. Luria-Bertani (LB) broth: 10 g NaCl, 10*g* Bacto-Tryptone (DifCo) 5*g* yeast extract (Difco).
4. Ampicillin (stocks of 50 mg/mL stored at –20°C).
5. Incubator shaker capable of maintaining a temperature of 37°C and rotation at 250 rpm.
6. Spectrometer capable of reading at 600 nm.
7. Disposable acrylic cuvets (Jencons Scientific).
8. IPTG (1 *M* stocks stored at –20°C).
9. Centrifuge capable of spinning up to 2 L of media at 9,000*g* and suitable centrifuge tubes.

2.2. Lysis of Cell Pellet

1. Lysis buffer: 50 m*M* Tris-HCl, pH 8.0, 0·1 *M* NaCl, 1 m*M* EDTA.
2. Lysozyme at 10 mg/mL (Sigma).
3. Diisopropyl fluorophosphate (DFP): 200 m*M* in isopropanol (Sigma).
4. Deoxycholic acid 40 mg/mL (Sigma).
5. Water bath at 37°C.
6. 1 *M* MgSO$_4$.
7. DNase I at 10 mg/mL (Sigma).
8. Centrifuge and tubes capable of centrifuging 30 mL at 20,000*g*.

2.3. Refolding TIMP-1

1. Solubilisation/denaturation buffer: 6 *M* GdHCl, 0·1 *M* NaCl, 20 m*M* DTT, 50 m*M* Tris-HCl, pH 8.0, 4.2 m*M* reduced glutathione, 0.2 m*M* oxidised glutathione (*see* **Note 1**)
2. Refolding buffer: 0.5 *M* GdHCl, 0·1 *M* NaCl, 50 m*M* Tris-HCl, pH 8.0, 2 m*M* reduced glutathione, 0.2 m*M* oxidised glutathione (*see* **Note 1**).
3. Pump capable of 10 mL/h flow rate (i.e., Pharmacia P1).
4 Magnetic stirrer and stirrer bar.
5. Cold room to maintain all equipment and reagents at 4°C.

2.4. Purification of TIMP-1

1. Anti-TIMP-1 monoclonal antibody coupled Sepharose (RRU-T3 antibody available from Cambio, Cambridge UK , CNBr activated Sepharose from Pharmacia).
2. Dilution buffer/column equilibration buffer: 50 m*M* Tris-HCl, pH 7.4, 0·1 *M* NaCl.
3. Column wash buffer: 50 m*M* Tris-HCl, pH 7.4, 1 *M* NaCl.
4. Elution buffer: column wash buffer containing 3 *M* sodium iodide.
5. Storage buffer: 50 m*M* Tris-HCl, pH 8·0, 0·1 *M* NaCl, 0.02% NaN$_3$.

3. Methods

The full-length human TIMP-1 cDNA was cloned into the pET11a vector as a BamH1- Nde1 fragment and transformed into HM174 cells. The protein was always found as inclusion bodies so the highest expressing clone with the correct N-terminal sequence was stored as a glycerol stock at –80°C.

3.1. Growing Cells for Lysis

1. Prepare an overnight culture by inoculating LB medium containing 50 mg/mL ampicillin (LB-amp) with cells from a glycerol stock.
2. Grow at 37°C in an incubator-shaker at 250 rpm.
3. Inoculate up to 0.5 L of LB-amp in each 2 L flask to 1% with the overnight culture.
4. Grow the cells at 37°C/200 rpm for approx 2 h, until the OD_{600} is 0·5.
5. Initiate protein expression by adding isopropyl-β-D-thiogalactopyranoside (IPTG) to a final concentration of 1 mM.
6. Grow the culture for a further 4 h.
7. Harvest the cells by centrifugation at 9,000g for 15 min at 4°C.

3.2. Cell Lysis

The cells are lysed using a modification of the method found in Sambrook et al. *(11)*.

1. Resuspend the cell pellet from 1 L of culture in 30 mL of ice-cold lysis buffer.
2. Incubate the cells on ice for 20 min, with occasional gentle mixing.
3. Protease activity is reduced by the addition of DFP to a final concentration of 2 mM.
4. Incubate the mixture for a further 10 min on ice.
5. To complete the lysis, add deoxycholic acid to a final concentration of 1.3 mg/mL.
6. Transfer the lysate to a water bath at 37°C and stir with a glass rod until it becomes viscous.
7. Digest the DNA (to reduce the viscosity) by adding DNase I to a final concentration of 80 µg/mL with $MgSO_4$ to 2 mM.
8. Incubate the lysate at room temperature until no longer viscous (about 30 min).
9. Spin down the insoluble portion, containing the TIMP-1 inclusion bodies by centrifugation at 20,000g for 20 min at 4°C.

3.3. Refolding of TIMP-1

1. Dissolve the insoluble pellet in solubilization/denaturation buffer at the equivalent of 30 mL per liter of cell culture.
2. Resuspend the pellet by pipeting it up and down and vortexing to break it up as far as possible.
3. Incubate the solution at 37°C on an incubator shaker set at 250 rpm for 2 h to ensure complete solubilization of the proteins.
4. Remove any remaining insoluble material by centrifugation at 20,000g for 20 min.
5. The fully denatured cell extract can be stored frozen at –20°C until it is required.
6. Prepare and pre-cool the refolding buffer to 4°C (*see* **Note 2**). To refold 30 mL of solubilised protein (from 1 L of cell culture) it is necessary to use 1 L of refolding buffer.

7. Transfer the solubilized protein into a plastic beaker with a stirrer bar on a magnetic stirrer.
8. Pump the renaturation buffer into the solubilized protein at 30 mL/h at 4°C whilst stirring gently (*see* **Note 3**).
9. Centrifuge the refolded mixture at 9,000*g* for 20 min at 4°C.

3.4. Purification of TIMP-1

A monoclonal antibody affinity column gives an excellent single step purification in this instance (*see* **Notes 4** and **5**). The details of this affinity purification have been described previously *(12)*.

1. Dilute the refolded protein 10-fold in column equilibration buffer. This is to protect the antibody from the potentially deleterious effects of residual guanidinium and redox reagents.
2. Pump the diluted protein solution onto a 76 mm × 170 mm monoclonal (anti-TIMP) IgG-Sepharose affinity column at 500 mL/h.
3. Wash off contaminating proteins with 3 vol of column wash buffer.
4. Elute the TIMP-1 with 600 mL of elution buffer, discarding the first 100 mL and retaining 500 mL as a single fraction.
5. Dialyze immediately into storage buffer.
6. Wash the column with 3 vol equilibration buffer.
7. Assess protein purity on 12% SDS-PAGE gels stained with Coomassie blue.
8. Assess inhibitory activity (*see* **Note 6**).

4. Notes

1. A key features that proved critical for the success of the protocol is the use of 6 *M* guanidinium chloride as the denaturant in preference to urea. It is also essential to have the correct "redox" potential in the buffer. A 10:1 ratio of reduced:oxidized glutathione is used here. If no redox agents are present, the yield is much reduced.
2. For the refolding reaction to be successful it is essential for all of the buffers to be precooled to 4°C. The use of warm buffers will significantly reduce the final yield.
3. It is important to add the renaturation buffer very slowly. Significant losses occur when the denaturant is removed too quickly. If small starting volumes are used, the flow rate should be reduced to give an overall time of 16–17 h.
4. This method is optimized for larger scale preps (of 50 mg), where it is important to elute the TIMP-1 in the minimum possible volume. On a smaller scale, or where the antibody-Sepharose has a higher binding capacity smaller columns and gentler elution buffers (low pH) may be preferred.
5. Since the initial preparation enriches the initially insoluble TIMP-1 and the refolding also appears to favor the solubility of the TIMP-1 over other proteins other purification methods such as ion exchange and gel filtration should also give a good final purification.

6. There are a number of assays available where the inhibitory activity of TIMP-1 can be assessed. Some of these are described in this vol. (e.g., *see* Chapters 22 and 31). In the case of the collagen diffuse fibril assay one unit of inhibitory activity will reduce the activity of 1 unit of collagenase by 50%. The expected specific activity should be 10,000–20,000 units/mg.

Acknowledgments

I would like to thank M. O'Hare for initial work in producing the TIMP-1 expression system. This work was supported by a grant from the Arthritis and Rheumatism Council.

References

1. Missiakas, D., Georgopoulos, C., and Raina, S. (1994) The *Escherichia coli* dsbC (xprA) gene encodes a periplasmic protein involved in disulphide bond formation. *EMBO J.* **13**, 2013–2020.
2. Van Der Wal, F. J., Ten-Hagen-Jongman, C. M., Oudega, B., and Luirink, J. (1995) Optimization of bacteriocin-release-protein-induced protein release by Escherichia coli: Extracellular production of the periplasmic molecular chaperone FaeE. *Applied Microbiology and Biotechnology* **44**, 459–465.
3. Williamson, R. A., Marston, F. A. O., Angal, S., Koklitis, P., Panico, M., Morris, H. R., Carne, A. F., Smith, B. J., Harris, T. J. R., and Freedman, R. B. (1990) Disulphide bond assignment in human tissue inhibitor of metalloproteinases (TIMP). *Biochem. J.* **268**, 267–274.
4. Caterina, N. C., Windsor, L. J., Yermovsky, A. E., Bodden, M. K., Taylor, K. B., Birkedal-Hansen, H., and Engler, J. A. (1997) Replacement of conserved cysteines in human tissue inhibitor of metalloproteinases-1. *J. of Biol. Chem.* **272**, 32,141–32,149.
5. Derman, A. I., Prinz, W. A., Belin, D., and Beckwith, J. (1993). Mutations that allow disulfide bond formation in the cytoplasm of Escherichia coli. *Science* **262**, 1744–1747.
6. Negro, A., Onisto, M., Grassato, L., Caenazzo, C., Garbisa, S. (1997) Recombinant human TIMP-3 from *Escherichia coli:* Synthesis, refolding, physico-chemical and functional insights. *Protein Engineering* **10**, 593-599.
7. Williamson, R. A., Natalia, D., Gee, C. K., Murphy, G., Carr, M. D., and Freedman, R. B. (1996) Chemically and conformationally authentic active domain of human tissue inhibitor of metalloproteinases-2 refolded from bacterial inclusion bodies. *Eur. J. Bio. Chem.* **241**, 476–483.
8. Guise, A. D., West, S. M., and Chaudhuri, J. B. (1996) Protein folding in vivo and renaturation of recombinant proteins from inclusion bodies. *Mol. Biotech.* **6**, 53–64.
9. Fischer, B. E. (1994) Renaturation of recombinant proteins produced as inclusion bodies. *Biotechnology Advances* **12**, 89–101.
10. Knauper, V., López-Otín, C., Smith, B., Knigh, G., and Murphy, G. (1996) Biochemical characterization of human collagenase-3. *J. Biol. Chem.* **271**, 1544–1550.

11. Maniatis, T., Fritsch, E. F., and Sambrook, J. (1992) *Molecular Cloning*. A Laboratory Manual. Cold Spring Harbour Laboratory Press, N.Y.
12. Hodges, D. J. Lee, D. C., Salter, C. J., Reid, D. G., Harper, G. P., and Cawston. T. E. (1994) Purification and secondary structural analysis of tissue inhibitor of metalloproteinases-1. *Biochem. Biophys. Acta.* **1208**, 94–100

14

Purification of MMPs and TIMPs

Ken-ichi Shimokawa and Hideaki Nagase

1. Introduction

Currently more than twenty different matrix metalloproteinases (MMPs) and four tissue inhibitors of metalloproteinases (TIMPs) have been identified *(1,2)* by cDNA cloning. This chapter describes methods to purify collagenases (MMP-1, MMP-8, and MMP-13), gelatinases (MMP-2 and MMP-9), stromelysins (MMP-3 and MMP-10), macrophage elastase (MMP-12), and TIMP-1 and TIMP-2 from natural sources.

2. Materials

2.1. Purification of proMMP-1

1. Sources: Serum-free conditioned medium of human fibroblasts (rheumatoid synovium, skin, or gingiva) cultured in the presence of phorbol myristate acetate (PMA) (50 ng/mL) and/or interleukin 1 (IL-1) (10 ng/mL) for 3–4 d.
2. Substrates: ^{14}C-labeled collagen or Mca-Pro-Gly-Leu-Dpa-Ala-Arg-NH$_2$ from Peptide International (Louisville, KY), Peptide Institute Inc. (Osaka, Japan) or Bachem AG (Bubendorf, Switzerland).
3. DEAE-cellulose (DE 52) from Whatman, UK.
4. Green A Dyematrex gel, Diaflo apparatus, and YM-10 membrane from Amicon Corp (Beverly, MA).
5. Affi-gel 10 from Bio-Rad (Hercules, CA).
6. Sephacryl S-200, concanavalin A (ConA)–Sepharose, and gelatin-Sepharose 4B from Pharmacia Biotech (Piscataway, NJ).
7. 0.2 *M* 4-Aminophenylmercuric acetate (APMA) solution: APMA from ICN Biochemistry (Cleveland, OH) is dissolved in 0.2 *M* NaOH and titrated with concentrated HCl to pH 11.0. Below pH 11.0 APMA precipitates at that concentration. The solution is stable at room temperature for at least 4 wk.

From: *Methods in Molecular Biology, vol. 151: Matrix Metalloproteinase Protocols*
Edited by: I. Clark © Humana Press Inc., Totowa, NJ

8. TNC buffer: 50 mM Tris-HCl (pH 7.5), 0.15 M NaCl, 10 mM CaCl$_2$, 0.02% NaN$_3$, 0.05% Brij 35.
9. TC buffer: 50 mM Tris-HCl (pH 7.5), 10 mM CaCl$_2$, 0.02% NaN$_3$, 0.05% Brij 35.
10. Anti-(human MMP-3) IgG-Affi-Gel 10 (see below for preparation).

2.2. Purification of Human proMMP-2

1. Sources: Conditioned medium of cultured fibroblasts. Fibroblasts from various species in culture constitutively produce proMMP-2. Other cell types such as epithelial cells, neoplastic cell-lines, macrophages in culture also produce this zymogen but the level may be variable.
2. Substrates: ^{14}C-labeled gelatin (heat-denatured collagen) or Mca-Pro-Leu-Gly-Leu-Dpa-Ala-Arg-NH$_2$ from Peptide International, Peptide Institute Inc., or Bachem AG.
3. Gelatin-Sepharose 4B and Sephacryl S-200 from Pharmacia Biotech.
4. Diaflo apparatus and YM-10 membrane from Amicon Corp.
5. TNC buffer: *see* **Subheading 2.1.**

2.3. Purification of Human proMMP-3

1. Materials required for purification are essentially same as for human proMMP-1. (*See* **Subheading 2.1.**).
2. Anti-(human MMP-1) IgG-Affi-Gel 10 and CM-cellulose (CM-52; Whatman).
3. Protein substrates: reduced, [^{3}H]carboxymethylated transferrin *(14)*, casein, or Azocoll (Calbiochem/Novabiochem, La Jolla, CA).
4. Synthetic substrates: Mca-Pro-Leu-Gly-Leu-Dpa-Ala-Arg-NH$_2$ or Mca-Arg-Pro-Lys-Pro-Val-Glu-Nva-Trp-Arg-Lys(Dnp)-NH$_2$ from Peptide International, Peptide Institute, Inc. or Bachem AG.

2.4. Purification of MMP-7

2.4.1. Purification from the Medium of CaR-1 Cells (3)

1. Sources: Conditioned medium of CaR-1 human rectal carcinoma cells (Japanese Cancer Research Cell Bank).
2. Substrates: Azocoll from Calbiochem/Novabiochem (La Jolla, CA) or Mca-Pro-Leu-Gly-Leu-Dpa-Ala-Arg-NH$_2$ from Peptide International, Peptide Institute, Inc., or Bachem AG.
3. DEAE-cellulose from Whatman.
4. Diaflo apparatus, YM-10 membrane, and Green A Dyematrex gel from Amicon Corp (Beverly, MA).
5. Sephacryl S-200 from Pharmacia Biotech.
6. TNC Buffer: *See* **Subheading 2.1.**
7. Buffer 7.1: 10 mM Tris-HCl (pH 8.0), 0.5 mM CaCl$_2$, 0.02% NaN$_3$, 0.05% Brij 35.
8. Buffer 7.2: 25 mM Sodium borate (pH 8.0), 0.15 M NaCl, 1 mM CaCl$_2$, 0.02% NaN$_3$, 0.05% Brij 35.

9. Buffer 7.3: 25 mM Sodium cacodylate (pH 6.5), 1 mM CaCl$_2$, 0.02% NaN$_3$, 0.05% Brij 35.

2.4.2. Purification from Postpartum Rat Uterus (4)

1. Sources: Rat uterus at 1 d postpartum.
2. Substrate: Same as in **Subheading 2.4.1.**
3. Ultrogel AcA54 from Sepracor (Marlborough, MA).
4. Blue-Sepharose and Chelating Fast Flow resin from Pharmacia Biotech.
5. Buffer 7.4: 50 mM Tris-HCl (pH 7.5), 0.15 M NaCl, 0.1 M CaCl$_2$.
6. Buffer 7.5: 50 mM Tris-HCl (pH 7.5), 0.2 M NaCl, 10 mM CaCl$_2$, 0.02% NaN$_3$, 0.05% Brij 35.
7. Buffer 7.6: 25 mM boric acid-NaOH (pH 8.0),1 mM CaCl$_2$, 0.05% Brij 35, 0.03% toluene.
8. Buffer 7.7: 25 mM sodium cacodylate (pH 6.5).

2.5. Purification of Human proMMP-8 and MMP-8

1. Source: human blood buffy coat (neutrophils)
2. Plasmatonin solution: 6% hydroxyethylstarch in 0.9% NaCl
3. Substrates: same as in **Subheading 2.1.**
4. Chelating-Sepharose Fast Flow, activated CH-Sepharose 4B, Q-Sepharose Fast Flow, Orange-Sepharose CL-6B, and Sephacryl S-300 from Pharmacia Biotech (Uppsala, Sweden).
5. Buffer 8.1: 20 mM Tris-HCl (pH 9.0), 5 mM CaCl$_2$, 0.5 mM ZnCl$_2$.
6. Buffer 8.2: 20 mM Tris-HCl (pH 8.5), 5 mM CaCl$_2$, 0.5 mM ZnCl$_2$, 0.02% NaN$_3$.
7. Buffer 8.3: 20 mM Tris-HCl (pH 7.5), 5 mM CaCl$_2$, 0.5 mM ZnCl$_2$, 0.1 M NaCl.
8. Buffer 8.4: 50 mM Tris-HCl (pH 7.5), 5 mM CaCl$_2$, 0.5 M NaCl, 0.5 mM ZnCl$_2$, 0.02% NaN$_3$.
9. Buffer 8.5: 20 mM Tris-HCl (pH 7.5), 5 mM CaCl$_2$, 0.5 M NaCl.
10. Buffer 8.6: 100 mM Tris-HCl (pH 9.0), 20 mM CaCl$_2$ 0.5 M NaCl.
11. Buffer 8.7: 600 mM Tris-HCl (pH 7.5), 20 mM CaCl$_2$, 0.5 M NaCl.
12. Benzamidine, diisopropyl fluorophosphate (DFP) and phenylmethyl sulfonylfluoride (PMSF) and Pro-Leu-Gly-hydroxamate HCl from Sigma (St. Louis, MO).

2.6. Purification of proMMP-9

1. Sources: The serum-free conditioned medium of U937 monocytic leukemia cells or HT1080 fibrosarcoma cells stimulated with PMA (50 ng/mL) *(5)*.
2. Substrates: same as in **Subheading 2.2.**
3. Gelatin-Sepharose, Con-A Sepharose, and Sephacryl S-200 from Pharmacia Biotech.
4. TNC buffer: *see* **Subheading 2.1.**

2.7. Purification of proMMP-10

1. Sources: the conditioned medium of human keratinocytes in culture stimulated with PMA, transforming growth factor α, epidermal growth factor or TNF *(6)*;

The conditioned medium of OSC-20 human oral squamous carcinoma cells stimulated with PMA *(7)*.

2. Substrates: same as in **Subheading 2.3.**
3. DEAE-cellulose (DE52) from Whatman.
4. Green A Dyematrex gel and YM-10 membrane from Amicon Corp.
5. Affi-Gel 10 from Bio-Rad.
6. Sephacryl S-200 from Pharmacia Biotech.
7. Anti-(human MMP-1) antibody (clone 78-12G8) and anti-(human MMP-3) antibody (clone 55-2A4) are from Fuji Chemical Industries, Inc. (Takaoka, Japan).
8. 0.2 M APMA: *see* **Subheading 2.1.**
9. TNC buffer: *see* **Subheading 2.1.**
10. Buffer 10.1: 50 mM Tris-HCl (pH 8.0), 5 mM CaCl$_2$, 0.05% Brij 35, 0.02% NaN$_3$.

2.8. Purification of MMP-12

1. Sources: the conditioned culture medium of mouse peritoneal macrophages treated with 2 µM colchicine for 48 h.
2. Brewer's thioglycollate medium from Difco Laboratories (Detroit, MI).
3. DME medium: Dulbecco's modified Eagle's medium, high glucose formulation from Gibco-BRL (Grand Island, NY).
4. DME–LH medium: DME medium containing 0.2% lactalbumin hydrolysate from Gibco-BRL.
5. Substrates: ^{3}H-labeled elastin, or Mca-Pro-Leu-Gly-Leu-Dpa-Ala-Arg-NH$_2$ from Peptide International, Peptide Institute Inc. or Bachem AG.
6. DEAE-Sephadex A-25 from Pharmacia Biotech.
7. Ultrogel AcA54 from LKB.
8. Buffer 12.1: 20 mM Tris-HCl (pH 7.6).
9. Buffer 12.2: 50 mM Tris-HCl (pH 7.6), 0.15 M NaCl, 50 mM CaCl$_2$, 0.02% NaN$_3$.

2.9. Purification of TIMP-1

1. Sources: the conditioned medium of fibroblasts (skin, tendon, gingiva, etc); human plasma.
2. MMP-1 or MMP-3.
3. Substrate for MMP-1 or MMP-3 (*see* **Subheading 2.1.** and **2.3.**).
4. Heparin-Sepharose CL-6B and DEAE-Sepharose from Pharmacia Biotech.
5. C4 Reverse-phase HPLC from Vydac (Hesperia, CA).
6. Anti-(human TIMP-1) IgG – Affi-Gel 10.
7. TCO buffer: 50 mM Tris-HCl (pH 7.5), 10 mM CaCl$_2$, 0.01% β-octylglucoside.
8. Buffer T1-1: 25 mM sodium cacodylate (pH 7.6), 10 mM CaCl$_2$, 0.5 M NaCl, 0.05% Brij 35, 0.02% NaN$_3$.
9. Buffer T1-2: 50 mM glycine-HCl (pH 3.0), 0.5 M NaCl.
10. Buffer T1-3: 25 mM sodium cacodylate (pH 7.2), 1 M NaCl, 0.05% Brij 35, 0.02% NaN$_3$.

2.10. Purification of TIMP-2

2.10.1. Method 1

1. Sources: the serum-free conditioned medium of fibroblasts (human uterine cervix, human gingiva).
2. MMP-1 or MMP-3.
3. Substrates for MMP-1 or MMP-3 (*see* **Subheadings 2.1.** and **2.3.**).
4. Gelatin-Sepharose and Sephacryl S-200 from Pharmacia Biotech.
5. TNC buffer: *see* **Subheading 2.1.**

2.10.2. Method 2

1. Source: the serum-free conditioned medium of human hepatoma HLE cells from Japanese Cancer Research Resource Bank.
2. MMP-1 or MMP-3.
3. Substrates for MMP-1 or MMP-3 (*see* **Subheadings 2.1.** and **2.3.**).
4. Cellulofine GCL-2000-m column from Chisso (Tokyo, Japan).
5. Reactive Red-agarose from Sigma.
6. Synchropak RP-4 column from SynChrom (Lafayette, IN).
7. Buffer T2-1: 10 m*M* Tris-HCl (pH 7.5), 0.5 *M* NaCl, 0.1% Brij 35.
8. Buffer T2-2: 20 m*M* Tris-HCl (pH 7.5), 0.01% Brij 35.

3. Methods

See **Notes 1,2,4** and **51** for general points.

3.1. Purification of Human proMMP-1 (see Note 3)

3.1.1. Assays

MMP-1 activity is measured, after activation of proMMP-1 with APMA, against synthetic substrates such as Mca-Pro-Leu-Gly-Leu-Dpa-Ala-Arg-NH$_2$ *(8)* (*see* the details of assay procedures in Chapter 31) and ^{14}C-labeled collagen diffuse fibril assay described by Cawston and Barrett *(9)* (*see* Chapter 22).

3.1.2. Activation of proMMP-1 (see **Notes 5** and **6**)

1. *Activation by APMA.* Add 0.2 *M* APMA stock solution to proMMP-1 (1–100 μg/mL) to a final concentration of 1 m*M*, and incubate the mixture at 37°C for 2 h. SDS/PAGE analysis shows 43-kDa and 41-kDa MMP-1. The 43-kDa form has Met68 and the 41-kDa has Val82 and Leu83 for their N-termini, respectively *(10)*. These forms have about 25% of the full activity of MMP-1.
2. *Activation by APMA in the presence of MMP-3 (Full-activation of proMMP-1).* Add an equimolar amount of MMP-3 and 1/200 vol of 0.2 *M* APMA stock solution to proMMP-1 in TNC buffer, and incubate the mixture at 37°C for 4 h. ProMMP-1 is activated to the 41 kDa form with Phe81 for its N-terminus under

these conditions, and it exhibits the full collagenolytic activity. Remove MMP-3 by anti-(MMP-3) IgG-Affi-Gel 10, if necessary.

3.1.3. Preparation of Anti-(Human MMP-3) IgG-Affi-Gel 10

First prepare the IgG from antiserum:

1. Add 0.5 vol of 60% (w/v) polyethylene glycol 8000 in 50 mM Tris-HCl (pH 7.8) to the specific antiserum raised against MMP-3.
2. Stand overnight at 4°C.
3. Centrifuge at 10,000g for 15 min, dissolve the pellet with 0.1 M Tris-HCl (pH 8.0), and dialyze the sample against the same buffer.
4. Apply the sample to a column of DEAE-cellulose equilibrated with 0.1 M Tris-HCl (pH 8.0) and wash the column until A$_{280nm}$ reaches the base line.
5. Combine unbound fractions (IgG fraction) and concentrate to about 10–15 mg/mL.
6. Dialyze the IgG fraction against 0.1 M NaHCO$_3$ (pH 8.5).

Then, couple the anti-(human MMP-3) IgG to Affi-Gel 10:

7. Shake the Affi-Gel 10 vial and make a uniform suspension.
8. Transfer 2 mL of slurry to a glass-fritted funnel and wash with 6 mL of cold deionized water.
9. Add 3 mL of IgG solution immediately to the gel cake of Affi-Gel 10.
10. Transfer the gel suspension to a plastic tube and gently mix the gel slurry on a rocker, shaker, or wheel for 4 h at room temperature.
11. Transfer the gel slurry to a glass-fritted funnel and wash with 6 mL of deionized water.
12. Suspend the gel cake into 6 mL of 0.1 M glycine-NaOH, pH 9.0 and gently mix for 2 h at room temperature.
13. Transfer the gel to a column and equilibrate with the desired buffer.

3.1.4. Purification Procedures

Step 1: DEAE-Cellulose Chromatography (*see* **Note 7**)

1. Concentrate the crude serum-free conditioned medium from cultured fibroblasts (500 mL) about 20-fold using Diaflo apparatus with an Amicon YM-10 membrane, and apply to a column of DEAE-cellulose (25 mL packed resin) equilibrated with TNC buffer without Brij 35.
2. Wash the column with the same buffer until A$_{280nm}$ reaches the base line, and collect unbound protein fractions, which contain proMMP-1, proMMP-2 and proMMP-3.

Step 2: Green A Dyematrex Gel Chromatography (*see* **Note 8**)

3. Apply the DEAE unbound fraction to a column of Green A Dyematrex gel (2.5 × 5 cm) equilibrated with TNC buffer without Brij 35. Most proteins in the

medium do not bind to the column, whereas three proMMPs (proMMPs-1, -2, and -3) bind.

4. Wash the column with TNC buffer, and elute proMMPs bound to the column in a stepwise manner with TC buffer containing 0.3 M NaCl and then 0.5 M NaCl.
5. Monitor protein concentrations in fractions by reading at A_{280nm}, and assay for MMP-1 activity.
6. Pool 0.5 M NaCl fractions which should contain proMMP-1.

Step 3: Gel-Filtration on Sephacryl S-200

7. Add 10% (w/v) Brij 35 to the 0.5 M NaCl fraction containing proMMP-1 (40 mL) to the final concentration of 0.05%; concentrate it to 1.5 mL with an Amicon YM-10 membrane, and then apply to a Sephacryl S-200 column (1.5 × 120 cm) equilibrated with TNC buffer.
8. Monitor MMP-1 activity and pool fractions containing proMMP-1. The yield of proMMP-1 from 500 mL of the conditioned medium is about 150–200 µg with recovery of about 45%.

*Step 4: Gelatin-Sepharose and Anti-MMP-3 Affinity Chromatography (see **Note 9**)*

9. Concentrate the Sephacryl S-200 fraction by Amicon YM-10 membrane to 2–3 mL and apply to a column of gelatin-Sepharose (5 mL) and a column of anti-(MMP-3) IgG-Affi-Gel 10 successively.
10. Collect unbound fractions of each chromatography. The final product exhibits a major 52-kDa form (unglycosylated) and a minor 56-kDa (10–20%; glycosylated) of proMMP-1 on SDS/PAGE (*see* **Note 11**).

Step 5: Separation of Glycosylated and Unglycosylated proMMP-1

11. Apply proMMP-1 preparation containing both glycosylated and unglycosylated forms to a column of ConA-Sepharose equilibrated with TC buffer containing 0.4 M NaCl.
12. Collect the unbound fraction (containing unglycosylated proMMP-1) and apply it to a gel filtration column of Sephacryl S-200 (1.5 × 110 cm) equilibrated with TC buffer containing 0.4 M NaCl (*see* **Note 10**).
13. Elute the glycosylated 56-kDa form from the Con A-Sepharose column with 1 M α-methyl-D-mannoside in TC buffer containing 0.4 M NaCl.
14. Monitor the protein peak by reading A_{280nm}.
15. Concentrate the fraction eluted with α-methyl-mannoside and apply it to gel filtration chromatography on Sephacryl S-200 as above.

3.1.5. Separation of 41-kDa MMP-1, 22-kDa Catalytic Domain and 25-kDa C-Terminal Domain of MMP-1

When proMMP-1 is incubated with 1 mM APMA at 37°C for >2 h, proMMP-1 converts to the 41-kDa active form (major), 25-kDa C-terminal domain, and 22-kDa catalytic domain (minor). The 22-kDa MMP-1 has peptidase activity but lacks collagenolytic activity. The 22-kDa catalytic domain of MMP-1 can be separated from the mixture by Green A Dyematrex gel chromatography.

1. Apply the mixture of the 41-kDa and 22-kDa MMP-1 in TNC buffer to a column of Green A Dyematrex gel (1 × 5 cm) equilibrated with TNC buffer. The 22-kDa catalytic domain of MMP-1 does not bind to the resin, whereas the 41-kDa active form and the 25-kDa C-terminal domain does.
2. Elute the bound protein with 0.5 M NaCl in TNC buffer. This fraction contains the 41-kDa MMP-1 and the 25-kDa C-terminal domain.
3. Concentrate the 0.5 M NaCl-fraction with an Amicon YM-10 membrane and apply it to a gel filtration column of Sephacryl S-200 (1.5 × 110 cm) equilibrated with TNC buffer. This step separates the 41-kDa MMP-1 and the 25-kDa C-terminal domain.

3.2. Purification of Human proMMP-2 (see Note 3)

3.2.1. Assays

MMP-2 activity is measured after activation with APMA using gelatin (denatured type I collagen) as a substrate (*see* Chapter 22) or synthetic substrates such as Mca-Pro-Leu-Gly-Leu-Dpa-Ala-Arg-NH$_2$ (*see* Chapter 31). Gelatin zymography (*see* Chapter 23) is also useful to detect both proMMP-2 and active form of MMP-2.

*3.2.2. Activation of proMMP-2 (see **Note 12**)*

1. Dilute 0.2 M APMA solution to 2 mM in TNC buffer.
2. Add an equal volume (10–50 µL) of 2 mM APMA to proMMP-2 preparation (1–5 µg/mL), incubate at 37°C for 45 min, and assay immediately for MMP-2 activity.

3.2.3. Purification Procedures

This method is based on Itoh et al. *(11)* and Ward et al. *(12)*.

*Step 1. Gelatin-Sepharose Chromatography (*see* **Notes 13** and **14**)*

1. Apply the conditioned culture medium (250 mL) of fibroblasts to a column of gelatin-Sepharose (1 × 8 cm) equilibrated with TNC buffer.
2. Wash the column with the same buffer, and then TNC buffer containing 1% (v/v) dimethyl sulfoxide (DMSO).
3. Elute the bound proMMP-2 with 5% DMSO in TNC buffer.
4. Read A$_{280nm}$, combine the protein peak and dialyze against TNC buffer for 18 h at 4°C.

Step 2: Gel Filtration on Sephacryl S-200

5. Concentrate the dialyzed material to about 1.5 mL by Diaflo apparatus with an Amicon YM-10 membrane and apply to a gel filtration column of Sephacryl S-200 (1.5 × 110 cm) equilibrated with TNC buffer.

6. Analyze fractions for proMMP-2 by SDS/PAGE (10% acrylamide) with reduction. Molecular mass of proMMP-2 is about 72 kDa. Some fractions contain proMMP-2 and TIMP-2 (21 kDa). Reduced fibronectin exhibits a protein band of 220,000, which is eluted close to the void volume of the column.
7. Pool fractions containing only proMMP-2. The proMMP-2-TIMP-2 complex is eluted between fibronectin and proMMP-2.

3.3. Purification of proMMP-3 (see Note 3)

3.3.1. Assays

MMP-3 activity is measured after activation of proMMP-3 with APMA using Azocoll *(13)*, casein, reduced [^{3}H]-carboxymethylated transferrin, [^{3}H]Cm-Tf) as protein substrates *(14)*, or synthetic substrates such as Mca-Pro-Leu-Gly-Leu-Dpa-Ala-Arg-NH$_2$ *(8)* and Mca-Arg-Pro-Lys-Pro-Val-Nva-Glu-Trp-Arg-Lys(Dnp)-NH$_2$ *(15)*.

3.3.2. Activation of proMMP-3

3.3.2.1. ACTIVATION BY APMA (*SEE* NOTE 15)

1. Add 0.2 *M* APMA stock solution to proMMP-3 solution (1–100 µg/mL) to a final concentration of 1.5 m*M* and incubate at 37°C for 22 h.

3.3.2.2. ACTIVATION BY CHYMOTRYPSIN (GENERATION OF THE FULL-LENGTH [45 KDA] MMP-3) (*SEE* NOTE 16)

1. Incubate proMMP-3 (100 µg/mL) and chymotrypsin (10 µg/mL, final concentration) at 37°C for 2 h.
2. Stop the reaction by adding 200 m*M* PMSF in dry isopropanol to a final concentration 1 m*M*.

3.3.3. Purification Procedures

3.3.3.1. METHOD 1

Step 1: DEAE-Cellulose Chromatography (see **Note 17***)*

1. Concentrate the crude conditioned culture medium of fibroblasts stimulated by PMA and/or IL-1 (500 mL) about 20-fold using Diaflo apparatus with an Amicon YM-10 membrane and apply it to a column of DEAE-cellulose (25 mL) equilibrated with TNC buffer (pH 8.0) without Brij 35.
2. Collect unbound fractions. All proMMPs in the medium are recovered in the unbound fraction.

Step 2: Green A Dyematrex Gel Chromatography

3. Apply the unbound fraction of DEAE-cellulose to a column of Green A Dymatrex gel (2.5 × 5 cm) equilibrated with TNC buffer.

4. Wash the column with TNC buffer extensively, and elute proMMP-3 with TC buffer containing 0.3 *M* NaCl and collect the eluted protein peak. (*see* **Note 18**).

Step 3: Gelatin-Sepharose Column Chromatography

5. Apply the 0.3 *M* NaCl fraction to a column of gelatin-Sepharose column (5 mL) equilibrated with TC buffer containing 0.3 *M* NaCl.
6. Wash the column with the same buffer and collect the unbound fraction.

Step 4: Gel-Filtration on Sephacryl S-200

7. Concentrate the proMMP-3 fraction to 1.5 mL using a YM-10 membrane and apply to a column of gel filtration on Sephacryl S-200 equilibrated with TC buffer containing 0.4 *M* NaCl.
8. Collect the main protein peak eluted in fractions corresponding to 50–60 kDa. ProMMP-3 exhibits a doublet of 57-kDa (major) and 59-kDa (minor) bands on SDS-PAGE.

Step 5: Further Purification of ProMMP-3 (see **Note 19**)

9. Apply the proMMP-3 fraction to a column of anti-(human MMP-1) IgG-Affi Gel 10 (2 mL) equilibrated with TNC buffer.
10. Wash the column with the same buffer and collect the unbound protein peak. The final product exhibits a doublet of proMMP-3 (57 and 59 kDa).

3.3.3.2. METHOD 2: IMMUNO-AFFINITY CHROMATOGRAPHY

This procedure has only two chromatographic steps, but requires antibody against MMP-3 *(16)*.

1. Apply the crude conditioned medium (500 mL) or tissue extracts to a column of Affi-Gel 10 coupled with polyclonal anti-(human MMP-3) (1.5 × 5 cm) equilibrated with TNC buffer without Brij 35.
2. Wash the column with the same buffer until A_{280nm} reaches the base line.
3. Elute proMMP-3 with TNC buffer containing 6 *M* urea without Brij 35 (*see* **Note 20**).
4. Pool fractions with a protein peak and dialyze it against 200 vol of TNC buffer without Brij 35 three times.
5. Concentrate the sample using Diaflo apparatus with an Amicon YM-10 membrane.
6. Apply it to a column of Sephacryl S-200 (1.5 × 10 cm) equilibrated with TC buffer containing 0.4 *M* NaCl.
7. Analyze fractions with the major protein peak by SDS/PAGE. Those fractions contain essentially homogeneous proMMP-3 of 57 and 59 kDa (*see* **Note 22**).

3.3.4. Separation of Glycosylated and Unglycosylated Forms of proMMP-3 (see **Note 21**)

1. Apply purified proMMP-3 consisting of unglycosylated 57-kDa and glycosylated 59-kDa forms to a column of Con A-Sepharose (1.5 × 5 cm) equilibrated with TC buffer containing 0.4 *M* NaCl.

2. Wash the column with the same buffer until A_{280nm} reaches the base line.
3. Collect the unbound fraction as unglycosylated proMMP-3.
4. Elute the bound glycosylated proMMP-3 with 1 M α-methyl-D-mannoside in the same buffer, and collect the protein peak.
5. Concentrate the unglycosylated proMMP-3 (unbound fraction) and the glycosylated proMMP-3 (bound fraction) separately; and apply each fraction to a gel filtration column of Sephacryl S-200 (1.5 × 110 cm) equilibrated with TC buffer containing 0.4 M NaCl.

3.3.5. Separation of Active MMP-3 and proMMP-3

1. Apply the mixture of proMMP-3 and active form of MMP-3 in TNC buffer to a column of Green A Dyematrex gel equilibrated with TNC buffer.
2. Wash the column with the same buffer until A_{280nm} reaches the base line.
3. Collect the unbound protein peak as active MMP-3 (45 kDa and 28 kDa).
4. Wash the column with TNC buffer then with TC buffer containing 0.3 M NaCl. Collect the protein peak as proMMP-3.

3.4. Purification of proMMP-7 (see Notes 23–25)

3.4.1. Assays

MMP-7 activity is measured after activation with 1 mM APMA using Azocoll, or synthetic peptides (e.g., Mca-Pro-Leu-Gly-Leu-Dpa-Ala-Arg-NH$_2$).

3.4.2. Activation of proMMP-7

1. Dilute 0.2 M APMA solution to 2 mM in TNC buffer
2. Add an equal volume (10–50 µL) of 2 mM APMA to proMMP-7 preparation and incubate the mixture at 37°C for 1 h.

3.4.3. Purification from the Medium of CaR-1 Cells

This method is based on Imai et al. *(3)*.

Step 1: DEAE-Cellulose Chromatography

1. Concentrate the serum-free conditioned culture medium of CaR-1 cells (500 mL) about 20-fold by Diaflo apparatus with an Amicon YM-10 membrane.
2. Apply the sample to a column of DEAE-cellulose (2.5 × 8.5 cm) equilibrated with TNC buffer.
3. Wash the column with the same buffer until A_{280nm} reaches the base line.
4. Pool unbound protein fractions.

Step 2: Green A Dyematrex Gel Chromatography

5. Apply the unbound fraction of DEAE-cellulose to a column of Green A Dyematrex gel (2.5 × 8 cm) equilibrated with TNC buffer.
6. Wash the column with TNC buffer until A_{280nm} reaches to the base line.
7. Elute the bound proMMP-7 with a linear gradient of NaCl (0.15 M–2.0 M) in the same buffer.

8. Monitor proMMP-7 activity against a synthetic substrate or Azocoll after activation of a 50–100 µL portion of each fraction with APMA. Alternatively, analyze the eluted fractions by SDS/PAGE (12.5% acrylamide). ProMMP-7 exhibits a protein band around 28 kDa.

9. Combine fractions containing proMMP-7, and dialyze the sample against buffer 7.1.

Step 3: DEAE-Cellulose Chromatography

10. Apply the dialyzed sample to a column of DEAE-cellulose (1 × 18.5 cm) equilibrated in buffer 7.1 and wash the column with the same buffer until A_{280nm} reaches the base line.

11. Combine unbound protein fractions and dialyze the pool against buffer 7.2.

Step 4: Zn^{2+}– Chelating Sepharose Fast Flow-Chromatography

12. Make a Zn^{2+}– chelating Sepharose Fast Flow column (0.7 × 8 cm) by passing through 10 mL of 0.2 M $ZnCl_2$.

13. Wash the column with 20 mL of buffer 7.2.

14. Apply the dialyzed sample to the Zn^{2+}– chelating Sepharose Fast Flow column and wash the column with buffer 7.2 until A_{280nm} reaches to the base line.

15. Elute the bound active MMP-7 with Buffer 7.3 containing 0.15 M NaCl.

16. Wash the column with the same buffer until A_{280nm} reaches to the base line.

17. Elute proMMP-7 with buffer 7.3 containing 1 M NaCl.

18. Identify fractions containing proMMP-7 by assaying MMP-7 activity after activation of a 50–100 µL portion with APMA or by SDS-PAGE. ProMMP-7 exhibits a band of 28 kDa on SDS-PAGE.

19. Pool fractions containing proMMP-7 and concentrate it to 1.5 mL with an Amicon YM-10 membrane.

Step 5: Gel Filtration on Sephacryl S-200

20. Apply the sample to a gel filtration column of Sephacryl S-200 (1.5 × 15 cm) equilibrated with TNC buffer containing 0.4 M NaCl.

21. Analyze fractions for proMMP-7 (28 kDa) by SDS-PAGE (12.5% acrylamide).

22. Collect fractions containing proMMP-7, and store at 4°C for a short term (up to 4 wk), or at –20°C for a longer term.

3.4.4. Purification from Postpartum Rat Uterus (see **Note 26**)

This method is based on Woessner and Taplin *(4)*. All purification steps should be carried out at 4°C unless otherwise mentioned.

Step 1: Extraction

1. Dissect rat uteri (12 g, 9 uterine tissues) at 1 d postpartum free of mesentery and cervix.

2. Rinse the tissue with saline, and mince and homogenize in 10 vol of 10 mM $CaCl_2$ containing 0.25% (w/v) Triton X-100 in a glass homogenizer.

3. Centrifuge the homogenate at 600g for 30 min.
4. Re-homogenize the precipitate in 10 vol of buffer 7.4; place this suspension in metal centrifuge tubes, heat at 60°C for 4 min, and then centrifuge at 17,500g for 30 min at 4°C. Save the supernatant for proMMP-7 purification.

Step 2: (NH4)2SO4 Fractionation

5. Add solid $(NH_4)_2SO_4$ to the uterine extract to 25% saturation; stand the solution for 30 min on ice and remove the precipitate by centrifugation at 20,000g for 40 min.
6. Add solid $(NH_4)_2SO_4$ to the supernatant to 75% saturation; stand the solution for 30 min on ice.
7. Collect the precipitate by centrifugation at 20,000g for 40 min and dissolve the pellet in 40 mL of buffer 7.5.

Step 3: Gel Filtration on Ultrogel AcA54

8. Apply the ammononium sulfate fraction to a column of Ultrogel AcA54 (5 × 50 cm) equilibrated with buffer 7.5. The latent enzyme emerges at about 680 mL, the volume between 630 and 730 mL.
9. Assay for MMP-7 activity with Azocoll or a peptide substrate after APMA activation; pool the fractions with proMMP-7. This position is close to molecular weight marker of 25 kDa.

Step 4: Blue-Sepharose Chromatography

10. Apply the sample from Ultrogel AcA54 to a column of Blue-Sepharose (1.5 mL) equilibrated with buffer 7.4 and washed with the same buffer until A_{280nm} reaches to the base line.
11. Elute the enzyme with a linear NaCl gradient (0.2–2.0 M) in buffer 7.5. The enzyme emerges at about 0.8 M NaCl. At this stage the enzyme is only about 5% pure.
12. Dialyze the sample from the Blue-Sepharose column against buffer 7.6.

Step 5: Zn^{2+}– Chelating-Sepharose Fast Flow Chromatography

13. Make Zn^{2+}– chelating Fast Flow column (1.5 mL) by passing through 10 mL of 0.2 M ZnCl$_2$.
14. Wash the column with 20 mL of buffer 7.6.
15. Apply the dialyzed sample to a column of Zn^{2+}– Sepharose Fast Flow, and wash the column with buffer 7.7.
16. Elute the enzyme with a linear NaCl gradient (0–0.5 M) in buffer 7.7. The enzyme is eluted at about 50 mM NaCl.
17. Dialyze the eluted peak against 25 mM Tris-HCl (pH 7.5), 5 mM CaCl$_2$, 0.02% NaN$_3$, 0.05% Brij 35.

*Step 6: Gel Filtration on Ultrogel AcA54 (see **Note 27**)*

18. Apply the dialyzed sample to a column of Ultrogel AcA54 (1.6 × 80 cm) equilibrated with buffer 7.5.

19. Store the sample as in **Subheading 3.4.3.**

3.5. Purification of proMMP-8 and MMP-8

The procedures are based on Engelbrecht et al. *(17)* and Knäuper et al. *(18)*.

3.5.1. Assays

Same as MMP-1 (*see* **Subheading 3.1.1.**).

3.5.2. Activation of proMMP-8

Same as MMP-1 (*see* **Subheading 3.1.2.**) except activation by MMP-3 does not require APMA *(19)*.

3.5.3. Preparation of Hydroxamate-Sepharose

1. Allow to swell 15g of activated CH-Sepharose 4B in 1 mM HCl and wash the swollen gel with 3 L of 1 mM Tris-HCl on a glass scintered funnel. This gives 45 mL of gel with a capacity of 5–7 μmol/mL.
2. Dissolve 180 mg of Pro-Leu-Gly-NHOH in 45 mL of 10 mM sodium bicarbonate (pH 8.0) and mix with the gel for 60 min at room temperature.
3. Wash the coupled gel with 0.1 M Tris-HCl (pH 8.0), 0.5 M NaCl and shake the gel gently in this buffer for 1 h at room temperature.
4. Wash the coupled gel three times, alternating with 0.1 M sodium acetate (pH 4.0), 0.5 M NaCl and with 0.1 M Tris-HCl (pH 8.0), 0.5 M NaCl.
5. Store the coupled gel at 4°C in 50 mM Tris-HCl (pH 7.5), 0.5 M NaCl, 10 mM CaCl$_2$, 0.02% NaN$_3$.

3.5.4. Purification of proMMP-8

Step 1: Preparation of Buffy Coat Extract

1. Add plasmatonin solution to buffy coat (2 L from 20 L of human blood) in a 1:4 (v/v) ratio.
2. Stand buffy coat for 75 min, and aspirate leukocytes in the supernatant.
3. Centrifuge the collected supernatant to pellet leukocytes at 150g for 20 min.
4. Suspend the pellet in 0.45% NaCl; add 6 vol of water and shake for 30 s for hypotonic hemolysis. Immediately after this process, add 2.4 vol of 3.5% NaCl to make the solution isotonic (0.9% NaCl).
5. Centrifuge the suspension at 150g for 10 min and wash the pellet with 0.9% NaCl and pellet the cells again by centrifugation at 150g for 10 min.
6. Suspend the cells in an equal volume of buffer 8.1 containing 2 mM benzamidine, 1 mM PMSF and 1 mM DFP and homogenize in a Virtis homogenizer.
7. Freeze-thaw the homogenate 5-times in dry ice/acetone and in a 35°C water bath.
8. Dilute the homogenate to approx 300 mL with buffer 8.1 containing 2 mM benzamidine, 1 mM PMSF and 1 mM DFP.

9. Centrifuge the cell homogenate at 48,000g for 30 min and save the supernatant as crude buffy coat extract.

Step 2: Zn^{2+}– Chelating Sepharose Fast Flow Chromatography

10. Make a column of Zn^{2+}– chelating Sepharose Fast Flow column (3.4 × 19 cm) by slowly passing through 200 mL of 0.2 M $ZnCl_2$, and wash the column with buffer 8.2.
11. Apply the crude buffy coat extract to the column of Zn^{2+}-chelating Sepharose Fast Flow column.
12. Wash the column with buffer 8.2 until the A_{280nm} reaches the base line. Collect the unbound protein peak.

*Step 3: Q-Sepharose Fast Flow Chromatography (see **Note 28**)*

13. Apply the flow-through fraction from the Zn^{2+}– chelating column to a column of Q-Sepharose Fast Flow (3.5 × 25 cm) equilibrated with buffer 8.2.
14. Wash the column until A_{280nm} reaches the base line and elute the bound proMMP-8 with a linear NaCl gradient (0–0.15 M; 2 × 600 mL) in buffer 8.2.
15. Collect the initial protein peak up to a conductivity of 4.5 mS/cm (low conductivity fraction) and fractions with conductivity 4.5–7.0 mS/cm (high conductivity fraction).

Step 4: Orange-Sepharose CL-6B Chromatography

16. Dialyze the low conductivity fractions from the Q-Sepharose column against buffer 8.3 and apply to a column of Orange-Sepharose CL-6B (4 × 25 cm) equilibrated in buffer 8.3.
17. Wash the column with buffer 8.3 until A_{280nm} reaches the base line, and elute the bound proMMP-8 with nonlinear NaCl gradient [0.1 M (600 mL) – 2 M (250 mL)] in buffer 8.3.
18. Pool fractions with MMP-8 activity after activation with APMA. ProMMP-8 elutes in fractions with conductivity around 35–50 mS/cm.

Step 5: Gel Filtration on Sephacryl S-300

19. Concentrate the sample from the Orange-Sepharose column to 1.5 mL using Diaflo apparatus with an Amicon XM-50 membrane, and apply it to a gel filtration column of Sephacryl S-300 (1.8 × 80 cm) equilibrated with buffer 8.4.
20. Analyze fractions for proMMP-8 by assaying MMP-8 activity after activation with APMA and by SDS/PAGE. The main protein peak contains proMMP-8 of 85 kDa on SDS/PAGE (*see **Notes 29–31***).

3.5.5. Purification of the Active Form of MMP-8

The method is based on Moore and Spilburg *(20)*.

1. Use the high conductivity fraction containing proMMP-8 from the Q-Sepharose column.

Step 1: Orange-Sepharose CL-6B Chromatography

2. The procedure is essentially identical to that of proMMP-8 (*see* **Subheading 3.5.4.**). Pool proMMP-8 fractions detected by peptidolytic activity after activation with APMA.
3. Concentrate the pool from the Orange-Sepharose CL-6B column to 100 mL by Diaflo apparatus with an Amicon XM50 membrane, and activate proMMP-8 in the sample by incubating with 1 mM HgCl$_2$ at 37°C for 1 h.

Step 2: Hydroxamate-Sepharose Affinity Chromatography

4. Dialyze the activated MMP-8 fraction against buffer 8.5 overnight at 4°C, and apply the sample to a column of hydroxamate (Pro-Leu-Gly-NHOH)-Sepharose (1.2 × 15 cm) equilibrated with the same buffer.
5. Wash the column with buffer 8.5 to remove all unbound proteins.
6. Elute MMP-8 with buffer 8.6.
7. Collect 6 mL fractions into tubes containing 2 mL of buffer 8.7. Monitor the MMP-8 peak by measuring peptidolytic activity, pool active fractions and dialyze against buffer 8.5.

Step 3: Gel Filtration on Sephacryl S-300

8. Concentrate the sample using Diaflo apparatus with an Amicon XM50 membrane to 1.5 mL and apply it to a column of Sephacryl S-300 (1.8 × 80 cm) equilibrated with buffer 8.5. The main protein peak contains active form of MMP-8 that exhibits a protein band of 64 kDa on SDS/PAGE (*see* **Notes 29–31**).

3.6. Purification of proMMP-9

3.6.1. Preparation of Conditioned Medium

1. Grow U937 (or HT1080) cells in Dulbecco's modified Eagle's medium (DMEM) containing 10% fetal calf serum.
2. After confluency, wash the cells with phosphate-buffered saline (PBS) three times and treat with PMA (50 ng/mL) in serum-free DMEM supplemented with 0.2% (w/v) α-lactalbumin hydrolyzate (LH).
3. Harvest the conditioned medium after 2–3 d and store at –20°C after removal of the cells and cell debris by centrifugation at 400g for 5 min.

3.6.2. Assays

MMP-9 activity is measured after activation with APMA using ^{3}H-gelatin (*see* Chapter 22), a synthetic substrate (e.g., Mca-Pro-Leu-Gly-Leu-Dpa-Ala-Arg-NH$_2$) (*see* Chapter 31) or directly applying to gelatin-zymography (*see* Chapter 23).

3.6.3. Activation of proMMP-9

1. Dilute 0.2 M APMA stock solution to 2 mM with TNC buffer.
2. Add an equal volume (50–100 µL) of 2 mM APMA to proMMP-9 prep (0.1–10 µg/mL) and incubate at 37°C for 24 h.

3.6.4. Purification Procedures

The method is based on Morodomi et al. *(5)*.

Step 1: Gelatin-Sepharose Chromatography

1. Apply the serum-free conditioned medium of U937 cells or HT1080 cells (500 mL) to a column of gelatin-Sepharose (1.5 × 15 cm) equilibrated with TNC buffer.
2. Wash the column with TNC buffer until A_{280nm} reaches the base line.
3. Elute the bound proMMP-9 with 5% (v/v) DMSO in TNC buffer.
4. Analyze the eluted fractions for proMMP-9 by direct application to gelatin zymography, or by detecting gelatinase activity after treating the sample with 1 mM APMA for 24 h at 37°C. ProMMP-9 is detected as M_r around 92,000 on gelatin zymography (*see* **Note 32**).
5. Pool the fractions containing proMMP-9 and dialyze against TNC buffer.

Step 2: Con A-Sepharose Affinity Chromatography

6. Apply the dialyzed sample to a column of Con A-Sepharose (3 mL) equilibrated with TNC buffer.
7. Wash the column with the same buffer until the A_{280nm} reaches the base line.
8. Elute the bound proMMP-9 with 1 M α-methyl-D-mannoside in TNC buffer (*see* **Note 33**).
9. Collect the protein peak eluted from the column and concentrate it to 1.5 mL using Diaflo apparatus with an Amicon YM-10 membrane.

Step 3: Gel Filtration Chromatography on Sephacryl S-200

10. Apply the concentrated proMMP-9 sample to a column of gel filtration chromatography on Sephacryl S-200 (1.5 × 110 cm) equilibrated with TNC buffer.
11. Analyze the fractions by SDS/PAGE and by gelatin zymography. The earlier fractions from this column contain the proMMP-9-TIMP-1 complex (92 kDa and 30 kDa bands) and the later fractions contain proMMP-9 (92 kDa) free of TIMP-1 (29 kDa) (*see* **Note 34**).

3.7. Purification of proMMP-10

3.7.1. Assays

Same as MMP-3 (*see* **Subheading 3.3.1.**).

3.7.2. Activation of proMMP-10

Same as for proMMP-3 (*see* **Subheading 3.3.2.**).

3.7.3. Purification Procedures

The method is based on Nakamura et al. *(7)*.

Step 1: DEAE-Cellulose Chromatography

1. Concentrate the conditioned media of OSC-20 cells 10-20-fold; apply the sample to a column of DEAE-cellulose (2.5 × 8.5 cm) equilibrated with TNC buffer without Brij 35.
2. Wash the column with the same buffer until A_{280nm} reaches the base line; pool unbound fractions.

Step 2: Green A Dyematrex Gel Chromatography (see **Note 35***)*

3. Apply the unbound sample to a column of Green A Dyematrex gel (2.5 × 8 cm) equilibrated with TNC buffer without Brij 35, and wash the column with TNC buffer.
4. Elute the bound proteins with a linear gradient of NaCl (0.15–2.0 *M*) in TC buffer.
5. Monitor the MMP-10 activity of the eluted fractions after activation with APMA.
6. Pool fractions with MMP-10 activity.

Step 3: Gelatin-Sepharose Chromatography (see **Note 36***)*

7. Apply the sample from the Green A Dyematrex gel to a column of gelatin-Sepharose (2.5 cm × 4 cm) equilibrated with TNC buffer.
8. Wash the column with TNC buffer until A_{280nm} reaches the base line and collect unbound fractions.

Step 4: Immunoadsorbent Chromatography on Anti-(Human MMP-1) IgG-Sepharose and on Anti-(MMP-3) IgG-Sepharose (see **Notes 37** *and* **38***)*

9. Apply the sample from the gelatin-Sepharose column to a column of Affi-Gel 10 coupled with anti-(human MMP-1) IgG (clone 78-12G8, Fuji Chemical Industries, Ltd.) (2 mL) and a column of Affi-Gel 10 coupled with anti-(human MMP-3) IgG (clone 55-2A4, Fuji Chemical Industries, Ltd.) (2 mL) equilibrated with TNC buffer. Collect flow-through fractions from each chromatography.

Step 5: DEAE-Cellulose Chromatography

10. Combine the flow-through fraction of the immunoadsorbent columns and dialyze it against buffer 10.1.
11. Apply the dialyzed sample to a column of DEAE-cellulose (1.5 cm × 12 cm) equilibrated with the same buffer.
12. Wash the column with the same buffer until A_{280nm} reaches to the base line; elute the bound proMMP-10 by a linear gradient of NaCl (0–0.5 *M*) in the same buffer.
13. Monitor MMP-10 activity after activation with APMA and combine active fractions.

Step 6: Gel Filtration on Sephacryl S-200

14. Concentrate the sample from the DEAE-cellulose column to 1.5 mL using Diaflo apparatus with an Amicon YM-10 membrane, and apply it to a gel filtration column of Sephacryl S-200 (1 × 95 cm) equilibrated with TC buffer containing 0.4 NaCl.
15. Monitor A_{280nm} and MMP-10 activity after activation with APMA, and analyze for proMMP-10 by SDS/PAGE. ProMMP-10 exhibits a protein band of 56 kDa on SDS/PAGE (*see* **Notes 39** and **40**).

3.8. Purification of MMP-12

The method is based on Banda and Werb *(21)*. MMP-12 is purified as an active enzyme.

3.8.1. Preparation of Mouse Macrophage-Conditioned Medium

1. Inject 1.0 mL of 3% Brewer's thioglycollate medium into mice intraperitoneally (*see* **Note 41**).
2. After 4 d, obtain peritoneal macrophage from peritoneal lavarge with PBS containing 100 U heparin/mL.
3. Wash cells with PBS and suspend in DME medium containing 10% (v/v) fetal calf serum.
4. Culture at a density of $1.5 \times 10^7 - 2 \times 10^7$ cells per 75 cm^2 flask, and allow cells to adhere for 2–4 h.
5. Wash cells with Hank's balanced salt solution twice.
6. Add 10 mL of DME-LH medium containing 2 μ*M* colchicine to each flask.
7. After 48 h culture, harvest the conditioned medium, which is centrifuged at 8,000*g* for 10 min, and store at –20°C.
8. Add fresh DME-LH medium containing 2 μ*M* colchicine and culture for 48 h. Repeat this process to obtain about 500–600 mL of conditioned medium.

3.8.2. Assays

MMP-12 activity is measured by measuring the amount of radioactivity released after incubation of the enzyme preparation with [^{3}H]-labeled elastin. MMP-12 hydrolyzes a synthetic substrate Mca-Pro-Leu-Gly-Leu-Dpa-Ala-Arg-NH$_2$ *(8)*, but this substrate is hydrolyzed by other MMPs. Therefore, at the initial stage of purification, the assay using [^{3}H]-labeled elastin is recommended.

3.8.3. Purification Procedures

All procedures should be carried out at 4°C.

Step 1: Dialysis and Centration of Conditioned Medium

1. Dialyze the pooled conditioned medium against three changes of 20 vol of 10 m*M* NH$_4$HCO$_3$, and freeze-dry the sample.
2. Dissolve the freeze-dried material with 4 mL of cold water and dialyze it against 4 liters of buffer 12.1.

Step 2: DEAE-Sephadex A-25

3. Apply the dialyzed condition medium to a column of DEAE-Sephadex A-25 (0.9×27 cm) equilibrated with buffer 12.1 and wash the column with the same buffer until A$_{280nm}$ reaches the base line.
4. Elute the bound protein with a linear gradient (0–0.7 *M*; 80 mL total volume) in buffer 12.1 and collect 2-mL fractions.

5. Assay for protein (A_{280nm}) and enzyme activity.
   [*Note*: Two major peaks of elastase are detected; one eluted in the flow-through fraction and a second peak eluted with conductivity 2–15 mS/cm.]
6. Pool two peaks of elastase activity separately (pool I and pool II, respectively).
7. Collect fractions with conductivity ranging from 15 to 30 mS/cm as pool III.
8. Freeze-dry pools I, II, and III separately; dissolve each pool into 1.8 mL of water, and add 0.2 mL of glycerol.
9. Apply each pool of MMP-12 (pool I-III) to a gel filtration column of Ultrogel AcA54 (1.5 × 87 cm) equilibrated with buffer 12.2. Three peaks of elastase activity can be resolved by this gel filtration (peak A, *Kav* 0.13; peak B, *Kav* 0.33; peak C, *Kav* 0.61). Pool I gives peaks A and B in a ratio of 1:4; pool II gives peaks A, B, and C in a ratio of 1:15:4; and pool III gives four peaks of elastase activity. The peak B from pool II gives the highest specific activity with the main protein of 21 kDa (*see* **Note 42**).
10. Freeze active fractions in aliquots at –70°C for storage.

3.9. Purification of TIMP-1

TIMP-1 is a glycoprotein of 29–30 kDa. Methods discussed below are based on Sudbeck et al. *(22)* and Cawston et al. *(23)*. The latter method is rapid but requires anti-(human TIMP-1) IgG - Sepharose.

3.9.1. Assay for TIMP Activity (see *Notes 43–45*)

1. Mix 45 µL of MMP-3 (1 µg/mL) in TNC buffer and various amount (5–45 µL) of the inhibitor sample; make the mixture to a final volume of 90 µL by adding TNC buffer.
2. As positive control, mix 45 µL of MMP-3 (1 µg/mL) in TNC buffer and 45 µL of TNC buffer in duplicate. This gives 100% MMP-3 activity.
3. As negative control, place 90 µL of TNC buffer in an assay tube in duplicate.
4. Stand mixtures at 37°C for at least 30 min.
5. Add 10 µL of 10 µ*M* Mca-Pro-Leu-Gly-Leu-Dpa-Ala-Arg-NH$_2$.
6. Incubate 30 min at 37°C.
7. Stop the reaction by adding 900 µL of 3% acetic acid.
8. Read fluorescence with excitation at 325 nm and emission at 393 nm.

3.9.2. Purification of TIMP-1 from the Conditioned Medium of Bovine Skin Fibroblasts

This method is based on Sudbeck et al. *(22)*.

Step 1: Heparin-Sepharose Chromatography.

1. Apply serum-free conditioned medium (3 L) of bovine fetal skin fibroblasts to a column of heparin-Sepharose (1.6 × 12 cm) equilibrated with TCO buffer.
2. Wash the column with TCO buffer until A_{280nm} reaches the base line, and elute the bound TIMP-1 with a 500 mL linear gradient (0–2 mg/mL) of dextran sulfate in TCO buffer (5 mL/fraction).

3. Monitor TIMP-1 activity in each fraction, and pool fractions containing MMP inhibitory activity.

Step 2: DEAE-Sepharose Column

4. Apply the pooled fractions from the heparin-Sepharose column to a column of DEAE-Sepharose (0.5 mL) equilibrated with TCO buffer.
5. Wash the column with the same buffer until A_{280nm} reaches to the base line. Collect the unbound protein peak.

Step 3: Concentration by Heparin-Sepharose Chromatography

6. Apply the flow-through fraction of the DEAE-Sepharose column to a column of heparin-Sepharose (0.5 mL) equilibrated with TCO buffer.
7. Wash the column with the same buffer until A_{280nm} reaches the base line; elute TIMP-1 with TCO buffer containing 2 M NaCl.
8. Monitor the protein peak by reading A_{280nm}.
9. Collect the protein peak and dialyze it against 0.1% trifluoroacetic acid (TFA) in water.

Step 4: C4 Reverse-Phase HPLC

10. Apply the dialyzed sample to a C4 reverse-phase column (Vydac, Hesperia, CA) equilibrated in 0.1% TFA; elute TIMP-1 with a linear gradient of 30–50% acetonitrile containing 0.1% TFA over 50 min.
11. Analyze components of the major peaks for TIMP-1 by SDS/PAGE, and pool the fractions containing TIMP-1. TIMP-1 exhibits a band around 29-30 kDa. Typically, about 200 µg of TIMP-1 is recovered from 3 liters of the conditioned medium (*see* **Note 46** and **47**).

3.9.3. Purification of TIMP-1 from Human Plasma

This method is based on Cawston et al. *(23)*.

1. Dialyze human plasma (400 mL) against 20–40 vol of buffer T1-1 at 4°C overnight twice.
2. Remove precipitates formed by centrifugation at 40,000g for 1 h and pass the supernatant through a column of Sepharose 6B (1.6 × 30 cm) equilibrated with buffer T1-1 to remove any plasma protein that binds to Sepharose nonspecifically.
3. Wash the column with buffer T1-1 and collect the flow-through fraction.

Anti-(Human TIMP-1) Affinity Chromatography

4. Apply the flow-through fraction to a column of anti-(human TIMP-1) IgG-Affi-Gel 10 (5 × 0.6 cm).
5. Wash the column extensively, elute the bound TIMP-1 with buffer T1-2 and collect 2-mL fractions; pool the fractions with a protein peak.
6. Concentrate the sample eluted from the affinity chromatography to 1.5 mL by Amicon Diaflo with a YM-10 membrane and apply the concentrated sample to a gel filtration column on Ultrogel AcA 44 (1.6 × 88 cm) equilibrated with buffer T1-3. A single peak of MMP inhibitory activity is eluted at a position corre-

sponding to M_r 30,500. About 160 µg of TIMP-1 is purified from 400 mL plasma (*see* **Note 46** and **47**).

3.10. Purification of TIMP-2

3.10.1. Method 1

This method is based on the isolation of the proMMP-2-TIMP-2 complex from the conditioned medium of fibroblasts and the dissociation of TIMP-2 from the complex *(11,12)* (*see* **Note 48**).

Step 1: Gelatin-Sepharose Chromatography

1. Apply the conditioned culture medium of human fibroblasts (2 L) to a column of gelatin-Sepharose (2.5 × 5 cm) equilibrated with TNC buffer without Brij 35.
2. Wash the column with TNC buffer containing 1% DMSO until A_{280nm} reaches the base line.
3. Elute the proMMP-2-TIMP-2 complex with 5% DMSO in TNC buffer without Brij 35.
4. Monitor A_{280nm} and pool the protein peak.

Step 2: Gel Filtration on Sephacryl S-200

5. Dialyze the sample against 200 volumes of TNC buffer and concentrate by Diaflo apparatus with an Amicon YM-10 membrane.
6. Apply the concentrated samples to a gel filtration column of Sephacryl S-200 (1.5 × 110 cm) equilibrated with TNC buffer.
7. Monitor A_{280nm} for protein and analyze the proMMP-2-TIMP-2 complex by SDS-PAGE (10% acrylamide).
8. Pool the fraction containing the proMMP-2-TIMP-2 complex.

Step 3: Dissociation of TIMP-2 from the proMMP-2-TIMP-2 Complex and Gel Filtration on Sephacryl S-200

9. Concentrate the pool from Sephacryl S-200 to 1 mL and add 1 mL of 0.4 *M* Glycine-HCl (pH 2.8), 50 m*M* EDTA.
10. Heat the solution at 45°C for 30 min.
11. Apply the sample to a column of gel filtration on Sephacryl S-200 (1.6 × 110 cm) equilibrated with TNC buffer containing 25 m*M* EDTA.
12. Monitor A_{280nm} for protein and the presence of TIMP-2 by SDS/PAGE (10% acrylamide).
13. Pool the fractions containing TIMP-2, and store the sample at –70°C in aliquots (*see* **Notes 49** and **50**).

3.10.2. Method 2

This is based on the method by Umenishi et al. *(24)* using the conditioned medium of HLE human hepatoma cells.

Step 1: (NH4)2SO4 Precipitation and Gel Filtration on Cellulofine GCL-2000-m

1. Add solid $(NH_4)_2SO_4$ to pooled serum-free conditioned medium of HLE cells (about 4 L) to a final saturation of 80% at 4°C, and allow the mixture to stand overnight at 4°C.
2. Collect the precipitate by centrifugation at 20,000g at 40 min.
3. Dissolve the pellet with 10 mL of buffer 1; centrifuge the solution at 20,000g for 5 min and use the supernatant for purification.

Step 2: Gel Filtration on Cellulofine GCL-2000-m

4. Apply the supernatant to a column of Cellulofine GCL-2000-m (2.6 × 98 cm) equilibrated with buffer T2-1.
5. Monitor A_{280nm} and MMP inhibition activity.
6. Pool fractions containing MMP inhibitory activity and dialyze against 200 vol of buffer T2-2.

Step 3: Reactive Red-Agarose Chromatography

7. Apply the dialyzed sample to a column of Reactive Red-agarose (1.5 × 7 cm) equilibrated with buffer T2-2.
8. Wash the column until A_{280nm} reaches the base line, and elute the bound materials with a linear gradient of NaCl (0–2 M; total 200 mL).
9. Monitor A_{280nm} and MMP inhibitory activity, and pool fractions with the inhibitor. TIMP-2 is eluted at NaCl concentration of 0.9–1.3 M.
10. Dialyze the sample against 0.05% TFA.

Step 4: Reverse Phase HPLC

11. Apply the sample in 0.05% TFA to reverse phase HPLC on a SynChropak RP-4 column (0.41 × 25 cm) equilibrated with 0.05% TFA.
12. Elute TIMP-2 with a linear gradient of acetonitrile (0–60%) in 30 mL of 0.05% TFA.
13. Monitor A_{280nm} and MMP inhibitory activity. TIMP-2 is eluted at an acetonitrile concentration of about 40%.
14. Analyze fractions containing TIMP-2 by SDS/PAGE (12.5% acrylamide). Later fractions of the inhibitory peak contains a small amount of TIMP-1 (*see* **Notes 49** and **50**).

4. Notes

1. Purification procedures can be carried out either at room temperature when the starting material is a conditioned culture medium. However, when a tissue extract is used, all purification steps should be at 4°C.
2. Many MMPs are stable as proenzyme. When activated, they often lose their C-terminal hemopexin domains. Therefore, it is generally recommended to purify MMPs as proenzyme, and activate them after purification of proMMPs.
3. Many connective tissue cells in culture stimulated with IL-1 or PMA produce proMMP-1, proMMP-2 and proMMP-3. These three zymogens can be purified

from the same conditioned medium. The recommended purification procedures of proMMP-1, proMMP-2, and proMMP-3 from the conditioned medium of IL-1-treated fibroblasts are as follows:

a. Concentrate the medium about 20-fold and apply to a column of DEAE cellulose as in **Subheading 3.1.4.** The flow through fraction contains proMMPs -1, -2, and -3.

b. Apply the flow through fraction to a column of gelatin-Sepharose as in **Subheading 3.2.3.** The flow through fraction contains proMMP-1 and proMMP-3. The bound proMMP-2 is eluted from the column with 5% DMSO in TNC buffer and further purified as in **Subheading 3.2.3.**

c. Apply the flow through fraction from the gelatin-Sepharose column to a column of Green A Dyematrex as in **Subheading 3.1.4.** Elute proMMP-3 and proMMP-1 with 0.3 M NaCl and 0.5 M NaCl in TC buffer, respectively.

d. Further purify proMMP-3 and proMMP-1 by gel filtration Sephacryl S-200. ProMMP-1 and proMMP-3 fractions may contain small amounts of proMMP-3 and proMMP-1. Because they are close in molecular mass, it is recommended to check contamination by Western blotting analysis. The contaminating proMMPs may be removed using immunoadsorbent chromatography using an appropriate antibodies.

4. Assays with synthetic substrates are convenient to measure MMP activities but most substrates are hydrolyzed by many MMPs. To purify a specific MMP, assays methods with protein substrates, Western blotting are often required.

5. Human proMMP-1 is activated by treatment with APMA in the presence or absence of MMP-3 *(10)*. On activation the proMMP-1 is processed to the active forms (41 kDa) which may undergo autolytic cleavage into the C-terminal domain (25 kDa) and catalytic domain (22 kDa). In the case of MMP-1 activated with APMA and MMP-3, it may be necessary to remove MMP-3 depending on the study. However, MMP-3 does not participate in interstitial collagen-degradation *(10)*.

6. When proMMP-1 is activated with APMA and MMP-3 the primary product is the 41 kDa MMP-1.

7. **Step 1**: DEAE-cellulose chromatography removes glycosaminoglycans and phenol red. This step is essential for reproducible chromatography of Green A Dyematrex gel in the next step.

8. **Step 2**: Green A Dyematrex gel chromatography. The majority of proMMP-2 and proMMP-3 are recovered in the 0.3 M NaCl fraction, whereas about 75% of proMMP-1 is recovered in the 0.5 M NaCl fraction.

9. The purified proMMP-1 fraction often contains a small amount of proMMP-2 and proMMP-3. ProMMP-2 may be removed by chromatography on a gelatin-Sepharose (Pharmacia) and ProMMP-3 on an anti-(MMP-3) IgG-Affi-Gel 10 affinity chromatography.

10. **Step 5**: Con A-Sepharose chromatography. The sample passed through or eluted from this column is contaminated with a small amount of Con A from the resin. Thus, it is essential to purify it further by gel filtration as indicated.

11. Store proMMP-1 at 4°C for short storage (up to 4 wk), but at –70°C for longer storage.

12. ProMMP-2 is rapidly activated by APMA, but it is also gradually inactivated by autodegradation. Therefore, when the enzyme and substrate are incubated for a long time (>2 h), it is recommended to incubate proMMP-2, the substrate to be tested and 1 mM APMA together.

13. Gelatin-Sepharose chromatography adsorbs proMMP-2, the proMMP-2-TIMP-2 complex, and fibronectin. These three components, (molecular mass of fibronectin [440,000], proMMP-2-TIMP-2 complex [95,000], and proMMP-2 [72,000]) are reasonably well separated by gel filtration on Sephacryl S-200.

14. When the conditioned medium of cytokine-stimulated cells is used it may also contain proMMP-9 and the complex of proMMP-9 and TIMP-1. These two components can be removed by Con A-Sepharose chromatography.

15. Activation of proMMP-3 by APMA results in formation of both 45-kDa and 28-kDa forms. The latter is the catalytic domain lacking the C-terminal hemopexin-like domain. Both are equally active against various substrates including extracellular matrix macromolecules *(25)*.

16. Activation of proMMP-3 by chymotrypsin generates primarily the 45-kDa form.

17. **Step 1**: DEAE-cellulose chromatography removes glycosaminoglycans and phenol red. This step is essential for reproducible chromatography of Green A Dyematrex gel in the next step.

18. About 50% of proMMP-3 is recovered in 0.3 M NaCl fraction of Green A Dyematrex gel chromatography. This fraction contains proMMP-2, which is effectively removed by the next gelatin-Sepharose step. About 75% proMMP-1 is recovered in 0.5 M NaCl. This fraction can be used for proMMP-1 purification (*see* **Subheading 3.1.4.**).

19. **Step 5** removes a small amount of proMMP-1 contamination in the proMMP-3 preparation. When an anti-(MMP-1) affinity column is not available, dialyze the sample against 20 mM Tris-HCl (pH 7.5), 10 mM CaCl$_2$ and apply to a column of CM-cellulose (1.5 × 5 cm) equilibrated with the same buffer and collect the unbound fraction *(26)*. ProMMP-3 is recovered in the unbound fraction. The CM-cellulose step may be used after **Step 1**, DEAE-cellulose chromatography.

20. ProMMP-3 is eluted from an immuno-affinity column with 6 M urea. Under these conditions proMMP-3 is stable and the majority (>90%) is recovered as a proenzyme.

21. Con A-Sepharose chromatography separates glycosylated and unglycosylated forms of proMMP-3, but each fraction after a Con A-Sepharose column contains a small amount of Con A from the column. This is removed by gel filtration on Sephacryl S-200.

22. ProMMP-3 is stable at 4°C at least for 4 wk. For a longer storage, it is recommended to store at –20°C.

23. ProMMP-7 stored at 4°C is stable for at least 4 wk.

24. ProMMP-7 stored at –20°C is activated to a 19-kDa form when it is thawed.

25. APMA-activated MMP-7 is stable at 23°C for at least 4 wk.

26. Rat uterine extract contains several MMPs. These enzymes also hydrolyze Azocoll and synthetic substrates. However, the molecule mass of proMMP-7 (28 kDa) is smaller than that of other proMMPs (>55 kDa). For example, gel filtration in step 3 separates gelatinases, which are eluted in the void volume.

27. Rat proMMP-7 in 25 mM Tris-HCl (pH 7.5), 5 m*M* CaCl$_2$, 0.02% NaN$_3$, 0.05% Brij interacts with Ultrogel AcA54 (step 6) and emerges later than normal (*Kav* = 0.65). The product after Step 6 is homogeneous and shows a single band of 28 kDa on SDS/PAGE after staining with silver. However, a minor band of 19 kDa may appear sometimes due to partial activation of proMMP-7.
28. ProMMP-8 from the Q-Sepharose anion exchange column (step 4) is found at conductivity up to 7.0 mS/cm, but only fractions with the lower conductivity (up to 4.5 mS/cm) can be purified as proMMP-8. Fractions with higher conductivity contain proteins that cannot be removed in the subsequent purification steps. Those fractions can be used separately for another purification or for purification of the active form of MMP-8 by hydroxamate-Sepharose affinity chromatography *(20)* after separate purification on Orange-Sepharose CL-6B.
29. Immediate freezing and storage at –70°C in aliquots is required to prevent spontaneous activation of proMMP-8 or autolytic degradation of both active and latent MMP-8.
30. Storage at –10 to –20°C results in a slow loss in activity.
31. Frequent freeze-thawing of enzyme samples results in loss of activity.
32. The major components eluted from the gelatin-Sepharose column are proMMP-9, proMMP-9-TIMP-1 complex and possibly a small amount of proMMP-2.
33. The sample eluted from the Con A-Sepharose column contains both proMMP-9 and the proMMP-9-TIMP-1 complex. The proMMP-9-TIMP-1 complex and proMMP-9 can be separated by gel filtration on Sephacryl S-200. Alternatively the proMMP-9-TIMP-1 complex may be removed by anti-(TIMP-1) IgG-Sepharose. In this case anti-TIMP-1 antibody must be checked to see whether it recognizes TIMP-1 in the complex. Polyclonal antibodies raised against reduced human TIMP-1 (e.g., TIMP-1 eluted after SDS/PAGE under reducing conditions) does not recognize the native TIMP-1.
34. The proMMP-9-TIMP-1 complex may be isolated by collecting earlier fractions of gel filtration (**step 3**). The treatment of the complex with APMA does not exhibit gelatinolytic activity against ^{14}C-labelled gelatin in solution *(27)*. Any gelatinolytic activity detected with the complex is indicative of the presence of free proMMP-9.
35. Green A Dyematrex gel chromatography (**step 2**). The majority of proteins do not bind to the column. The bound proMMP-10 may also be eluted from the column with 0.3 *M* NaCl in TC buffer.
36. Gelatin-Sepharose chromatography (**step 3**) removes proMMP-2 and proMMP-9. ProMMP-10 is recovered in the unbound fraction.
37. Anti-(human MMP-1) IgG-Sepharose and anti-(human MMP-3) IgG-Sepharose remove proMMP-1 and proMMP-3, respectively. Polyclonal antibodies against those proteins can also be used.
38. MMP-3 and MMP-10 are about 73% identical in sequence, but sheep anti-(human MMP-3) IgG does not crossreact with proMMP-10.
39. The final product has a major species of 56 kDa but it may contain minor protein bands of 47 kDa, 24 kDa and 22 kDa. Both 47 kDa and 24 kDa species are activated forms of MMP-10.

40. About 400 µg of proMMP-10 may be obtained from 600 mL of the conditioned medium of OSC-20 human oral squamous carcinoma cells.
41. About 1.5×10^7 exudate cells are obtained from each mouse injected with 3% Brewer's thioglycollate medium. About 3×10^8 cells are required for a complete enzyme purification.
42. The zymogen of MMP-12 is estimated to be about 56 kDa *(28)*, but proMMP-12 is readily activated during purification. Only the active form of 21 kDa is purified by this method.
43. MMP-1 and MMP-3, which may be obtained relatively easily, are recommended to use for TIMP inhibition assay, although any MMP, except MT1-MMP (MMP-14), may be used as a target enzyme. MMP-2 is not a good enzyme for routine detection of TIMP as it autodegrades after activation.
44. The concentration of target enzyme should be greater than 20 nM when the quantity of TIMP-1 is determined.
45. To determine the amount of TIMP, inhibition of MMP has to be 25–75%. When inhibition is >90%, dilute the inhibitor sample several-fold and reassay.
46. The $A^{1\%}_{280nm,cm}$ value of human TIMP-1 is 9.14.
47. TIMP-1 is very stable; >95% activity is recovered after heating the solution of 10 µg/mL at 95°C for 15 min.
48. TIMP-2 in the conditioned medium of fibroblasts exists primarily as a complex with proMMP-2. However, the medium usually contains free proMMP-2. The material eluted from the gelatin-Sepharose column is a mixture of the proMMP-2-TIMP-2 complex and proMMP-2, which are separated by gel filtration on Sephacryl S-200.
49. The $A^{1\%}_{280nm,cm}$ value of human TIMP-2 is 18.0.
50. Like TIMP-1, TIMP-2 is very stable.
51. After activation and during storage MMPs often undergo autodegradation. It is therefore important to titrate the active site of an activated MMP. This is accomplished by using TIMP-1, TIMP-2 or α_2-macroglobulin ($\alpha_2 M$) which form a 1:1 complex with the enzyme. When $\alpha_2 M$ is used, a large protein substrate must be used. The following procedure is recommended
    a. Mix 10 µL of the MMP solution (about 100 nM) with 10 µL of 0, 20, 40, 60 …200 nM TIMP-1, TIMP-2, or $\alpha_2 M$.
    b. Incubate at 37°C for 1 h.
    c. Mix a 10-µL portion of the mixture with 10 µL of an appropriate substrate for the MMP and measure the residual enzyme activity.
    d. Plot the enzyme activity against the inhibitor concentration.
    e. Draw a linear line through the plots and determine, or extrapolate, the inhibitor molarity that gives zero activity of the enzyme. This molarity of inhibitor is equal to that of the enzyme solution.

Acknowledgments

We thank Linda Chung for useful comments. This work is supported by NIH grants AR39189 and AR40994.

References

1. Massova, I., Kotra, L. P., Fridman, R., and Mobashery, S. (1998) Matrix metalloproteinases - structures, evolution, and diversification. *FASEB J.* **12**, 1075–1095.
2. Gomez, D. E., Alonso, D. F., Yoshiji, H., and Thorgeirsson, U. P. (1997) Tissue inhibitors of metalloproteinases—structure, regulation and biological functions. *Eur. J. Cell. Biol.* **74**, 111–122.
3. Imai, K., Yokohama, Y., Nakanishi, I., Ohuchi, E., Fujii, Y., Nakai, N., and Okada, Y. (1995) Matrix metalloproteinase 7 (matrilysin) from human rectal carcinoma cells. Activation of the precursor, interaction with other matrix metalloproteinases and enzymic properties. *J. Biol. Chem.* **270**, 6691–6697.
4. Woessner, J. F., Jr. and Taplin, C. J. (1988) Purification and properties of a small latent matrix metalloproteinase of the rat uterus. *J. Biol. Chem.* **263**, 16,918–16,925.
5. Morodomi, T., Ogata, Y., Sasaguri, Y., Morimatsu, M., and Nagase, H. (1992) Purification and characterization of matrix metalloproteinase 9 from U937 monocytic leukemia and HT1080 fibrosarcoma cells. *Biochem. J.* **285**, 603–611.
6. Windsor, L. J., Grenett, H., Birkedal-Hansen, B., Bodden, M. K., Engler, J. A., and Birkedal-Hansen, H. (1993) Cell type-specific regulation of SL-1 and SL-2 genes. Induction of the SL-2 gene but not the SL-1 gene by human keratinocytes in response to cytokines and phorbolesters. *J. Biol. Chem.* **268**, 17,341–17,347.
7. Nakamura, H., Fujii, Y., Ohuchi, E., Yamamoto, E., and Okada, Y. (1998) Activation of the precursor of human stromelysin 2 and its interactions with other matrix metalloproteinases. *Eur. J. Biochem.* **253**, 67–75.
8. Knight, C. G., Willenbrock, F., and Murphy, G. (1992) A novel coumarin-labelled peptide for sensitive continuous assays of the matrix metalloproteinases. *FEBS Lett.* **296**, 263–266.
9. Cawston, T. E. and Barrett, A. J. (1979) A rapid and reproducible assay for collagenase using [1-^{14}C]acetylated collagen. *Anal. Biochem.* **99**, 340–345.
10. Suzuki, K., Enghild, J. J., Morodomi, T., Salvesen, G., and Nagase, H. (1990) Mechanisms of activation of tissue procollagenase by matrix metalloproteinase 3 (stromelysin). *Biochemistry* **29**, 10,261–10,270.
11. Itoh, Y., Binner, S., and Nagase, H. (1995) Steps involved in activation of the complex of pro-matrix metalloproteinase 2 (progelatinase A) and tissue inhibitor of metalloproteinases (TIMP)-2 by 4-aminophenylmercuric acetate. *Biochem. J.* **308**, 645–651.
12. Ward, R. V., Hembry, R. M., Reynolds, J. J., and Murphy, G. (1991) The purification of tissue inhibitor of metalloproteinases-2 from its 72 kDa progelatinase complex. Demonstration of the biochemical similarities of tissue inhibitor of metalloproteinases-2 and tissue inhibitor of metalloproteinases-1. *Biochem. J.*, **278**, 179–187.
13. Gunja-Smith, Z., Nagase, H., and Woessner, J. F., Jr. (1989) Purification of the neutral proteoglycan-degrading metalloproteinase from human articular cartilage tissue and its identification as stromelysin matrix metalloproteinase-3. *Biochem. J.* **258**, 115–119.

14. Nagase, H. (1995) Human stromelysins 1 and 2. *Methods Enzymol.* **248**, 449–470.
15. Nagase, H., Fields, C. G., and Fields, G. B. (1994) Design and characterization of a fluorogenic substrate selectively hydrolyzed by stromelysin 1 (matrix metalloproteinase- 3). *J. Biol. Chem.* **269**, 20,952–20,957.
16. Ito, A. and Nagase, H. (1988) Evidence that human rheumatoid synovial matrix metalloproteinase 3 is an endogenous activator of procollagenase. *Arch. Biochem. Biophys.* **267**, 211–216.
17. Engelbrecht, S., Pieper, E., Macartney, H. W., Rautenberg, W., Wenzel, H. R., and Tschesche, H. (1982) Separation of the human leucocyte enzymes alanine aminopeptidase, cathepsin G, collagenase, elastase and myeloperoxidase. *Hoppe Seylers Z. Physiol. Chem.* **363**, 305–315.
18. Knäuper, V., Kramer, S., Reinke, H., and Tschesche, H. (1990) Characterization and activation of procollagenase from human polymorphonuclear leucocytes. N-terminal sequence determination of the proenzyme and various proteolytically activated forms. *Eur. J. Biochem.* **189**, 295–300.
19. Knäuper, V., Wilhelm, S. M., Seperack, P. K., DeClerck, Y. A., Langley, K. E., Osthues, A., and Tschesche, H. (1993) Direct activation of human neutrophil procollagenase by recombinant stromelysin. *Biochem. J.*, **295**, 581–586.
20. Moore, W. M. and Spilburg, C. A. (1986) Purification of human collagenases with a hydroxamic acid affinity column. *Biochemistry* **25**, 5189–5195.
21. Banda, M. J. and Werb, Z. (1981) Mouse macrophage elastase. Purification and characterization as a metalloproteinase. *Biochem. J.* **193**, 589–605.
22. Sudbeck, B. D., Jeffrey, J. J., Welgus, H. G., Mecham, R. P., McCourt, D., and Parks, W. C. (1992) Purification and characterization of bovine interstitial collagenase and tissue inhibitor of metalloproteinases. *Arch. Biochem. Biophys.* **293**, 370–376.
23. Cawston, T. E., Noble, D. N., Murphy, G., Smith, A. J., Woodley, C., and Hazleman, B. (1986) Rapid purification of tissue inhibitor of metalloproteinases from human plasma and identification as a gamma-serum protein. *Biochem. J.* **238**, 677–682.
24. Umenishi, F., Umeda, M., and Miyazaki, K. (1991) Efficient purification of TIMP-2 from culture medium conditioned by human hepatoma cell line, and its inhibitory effects on metalloproteinases and in vitro tumor invasion. *J. Biochem.* (Tokyo) **110**, 189–195.
25. Okada, Y., Nagase, H., and Harris, E. D., Jr. (1986) A metalloproteinase from human rheumatoid synovial fibroblasts that digests connective tissue matrix components. Purification and characterization. *J. Biol. Chem.* **261**, 14,245–14,255.
26. Lark, M. W., Walakovits, L. A., Shah, T. K., Vanmiddlesworth, J., Cameron, P. M., and Lin, T. Y. (1990) Production and purification of prostromelysin and procollagenase from IL-1 β-stimulated human gingival fibroblasts. *Connect. Tissue Res.* **25**, 49–65.
27. Ogata, Y., Itoh, Y., and Nagase, H. (1995) Steps involved in activation of the pro-matrix metalloproteinase 9 (progelatinase B)-tissue inhibitor of metalloproteinases-1 complex by 4-aminophenylmercuric acetate and proteinases. *J. Biol. Chem.* **270**, 18,506–18,511.

28. Shapiro, S. D., Kobayashi, D. K., and Ley, T. J. (1993): Cloning and characterization of a unique elastolytic metalloproteinase produced by human alveolar macrophages. *J. Biol. Chem.* **268**, 23,824–23,829

III

DETECTION OF MMPS AND TIMPS

Monitoring MMP and TIMP mRNA Expression by RT-PCR

Howard Wong, Huong Muzik, Lori L. Groft, Marc A. Lafleur, Charles Matouk, Peter A. Forsyth, Gilbert A. Schultz, Steven J. Wall, and Dylan R. Edwards

1. Introduction

Reverse Transcriptase-Polymerase Chain Reaction (RT-PCR) is a sensitive and rapid method to monitor the expression of specific mRNAs. Here we describe two approaches to quantification of steady-state levels of Matrix Metalloproteinase (MMP) and Tissue Inhibitor of Metalloproteinase (TIMP) mRNAs using RT-PCR. The first is a modified RT-PCR protocol called the "primer-dropping" method *(1)*, which involves the simultaneous amplification of the specific MMP and TIMP target mRNAs (which are expressed in variable amounts in human cells) and that of an internal standard (glyceraldehyde phosphate dehydrogenase, GAPDH) which is expressed at constant levels. The internal standard provides a means to monitor the RT-PCR reaction efficiencies and to normalize the reaction products. The second approach uses coamplification of the specific target with a multi-MMP competitor cDNA within the same tube *(2)*.

The RT-PCR protocol involves two major steps: (a) the semiquantitative PCR amplification of cDNA copies of the target mRNA and (b) the detection, quantification, and analysis of the PCR products. The amplification step is further composed of two procedures. RNA harvested from tissues or tissue culture cells is first converted to cDNA using the retroviral enzyme reverse transcriptase, followed by PCR amplification of the cDNA using the thermostable *Taq* DNA polymerase. The amplification is carried out in a controlled manner (by regulating the number of amplification cycles for each coamplified target) so that the products do not reach "plateau" or "saturating" levels where

From: *Methods in Molecular Biology, vol. 151: Matrix Metalloproteinase Protocols*
Edited by: I. Clark © Humana Press Inc., Totowa, NJ

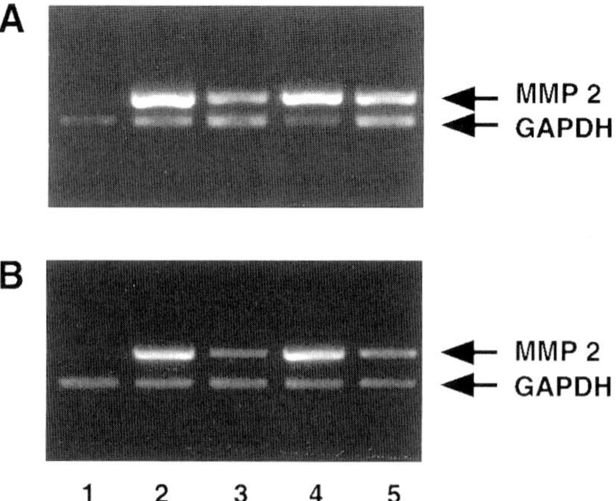

Fig. 1. Equalization of the GAPDH Signal Intensity. (**A**) Primer-dropping RT-PCR of MMP-2 (30 PCR cycles) and GAPDH (23 PCR cycles) before and (**B**) after equalization of the GAPDH signal between samples. Lane 1, normal brain tissue. Lanes 2–5, human brain tumor specimens.

the reaction is no longer quantitative. By maintaining sub-saturating levels of amplification, the products of the PCR reaction remain proportional to the amount of mRNA target in the original sample *(3,4)*. When testing a new set of primers, the range of PCR cycles to quantitatively amplify the target is determined empirically by performing a "range finding" PCR experiment.

In the examples shown in **Figs. 1** and **2**, comparisons of MMP mRNA expression levels among human brain tumor specimens versus normal brain tissue were performed after equalizing the levels of the GAPDH internal standard among the samples. The analysis and quantification of the specific target signal is achieved by scanning densitometry of photographic images of the PCR-products that have been run on agarose gels. In summary, RT-PCR is a powerful and reliable method for the rapid analysis of MMP and TIMP mRNA gene expression in clinical material. Indeed the RNA extraction, RT-PCR, detection and quantification of several specimens can be completed in as little as a single day.

We describe in detail the methodology used in both the primer-dropping and internal competitor procedures. Aspects that are common to both, such as RNA extraction and gel quantification, are described in detail in the first method.

2. Materials

2.1. Primer Dropping RT-PCR

1. Tissue homogeniser.
2. Guanidinium solution D: 4 *M* guanidinium isothiocyanate, 25 m*M* sodium citrate pH 7.0, 0.5% sarcosyl, 0.1 *M* 2-mercaptoethanol (added fresh).

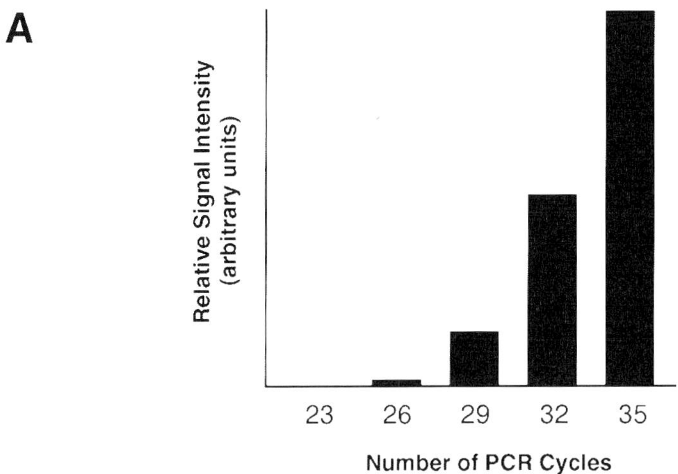

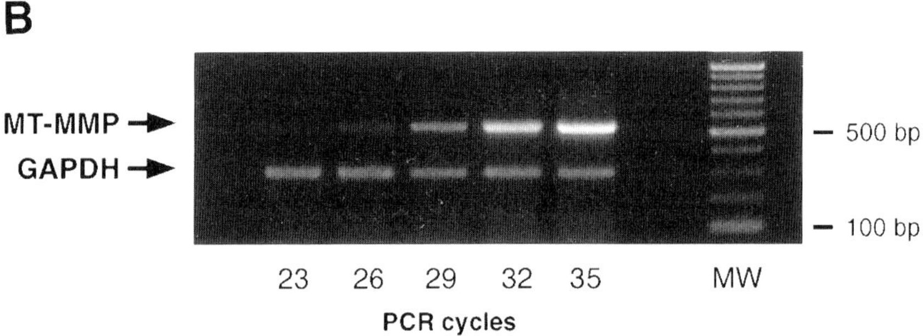

Fig. 2. PCR cycle range finding and quantification. (**A**) Quantification of MT1-MMP PCR product signals using the NIH Image densitometry program for the experiment shown in panel B. (**B**) Range finding experiment for MT1-MMP from a human brain tumor specimen. GAPDH was amplified for 22 cycles while MT1-MMP was amplified for 23, 26, 29, 32, 35 cycles. The semiquantitative PCR cycle number for MT1-MMP was chosen to have a product intensity that is similar to the intensity of GAPDH product signal. Thus for MT1-MMP, coamplification (with GAPDH at 22 cycles) at about 29 cycles appears appropriate.

3. 2 *M* sodium acetate, pH 4.0.
4. Water-saturated phenol.
5. 70% ethanol.
6. Isopropanol.
7. Chloroform: isoamylalcohol (24:1).
8. DEPC-water or distilled water.

9. 10X PCR buffer A: 100 mM Tris-HCl, pH 9.0, 500 mM KCl, 15 mM magnesium chloride.
10. 2 mM and 10 mM dNTPs mix (dATP, dCTP, dGTP, TTP).
11. Oligonucleotide primers for MMPs, TIMPs and the internal control GAPDH (all at 10 pmol/μL). Primers for the MMP and TIMP PCR products are designed to be of different size than the GAPDH product so they can be separately identified following electrophoresis. Most primers are 20–24 nucleotides long and have a GC content of about 50–60%.
12. Reverse transcriptase (RT)-master mix A (volumes per sample): 2 μL of 10X PCR buffer A, 2 μL of 10 mM dNTP mix, 2 μL of 100 pmol/μL N6-random oligonucleotide (Pharmacia), 0.5 μL of RNAguard (placental RNasin, Pharmacia), 12 μL ddH$_2$O.
13. MuLV-Reverse transcriptase (e.g., Superscript I, Gibco-BRL).
14. PCR-master mix A (volume per sample): 5 μL of 10X PCR buffer A, 2 μL of 2 mM dNTP mix, 2 μL 5' primer (10 pmol/μL), 2 μL 3' primer (10 pmol/μL), 0.5 μL *Taq* polymerase (2.5 U) and 36.5 μL ddH$_2$O.
15. PCR-master mix B (volume per sample): 5 μL of 10X PCR buffer A, 2 μL of 2 mM dNTP mix, 2 μL 5' primer (10 pmol/μL), 2 μL 3' primer (10 pmol/μL), and 37 μL ddH$_2$O.
16. *Taq* DNA polymerase (Pharmacia, or any other manufacturer).
17. Thermocycler and PCR tubes.

2.2. Quantitative RT-PCR Using a Multi-Competitor Standard

1. Multi-competitor plasmid e.g., pQM1.
2. 5X Transcription buffer: 200 mM Tris-HCl, pH 8.0, 40 mM MgCl$_2$, 250 mM NaCl, 10 mM spermidine.
3. Transcription mix (per sample): 5 μL of 5X transcription buffer, 1 μL of 0.75 M DTT, 1 μL of each 10 mM rNTP, 1 μL of 10,000 U/mL T7 polymerase, 1 μg DNA template, RNase-free water to 25 μL.
4. 10X RT buffer: 500 mM Tris-HCl, pH8.3, 750 mM KCl, 30 mM MgCl$_2$.
5. RT master mix B (volume per sample): 10 μL of 10X RT buffer, 5 μL of 10 mM each dNTP, 10 μL 0.1 M DTT, 1 μL of RNAsin (50 U/mL), 2 μL 100 pmol/μL N6-Random oligonucleotides, 2 μL of MuLV-Reverse transcriptase (400 U, Gibco-BRL), RNase-free water to 100 μL – volume of sample and competitor (i.e., so that when sample and competitor are added, final volume is 100 μL).
6. 10X PCR buffer B: 100 mM Tris-HCl, pH 8.4, 50 mM KCl, 15 mM MgCl$_2$.
7. PCR master mix C (volume per sample): 5 μL of 10X PCR buffer B, 1 μL of each 10 mM dNTP, 2.5 μL 5' primer (10 pmol/μL), 2.5 μL 3' primer (10 pmol/μL), 0.5 μL *Taq* polymerase (2.5 U), distilled water to 45 μL.

3. Methods

3.1. Primer-Dropping RT-PCR

3.1.1. RNA Extraction

Total RNA extraction from human brain tumors, normal brain tissues, and brain tumor cell lines is performed using a modified acid guanidinium isothiocyanate method *(5)*. However, other RNA extraction protocols that produce high quality RNA are also acceptable. One should include control tissue(s) or cell line(s) that can be included in the RT and PCR assays (*see* **Note 1**).

1. Grind fresh or freshly frozen brain tissue (~100 mg tissue) using a cold mortar and pestle. Liquid nitrogen is added periodically to the sample to keep the tissue frozen. Transfer the sample to a 12 mL polypropylene tube.
2. For tissue culture cells in plates, remove the media and add 1.8 mL guanidinium solution D with gentle pipeting (using a 1 mL pipet) to lyse the cells and shear the DNA (to reduce the viscosity). Transfer the sample to a 12 mL polypropylene tube and go to **step 4**.
3. For tissue samples, add 1.8 mL guanidinium solution D and homogenize the sample for 10 s using a tissue homogenizor (Note: clean the homogenizor probe thoroughly after processing each sample).
4. Add 100 µL 2 *M* sodium acetate (pH 4) and vortex briefly.
5. Add 2 mL water-saturated phenol and vortex briefly.
6. Add 0.2 mL chloroform: isoamyl alcohol mixture and vortex briefly.
7. Incubate on ice for 15 min.
8. Centrifuge for 10 min at 10,000g and transfer the aqueous phase to another 12 mL tube.
9. Add 1 vol of isopropanol and cool at –70°C for 10–15 min (or at –20°C for 1 h).
10. Centrifuge 30 min at 10,000g at 4°C (to pellet the RNA) and discard the supernatant.
11. Redissolve the whitish RNA pellet in 250 µL of guanidinium-solution D, transfer the solution to an Eppendorf tube, add 250 µL isopropanol and mix by inverting several times.
12. Incubate at –70°C for 10–15 min (or at –20°C for 1 h).
13. Centrifuge in a microfuge for 8 min at 6,000g (approx 8,000 rpm in a table top microfuge) and discard the supernatant.
14. Wash the RNA pellet with 70% ethanol and centrifuge for 3 min in a microfuge at 6,000g (8,000 rpm).
15. Carefully remove the ethanol and let the RNA pellet air dry for 5 min (or less).
16. Resuspend the RNA in a small volume of DEPC-water or distilled water (for example 20–50 µL).
17. Determine the RNA quantity and quality using a spectrophotometer. The 260 nm: 280 nm ratio should be between 1.7 and 2.0 (85% to 100% purity). If the purity is less than 80%, repeat the RNA extraction.
18. Dilute the RNA to approx 1–2 µg per µL.
19. Store RNA samples at –70°C until required.

3.1.2. Reverse Transcriptase (RT) Protocol for Synthesis of Template cDNA

1. Prepare RT-master mix A allowing for one extra sample.
2. Aliquot 18 µL of RT-master mix A into each PCR tube.
3. Add 1–4 µL total RNA (1–4 µg, depending on the quality/purity, *see* **Note 2**) to the appropriate tube.

4. Add 1 µL MuLV-Reverse-Transcriptase (200 U) and cap the tube. Final volume should be ~20 µL.
5. Place the tubes into a thermocycler and start the RT program (10 min at 20°C, 50 min at 42°C, 5 min at 95°C).
6. Store RT products (cDNA) at +4 or –20°C until required.

3.1.3. Basic Primer-Dropping PCR Protocol

The primer-dropping method involves the coamplification of the specific target (i.e., MMP and TIMP) and internal control GAPDH cDNAs in the same tube. However, both targets are not (usually) amplified throughout the entire PCR run. Each target is amplified using a distinct and predetermined number of PCR cycles. This then provides a means to control the overall amplification level within a reaction tube to minimize PCR product competition (*see* **Note 3**). The coamplification of the two targets is carried out by first starting the PCR using primers for the MMP or TIMP targets (which are present at relatively low levels within cells) to facilitate amplification up to some predetermined level. At which time the GAPDH primers, which have been withheld up to this point, are then dropped into the reaction to initiate amplification of the GAPDH target (which is present at high levels). For the balance of the PCR run, both targets are amplified simultaneously (both targets will now compete more or less equally for PCR substrates) in a semiquantitative manner.

1. Prepare enough PCR-master mix A (containing 5' and 3' primers for a MMP or TIMP target) allowing for one extra sample (*see* **Note 4**).
2. Aliquot 48 µL of PCR-master mix A into each PCR tube. Add 2 µL of template cDNA (i.e., RT products) to the appropriate PCR tube (*see* **Note 5**). (*See* below for more information regarding the precise amount of cDNA to use in subsequent PCRs.)
3. Place the PCR tubes into the thermocycler and start PCR program. Set the temperature cycles as follows: denaturation at 94°C for 1 min, annealing at 55°C for 30 s and elongation at 72°C for 1 min. Refer to **Table 1** for the appropriate number of PCR cycles to run for the specific MMP or TIMP targets.
4. Hot start method: Use PCR-master mix B (no *Taq*); wait until the thermocycler temperature reaches 85–94°C during the first denaturation step and press "PAUSE/STOP." During the pause, remove the tubes individually and add 0.5 µL (2.5 U) of *Taq* polymerase directly into the reaction mixture. Recap the tube and immediately replace into the thermocycler. After addition of *Taq* to all PCR tubes, press "START" to continue the PCR program (*see* **Note 6**).
5. With 22 cycles left to run, PAUSE/STOP the thermocycler when the temperature reaches 85–94°C and remove the PCR tubes from the thermocycler (again one at a time) and add 4 µL of a mixture of 5' and 3' GAPDH internal control primers directly into the reaction mixture. (Note: prepare enough GAPDH primer mixture ahead of time.) Following the addition of the GAPDH internal control primers to each tubes, press START to complete the remaining 22 cycles of the PCR program (*see* **Note 7**).

6. Following completion of the PCR amplification, load 10 µL of PCR products into a 1.8 to 2% agarose gel containing 0.2 µg/mL ethidium bromide and separate the two PCR products (target and GAPDH) by electrophoresis. Photograph the gel under UV illumination using Polaroid film.

7. Store PCR products at +4 or –20°C.

3.1.4. Equalizing the Final GAPDH Levels Between the Samples

In virtually all cases, the signal intensity of the GAPDH PCR-products will differ (sometimes significantly) from sample to sample because of uncontrollable variations in the quality and/or quantity of the RNA substrate. So the level of final GAPDH PCR-products are adjusted so they are equivalent (or nearly so) in each of the samples of a particular PCR run. Adjustments to equalize the GAPDH signals can be accomplished in three ways (in the order of preference) by altering the amount of: (1) PCR product loaded into the agarose gel, (2) RT product used in the PCR, and (3) RNA used in RT reaction. For slight differences in GAPDH abundance, adjusting the amount of PCR products loaded into the gels is usually sufficient. For very large differences, it is sometimes necessary to repeat the cDNA synthesis using more or less RNA in the RT reaction. **Figure 1** shows an example of an RT-PCR experiment before and after making adjustments to equalize the GAPDH signal levels (*see* **Note 8**).

3.1.5. Quantification

The Polaroid image of the agarose gel is captured by computerized digital imaging scanning and densitometry performed using the NIH Image program or the 1D gel analysis program of UVP Products, Cambridge. The GAPDH internal control signals are normalized first to some common value (for example, a value of 1). The specific-MMP or TIMP signal values are multiplied by the factor that the GAPDH differs from the common value. The values can then be plotted on graphs as shown in **Fig. 2A**.

3.1.6. Range Finding

The PCR cycle numbers for the MMP and TIMP primers shown in **Table 1** were determined empirically (in preliminary titration experiments) to coamplify with the GAPDH primers such that the overall amplification remains in the exponential phase of the reaction and thus produce products that are semiquantitative. If a different set of primers than those herein described are to be tested, the compatible PCR cycle numbers (that can be coamplified with GAPDH in a quantitative manner) can be determined in a "range finding" experiment. In this experiment, the number of PCR cycles used to amplify GAPDH is kept constant whereas the number of amplification cycles for the new set of primers is incrementally varied. We chose the cycle number(s) where

Table 1
MMP and TIMP Primer Sequences, PCR Product Size and PCR Cycle Numbers[a]

Target gene	Primer Sequence (5´– 3´)	Target accession number	Position	Product size (bp)	PCR cycle no.
MMP-2	GGCCCTGTCACTCCTGAGAT	J03210	1337–1356	474	28–30
	GGCATCCAGGTTATCGGGGA		1810–1791		
MMP-9	TGGACGATGCCTGCAACGTG	NM 004994.1	1554–1573	455	32-34
	GTCGTGCGTGTCCAAAGGCA		2008–1989		
TIMP-1	AGCGCCCAGAGAGACACC	X03124.1	33–50	670	25–26
	CCACTCCGGGCAGGATT		702–686		
TIMP-2	GGCGTTTTGCAATGCAGATGTAG	NM 003255.1	378–400	497	30–32
	CACAGGAGCCGTCACTTCTCTTG		874–852		
TIMP-3	CTTCTGCAACTCCGACATCGTG	NM 000362.1	389–410	459	22–23
	TGCCGGATGCAGGCGTAGTGTTT		847–825		
TIMP-4	AATCTCCAGTGAGAAGGTAGTTCC	NM 003256.1	212–235	512	31–32
	CGATGTCAACAAACTCCTTCCTGA		723–700		
MT1-MMP	GCCCATTGGCCAGTTCTGGCGGG	NM 004995.1	1178–1200	530	29–30
	CCTCGTCCACCTCAATGATGATC		1707–1685		
MT2-MMP	CGAGAACTGGCTGCGGCTTTATGG	NM 002428.1	207–230	794	33–35
	TGGGGAGAGGCTGGGTAGGCTGTG		1000–977		
MT3-MMP	GGAGATACCCATTTTGACTCAG	NM 005941.1	764–785	478	29–30
	ATCCATCCATCACCCTGTTGTT		1241–1220		

MT4-MMP	CGGTGTGCGGGAGTCTGTGTCT	NM 004141.1	862–883	498	29–30
	TGCCTCCTGTGGTCATCGTAGC		1359–1338		
MT5-MMP	ATGTTTCTCTTTAAGGATCGCT	NM 006690.1	1180–1201	428	40–45
	GTGTAATATCCTTCCTTGCTGA		1607–1586		
mMMP-2	CACCTACACCAAGAACTTCC	NM 008610.1	1318–1337	333	ND
	AACACAGCCTTCTCCTCCTG		1650–1631		
mMMP-9	TTGAGTCCGGCAGACAATCC	NM 013599.1	1579–1598	433	29–30
	CCTTATCCACGCGAATGACG		2011–1992		
mTIMP-1	CGCAGATATCCGGTACGCCTA	X04684	343–363	354	ND
	CACAAGCCTGGATTCCGTGG		696–677		
mTIMP-2	CTCGCTGGACGTTGGAGGAA	NM 011594.1	526–545	309	ND
	CACGCGCAAGAACCATCACT		834–815		
mTIMP-3	CTTGTCGTGCTCCTGAGCTG	L27424.1	330–349	244	ND
	CAGAGGCTTCCGTGTGAATG		573–555		
mTIMP-4	AATATCCAGTGAGAAGGTAGTCCC	na	na	511	30–35
	TGATGTCAACGTACTCCTTCCGG				
GAPDH	CGGAGTCAACGGATTTGGTCGTAT	M33197	78–101	307	22–23
	AGCCTTCTCCATGGTGGTGAAGAC		384–361		

[a]mMMP and mTIMP represents murine primer sequences; ND = not done for primer dropping RT-PCR; na = not available.

313

the signal intensity of the new PCR-product closely matches that of GAPDH. As an illustration of this process, **Fig. 2B** shows a range finding experiment for MT1-MMP primers.

1. Prepare enough PCR-master mix B including the new primers (20 pmol each 5' and 3' for each tube) for six tubes. In this case the master mix should include 6–12 µL of cDNA from a single RT reaction from tissue/cells that are known to express the target mRNA (preferably at levels that are in the middle range of expression); alter the volume of ddH$_2$O to take this into account.
2. Aliquot 48 µL of this PCR-master mix (i.e., including cDNA from RT) into five PCR tubes.
3. Set the thermocycler for 35 cycles (using the same cycling profile described above).
4. Place one tube into the thermocycler and start the PCR cycle program and add *Taq* using the hot start method.
5. Insert the other tubes after every third PCR cycle until there are none left (i.e., 32, 29, 26, 23 cycles remaining), each time adding *Taq* polymerase to the newly inserted tube.
6. When the thermocycler has 22 cycles remaining, PAUSE/STOP the thermocycler and add the GAPDH primers into each tube (as described above) and resume the PCR program.
7. Separate the PCR products by electrophoresis and photograph the gel.
8. Look for the signal intensity of the target that is similar to that of GAPDH (within one or two cycles) and use that cycle number for the new set of primers.

3.2. Quantitative RT-PCR Using a Multi-Competitor Standard

This method involves the coamplification of specific target and competitor cDNAs within the same tube. Both targets are amplified by the same primers, however the product amplified from the competitor differs in size from that of the sample, and so can be distinguished when run on a 2% agarose gel. By varying the amount of competitor although keeping the sample concentration constant it is possible to determine the amount of target mRNA within your starting RNA sample. This is the point where sample and competitor band intensities are equal (after normalizing for amplicon length, *see* **Note 11**).

The competitor RNA is derived from the pQM1 plasmid (for mouse sequences). This plasmid has two blocks of sequence corresponding to 5' and 3' primer sites for the indicated MMPs and other genes such as β-actin, TNFα and GAPDH constructed either side of a short linker region *(2)*. In **Fig. 3**, the upper block contains the sequence information corresponding to 5'-(forward) primers for murine MMP-3, -10, -11, -2, -9, -13, -7, -12 and -14. The lower block has the 3'-(reverse) primer sequences laid out in the same order. **Table 2** details these primer sequences. A "sense" copy of this region is generated using T7 RNA polymerase, and then varying concentrations of this synthetic RNA

	Distance from 3' end	Length of cellular cDNA	Length of standard cDNA	C content of cellular cDNA	C content of standard cDNA
Stromelysin 1	75	204	292	81	141
Stromelysin 2	28	235	291	91	142
Stromelysin 3	33	216	288	113	140
72kDa gelatinase	27	207	288	93	139
92kDa gelatinase	63	209	288	93	137
Collagenase 3	26	237	288	92	139
Matrilysin	35	215	288	97	138
MME	45	215	291	86	136
MT-MMP	62	221	292	134	141
B-actin	35	207	292	99	136
G3PDH	39	198	292	97	138
TNF	33	203	292	105	135

LINKER

Stromelysin 1
Stromelysin 2
Stromelysin 3
72kDa gelatinase
92kDa gelatinase
Collagenase 3
Matrilysin
MME
MT-MMP
B-actin
G3PDH
TNF

Fig. 3. Organization of the mouse MMP quantification amplicon in the vector pQM1. The figure shows the order of the target sequences for various MMP and other genes, and the sizes of the corresponding products from both the competitor and the natural RNA templates.

Table 2
Mouse MMP Primer Sequences Used
for the Multi-Competitor RT-PCR Method

Name	Sequence (5' – 3')
MMP-3	GAAGAAGATCGATGCTGCCA AGATCCACTGAAGAAGTAGAG
MMP-10	GCTGTTTTTGAAAAGGAGAAGAA AGTATGTGTGTCACCGTCCT
MMP-11	AGACTATTGGCGTTTCCACC AAAGAAGTCAGGACCTACGG
MMP-2	TGCAGGAGACAAGTTCTGGA GCTTCCAAACTTCACGCTCT
MMP-9	AAGGCTCTGCTGTTCAGCAA GTCCACCTTGTTCACCTCAT
MMP-13	GCGGTTCACTTTGAGAACAC GCATGACTCTCACAATGCGA
MMP-7	CTGGGCCAGGCCTAGGC TGCCTGCAATGTCGTCCTTT
MMP-12	GTCTTTGACCCACTTCGCC ACGGAACAGGGGGTCATATT
MMP-14	AGAAGCTGAAGGTAGAGCCA TGAAGAAGAAGACAGCGAGG

(generally ranging from 1–100 pg) are added to a known amount (generally 1 µg) of total RNA from the cell or tissue source.

1. It is necessary either to generate a multicompetitor plasmid customized to the genes of interest, or obtain a ready-made construct *(2)*. The mouse quantification vector pQM1 (*see* **Fig. 3**) that we have used was obtained from British Biotechnology, Oxford, UK. The competitor plasmid was constructed in the same manner as the rat quantification vector previously described in **ref. *2***. (*see* **Note 12**).
2. Prepare competitor RNA from the quantification vector cDNA: Prepare the transcription mix with 1 µg template DNA and incubate at 37°C for 30 min. Treat

with 10 U of DNase 1 per 1 µg DNA template used; extract with phenol:chloroform, precipitate with sodium acetate and quantitate RNA.

3. Prepare RT-master mix B allowing for one extra sample.
4. Aliquot RT-master B mix into each PCR tube.
5. Add 1 µg of total RNA (sample) and one concentration of competitor (from 1 pg to 100 pg) to the appropriate tube.
6. Place tubes into a thermocycler and start the RT program (37°C for 60 min).
7. Store RT products (cDNA) at –20°C until required.
8. Prepare PCR-master mix C allowing for one extra sample
9. Aliquot 45 µL PCR-master mix C into each tube.
10. Add 5 µL cDNA (RT product) to the appropriately labeled PCR tube.
11. Place tubes into the thermocycler and start the program (one cycle at 94°C for 1 min; then 26 cycles of: 95°C for 30 s, 57°C for 30 s, 72°C for 2 min; then one cycle at 72°C for 10 min) (*see* **Note 13**).
12. Following completion of the PCR amplification, load 10 µL of the PCR products into a 2% agarose gel containing 0.2 mg/mL ethidium bromide. Separate the two PCR products (target and competitor) by electrophoresis and scan the gel using "Grab-it" Version 2.5 (UVP Products, Cambridge, UK) or equivalent (*see* **Fig. 4A**).
13. Store PCR products at –20°C.
14. Densitometry of the scanned gel can be performed using the 1D gel analysis program (UVP Products, Cambridge, UK) or as described for the primer-dropping method. The ratios of competitor to sample band densities are calculated, after correcting densities for band length, and plotted against competitor concentration (*see* **Fig. 4B**). The amount of sample mRNA present in the original 1 µg total RNA can now be calculated from this graph; this is the point where the competitor/sample ratio is 1. This value can be used for comparison between different samples. Alternatively the transcript copy number can be calculated from this value and can be used for comparison between samples.

4. Notes

1. When extracting RNA from the test specimens, always include one or more external standard controls such as normal tissue or cells. In addition, when possible, we prefer to run controls that provide a low and a high level of target. Because the efficiency of PCR reaction can vary considerably between runs, the external controls will provide information regarding the efficiency of the PCR reaction.
2. In our experience, RNA quality (and to a lesser degree, quantity) is the most important factor for successful RT-PCR. Consistently high quality RNA will eliminate many of the steps necessary for optimization of the reactions. However, it is not often possible to control the quality of the clinical samples due to differences in sample handling procedures and thus the RT-PCR methodology will require some adjustments to obtain satisfactory results. For poor quality RNA, add more to the RT reaction and reduce the volume of the water accordingly.

A

Competitor (288nucleotides) -   100pg   50pg   25pg   10pg   1pg

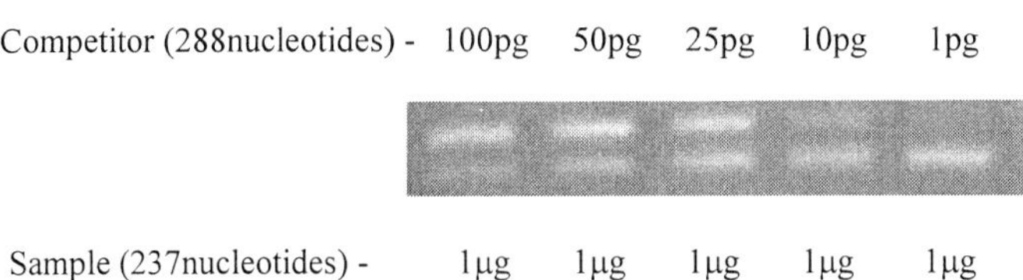

Sample (237nucleotides) -   1µg   1µg   1µg   1µg   1µg

B

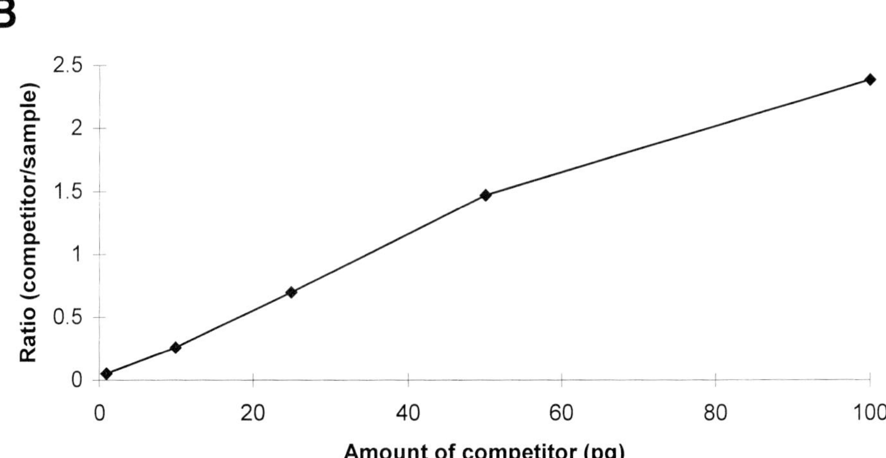

Fig. 4. Competitor quantification of mouse collagenase-3 in total RNA from mouse skin. (**A**) 2% agarose gel showing separation of competitor (upper) and sample (lower) RT-PCR bands. The competitor is the mouse quantitation vector pQM1, while the sample is total RNA isolated from mouse skin. Primers used were for mouse collagenase-3. (**B**) Graph showing the competitor/sample ratio (after amplicon length correction) plotted against amount of competitor added. This indicates that the mouse skin has approx 35 pg collagenase-3 mRNA per µg total RNA.

3. Simultaneous amplification of more than one target is usually accompanied by a competitive event for substrates of the PCR reaction causing an attenuation of one of the products. For example, if both sets of primers (GAPDH and the MMP or TIMP) were included at the beginning of the PCR run, the GAPDH product would dominate the reaction because of its high abundance in cells (which provides it a competitive advantage). Indeed, GAPDH is often expressed at up to 1000-fold higher levels than most other transcripts and will require 21 to 23 cycles (depending upon the RNA quality/quantity) for detection on gels, and yet remain within the exponential phase (i.e., quantitative range) of the amplification curve. In contrast, MMP and TIMP mRNAs are expressed at considerably lower levels

than GAPDH and thus require greater numbers of PCR cycles (e.g., 23–34 cycles) for detection at similar levels. Optimal PCR cycle numbers to run the MMP and TIMP primer sets have been determined empirically by range-finding or titration experiments (as described in the text) and are provided in **Table 1.**

4. For ease of handling, we suggest (at least at the beginning) working with a limited number of samples (e.g., 3–8). After becoming more experienced, the number of samples can increase, but it is not advisable to assay more than about 14–16 samples in a single PCR run because of the increase in complexity in equalizing the GAPDH signals among the samples (see below regarding equalizing the GAPDH levels).

5. For thermocyclers that do not have heated lids, we use PCR tubes supplied by Sarstedt. These tubes have inserts that reduce the interior tube volume and thus do not require mineral oil overlays. Furthermore, their larger size provides for ease of handling during the hot-start and primer-dropping procedures. With many thermocyclers the lid will not close completely due to the larger size of the Sarstedt tubes but we have not encountered any difficulties with this arrangement. For thermocyclers with heated lids and smaller tubes, such as the fragile thin-walled 0.2 mL tubes, primer-dropping PCR can still be performed, but tubes should be opened carefully to prevent damage to the tubes and to inhibit possible cross contaminating aerosols.

6. Performing hot-start PCR increases the specificity of primer-mediated amplification and thus reduces spuriously primed PCR products.

7. Although the primer-dropping method involves the opening of PCR tubes during the amplification step, we have not experienced cross contamination due to aerosols. This could be due to several factors. First, because amplification of RNA targets requires significantly fewer PCR cycles (because cells possess multiple RNA copies) than genomic DNA amplification, minute levels of possible aerosol contamination are not detected at this low level of amplification. Second, the PCR tubes are opened during the early stages of the PCR run and hence contain extremely low levels of potentially contaminating PCR products. However, it is absolutely essential to discard the pipet tips following each operation to prevent carry over contamination by this method.

8. Once the optimal amount of target cDNA (to produce similar levels of GAPDH PCR-product in all the samples) has been determined for a set of samples, the same volumes can be used for the analysis of any other mRNA that may be expressed in the sample cells.

9. We have found that some specimens express MMP and TIMP mRNAs that are significantly below the level of detection whereas others express so much that the final products have reached plateau and therefore are outside the quantifiable range. In this case, the number of PCR cycles can be either increased or decreased, respectively, to bring the products into a range that can be visualized and measured.

10. Proper PCR technique should be used throughout the procedures. That is, be sure to discard all pipet tips after each use. If available, use aerosol-resistant tips.

Keep the final PCR products away from the PCR set-up area to reduce the possibility of cross contamination. Finally, it is important to use dedicated pipet for sample preparation and for PCR product analysis.

11. Band intensity is dependent upon amplicon length. Therefore the intensity values need to be normalized for their size—divide intensity by amplicon length.

12. Generation of the competitor was achieved by stepwise annealing and ligation of oligonucleotides correponding to the indicated genes. A 10bp linker region was inserted between the forward and reverse primers to create a competitor which when amplified produces an amplicon differing from target by 55–90bp. A poly A^+ tail was inserted at the 3' end to allow the RNA synthesized from this region to be used in oligo-dT primed RT reactions as well as the random hexamer-primed procedure that we have described. The competitor was amplified by PCR and cloned into a pGEM-T vector (Promega) downstream of a T7 promoter.

13. The number of cycles used was determined by preliminary range finding experiments to determine where the reaction is still within the exponential phase.

References

1. Wong, H., Anderson, W. D., Cheng, T., and Riabowol, K. T. (1994) Monitoring messenger RNA expression by polymerase chain-reaction—the primer dropping method. *Anal. Biochem.* **223**, 251–258.

2. Wells, G. M. A., Catlin, G., Cossins, J. A., Mangan, M., Ward, G. A., Miller, K. M., and Clements J. M. (1996) Quantitation of matrix metalloproteinases in cultured rat astrocytes using the polymerase chain reaction with a multi-competitor cDNA standard. *Glia.* **18**, 332–340.

3. Wang, A. M., Doyle, M. V., and Mark, D. F. (1989) Quantitation of messenger RNA by the polymerase chain reaction. *Proc. Natl. Acad. Sci.* **86**, 9717–9721.

4. Robinson, M. O. and Simon, M. I. (1991) Determining transcript number using the polymerase chain-reaction—PGK-2, MP2, and PGK-2 transgene messenger RNA levels during spermatogenesis. *Nucleic Acid Res.* **19**, 1557–1562.

5. Chomczynski, P. and Sacchi, N. (1987) Single-step method of RNA isolation by acid guanidinium thiocyanate phenol chloroform extraction. *Anal. Biochem.* **162**, 156–159.

16

Measuring Transcription of Metalloproteinase Genes

Nuclear Run-Off Assay vs Analysis of hnRNA

Anne M. Delany

1. Introduction

Many metalloproteinase genes are transcriptionally regulated by differentiation, growth factors, cytokines, and hormones. Nuclear run-off assays and analysis of heterogeneous nuclear RNA (hnRNA) levels are two methods for studying changes in gene transcription. In this chapter these two techniques will be outlined and compared.

1.1. Nuclear Run-Off Assay

A nuclear run-off assay is probably the best method for measuring gene transcription. In this procedure, nuclei are isolated from cells at a particular stage of differentiation or from cells treated with or without a test agent. Transcripts that were initiated at the time the nuclei were harvested are labeled in vitro using [^{32}P]-UTP. The labeled RNA is purified and hybridized to immobilized cDNA corresponding to the gene(s) of interest, and detected by autoradiography. The major advantage to this procedure is that it detects only changes in transcriptional activity, not changes in processing of RNA or RNA stability *(1,2)*.

Although the nuclear run-off assay appears to be the "gold standard" for measuring gene transcription, there are several drawbacks to the procedure. First, since the assay requires a large number of cells as starting material, it can be expensive or impossible to perform if cells or tissue are very limited. Second, the assay requires the manipulation of a significant amount of radioactivity, which can often lead to contamination of laboratory equipment. Third, it may be difficult to evaluate small changes (i.e., twofold) in gene transcription or measure the activity of genes with a low level of transcription. Because

From: *Methods in Molecular Biology, vol. 151: Matrix Metalloproteinase Protocols*
Edited by: I. Clark © Humana Press Inc., Totowa, NJ

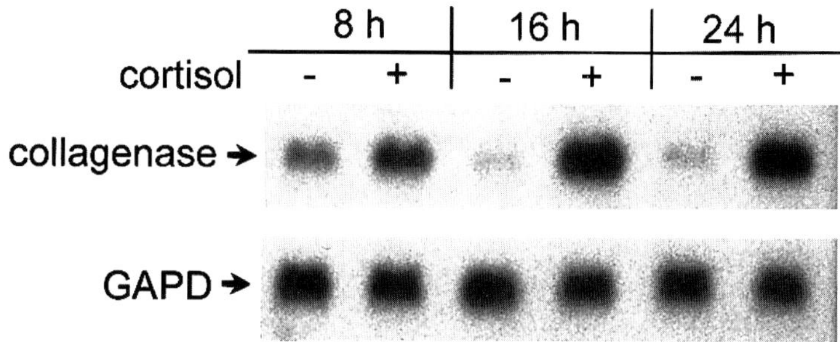

Fig. 1. Cortisol increases collagenase-3 mRNA in primary rat osteoblastic cells. Northern blot analysis of total RNA from cultures treated for 8, 16, or 24 h with or without 1 μ*M* cortisol. The blot was probed for rat collagenase-3 in the *top panel* and for rat GAPD in the *bottom panel*.

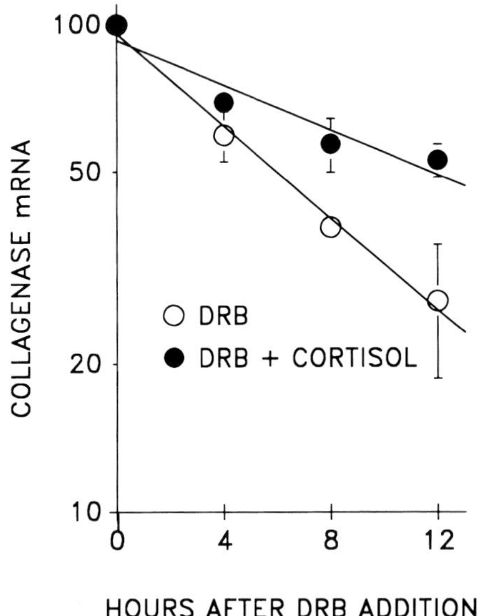

Fig. 2. Cortisol stabilizes collagenase-3 mRNA in transcriptionally arrested primary rat osteoblastic cells. Cultures were exposed to 1 μ*M* cortisol or to control medium for 4 h prior to the addition of 75 μ*M* DRB, and RNA was isolated at selected times after DRB addition. Collagenase-3 mRNA levels were determined by Northern blot analysis. Values are mean ± standard error for triplicate cultures. The slope for DRB = –0.052, and the slope for DRB + cortisol = –0.022 (significantly different, $p < 0.01$)

of these disadvantages, the analysis of hnRNA by reverse transcriptase-polymerase chain reaction (RT-PCR) may be considered.

1.2. hnRNA Analysis by RT-PCR

hnRNA represents newly transcribed, unspliced RNA and is generally a very small percentage of total RNA. hnRNA levels are affected by the rate of gene transcription, RNA processing, and nuclear RNA stability. Analysis of hnRNA can be a reasonable surrogate for the nuclear run-off assay if the treatment of interest does not affect RNA processing or nuclear RNA stability *(3,4,5)*. Reports of growth factors, cytokines, and hormones that differentially affect RNA processing are limited. However, there is increasing evidence for the differential regulation of RNA stability by these factors, and agents that affect cytoplasmic RNA stability may also affect stability of hnRNA (*6*, Delany, unpublished observations).

One example of this is seen in primary rat osteoblastic cells, where the glucocorticoid cortisol stimulates collagenase-3 mRNA levels seven- to tenfold after 16 or 24 h of treatment *(7)* (**Fig. 1**). To examine the effect of cortisol on collagenase-3 mRNA stability, the RNA polymerase II inhibitor 5, 6-dichlorobenzimidizole riboside (DRB) was used. Cells were cultured in control medium or in the presence of cortisol for 4 h, then exposed to DRB for up to 12 h (**Fig. 2**). In the presence of DRB alone, the half-life of collagenase-3 mRNA is ~6 h, but in the presence of DRB and cortisol collagenase-3 mRNA half-life is ~12 h. This experiment shows that cortisol significantly increases the stability of collagenase-3 mRNA in osteoblasts. Nuclear run-off assays show that cortisol does not affect or modestly decreases collagenase-3 gene transcription, and these results are supported by data obtained from the transient transfection of osteoblasts with a collagenase-3 promoter-luciferase gene construct (**Figs. 3** and **4**) *(7)*. However, analysis of collagenase-3 hnRNA by RT-PCR shows that cortisol increases collagenase-3 hnRNA two- to threefold (**Fig. 5**). Given these data, it is likely that collagenase-3 hnRNA is also stabilized by cortisol and, in this case, analysis of hnRNA can not be used as a surrogate for the nuclear run-off assay.

A preliminary experiment evaluating the effect of a test agent on the stability of a specific RNA will help in choosing a method for measuring gene transcription *(5,7)*. If analysis of hnRNA by RT-PCR is chosen, it has the advantage of using very little starting material. In addition, this method can be sensitive and little radioactivity is required. Once the RT-PCR assay is set up, data can be generated easily, rapidly, and reproducibly.

2. Materials

2.1. Nuclear Run-Off

All reagents, tubes, tips, and equipment should be as RNase-free as possible (*see* **Note 1**).

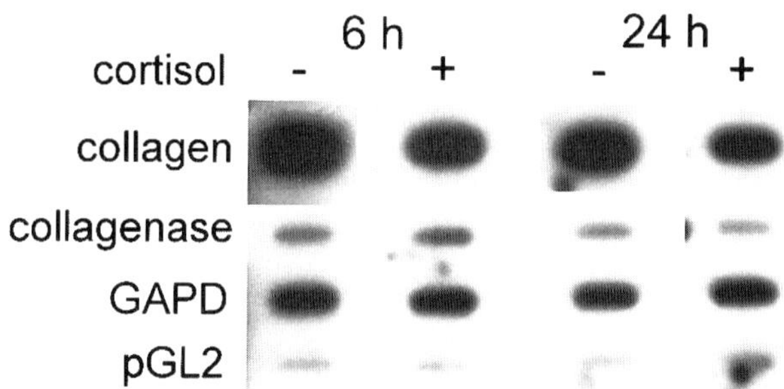

Fig. 3. Nuclear run-off assay on nuclei isolated from primary rat osteoblastic cells cultured in the presence or absence of 1 μM cortisol for 6 or 24 h. In vitro ^{32}P-UTP labeled RNA was hybridized to immobilized cDNA for α1(I) collagen, glyceralde-hyde-3-phosphate dehydrogenase (GAPD), and collagenase-3. pGL2-Basic vector DNA was used as a control for nonspecific hybridization.

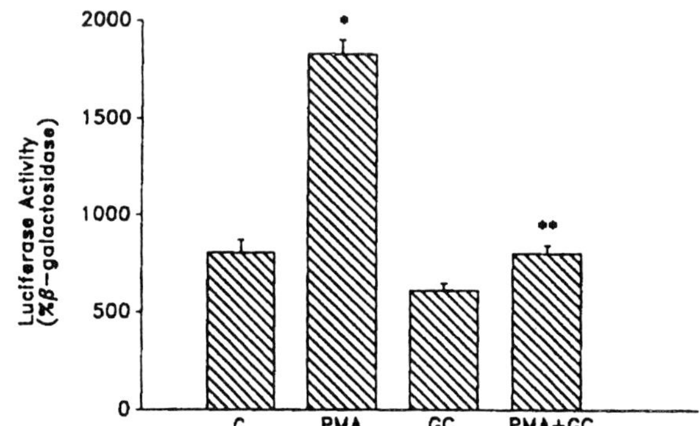

Fig. 4. Rat collagenase-3 promoter activity in transiently transfected rat osteoblastic cells treated for 6 h with cortisol at 1 μM and PMA at 0.1 μM. A 2.1 kb fragment of the rat collagenase-3 promoter was used to drive expression of the luciferase gene in the promoterless reporter plasmid pGL2-Basic. Osteoblastic cells were transiently trans-fected with the collagenase-3 promoter construct and a plasmid that constitutively expresses the β-galactosidase gene. Transfected cells were treated for 6 h with control medium (C), or with medium containing cortisol (GC), phorbol myristate acetate (PMA) or PMA plus cortisol. Luciferase activity was normalized to β-galactosidase activity. *, significantly different from C, $p < 0.01$; **, significantly different from PMA, $p = 0.01$.

1. Refrigerated low speed centrifuge; slot blot or dot blot apparatus.
2. Dounce homogenizers, 30 mL type, siliconized and baked.
3. Phosphate buffered saline (autoclaved).
4. RLW (rinsing/lysing/washing) solution: 10 m*M* Tris-HCl, pH 7.5, 10 m*M* NaCl, 3 m*M* MgCl$_2$ (autoclaved).

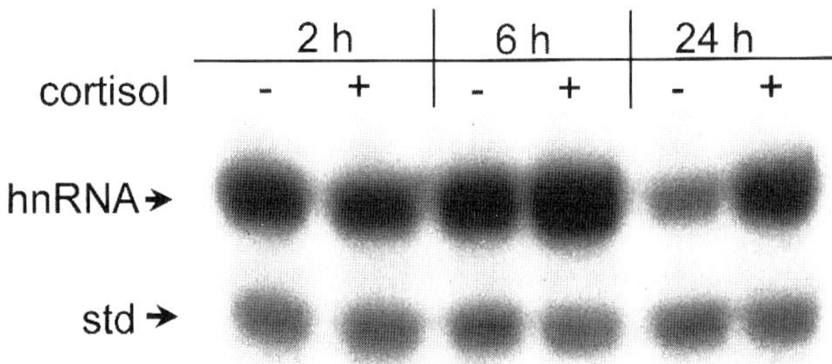

Fig. 5. RT-PCR analysis of collagenase-3 hnRNA in primary rat osteoblastic cells cultured in the presence or absence of 1 μM cortisol for 2, 6, or 24 h. 1 µg of total RNA was treated with DNase I, reverse transcribed using MMLV-reverse transcriptase, and amplified by PCR using a sense primer corresponding to nucleotides 4–23 of exon 1 and nucleotides 61–80 of intron 1, yielding a 186 base pair product. The ~150 base pair standard was synthesized by low stringency annealing PCR using the rat collagenase hnRNA primer set and pGL2-Basic as a template. The newly synthesized cDNA and 0.05 attomole standard DNA were PCR amplified in the presence of α-$[^{32}P]$-dCTP using 24 cycles of 94°C for 1 min, 59°C for 1 min, and 72°C for 1 min. Aliquots of the reaction products were fractionated on a 6% polyacrylamide gel containing 100 mM Tris-borate and 1 mM EDTA, and visualized by autoradiography.

5. RLW containing 0.5% Nonidet-P40 (filtered).
6. Nuclei storage buffer: 5 mM MgCl$_2$, 10 mM Tris-HCl, pH 7.5, 0.5 M sorbitol, 2.5% Ficoll (MW~400,000), 50% glycerol, 5 mM DTT. Make this with RNase-free reagents, as it will be too viscous to filter. Add the DTT fresh at the time of use.
7. 2× nuclei labeling mix: 20 mM Tris-HCl, pH 8.0, 5 mM MgCl$_2$, 300 mM KCl, 10 mM DTT, 1 mM each ATP, CTP, and GTP, 20% glycerol. Make this with RNase-free reagents. Add the DTT fresh at the time of use.
8. 10× proteinase K mix: 20 mg/mL proteinase K, 100 mM Tris-HCl, pH 8.0, 50 mM EDTA, 5% SDS. Make this fresh at the time of use.
9. Water saturated (acid) phenol; chloroform:isoamyl alcohol (49:1).
10. 2 M sodium acetate, pH 4.0; 3 M sodium acetate, pH 5.3.
11. 10 M ammonium acetate; 10 mM EDTA.
12. α-$[^{32}P]$-UTP (800 Ci/mM).
13. Yeast tRNA (10 mg/mL), RNase-free DNase I, RNasin (Promega) or another RNase inhibitor.
14. Nylon or nitrocellulose membrane, cDNAs.
15. Buffer saturated phenol.

2.2. hnRNA Analysis by RT-PCR

1. DNA thermocycler.
2. Appropriate primers, target RNA isolated by any standard technique (*see* **Notes 1** and **5**).

3. Amplification grade DNase I, *Taq* polymerase, RNasin (or another RNase inhibitor), MMLV-reverse transcriptase.
4. 10X DNase I buffer: 200 m*M* Tris-HCl, pH 8.4, 20 m*M* MgCl$_2$, 500 m*M* KCl (usually supplied with the DNase I).
5. 10× *Taq* polymerase buffer: 200 m*M* Tris-HCl, pH 8.4, 500 m*M* KCl (usually supplied with the *Taq*).
6. 10 mM dNTP mix: 10 m*M* each of dCTP, dATP, dGTP, dTTP.
7. 100 m*M* DTT.
8. 50 m*M* MgCl$_2$.
9. α-[^{32}P]-dCTP or α-[^{32}P]-dATP (any specific activity).
10. Typical reagents for polyacrylamide gel electrophoresis and agarose gel electrophoresis.

3. Methods

3.1. Nuclear Run-Off

3.1.1 Preparation of cDNAs Immobilized on Membranes (see **Note 2**)

1. Clean the slot blot or dot blot apparatus very well, to remove any DNA or RNA contamination. We use Absolve (DuPont).
2. Linearize cDNAs with the appropriate restriction enzymes. Extract the DNA with buffer saturated phenol:chloroform:isoamyl alcohol, and ethanol precipitate. Cut enough DNA so that approx 1 μg of insert DNA will be used per slot. Cut extra DNA to allow for loss in the extraction and precipitation.
3. Resuspend the DNA in TE and denature it by adding an equal volume of 0.5 *M* NaOH/1 *M* NaCl. Incubate at room temperature for 10 min and then put it on ice. Dilute the DNA to the desired concentration with 0.1X SSC/125 m*M* NaOH.
4. Equilibrate the nylon or nitrocellulose membrane with buffer according to the manufacturer's directions. We use the nylon membrane GeneScreen Plus (DuPont) for Northern blot analysis and nuclear run-off assays. Assemble the blotting apparatus and slot or dot blot the DNA as instructed by the manufacturer. Neutralize the membrane with 0.5 *M* Tris-HCl, pH 7.5/0.5 *M* NaCl, and fix the DNA to the membrane by UV-cross linking or as instructed by the manufacturer. The membrane can be stored dry for several months.

3.1.2. Isolation of Nuclei

The procedure outlined here is for adherent cultured cells, but it can be easily modified for use with nonadherent cells (*1,2,8,9*). In general, the best results are obtained when at least 5 million nuclei are used in the reaction, so prepare a comparable number of cells. It is best to keep the cells cold while the nuclei are being harvested, either by keeping the flasks or plates on ice or by performing the procedure in a cold room. Solutions and Dounce homogenizers should be prechilled on ice, and the nuclei should be transported on ice to the refrigerated centrifuge. It is important to work quickly to get the nuclei from the cells and into the −70°C freezer in as little time as possible.

1. Wash the cells twice with PBS, followed by one wash with RLW buffer. Dab the edge of the plate on a towel to remove as much of the RLW as possible.
2. Pipet on RLW containing NP-40 onto each plate and scrape thoroughly, pooling into Dounce homogenizers. Use enough solution so that the total volume of the harvested cells is 25–30 mL.
3. Dounce on ice. The number of passes needed and which pestle to use will depend on the type of cells (*see* **Note 3**.) Check the lysis of the cells under the microscope. Mix a droplet of trypan blue and a droplet of nuclei on a slide; the nuclei should look like blue ovals. It is important to lyse the cells but not the nuclei, so Dounce "less rather than more." If the nuclei have some clinging cytoplasmic debris on them, this will often clean up when the nuclei are washed.
4. Pour the homogenate into 50 mL sterile culture tubes and pellet the nuclei gently at 4°C (~5 min at ~500–1000g). It is important not to squash the nuclei. The supernatant from this step contains the cytoplasmic RNA.
5. Pipet off the supernatant, saving some for the isolation of cytoplasmic RNA, if desired (*see* **Note 4**). To isolate the cytoplasmic RNA, immediately add 1 mL 10× proteinase K to every 10 mL of supernatant saved. Incubate at 55°C for ~1 h, this can then be frozen at –70°C and extracted later *(8,9)*. For the extraction, add 1/10 volume of 2 *M* sodium acetate, pH 4.0 and an equal vol of 3 parts water saturated phenol, 1 part chloroform:isoamyl alcohol. Mix well and centrifuge at 10,000g for 15 min at 4°C to separate the phases. Using the water saturated (acid) phenol will make sure that any contaminating DNA is left in the organic phase. Remove the aqueous phase and add 1/10 vol of 3 *M* sodium acetate and an equal volume of isopropanol. Incubate at –20°C for several hours and pellet the RNA at 4°C for 30 min at ~10,000g. Resuspend the pellet in ~300 µL 10 m*M* EDTA and reprecipitate in a microfuge tube using 1/10 vol of 3 *M* sodium acetate and 2 vol of ethanol. Incubate at –20°C for several hours to overnight and pellet the RNA in the 4°C microfuge. Wash the RNA with 70% ethanol and resuspend in water or buffer as desired.
6. Gently resuspend the nuclear pellet in 10 mL RLW. Pellet the nuclei gently at 4°C, as in **step 4**. This washes out the detergent and helps clean up the nuclei.
7. Gently resuspend the nuclei in ~500 µL nuclei storage buffer plus ~10 µL RNasin (or your favorite RNase inhibitor). Store at –70°C as soon as possible. If the nuclear pellet is big and difficult to resuspend, use more buffer.

3.1.3. Run-Off Reactions and Hybridization

1. Prehybridize the membranes in plastic heat sealable bags at 42°C. For hybridization of GeneScreen Plus membranes we use 50% formamide, 5X SSPE, 5X Denhardt's solution, 0.4% SDS, and 100 µg/mL salmon sperm DNA. Use a small volume of solution, approx 4 mL per filter strip.
2. Thaw nuclei on ice and count an aliquot of each. It is ideal to start with approximately equal numbers of nuclei in each reaction, and to use as many as possible.
3. Aliquot the nuclei into microfuge tubes with screw caps and sealing rings. The use of these tubes helps contain the radioactivity in subsequent steps. Pellet

the nuclei gently in the 4°C microfuge (3–4 min at 500g). Pipet off and discard the supernatant.

4. To the nuclei, add an approximately equal volume of the 2X labeling mix plus 10 µL RNasin and pipet gently to resuspend the nuclei. It is sometimes necessary to use a pipet tip with the end cut off to avoid shearing the nuclei. For a typical osteoblast experiment, the original volume of the nuclei can range from 250–500 µL. Add 250 µCi [^{32}P]-UTP (25 µL, 800Ci/mmol) to each tube and mix gently. Incubate at room temperature for 30 min.

5. Add at least 1000 U of RNase-free DNase I per tube and mix well. Incubate at 37°C for 15 min, mixing again half way through the incubation. The reaction mixture should become much less viscous. If not, add additional DNase I and incubate for an additional 10 min. We use Gibco DNase I, Cat. no. 18047-019.

6. Add 20 µL 10 mg/mL yeast tRNA to each tube to use as a carrier.

7. Add 1/10 reaction volume of 10X proteinase K mix and incubate at 55°C for 30 min, mixing halfway through the incubation. After proteinase K treatment the reaction mix should no longer be viscous. If necessary, add more proteinase K and incubate an additional 10 min.

8. To each reaction, add 1/10 vol of 2 M sodium acetate, pH 4.0, and an equal volume of three parts water saturated (acid) phenol, one part chloroform:isoamyl alcohol. The use of acid phenol will make any residual DNA segregate into the organic phase. Mix well and centrifuge 4–5 min to separate the phases.

9. Re-extract the aqueous phase with an equal volume of chloroform:isoamyl alcohol.

10. Remove the aqueous phase to a fresh tube and precipitate the RNA by adding ammonium acetate to a final concentration of 2.5 M (use a 10 M stock) and 2 vol of ethanol. The use of ammonium acetate decreases the co-precipitation of the unincorporated nucleotides. Incubate at –70°C for 30–60 min, then pellet the RNA at 4°C for 20–30 min. The pellet should be easy to see due to the carrier tRNA. Remove the supernatant with care, as it contains most of the unincorporated [^{32}P]-UTP.

11. Resuspend the RNA pellet in ~300 µL 10 mM EDTA and reprecipitate with ammonium acetate (2.5 M final) and two volumes of ethanol. Incubate at –70°C and pellet the RNA as in **step 10**.

12. Resuspend the final pellet in ~150 µL 10 mM EDTA. Count an aliquot of each sample in the scintillation counter. Heat the RNAs at 95°C for 1 min then cool on ice before adding them to the hybridization bags. Try to use approximately the same number of counts of labeled RNA per bag, but use as many counts as possible.

13. Allow the hybridization to proceed for 48–72 h, then wash the blots using the same stringency conditions as those normally used for Northern blots. For example, we use two 15 min washes at room temperature in 2X SSC/0.1% SDS, followed by two 15 min washes at 65°C in 1X SSC/0.1 % SDS. It may be necessary to shorten the time of the highest stringency wash if survey of the membranes with a Geiger counter suggests that there are not many counts. Expose the membranes to X-ray film, but keep the membranes moist so that additional washes can be performed after the first exposure if there is too much background.

3.2. hnRNA Analysis by RT-PCR

3.2.1. Primer Design

Since hnRNA is unspliced RNA, it is optimal to use a primer set in which one primer corresponds to intronic sequence and one primer corresponds to exonic sequence. If this information is not available, it may be necessary to sequence into an intron of a genomic clone. It is not necessary to sequence very far into the intron; a hundred bases or so should be sufficient. Design the primers so that the amplification product is between 150–500 base pairs. The primers should be designed taking into consideration the major rules of PCR primer design. For example, the 2 primers should have similar melting temperatures, should not be self complementary or complementary to each other (*see* **ref. 10** for a discussion of primer design).

3.2.2. RT-PCR Reactions

The reaction volumes given here are for 1 RNA sample. For multiple RNA samples, make a master mix of the buffers and enzymes and add an aliquot of this mix to each sample tube. This decreases the variability between the reactions and the potential for contaminating stock solutions. Each RNA sample will be treated with DNase I, to remove any contaminating genomic DNA. The reverse transcription reaction, performed using the 3' primer, will make a corresponding cDNA strand. Following addition of the 5' primer, PCR will amplify a DNA fragment corresponding to a portion of the hnRNA of interest.

1. DNase treatment. We use amplification grade DNase I (Gibco Cat. no. 18068-015). In a 0.5 mL microfuge tube, add:

RNA	8 µL (0.5 or 1 µg total)
10X DNase I buffer	1 µL
DNase I	1 µL
100 mM DTT	0.2 µL
RNasin	0.2 µL

Incubate 15 min at room temperature, then inactivate the enzyme by adding 1 µL 20 mM EDTA and incubating at 65°C for 10 min.

2. Reverse transcriptase reaction. To the DNase treated RNA sample, add:

100 mM DTT	2 µL
10X PCR buffer	2 µL
10 mM dNTP mix	2 µL
50 mM MgCl$_2$	1.2 µL
15 µM 3' primer	1 µL
RNasin	0.5 µL
MMLV-RT	1 µL

Incubate at room temperature for 10 min, then incubate at 42°C for 30 min. Inactivate the RT by incubation at 95°C for 10 min.

3. PCR reactions (*see* **Note 6**). On ice, add to the reverse transcribed sample above:

10× *Taq* buffer	8 μL
50 m*M* MgCl$_2$	1.2 μL
15 μ*M* 5' primer	1 μL
internal standard	1 μL (*see* **Subheading 3.2.4.**)
[^{32}P]-dCTP or dATP	0.5 μL or less
Taq polymerase	0.4 μL
water	67 μL

The radioactive nucleotide is added as a tracer, so very little is needed, and it need not be the freshest batch. The annealing temperature of the PCR reaction will depend on the melting temperature of the primers and the PCR reaction conditions may be modified to decrease false priming. An example PCR is 22 main cycles of 94°C, 1 min; 59°C, 30 s; 72°C, 1 min.

3.2.3. Electrophoresis of PCR Products

Mix an aliquot (10–20 μL) of the PCR products with gel loading buffer and fractionate on a polyacrylamide gel (denaturing or native). We use 8% polyacrylamide/6 *M* urea in 1X TBE (100 m*M* Tris, 90 m*M* boric acid, 1 m*M* EDTA) buffer. DNA fragments (i.e., pGEM standards or 1 kb ladder), labeled with γ-[^{32}P]-ATP using T4 DNA kinase, can be used to determine the molecular weight of the products. After electrophoresis, the gel can be dried without fixing and exposed to X-ray film. The product bands can be quantitated by densitometry or by scintillation counting.

3.2.4. Preparation of Internal Standard DNA

An internal standard can be used to determine if the efficiency of the PCR was the same for all of the samples tested (*see* **Note 7**). The standard is an exogenously added DNA fragment, with a size different from that of the expected PCR product, which can be coamplified with the same primer set used for the PCR. Low stringency primer annealing is a simple and inexpensive method for generating internal standards *(11)*. To do this, use the primer set that amplifies the hnRNA of interest to amplify nonspecific cDNA or vector sequences using low stringency annealing conditions. For example, we have used the promoterless luciferase reporter vector pGL2-Basic as a template and have used the amplification protocol of 40 main cycles 94°C, 1 min, 45°C, 30 s, 72°C 1 min. Electrophoresis of the PCR products on an agarose gel should reveal multiple bands. Excise a band that is 100–200 bases longer or shorter than the hnRNA band of interest. Reamplify an aliquot of this band using high stringency annealing conditions to produce the standard DNA. Purify the stan-

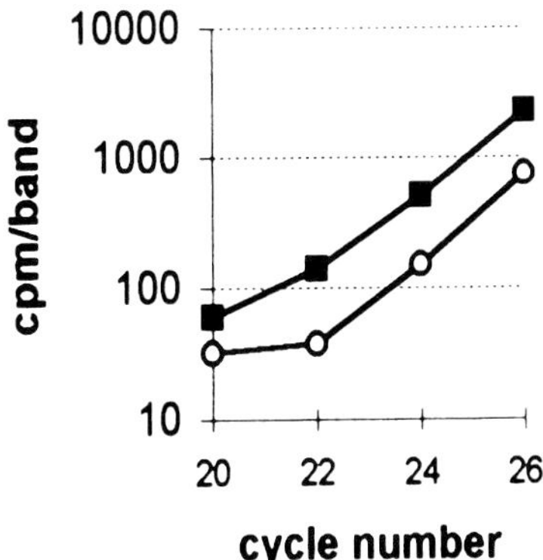

Fig. 6. Effect of PCR cycle number on the generation of rat collagenase-3 hnRNA and standard DNA products. Replicate tubes containing reverse transcribed RNA (1 µg) from rat osteoblastic cells and rat collagenase-3 hnRNA standard were PCR amplified in the presence of α-[^{32}P]-dCTP using 20–26 cycles of 94°C for 1 min, 59°C for 1 min, and 72°C for 1 min. Aliquots of reaction products were fractionated on a 6% polyacrylamide gel containing 100 mM Tris-borate and 1 mM EDTA, visualized by autoradiography, and quantitated by scintillation counting. ■, hnRNA; ○, standard.

dard by conventional means and quantitate it by DNA fluorimetry or by comparison to mass standards on an agarose gel, and calculate the concentration of the DNA standard in moles.

Perform a pilot experiment to determine the amount of standard to use. The amount of standard should be similar to the amount of input cDNA from a typical RT reaction. This is probably in the range of attomoles (1 attomole = 10^{-18} mole). Perform multiple RT reactions using the same RNA sample as starting material. Set up the subsequent PCR reactions containing different dilutions of internal standard. The amount of standard that gives a band intensity similar to that of the target hnRNA should be used for future experiments.

In addition, it is also important that the number of cycles used for amplification is within the linear range for both the standard and a typical RNA sample. This is done by preparing several identical PCR reactions and removing them from the thermocycler after different numbers of cycles. A plot of band intensity or cpm (on a log scale) vs cycle number should be linear for both the standard and the hnRNA (*see* **Fig. 6**).

4. Notes

1. Use the typical precautions for working with RNA. These include wearing gloves when working, having a set of stock reagents, tubes and tips dedicated to RNA

work only, and using DEPC-treated water to prepare reagents. A good discussion of RNases and laboratory hygiene can be found in **ref. 8**.

2. The membranes for the nuclear run-off should contain several controls as well as the cDNAs for the genes of interest. One or more loading controls should always be included; most popular are "house-keeping" genes or other genes not affected by the treatment of interest. As a control for nonspecific hybridization, linearized, empty cloning vectors are often used. In addition, it is good practice to include a positive control, namely cDNA for a gene known to be transcriptionally regulated by the treatment of interest. These controls make the experiment interpretable. For example, in the experiment shown in **Fig. 3**, GAPD was used as the loading control, the empty vector pGL2-Basic was used as the control for nonspecific hybridization, and $\alpha 1(I)$ collagen was used as the positive control, as cortisol is known to repress collagen gene transcription.

3. It is a good idea to do a test experiment with a small number of cells to determine the best conditions (tight or loose pestle, number of passes needed) for isolating nuclei from a particular cell type. As an example, we homogenize primary rat osteoblasts using 4 passes with the tight pestle. For the osteoblastic cell line MC3T3, we use 6–8 passes with the loose pestle, because these cells make abundant extracellular matrix, making them difficult to homogenize with the tight pestle.

4. The cytoplasmic RNA from a nuclear run-off can be a good control. If the nuclear run-off assay does not show the anticipated results, then Northern blot analysis of the cytoplasmic RNA could reveal whether the treatment worked or if the cells responded as expected. It is not much work to add proteinase K, EDTA, and SDS to the supernatant and start the incubation. The RNA can be stored in the freezer until the results of the nuclear run-off assay are known. It is a nice luxury to be able to throw the RNA away later if it is not needed.

5. The RNA to be used for RT-PCR can be prepared by any standard method. It is wise to visualize a small amount of the RNAs to be used on an agarose gel, to ensure that the RNA is intact and that the samples have been quantitated accurately. Only 0.5–1 µg of RNA is needed for each reaction, however, do not dilute an entire RNA stock as these diluted RNAs tend to be unstable.

6. Optimize the PCR conditions so that only the product of interest is detected. It may be necessary to adjust the concentration of Mg^{2+}, annealing temperature, or cycle number. If adjustment of the reaction conditions still does not give the desired result, it may be necessary to redesign one of the primers. Add the internal standard only after the RT-PCR itself has been optimized. This will decrease the number of variables in the PCR optimization phase.

7. If the input RNAs used for this procedure are well quantitated and of similar quality, one can expect to obtain reproducible results that reflect the amount of hnRNA present in the starting material. As with all experiments, it is best to obtain the same results with several different RNA preparations. However, it is important to remember that the protocol outlined here is not strictly quantitative in that it cannot be used to determine number of cDNA molecules or ng of cDNA in a sample. Competitive PCR, where several concentrations of competitor DNA

are used, is more appropriate for that purpose. Good discussions of competitive PCR can be found in **refs.** *12* and *13*.

Acknowledgments

This work was supported by National Institutes of Health grants DK 45227 and DK 09038. Special thanks to Drs. Zhenjian Du, Samuel Varghese, and Matthew Vincenti for reviewing the manuscript.

References

1. Greenberg, M. E. and Ziff, E. B. (1984) Stimulation of 3T3 cells induces transcription of the c-fos proto-oncogene. *Nature (London)* **311**, 433–438.
2. Groudine, M., Pertz, M., and Weintraub, H. (1981) Transcriptional regulation of hemoglobin switching on chicken embryos. *Mol. Cell. Biol.* **1**, 281–288.
3. Lipson, K. E. and Baserga R. (1989) Transcriptional activity of the human thymidine kinase gene determined by a method using the polymerase chain reaction and an intron-specific probe. *Proc. Natl. Acad. Sci. USA* **86**, 9774–9777.
4. Buttice G., and Kurkinen, M. (1993) A polyomavirus enhancer A-binding protein-3 site and Ets-2 protein have a major role in the 12-*O*-tetradecanoylphorbol-13-acetate response of the human stromelysin gene. *J. Biol. Chem.* **268**, 7196–7296.
5. Elferink, C. J. and Reiners, J. J., Jr. (1996) Quantitative RT-PCR on CYP1A1 heterogeneous nuclear RNA: a surrogate for the in vitro transcription run-on assay. *Biotechniques* **20**, 470–477.
6. Määttä, A., and Penttinen R. P. K. (1994) Nuclear and cytoplasmic α1(I) collagen mRNA-binding proteins. *FEBS Letts* **340**, 71–77.
7. Delany, A. M., Jeffrey, J. J., Rydziel S., and Canalis E. (1995) Cortisol increases interstitial collagenase expression in osteoblasts by post-transcriptional mechanisms. *J. Biol. Chem.* **270**, 26,607–26,612.
8. Sambrook, J., Fritsch, E. F., and Maniatis T. (1989) *Molecular Cloning. A Laboratory Manual.* Cold Spring Harbor Laboratory Press, Cold Spring Harbor, New York.
9. Ausubel, F. M., Brent, R., Kingston, R. E., Moore, D. D., Seidman, J. G., Smith J. A., Struhl, K. (1997) *Current Protocols in Molecular Biology.* John Wiley and Sons, Inc., New York.
10. Rychlik, W. (1993) Selection of primers for polymerase chain reaction, in *PCR Protocols: Current Methods and Applications* (White, B. A., ed.) pp. 31–40, Humana Press, Totowa, NJ.
11. Förster, E. (1994) Rapid generation of internal standards for competitive PCR by low-stringency primer annealing. *Biotechniques* **16**, 1006–1008.
12. Souazé F., Ntodou-Thomé, A., Tran, C. Y., Rostène, W., and Forgez, P. (1996) Quantitative RT-PCR: limits and accuracy. *Biotechniques* **21**, 280–285.
13. Gilliland, G., Perrin, S., and Bunn, H. F. (1990) Competitive PCR for analysis of mRNA. In *PCR Protocols: A Guide to Methods and Applications* (Innis, M. A., Gelfand, D. H., Sninsky, J. J., White T. J., eds.) pp. 60–69, Academic Press, San Diego, CA.

17

In Situ Hybridization for Metalloproteinases and Their Inhibitors

Tina L. Hurskainen and Suneel S. Apte

1. Introduction

1.1. The Role of In Situ *Hybridization in Metalloprotease and TIMP Biology*

The imminent completion of the human and mouse genome projects will reveal all metalloproteases and metalloprotease inhibitors such as those belonging to the MMP, TIMP, ADAM, and ADAMTS *(1)* families. At the present time, there are over 65 published genes in these families and it is very likely that their total number will be much higher. Considerable effort will subsequently be required to undertake a systematic examination of the expression, function, and regulation of these genes.

In situ hybridization (ISH) is invaluable in understanding tissue-specific gene expression and gene regulation within a spatial context and at a resolution that is not possible by any other method. While northern analysis and RT-PCR are amenable to quantitation, they are bulk methods, and cannot pinpoint the precise cell type which is expressing a gene. This is desirable information in organs which contain a heterogeneous cell population, such as the brain, bone marrow, lung, and so on, but less so in relatively monotypic tissues such as cartilage and muscle.

Localization of mRNA expression has been pivotal in developing a hypothesis of function and examining this in the context of a knock-out mouse or a naturally occurring gene mutation. For example, *Mmp9* is highly expressed by cells of the monocyte-macrophage lineage, particularly osteoclasts, in primary metaphyseal trabeculae, suggesting a role in cartilage resorption during endochondral ossification *(2)*. Targeted inactivation of *Mmp9* has demonstrated that

From: *Methods in Molecular Biology, vol. 151: Matrix Metalloproteinase Protocols*
Edited by: I. Clark © Humana Press Inc., Totowa, NJ

this is indeed the case, since MMP-9 deficient mice have delayed resorption of hypertrophic cartilage *(3)*. *Mmp14* (encoding mouse MT1-MMP or MMP-14) is expressed in osteoclasts, osteoblasts, and dense connective tissues such a joint capsules, ligaments, and tendons *(4,5)*. Consequently *Mmp14* inactivation in transgenic mice *(6,7)* leads to severe anomalies of skeletal development. Elucidation of the expression pattern has thus been helpful in understanding the mechanism underlying the phenotype of these two knock-out mice, the only ones among MMP or TIMP gene knockouts to show a developmental abnormality.

ISH may be performed on cells, tissue sections, or whole mounts of embryos or specific organs. Whole mount studies are useful for studying genes with a role in pattern formation and are not generally used in studies of metalloproteinases and their inhibitors. In this chapter we provide a simple annotated nonisotopic method which is optimized for the use of digoxigenin (dig)-labeled cRNA probes on tissue sections, but may easily be adapted for other material. cRNA probes are preferred over double-stranded cDNA probes because of their greater specificity; in this method, antisense cRNA provides experimental information and the sense cRNA acts as a negative control. Double-stranded cDNA probes may self-anneal and less probe is then available for hybridization to target mRNA. The greater strength of cRNA-mRNA associations allows hybridization to be performed at much higher stringency and specificity. Sense and antisense oligonucleotide probes may also be used in a similar fashion *(8)*.

1.2. Caveats in the Use of ISH

The presence of mRNA implies but does not always predict that the respective protein will be present. Many MMPs are secreted as proenzymes and thus the presence of mRNA or even protein does not necessarily imply activity. In some instances, mRNA may be present in one cell type and the protein in another *(9)*. This is an apparent discrepancy; it may be explained by the uptake of secreted protein by another cell type via cellular receptors. In the case of transmembrane metalloproteases there may be alternatively spliced soluble forms or sheddable extracellular domains that may show a different immuno-histological localization from the mRNA. Although attempts have been made to derive quantitative data from ISH, such data is meaningless at worst and semiquantitative at best, and detracts from the real value of ISH as a descriptive tool for gene regulation analyses.

1.3. Isotopic ISH (IISH) or Nonisotopic ISH (NISH)?

Proponents of IISH argue that NISH is less sensitive and is not useful for detecting low levels of expression. On the other hand, NISH does offer a fast and simple technique that relieves the investigator of two major drawbacks of IISH i.e., the use of radioisotope and the need for autoradiography. Another practical advantage of NISH is that dual dark field/bright field images are not required for visualization of the signal and tissue morphology. Bright field images of NISH make for more informative and elegant illustrations and presentations. NISH is rapid; the NISH protocol we describe here may be completed in as little as two whole days, more usually three. IISH may require a period of days to weeks depending on which isotope is used and how high the levels of probe labeling and gene expression are. NISH is cheaper, since autoradiographic emulsion and radioisotope used for ISH are more expensive than the labeling and detection reagents used in NISH and have a shorter shelf life. Additionally, NISH has the advantage of providing accurate subcellular localization of mRNA which may not be provided by IISH because of the scatter of radiation. Subcellular localization of mRNA is particularly useful in studies of polarized cells as we have shown recently for MMP-19 (Hurskainen, T. and Apte, S.S., unpublished data).

2. Materials

Unless specified, reagents are from Sigma or Fisher. Solution volumes may be scaled up or down as required. Reagents with a long shelf life such as DEPC-water and PBS may be stored at ambient temperatures. Where required, solutions can be filtered with a 0.2 μm filter. Refer to manufacturer's material safety data sheets (MSDS) for detailed information on hazards of individual chemicals.

1. 8% Paraformaldehyde (PFA) stock (*Caution:* PFA is toxic and must be handled in a fume hood and disposed of correctly): add 8 g of PFA powder to 80 mL of heated distilled water (approx 80°C), add one or more drops of 10 N NaOH to dissolve the paraformaldehyde and equilibrate pH to about 7.4, stir until dissolved, make up to 100 mL, and store at 4°C. Good for 2–3 mo.
2. Fixative (4% PFA in PBS): mix equal vol of 8% PFA and 2X PBS. Store at 4°C. Good for one month.
3. 10X Phosphate buffered saline (PBS): 80 g NaCl, 2 g KCl, 11.5 g Na_2HPO_4, 2 g KH_2PO_4 per liter. Store at 4°C. Dilute to 1X PBS with distilled water, and sterilize by autoclaving.
4. Diethylpyrocarbonate (DEPC)-treated water (*Caution:* DEPC is toxic and must be handled in a fume hood): add 1 mL of DEPC to 1 L of water in a glass bottle, shake well and let stand overnight in fume hood. The next morning, autoclave and store at room temperature. We re-use the same bottles for successive DEPC treatments without washing. Overnight incubation with the DEPC is essential for inactivation of ribonucleases; autoclaving subsequently destroys the DEPC.
5. 0.2 N HCl: add 0.5 mL concentrated HCl (approx 12 N) to 30 mL distilled water. Use a plugged pipet tip to transfer conc. HCl. Make just before use at ambient temperature. Do not add water to concentrated acid.

6. Acetylation mix (*Caution:* Acetic anhydride is toxic and must be handled in a fume hood). To make 50 mL of acetylation mix, add 660 µL of triethanolamine and 125 mL of acetic anhydride to 50 µL of distilled water. Make immediately prior to use (acetic anhydride is unstable) and incubate slides in this solution with constant stirring or rocking at room temperature.

7. 20X SSC: 3 M NaCl, 300 mM sodium citrate, pH adjusted to 7.0 with 1 M citric acid.

8. Hybridization solution (*Caution:* Formamide is toxic and must be handled in a fume hood and disposed of correctly; SDS powder is a respiratory irritant if inhaled). 50% formamide (Fluka, molecular biology grade), 10 mM Tris-HCl, pH 7.6, 200 mg/mL tRNA (Nuclease free, Sigma), 1X Denhardt's solution, 10% dextran sulfate, 600 mM NaCl, 0.25% sodium lauryl sulphate (SDS, enzyme grade), 1 mM EDTA (tetrasodium salt). Store at –20°C in 1 mL aliquots.

9. Denhardt's solution: 0.02% (w/v) Ficoll, 0.02% (w/v) polyvinylpyrrolidone and 10 mg/mL RNAse free bovine serum albumin. Make as a 50X solution, aliquot, and store at –20°C.

10. TNE: 10 mM Tris-HCl, pH 7.5, 0.5 M NaCl, 1 mM EDTA. Filter and store at room temperature.

11. Substrate mix: Nitro blue tetrazolium chloride (NBT) stock is 75 mg/mL (in 70% dimethylformamide) and 5-bromo-4-chloro-3-indolyl phosphate (BCIP) stock is 50 mg/mL (in dimethylformamide). Store at –20°C. (NBT and BCIP are purchased from Roche Biosciences and are also available as a premixed stock). Levamisole (Sigma) stock is 1 M in water and is stored at –20°C. Levamisole is an inhibitor of many isoforms of alkaline phosphatase, other than that which is tagged to the antibody used here. It is used as an inhibitor of endogenous alkaline phosphatase. Working substrate solution is prepared fresh from the above stocks by mixing 4.5 µL of NBT and 3.5 µL of BCIP with 1.0 µL of levamisole in 1.0 mL of DIG3 buffer.

12. DIG 1 solution: 100 mM Tris-HCl, pH 7.5, 150 mM NaCl. Filter after preparation.

13. DIG 3 solution: 100 mM Tris-HCl, pH 9.5, 100 mM NaCl, 50 mM MgCl$_2$. Make the Tris plus NaCl solution first, adjust pH to 9.5, then add MgCl$_2$, otherwise the MgCl$_2$ will precipitate out. Filter after preparation. This solution is stable for 6 months at ambient temperature. Note that pH 9.5 is critical for optimal activity of alkaline phosphatase.

14. Plastic peel away molds (Polysciences).

15. Superfrost Plus slides (Fisher).

16. Tissue embedding medium, e.g., OCT (Miles Scientific).

17. PAP pen or grease pencil.

18. Methyl green (Sigma).

19. Aqua Polymount cover slip mounting medium (Polysciences Inc).

20. Proteinase K (nuclease free, Roche Biosciences): 10 µg/mL in 10 mM Tris-HCl, pH 7.4, 1 mM calcium chloride. This solution should be prepared fresh and warmed to 37°C before use.

21. RNAse A (Roche Biosciences).

22. Digoxigenin-labeled UTP (as Dig-labeling mix, Roche Biosciences).
23. RNA polymerases and restriction enzymes (Promega, New England Biolabs or Roche Biosciences).

3. Methods

3.1. Tissue Fixation

1. Fix as required in 4% paraformaldehyde in PBS at 4°C without agitation (*see* **Note 1**).

3.2. Sectioning

1. Cryosections or paraffin sections of 5–10 µm thickness may be used (*see* **Note 2**).

3.3. Preparation of cDNA Template (see Note 3)

1. Cut 5–10 µg of plasmid DNA with a suitable restriction enzyme to generate a blunt or 3' recessed end.
2. Electrophorese the entire digest on a preparative gel to confirm completeness of the digest and cut out the band. It is very important to ensure that there is no circular DNA remaining at the end of the digestion. If there is difficulty in getting complete digests, cut less DNA in a larger final volume and add 1–10 U more enzyme after 2 h of digestion. Digest for another hour.
3. Elute the DNA from the gel slice using the Gene Clean kit (BIO 101). The protocol recommended by the manufacturer is modified so that DNA is eluted with 500 µL sodium iodide and 15 µL of glass milk. DNA is eluted into 10 µL DEPC-treated water. Assuming at least 75% recovery, this will be sufficient for 5–10 transcription reactions.

3.4. In Vitro Transcription

1. Mix the following in a microcentrifuge tube (1.5 mL) in a total vol of 20 µL: 2 µL each of 10X transcription buffer and Dig-labeling mix, and 1 µg of linearized DNA template.
2. Make up the volume with DEPC-treated water and add 1 µL of the appropriate RNA polymerase (T7,T3 or SP6, purchased from Promega or Roche Biosciences).
3. Incubate for 1.5 h at 37°C.
4. Electrophorese 1 µL of the reaction added to 1 µL of 10X loading buffer (RNAse free) and 8 µL of DEPC-treated water. Use a 1% agarose minigel containing ethidium bromide to visualize the template and transcript (*see* **Note 4**).
5. Add 2 µL of DNAse I and incubate for 15 min at 37°C to destroy the template.

3.5. Ethanol Precipitation of Probe

1. To the transcription reaction, add 2 µL of 3*M* sodium acetate (pH 5.2, made in DEPC-treated water), 2 µL of yeast tRNA (RNAse free, 20 mg/mL in DEPC-treated water, used as RNA co-precipitant) and 200 µL of ice-cold absolute ethanol (dedicated for RNA work).

2. Vortex lightly, and keep at –70°C to –80°C for 20 min.
3. Centrifuge at 14,000*g* (in a microfuge) for 15 min at 4°C.
4. Discard the ethanol by inverting the tube and add 200 µL of 70% ethanol (diluted with DEPC-treated water).
5. Centrifuge again for 5 min.
6. Discard the supernatant by inverting, allow the pellet to dry and suspend in 50 µL of DEPC-water. Store at –70°C.

3.6. Hybridization (see Note 5)

Slides are prepared for hybridization as follows:

1. Immerse slides in 0.2 *N* hydrochloric acid for 15 min at ambient temperature. Incubate (wash) slides twice in PBS (5 min each).
2. Treat slides with freshly prepared proteinase K (10 µg/mL) warmed to 37°C prior to use. Incubate (wash) slides twice in PBS (5 min each) (*see* **Note 6**).
3. Postfixation: Immerse slides in 4% PFA for 5 min at room temperature (*see* **Note 7**).
4. Acetylation: Immerse slides in acetylation mix for 15 min at ambient temperature. Incubate (wash) slides twice in PBS (5 min washes each) (*see* **Note 8**).
   Note: steps 3 and **4** should be carried out in an externally vented, fume hood and the user should use appropriate personal protection such as a lab coat, gloves and safety glasses.
5. Dehydrate the sections successively in 70%, 90%, and two changes of 100% ethanol (2 min in each change). Allow the slides to dry completely, sections up. Use a PAP pen or grease pencil to draw circles around each section and create wells.
6. Hybridization (*see* **Note 9**). Preincubate aliquots of the hybridization fluid at 80°C for 10 min. Add probe to the desired volume of hybridization fluid (usually 150 µL/section). Vortex vigorously to disperse the probe, heat for 3 min more, and add to preheated slides (80°C). Hybridize overnight at 50–55°C. We use 52°C for most probes. The temperature of hybridization may be titrated as indicated for different probes.

3.7. Stringency Washes and Ribonuclease Treatment

1. Pour off the hybridization fluid and rinse briefly in 5X SSC at 50–60°C.
2. Place the slides in 2X SSC/50% formamide for 30 min at 56°C. (*Caution:* Use formamide in a fume hood, particularly when it is heated, and collect the waste for proper disposal).
3. Immerse the slides in TNE at 37°C for 10 min, then in TNE with RNAse A (10 µg/mL) at 37°C for 30 min (*see* **Note 11**).
4. Place the slides in TNE at 37°C for 10 min to remove excess RNAse A.
5. Carry out stringency washes next, as follows: In 2X SSC at 50–60°C for 20 min (one wash) and 0.2X SSC at 50–60°C for 20 min (two washes).

3.8. *Immunohistochemical Detection of DIG and Counterstaining*

1. Place slides in DIG1 buffer at room temperature for 5 min and then incubate with 1.5% (w/v) blocking reagent (Roche Biosciences) in DIG1 buffer for 30 min (*see* **Note 12**).
2. Following this incubation, dip the slides once or twice in another change of DIG1 buffer and place 1:500 to 1:1000 anti-Dig antibody (sheep anti-Dig antibody, Fab fragment, alkaline phosphatase conjugated, Roche BioSciences) diluted in DIG1 buffer in each well.
3. Return the slides to the humid box and incubate for an hour at room temperature (note that this incubation may also be carried out overnight, but at 4°C).
4. Wash the antibody off with two rinses in DIG1 buffer (15 min each) followed by immersion in DIG3 buffer (5 min) to allow equilibration of sections for substrate deposition.
5. Prepare working substrate solution and add 100–300 µL of substrate/slide. Monitor continuously for development of magenta color. We find that genes with high level, specific expression (such as the *col2a1* gene expressed specifically in cartilage) develop a color product within one hour. Most other genes require incubation periods ranging from 3–18 h.
6. Rinse the slides in Tris-EDTA buffer (pH 7.5) or in distilled water when the desired signal/noise ratio is reached, then rinse in distilled water.
7. Counterstain sections in 0.3% (w/v) methyl green in distilled water (15–60 s), then rinse in cold running tap water to remove excess stain or until desired stain intensity is achieved.
8. Blot the slides dry with a paper towel and mount coverslips with warmed (to 45°C) Aqua Polymount (Polysciences). Remove air-bubbles. Coverslips may be held in place with nail-varnish.

4. Notes

1. *Procuring tissue sections:* Although *in situ* hybridization has been successfully performed on decades old archival tissue, it is best to procure and process tissue specifically for this purpose and to fix it immediately to be assured of mRNA preservation. In general, mRNA preservation is inconsistent in whole mouse embryos older than 15.5 d or in pieces of tissue greater than 0.5 cm thick in their smallest dimension. It is recommended that older embryos have multiple punctures or slits made in the skin to allow penetration of fixative. Fresh 4% paraformaldehyde in PBS should be made just prior to fixation and tissue immersed in at least 10 vol of fixative at 4°C. Agitation is not necessary; vigorous agitation may damage delicate tissues. The duration of fixation should be adjusted as required to optimize mRNA preservation and to permit probe access to the mRNA. Under-fixation may result in loss of tissue architecture during the ISH protocol whereas over-fixation may diminish the signal intensity. As a rule of thumb, overnight fixation (18 h) for tissue not less than 3 cubic mm but not larger than 1 cubic cm is effective. Very small pieces of tissue may require short fixation times.

2. *Tissue embedding and sectioning:* Following fixation, the tissue may be processed for cryosections or embedded in paraffin. Paraffin embedding is most reproducible in an automated tissue processor, allowing well controlled processing and vacuum infiltration prior to embedding in paraffin.

   For cryosections, following fixation, the tissue is rinsed briefly in PBS, then immersed in 20% w/v sucrose in PBS (autoclaved, stored at 4°C) until it sinks (usually 24–48 h). This provides the tissue with a buoyant density similar to most commercially available freezing media and facilitates sectioning. Freezing is done in plastic peel-away molds. Tissue is placed in the desired orientation, freezing medium (e.g., OCT compound) poured around it (avoiding formation of air bubbles) and frozen rapidly by immersion in liquid nitrogen or in a slurry of dry ice and ethanol. We have frozen tissue simply by placing the blocks on the shelf of a –80°C freezer without detriment of mRNA or appearance of freeze artefact. These blocks of frozen tissue are now ready for sectioning or may be stored at –80°C in an airtight container such as sealable freezer bags to prevent dehydration. We have successfully used tissue stored for up to one year. Note that tissues such as muscle and liver are prone to freeze artefact and should be frozen rapidly.

   Cryosections should be taken on Superfrost Plus slides (Fisher) and air-dried for up to 3 h. Dried sections may be stored at –70°C in an airtight box for up to 3 wk. We find that signal decreases after this time. Prior to use, slides are allowed to warm to room temperature and dry again. A PAP pen or grease pencil may be used to encircle the section thus allowing retention of fluids used for each step. Paraffin sections should be taken on Superfrost Plus slides and stored in a clean box at room temperature. They will last indefinitely. If sections are taken on the lower half of each slide, smaller volumes of all reagents and washes are required during *in situ* hybridization.

3. *Generating cloned cDNA templates for ISH probes:* cRNA probes are generated by in vitro transcription using prokaryotic RNA polymerases and cloned cDNA fragments. We describe at length in this section the requirements for and procurement of cloned DNA fragments. An alternative is to use PCR generated templates containing RNA polymerase promoter sequences.

   New genes: the first step is to clone cDNA fragments of a suitable size into a vector suitable for generation of labeled riboprobes. Inserts ranging from 200 bp to 900 bp work well. Shorter fragments may not be labeled sufficiently and longer fragments may not permeate into cells, although this problem may be circumvented by hydrolysis of long probes. The vectors used for transcription should have a versatile multiple cloning site (MCS), which should be immediately flanked by RNA polymerase binding sites (T3, T7, or SP6). Although most of the workhorse vectors fulfill this requirement, we have had the best results with pBluescript (Stratagene) or pLitmus28 (New England Biolabs). Some vectors commonly used for T/A cloning of PCR products have given us problems with transcription or with specificity although we have successfully used pT7Blue (Invitrogen). Although T/A cloning is very convenient, the use of the pT7Blue is complicated by the fact that the fragment needs to be cloned into both sense and antisense orientations owing to the presence of a single T7 polymerase site. pLitmus28 offers the advantage of transcribing both strands with a single (T7) RNA polymerase.

We recommend cloning and transcribing more than one nonoverlapping cDNA fragment for the following reasons. Some probes may give excess background and some may cross-react with other members of the gene family. Cross hybridization is more of a problem with repetitive sequences such as collagen coding sequences, rather than metalloproteinases and TIMPs, but it is a good idea to use a cDNA fragment from the untranslated regions as one of the new probes to be tested. The cDNA fragment to be used must contain minimum flanking sequences from the vector in which it was originally cloned, since these sequences may diminish transcriptional efficiency or contain additional polymerase binding sites. Some cDNA fragments may not transcribe very well owing to high G/C content or complex secondary structure. Newly cloned cDNA fragments and cDNA clones obtained from other investigators must be sequenced from both ends to confirm their identity.

'Old' genes and probes from dBEST: published cDNA clones representing cognate genes are generally made freely available by many investigators together with a detailed restriction map and if available, with the nucleotide sequence of the insert. Using these has the advantage of a tested reagent. Sequencing or restriction mapping is strongly recommended to confirm the identity and orientation of any cDNA clone thus obtained.

A recent alternative is to purchase clones from the IMAGE (**I**ntegrated **M**olecular **A**nalysis of **G**enomes and their **E**xpression, URL: http://www.bio.llnl.gov/ bbrp/image/image.html) consortium after first identifying them in dBEST (URL: http://www.ncbi.nlm.nih.gov/dbEST/) or UniGene (URL: http:// www.ncbi.nlm.nih.gov/UniGene/). Over the past several years, numerous investigators (who form the consortium) have deposited cDNA clones and their partial sequences into the database of expressed sequence tags (dBEST). This public domain subdatabase of GenBank contains >500,000 cDNA sequences, and is annotated with the corresponding cDNA clones which are readily available. Screening of dBEST with any cognate cDNA sequence from the MMP, ADAM, ADAM-TS, and TIMP family reveals the existence of large numbers of independent clones encoding various portions of that gene. Sequences from the 3'-untranslated region predominate and several of these clones are 1 kb or less in size. Furthermore, many of these clones are in the pBluescript and pT7T3 vectors. These three feature make these clones suitable for generation of riboprobes for in situ hybridization.

cDNA clones are selected by a dBEST search using the Basic Local Alignment Search Tool (BLAST, URL: http://www.ncbi.nlm.nih.gov/BLAST/) servers at the National Center for Biotechnology Information/National Library of Medicine (NCBI/NLM, URL: http://www.ncbi.nlm.nih.gov/). Selected clones can be purchased from any one of four commercial distributors (e.g., Genome Systems Inc.) for a nominal fee; their use is royalty-free. The only characterization that will need to be performed prior to their use is determination of sequence and orientation of the insert in relation to the T3 and T7 polymerase primers.

The advantages of purchasing IMAGE clones are many. Since the clones are commercially available, their use is unrestricted. They may often be quite well characterized, of suitable size, and in suitable vectors. Furthermore, where the gene sequence is known in one species e.g., human, it is usually quite a straight-forward matter to identity orthologous ESTs from another species such as mouse or rat, based on percent sequence identity (at both protein and nucleotide levels). There are also a number of disadvantages inherent in the use of IMAGE clones. The inserts may be too large and may need to be sequenced in their entirety, a. to verify their identity; b. to ascertain the complete sequence of the clone; c. to ensure absence of chimerism and d. to identify suitable restriction sites for subcloning of smaller fragments. There is a substantial error rate in the provision of the clones (owing to errors during plating of arrays, harvesting the clone or cross-contamination) and in some instances clones harbor infection with a lytic T1 phage. Complete reliance should not be placed on the identity provided for the IMAGE clones within the database, since in many instances they may be assigned based on partial similarity to cognate genes. Clearly, there are many drawbacks to the use of IMAGE clones, but we have found them to be a rich resource for gene discovery and probe acquisition for multiple applications.

4. The Dig-labeled RNA migrates slower than expected because of the incorporated Dig. In other words, the size of the probe will be judged to be 20–25% larger than the transcribed cDNA insert. In some instances, the band may be quite broad and occasionally it may be shorter than the template because of premature arrest of transcription. Both the template and primer should be visible on the gel and the transcript should be approx 10-fold brighter than the template (suggesting the generation of approx 20 μg of cRNA probe. Note that if the reaction product is electrophoresed *after* DNAse treatment, the template will not be visible.

5. **Steps 1–5** are carried out by immersing slides in Coplin jars. Jars with 5 or 10 slide capacity may be used. The capacity may be doubled by arranging slides back to back in each slot of the Coplin jar. Purchase jars with lids so that you will be protected from chemical fumes and to prevent evaporation of reagents. Volumes used in each step may be minimized by taking sections in the bottom third of each slide.

6. This is a key step in the protocol. The purpose of this step is to open up tissue for improved accessibility of probe. We usually incubate tissue for 15 min. However, the optimal time of incubation needs to be adjusted depending on the fixation conditions and considerable titration may be required. Tissue that has been lightly fixed will be more susceptible to destruction by proteinase K. Over-digestion with proteinase K leads to labeling of nuclei and loss of morphological detail. In our experience 7.5- and 9.5-day-old embryos need approx half the treatment time of 14.5-day-old embryos.

7. The purpose of this step is to post-fix the sections to protect them from further morphological disruption following proteolytic treatment.

8. Acetylation of sections prevents nonspecific binding of probe and decreases background.

9. We do not prehybridize our slides. It is important to preheat the hybridization fluid and slides. Note also that we do not use coverslips. We place the slides on rails constructed of parallel glass rods stuck to the bottom of Tupperware boxes with silicone rubber. A few mL of 50% formamide (v/v) in water are placed in the tray and the lid closed to maintain humidity. Slides are preheated in the box and the hybridization mix placed in the marked wells. It is important to cover each section, but not to flood the well. If too much mix is used the surface tension cannot hold the drop and hybridization fluid will spill out of the well.

10. Place 5X SSC, 2X SSC, 2X SSC/50% formamide, and 0.2X SSC at 50–60°C and TNE at 37°C overnight. This allows equilibration to required temperature prior to stringency washes. Shaking is not required.

11. Separate glassware should be kept for this step and for all subsequent steps to prevent RNAse contamination during subsequent runs of ISH.

12. Start dissolving the blocking reagent about 2 h prior to use because it dissolves very slowly. We make it fresh in a volume of 50 mL.

13. *RNA hygiene in ISH.* The seemingly ubiquitous presence of RNA degrading enzymes and their notorious stability makes it necessary to safeguard against degradation of the probe and target mRNA sequences during the ISH procedure. In general, RNA hygiene is more pertinent to probe preparation than the target mRNA. Solutions used for in vitro transcription should be prepared in DEPC-treated water using clean, baked glassware or autoclaved disposable plastic ware. It is not necessary to keep a separate set of pipets for RNA use. Use gloves throughout the ISH procedure. Once the target RNA has been fixed, it is quite stable and under the conditions used for hybridization, loss of target mRNA is unlikely (the hybridization mix contains formamide). Glassware used for dipping of sections for hybridization may be baked, but we have found that even this is not necessary as long as the glassware used prior to RNAse A step is kept separately from that used for the post-RNAse A steps. During the washes, the RNA-RNA hybrids are safe against RNAse attack, and after the RNAse treatment step, protection against RNAse is irrelevant. We ensure that any glassware used after the RNAses washes is not used for any of the prior steps. If needed, the glassware used for the pre-RNAse steps can be baked at 250°C for four hours (put a cautionary note on the oven so no one gets burnt and do not use autoclave tape because it will char). Where specified in the protocol, DEPC-treated water is prescribed, but otherwise molecular biology grade deionized distilled water is adequate. We do not use special pipets, but do use prepacked autoclaved sterile pipet tips and microfuge tubes. We recommend keeping a separate apparatus for electrophoresis of the template and for analysis of in vitro transcription product.

Acknowledgments

Support for this chapter was provided by NIH AR44436 and a Biomedical Sciences Grant from the Arthritis Foundation, by the Cleveland Clinic Foundation to S. Apte and by an award from the Academy of Finland to T. Hurskainen. We thank Mrs. Judy Christopher for secretarial assistance. We are grateful to Dr. Naomi Fukai, Harvard Medical School, for introducing us to the NISH techniques we have described above.

References

1. Hurskainen, T. L., Hirohata, S., Seldin M. F., and Apte S. S. (1999) ADAM-TS5, ADAM-TS6 and ADAM-TS7, novel members of a new family of zinc metalloproteases (ADAM-TS, A Disintegrin And Metalloprotease domain with ThromboSpondin type I motifs). General features and genomic distribution of the ADAM-TS family". *J. Biol. Chem.* **274**, 25,555–25,563.
2. Reponen, P., Sahlberg, C., Munaut, C., Thesleff, I., and Tryggvason, K. (1994). High expression of 92-kD type IV collagenase (gelatinase B) in the osteoclast lineage during mouse development. *J. Cell Biol.* **124**, 1091–1102.
3. Vu T. H., Shipley, J. M., Bergers, G., Berger, J. E., Helms, J. A., Hanahan, D., Shapiro, S. D., Senior, R. M., and Werb, Z. (1998). MMP-9/gelatinase B is a key regulator of growth plate angiogenesis and apoptosis of hypertrophic chondrocytes. *Cell*, **93**, 411–422.
4. Apte, S. S., Fukai N., Beier D. R., and Olsen, B. R. (1997) The matrix metalloproteinase-14 (MMP-14) gene is structurally distinct from other MMP genes and is co-expressed with the TIMP-2 gene during mouse embryogenesis. *J. Biol. Chem.* **272**, 25,511–25,517.
5. Kinoh, H, Sato, H., Tsunezuka, Y., Takino, T., Kawashima, A., Okada, Y., and Seiki, M. (1996) MT-MMP, the cell surface activator of proMMP-2 (progelatinase A), is expressed with its substrate in mouse tissue during embryogenesis. *J Cell Sci*, **109**, 953–959.
6. Zhou, Z., Apte, S. S., Wang, J., Rauser, R., Baaklini, G., Soininen, R., and Tryggvason, K. Abnormal endochondral ossification, dwarfism and early death in mice deficient in membrane-type matrix metalloproteinase 1 (MMP-14) (submitted for publication).
7. Holmbeck, K., Bianco, P., Caterina, J., Yamada, S., Kromer, M., Kuznetsov, S. A., Mankani, M., Robey, P. G., Poole, A. R., Pidoux, I., Ward, J. M., and Birkedal-Hansen, H. MT1-MMP Deficient Mice Develop Dwarfism, Osteopenia, Arthritis, and Generalized Connective Tissue Disease Because of Inadequate Collagen Turnover (submitted for publication).
8. Crabb, I. D., Hughes S. S., Hicks, D. G., Puzas, J. E., Tsao, G. J., and Rosier, R. N. (1992) Nonradioactive *In situ* hybridization using digoxigenin-labeled oligonucleotides. Applications to musculoskeletal tissues. *Am J Pathol.* **141**, 579–589.
9. Polette, M., Gilbert, N., Stas, I., Nawrocki, B., Noel, A., Remacle, A., Stetler-Stevenson, W. G., and Birembaut, P. (1994) Gelatinase A expression and localization in human breast cancers. An in situ hybridization study and immunohistochemical detection using confocal microscopy. *Virchows Arch.* **424**, 641–645.

18

Use of EIA to Measure MMPs and TIMPs

Noboru Fujimoto and Kazushi Iwata

1. Introduction

The activity of matrix metalloproteinases (MMPs) is closely regulated by tissue inhibitors of metalloproteinases (TIMPs) under pathophysiological conditions. The quantitative imbalance between MMPs and TIMPs in tissues and body fluids is considered to cause tumor invasion, metastasis, and various connective tissue diseases such as arthritis and periodontitis *(1)*. MMPs and TIMPs have been identified and purified from various samples such as cell-conditioned media, tissue homogenates, sera, and so on. Furthermore, monoclonal antibodies against MMPs and TIMPs have been also prepared, and enzyme immunoassays (EIAs), which can quantitate MMPs and TIMPs in specimens, have been established. A two-step sandwich EIA for human MMP-3 was prepared using the combination of a monoclonal antibody as the solid-phase with rabbit polyclonal antibody and horseradish peroxidase (HRP)-labeled donkey antirabbit gamma globulin as the second phase *(2)*. EIAs for human MMP-9 were developed using MMP-9 specific polyclonal antibody *(3,4)* or monoclonal antibodies *(5)*. For quantitative determination of MMP-8, a two-step sandwich EIA was developed using polyclonal antibody *(6)*. We also established one-step sandwich EIAs for MMPs and TIMPs using two kinds of monoclonal antibodies *(7–16)*.

In this chapter we describe the methodology of sandwich EIAs using enzyme-labeled antibody such as IgG-HRP or Fab'-HRP. This system is a solid-phase EIA based on a one-step sandwich immunoreaction. The antigen is allowed to react with an antibody immobilized on the solid phase and enzyme-labeled antibody directed against distinctly different antigenic site on the same antigen, resulting in the antigen being sandwiched between the solid-phase and HRP-labeled antibodies. After removal of the unbound HRP-labeled antibody,

From: *Methods in Molecular Biology, vol. 151: Matrix Metalloproteinase Protocols*
Edited by: I. Clark © Humana Press Inc., Totowa, NJ

the HRP bound to the solid phase is then incubated with the enzyme substrate and color reagent. The HRP activity is proportional to the amount of antigen, and thus the concentration of antigen can be determined.

2. Materials

2.1. Antibodies

1. Anti-MMP-2 (clone 75-7F7), anti-MMP-3 (clone 55-2A4), anti-MMP-13 (clone 181-15A12), anti-TIMP-1 (clone 7-6C1) and anti-MMP-2 (clone 67-4H11) antibodies used in these EIAs are sold by Oncogene Research Products, a product line of Calbiochem (Boston, USA).

2.2. Antigens

1. MMP-1, 2, 3, 7, 9, and 13 can be purified from the conditioned medium of human skin fibroblasts (CCD-41SK), normal human skin fibroblast cells (NBIRGB), CaR1 human rectal carcinoma cells, human fibrosarcoma (HT1080), NSO mouse myeloma cells transfected with MMP-13 cDNA; MMP-8 and TIMP-2 can be purified from human placenta, and the TIMP-1 purified from human serum to use as a standard in respective EIA.

2.3. General Buffers

1. Buffer A: 0.1 M sodium phosphate (pH 6.0).
2. Buffer B: 0.1 M sodium phosphate (pH 6.5).
3. Buffer C: 0.1 M sodium phosphate (pH 7.0).
4. Buffer D: 0.1 M sodium phosphate (pH 7.5).

2.4. Enzyme-Labeled Antibody

2.4.1. Preparation of IgG-HRP

1. IgG solution: 1.7 mg (10.8 nmol) of IgG in 500 μL of Buffer B (*see* **Note 1**).
2. S-acetylmercaptosuccinic anhydride solution: 0.19 mg (1,080 nmol) of S-acetylmercaptosuccinic anhydride (Wako Pure Chemicals) in 10 μL of N, N'-dimethylformamide (DMF). This reagent is prepared immediately before use.
3. HRP solution: 4 mg (100 nmol) of HRP (grade 1, Boehringer Mannheim) in 0.3 mL of Buffer C.
4. N-ε-maleimidecaproyloxy succinimide (EMCS) solution: 1.6–3.2 mg (4,800–9,600 nmol) of EMCS (Wako Pure Chemicals) in 30 μL of DMF. This reagent is prepared immediately before use.

2.4.2. Preparation of Fab'-HRP

1. IgG solution: 10 mg purified monoclonal antibody (IgG) in 1 mL of 0.1 M sodium acetate buffer (pH 4.0–4.5). This solution is dialyzed before use (*see* **Note 2**).
2. Pepsin solution: 0.2 mg pepsin (from porcine gastric mucosa, Boehringer Mannheim) in 0.1 M sodium acetate buffer (pH 4.0–4.5). This solution is prepared immediately use.

3. F(ab')$_2$ solution: 1–3 mg of the F(ab')$_2$ in 0.45 mL of Buffer A.
4. HRP-maleimide solution: 1.7 mg HRP-maleimide (43 nmol) in 0.3 mL of Buffer A.

2.5. Preparation of Standards Using Antibody-Sepharose 4B

2.5.1. Preparation of Antibody-Sepharose 4B

1. React 10–30 mg of monoclonal antibody in 0.1 M sodium borate buffer (pH 8.0) containing 0.5 M NaCl (coupling buffer) with 1 g CNBr-activated Sepharose 4B (Pharmacia) (previously treated with 200 mL of 1 mM HCl) and shake gently at room temperature for 2 h or at 4°C for 16 h.
2. Block unreacted sites with 0.2 M glycine-NaOH buffer (pH 8.0).
3. Wash sequentially with the coupling buffer, 0.1 M sodium acetate buffer (pH 4.0) and coupling buffer again, then store the matrix in Buffer C containing 0.1% NaN$_3$ at 4°C. This reagent is stable for more than 6 mo under these conditions.

2.6. One-Step Sandwich EIAs for MMPs and TIMPs

2.6.1. Microtitration Plate Coated with Antibody

1. Coat microtitration plates with 100 μL of purified mouse monoclonal antibody at an adequate concentration (more than 5 μg/mL) in Buffer D at 4°C for 24 h.
2. Wash twice with 10 mM sodium phosphate buffer (pH 7.0) containing 0.1 M NaCl (wash buffer)
3. Block the microplates with 10 mg/mL BSA in Buffer D at 4°C for 24 h and wash with wash buffer just before use. Such plates sealed with a plate sealer are stable at 4°C for more than 2 mo.

2.6.2. Bead Coated with Antibody

1. Place polystyrene beads into 100 μg/mL of the purified monoclonal antibody in Buffer D containing 0.1% (w/v) NaN$_3$ at 4°C for 24 h.
2. Wash the beads 5 times with 10 mM sodium phosphate buffer (pH 7.0) and immerse in 10 mM sodium phosphate buffer (pH 7.0) containing 1 mg/mL BSA, 1 mg/mL NaCl, and 10 mg/mL chlorhexidine and store at 4°C just until use. The antibody-coated beads are stable at 4°C for more than 6 mo.

2.6.3. Reagents for EIA

1. HRP-labeled antibody: IgG-HRP or Fab'-HRP in Buffer B containing 1 mg/mL BSA and 20 μg/mL chlorhexidine. This reagent is stable at 4°C for more than 6 mo.
2. Assay buffer (*see* **Note 3**): 10 mM sodium phosphate buffer (pH 7.0) containing 0.1 M NaCl, 5 mg/mL BSA, 20 mg/mL horse serum, and 0.02 mg/mL Fungizone for MMP-1 EIA; 10 mM sodium phosphate buffer (pH 7.0) containing 0.1 M NaCl, 10 mg/mL BSA, and 10 mM ethylenediamine tetraacetic acid (EDTA) for MMP-2, 7, 9, and TIMP-2 EIAs; 10 mM sodium phosphate buffer (pH 7.0) containing 0.1 M NaCl and 10 mg/mL BSA or horse serum for MMP-3 EIA; 30 mM sodium phosphate buffer (pH 7.0) containing 0.1 M NaCl, 10 mg/mL BSA,

10 m*M* EDTA, and 20 m*M* benzamidine for MMP-8 EIA; 10 m*M* sodium phosphate buffer (pH 7.0) containing 0.1 *M* NaCl, 30 mg/mL BSA, and 10 m*M* EDTA for MMP-13 EIA; 10 m*M* sodium phosphate buffer (pH 7.0) containing 0.1 *M* NaCl, 10 mg/mL BSA, and 0.1% (w/v) Tween 20 for TIMP-1 EIA. All assay buffers are stable at 4°C for more than 6 mo.

3. Wash buffer: 10 m*M* sodium phosphate buffer (pH 7.0) containing 0.1 *M* NaCl for MMP-1, 2, 9, and TIMP-2 EIAs; 5 m*M* sodium phosphate buffer (pH 7.0) containing 50 m*M* NaCl for MMP-3 EIA; 10 m*M* sodium phosphate buffer (pH 7.0) containing 0.1 *M* NaCl and 0.1% (w/v) Tween 20 for MMP-7, 13, and TIMP-1 EIAs; 0.9% (w/v) NaCl for MMP-8 EIA. All wash buffers are stable at 4°C for more than 6 mo.

4. Color reagents: 0.5–2 mg/mL OPD (Sigma) in 38 m*M* citric acid, 62 m*M* sodium phosphate buffer (pH 4.9) containing 0.02% (v/v) hydrogen peroxide are used for MMP-1, 2, 7, 8, 9, 13, and TIMP-1 and -2 EIAs, and 0.25 mg/mL tetramethyl-benzidine (TMB) and 75 µg/mL hydrogen peroxide for the MMP-3 EIA. These color reagents are prepared immediately prior to use. OPD is carcinogenic and TMB is noncarcinogenic.

5. Stop solution: 1 *M* sulfuric acid for MMP-1, 2, 7, 8, 9, 13, and TIMP-1 and -2 EIAs and 0.875 *M* sulfuric acid for the MMP-3 EIA.

6. Standard solution: MMPs or TIMPs are diluted to an appropriate concentration with their respective assay buffer and stored at less than –20°C.

7. Samples such as body fluids, cell-conditioned medium, tissue homogenates, etc. (*see* **Note 4**) are stored at –40°C and usually assayed within 3 mo.

3. Methods

3.1. Enzyme-Labeled Antibody

3.1.1. Preparation of IgG-HRP

The purified monoclonal antibody is conjugated with HRP using S-acetyl-mercaptosuccinic anhydride *(17)*.

1. Add the S-acetylmercaptosuccinic anhydride solution to the IgG solution, and incubate at 30°C for 30 min.

2. Add 100 µL of 0.1 *M* Tris-HCl buffer (pH 7.0) containing 0.1 *M* EDTA.

3. Add 100 µL of 1 *M* NH$_2$OH (pH 7.0).

4. Incubate the mixture at 30°C for 5 min.

5. Apply to a Sephadex G-25 column (1.0 × 45 cm) equilibrated with Buffer A containing 5 m*M* EDTA. Pool fractions containing the mercaptosuccinylated IgG and concentrate in a collodion bag (Sartorius, GmbH).

6. Add EMCS solution to 0.3 mL of Buffer C containing 4 mg (100 nmol) of HRP and incubate at 30°C for 1 h.

7. Apply to a Sephadex G-25 column (1.0 × 45 cm) equilibrated with Buffer A. Pool fractions (HRP-maleimide) showing absorbance at 403 nm and concentrate in a collodion bag at 4°C. Do not use NaN$_3$ as a preservative since it inactivates HRP and accelerates the decomposition of maleimide groups.

8. Add 1.6 mg (10 nmol) of the mercaptosuccinylated IgG to 2.0 mg (50 nmol) of the HRP-maleimide, and incubate at 4°C for 20 h. The final concentration of the IgG and HRP is 50–100 and 250–500 µ*M*, respectively.
9. Apply the mixture to an Ultrogel AcA 44 (LKB, France) column (1.5 × 80 cm) equilibrated with Buffer B. Measure the absorbance at 403 and 280 nm and pool fractions corresponding to IgG-HRP.
10. The molar ratio of HRP to IgG in the conjugate can be calculated from the absorbance at 403 nm and 280 nm of the conjugate by use of the extinction coefficients of HRP (2.275 g^{-1}•L•cm^{-1} at 403 nm and 0.73 g^{-1}•L•cm^{-1} at 280 nm) and IgG (1.4 g^{-1}•L•cm^{-1} at 280 nm) and the molecular weights of HRP (40,000) and IgG (155,000). The conjugate having a molar ratio of 2–4 are used in the EIA system.

3.1.2. Preparation of Fab'-HRP

1. Dialyze 10 mg of purified monoclonal antibody (IgG) in 1 mL against 0.1 *M* sodium acetate buffer (pH 4.0–4.5) at 4°C.
2. Add the pepsin solution to the IgG solution, and incubate the mixture is at 37°C for 15–24 h. The incubation time varies according to the characteristics of mouse monoclonal antibody (IgG) used.
3. Adjust the pH of the digested IgG solution to 7 with 0.1 *M* NaOH, and apply the solution to an Ultrogel AcA 44 column (1.5 × 45 cm) equilibrated with Buffer C. Collect F(ab')$_2$ fragments and concentrate by ultrafiltration using a UK10 membrane filter.
4. For the reduction of the purified F(ab')$_2$ *(18)*, add 50 µL of freshly prepared 0.1 *M* 2-mercaptoethylamine in Buffer A containing 5 m*M* EDTA to 1–3 mg of the F(ab')$_2$ in 0.45 mL of Buffer A, and incubate at 37°C for 1.5 h.
5. Apply the mixture to a Sephadex G-25 column (1.0 × 45 cm) equilibrated with Buffer A containing 5 m*M* EDTA. Collect Fab' fragments and concentrate by ultrafiltration using a UK10 membrane filter.
6. Conjugate 2 mg Fab' (43 nmol) with 1.7 mg HRP-maleimide (43 nmol) as described in **Subheading 3.1.1.** The final concentration of the Fab' and HRP is 50–100 µ*M*, respectively.
7. Incubate the mixture at 4°C for 20 h and apply to an Ultrogel AcA 44 column (1.5 × 80 cm) equilibrated with Buffer B. Measure the absorbance at 403 and 280 nm and pool fractions corresponding to Fab'-HRP. The approximate molar ratio of HRP to Fab' can be calculated by use of the extinction coefficients of HRP and Fab' (1.4 g^{-1}•L•cm^{-1} at 280 nm) and the molecular weights of HRP and Fab' (46,000). Use the conjugate having a 0.8–1.2 molar ratio in the EIA system.

3.2. Preparation of Standards

Here we introduce the purification of MMPs and TIMPs by affinity column chromatography using monoclonal antibodies against MMPs and TIMPs *(9,11,14,15)*. For other means of purification of MMPs and TIMPs, see the appropriate section in this book.

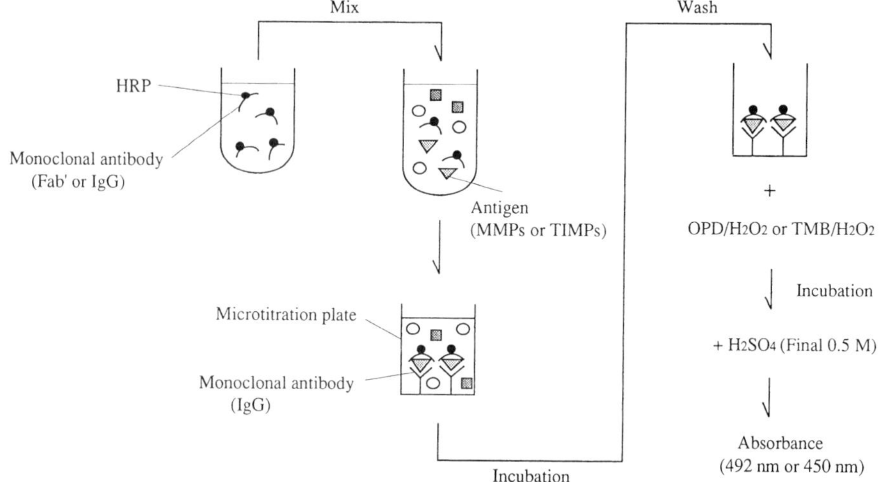

Fig. 1. Schematic representation of one-step sandwich EIA. A microtitration plate or bead was used as the solid. ○ and ■ depict extraneous antigens.

1. Dialyze the sample containing MMPs or TIMPs against 30 m*M* Tris-HCl buffer (pH 7.5) containing 0.1 *M* NaCl, 5 m*M* CaCl$_2$, and 0.05% (w/v) Brij 35.
2. Apply the dialyzed sample to a monoclonal antibody-Sepharose 4B column equilibrated with the same buffer. Wash the column with the same buffer.
3. Elute the bound proteins with 0.2 *M* glycine-HCl buffer (pH 2.5) or 3 *M* KSCN.
4. Dialyze the eluted pool against 30 m*M* Tris-HCl buffer (pH 7.5) containing 0.1 *M* NaCl, 5m*M* CaCl$_2$, and 0.05% (w/v) Brij 35, then concentrate the eluate by ultra-filtration using a UK 10 membrane filter.

Purity of the enzyme or inhibitor is verified by SDS-PAGE as described by Laemmli and Favre *(19)*. The protein concentration of MMPs and TIMPs is determined by micro BCA protein assay reagent using BSA as a standard protein according to the instructions of Pierce.

3.3. Procedure of One-Step Sandwich EIAs for MMPs and TIMPs

3.3.1. Plate Method (see Notes 3 and 5)

1. Mix an aliquot of standard MMP or specimen (*see* **Notes 4** and **6**) with 100 µL of assay buffer containing 0.1–2 µg/mL of the HRP-labeled antibody in plastic tubes or vinyl plates.
2. Transfer a 100-µL aliquot of the solution to each well of a microtitre plate coated with antibody, and incubate the plate at room temperature for the time indicated in **Table 1**, without shaking.
3. Wash each well with the respective wash buffer, then incubate with the color reagent at room temperature for 15–30 min.

4. Stop the reaction by adding 100 μL of stop solution, and read the absorbance at 492 nm using a microplate reader (**Table 1, Fig. 1**).

3.3.2. Bead Method

The MMP-3 EIA employs polystyrene beads as a solid phase *(9)*.

1. Incubate a 50-μL aliquot of standard proMMP-3 or of specimen in the assay buffer with the antibody-coated beads and 300 μL of Fab'-HRP in 30 mM sodium phosphate buffer containing 0.1 M NaCl and 10 mM EDTA for 1 h at room temperature.
2. Wash the beads 3 times with the wash buffer and then incubate with the color reagent for 30 min at room temperature.
3. Stop the reaction by adding 1.4 mL of stop solution, and read the absorbance at 450 nm in a micro-flow spectrophotometer (**Table 1**).

3.3.3. Calculations

1. Calculate the net absorbance value for each standard by subtracting the value for the buffer solution (zero standard) from the value for the individual dilution.
2. Plot the net absorbance as ordinate against the standard concentration as abscissa on graph paper and draw a smooth curve that fits the plotted points.
3. Using the net absorbance value for a specimen, read the concentration off of the standard curve. When a diluted specimen is used, multiply the value obtained by the dilution factor.

4. Notes

1. IgG or Fab' is used as the HRP-labeled antibody in our one-step sandwich EIAs. The preparation of IgG-HRP is relatively simple compared to that of Fab'-HRP, although IgG-HRP generally shows a higher background. The sensitivity of the EIA depends on the ratio of specific binding (absorbance of top standard) to nonspecific binding (zero standard). Therefore, at the present we usually use Fab'-HRP, which shows low nonspecific binding.
2. Regarding the preparation of Fab' from mouse IgG, the optimal pH of the buffer and incubation time in addition to pepsin concentration must be determined, because the digestion pattern of individual mouse monoclonal antibodies obtained with pepsin is different. Normally, we employ IgG, especially IgG1 type, since sufficient examination for the preparation of Fab' from IgA or IgM has not been performed yet.
3. We have selected a one-step sandwich EIA using two monoclonal antibodies to increase the sensitivity, specificity, and precision of the assay. The characteristics of MMP and TIMP EIAs are summarized in **Table 2**. The characteristics of these EIAs depend on those of the antibodies used. The MMP-1 EIA does not recognize the enzyme treated with EDTA, because the monoclonal antibody used is considered to recognize a higher order structure of the enzyme and not to recognize the MMP-1 whose structure is altered by EDTA. On the other hand, the

Table 1
Procedures for MMPs and TIMPs EIAs

	MMP-1	MMP-2	MMP-3	MMP-7	MMP-8	MMP-9	MMP-13	TIMP-1	TIMP2
Solid	Plate	Plate	Bead	Plate	Plate	Plate	Plate	Plate	Plate
HRP labeled	Fab'	IgG	Fab'	IgG	IgG	IgG	Fab'	Fab'	IgG
Color reagent	OPD	OPD	TMB	OPD	OPD	OPD	OPD	OPD	OPD
Sample or standard volume	25 μL	10 μL	50 μL	75 μL	20 μL	10 μL	20 μL	10 μL	10 μL
HRP-labeled antibody	150 μL	100 μL	300 μL	175 μL	100 μL	100 μL	100 μL	150 μL	100 μL
Immunoreaction									
Volume/assay	100 μL	100 μL	350 μL	100 μL	100 μL	100 μL	100 μL	100 μL	100 μL
Temperature	r.t.	r.t.	r.t.	4°C	r.t.	r.t.	4°C	r.t.	r.t.
Time	30 min	1 h	1 h	24 h	1 h	1 h	20 h	30 min	1 h
Wash	4 times	3 times	3 times	3 times	3 times	3 times	4 times	3 times	3 times
Color reaction									
Temperature	r.t.	r.t.	r.t.	r.t.	r.t.	r.t.	r.t.	r.t.	r.t.
Time	15 min	20 min	30 min	15 min	20 min	30 min	30 min	20 min	20 min
Stop solution	100 μL	100 μL	100 μL	100 μL	100 μL	100 μL	100 μL	100 μL	100 μL
Absorbance	492 nm	492 nm	450 nm	492 nm	492 nm	492 nm	492 nm	492 nm	492 nm
References	*7*	*8*	*9*	*10*	*11*	*12*	*13*	*14*	*15*

Table 2
Characterizations of MMPs and TIMPs EIAs[a]

	MMP-1	MMP-2	MMP-3	MMP-7	MMP-8	MMP-9	MMP-13	TIMP-1	TIMP2
Sensitivity (ng/mL)	0.3	0.24	9.2	0.16	0.34	0.24	0.08	0.15	1.6
(pg/assay)	6	2.4	460	1.5	5.7	2.2	1.3	1.5	16
Precision									
Intra-assay (CV%)	1.9–3.5	3.6–6.0	4.5–7.3	1.0–5.0	5.2–7.5	1.8–3.9	4.5–6.9	1.2–4.4	1.5–3.2
Inter-assay (CV%)	2.2–4.8	4.4–7.0	6.0–8.0	7.0–9.8	4.2–9.2	1.8–6.2	4.8–7.1	2.8–4.3	5.6–8.7
Recovery (%)	103 ± 1.2	103 ± 6.6	100 ± 5.4	92.4 ± 9.9	96.0 ± 9.0	100 ± 8.2	86 ± 100	99.4 ± 2.7	91.0 ± 5.1
Immunoreactivity									
Proform	100%	100%	100%	100%	100%	100%	100%		
Active form	125	0	29~47	0	135	~4	100		
Complex form									
with TIMP-1	48	(0)	54~62	(0)	~6	(pro) 100 (active) ~5	50		
with TIMP-2	~3	(pro) 100 (active) (0)	33~53	(0)	~6	~3	N.T.		
Free form								100%	100%
Complex form									
with MMP-1								100	350
with MMP-2								96	330 (pro) ~5
with MMP-3								105	105
with MMP-9								98 (pro) 92	300
Normal levels									
Human sera (ng/mL)	8.5 ± 5.2 ($n = 120$)	570 ± 118 ($n = 213$)	32 ± 16 ($n = 118$)	10.7 ± 18.8 ($n = 80$)	N.T.	N.T.	<0.13 ($n = 50$)	175 ± 39 ($n = 81$)	56 ± 20 ($n = 127$)
Human plasma (ng/mL)	N.T.	N.T.	N.T.	N.T.	1.5–28 ($n = 197$)	38 ± 13 ($n = 50$)	N.T.	69 ± 13 ($n = 37$)	50 ± 15 ($n = 39$)
References	7	8	9	10	11	12	13	14	15

[a]N.T. not tested

sensitivity of the MMP-7 EIA in the absence of EDTA was 640-fold lower than that in the presence of EDTA *(10)*. Requirement for EDTA-treatment of proMMP-7 in the immunoreaction can be explained by the fact that the antibody used in the assay was developed by immunization with proMMP-7 treated with EDTA. In MMP-2, 7 and 9 EIAs, only the proform and not the active form of the enzyme is recognized, because one of the monoclonal antibodies used recognizes the propeptide region of the respective enzyme. MMP-2 and 9 EIAs also detect the proMMP-2 and 9 complexed with TIMP-2 and TIMP-1, respectively. For measurement of the active form in addition to that of the proform of the enzyme, a competitive EIA using one antibody that reacts with the central or C-terminal region of these enzymes may be adequate.

4. Regarding the blood sample applied to EIAs, sera are generally used rather than plasma. However, it has been reported that MMP-8 and 9 are secreted by neutrophils in human blood *(4)*. We also confirmed that the secretion of these enzymes into serum but not into plasma depended on the temperature and time that transpired between blood sampling and centrifugation. Therefore, plasma prepared by use of EDTA or other agent as a specimen should be used to determine the exact value in the measurement of MMP-8 and 9. The concentration of TIMP-1 in human plasma was about three-fold lower than that in human sera, although the level of TIMP-2 was almost the same between the two. It has been reported that purified human platelets contain TIMP-1 that is immunologically, functionally, and chromatographically identical to that produced by human skin fibroblasts *(20)*. The difference of TIMP-1 concentration may be attributed to release of TIMP-1 from platelets during platelet activation.

5. A one-step sandwich EIA has the disadvantage of the prozone phenomenon, which occurs when a large amount of antigen is present compared to the amount of antibodies used in the first immunoreaction. To avoid this phenomenon, a two-step sandwich EIA can be applied to measure the antigen. However, the sensitivity of a two-step sandwich assay is lower than that of a one-step sandwich assay using the same antibodies. The appropriate EIA system should be selected based on the level of antigens in the specimens.

6. Sample dilution and recovery tests should be employed on specimens such as human sera or plasma, because EIAs are often affected by proteins in specimens. This is important to determine the exact concentration of antigens in specimens. Good recovery and linearity of values measured by these EIAs can be obtained with more than 3.3-fold dilutions of samples such as sera and plasma. The intra-assay, inter-assay ($n = 8$–10, respectively), and recovery test using samples are carried out to ensure the quality. CVs of less than 15% and recoveries of $100 \pm 15\%$ are acceptable.

References

1. Birkedal-Hansen, H., Moore, W. G. I., Bodden, M. K., Windsor, L. J., Birkedal-Hansen, B., Decarlo, A., and Engler, J. A. (1993) Matrix metalloproteinases: a review. *Oral Biol. Med.* **4**, 197–250.

2. Cooksley, S., Hipkiss, J. B., Tickle, S. P., Holmes-Ievers, E., Docherty, A. J. P., Murphy, G., and Lawson, A. D. G. (1990) Immunoassays for the detection of human collagenase, stromelysin, tissue inhibitor of metalloproteinases (TIMP) and enzyme-inhibitor complex. *Matrix* **10**, 285–291.

3. Ballin, M., Gomez, D. E., Sinha, C. C., and Thorgeirsson, U. P. (1988) Ras oncogene mediated induction of a 92 kDa metalloproteinase: strong correlation with the malignant phenotype. *Biochem. Biophys. Res. Commun.* **154**, 832–838.

4. Kjeldsen, L., Bjerrum, O. W., Hovgaard, D., Johnson, A. H., Sehested, M., and Borregaard, H. (1992) Human neutrophil gelatinase: a marker for circulating blood neutrophils. Purification and quantitation by enzyme linked immunosorbent assay. *Eur. J. Haematol.* **49**, 180–191.

5. Zucker, S., Lysik, R. M., Zarrabi, M. H., and Moll, U. (1993) Mr 92,000 type IV collagenase is increased in plasma of patients with colon cancer and breast cancer. *Cancer Res.* **53**, 140–146.

6. Bergmann, U., Michaelis, J., Oberhoff, R., Knäuper, V., Beckmann, R., and Tscheshe, H. (1989) Enzyme linked immunosorbent assay (ELISA) for the quantitative determination of human leukocyte collagenase and gelatinase. *J. Clin. Chem. Clin. Biochem.* **27**, 351–359.

7. Zhang, J., Fujimoto, N., Iwata, K., Sakai, T., Okada, Y., and Hayakawa, T. (1993) A one-step sandwich enzyme immunoassay for human matrix metalloproteinase 1 (interstitial collagenase) using monoclonal antibodies. *Clin. Chim. Acta.* **219**, 1–14.

8. Fujimoto, N., Mouri, N., Iwata, K., Ohuchi, E., Okada, Y., and Hayakawa, T. (1993) A one-step sandwich enzyme immunoassay for human matrix metalloproteinase 2 (72-kDa gelatinase/type IV collagenase) using monoclonal antibodies. *Clin. Chim. Acta.* **221**, 91–103.

9. Obata, K., Iwata, K., Okada, Y., Kohrin, Y., Ohuchi, E., Yoshida, S., Shinmei, M., and Hayakawa, T. (1992) A one-step sandwich enzyme immunoassay for human matrix metalloproteinase 3 (stromelysin-1) using monoclonal antibodies. *Clin. Chim. Acta.* **211**, 59–72.

10. Ohuchi, E., Azumano, I., Yoshida, S., Iwata, K., and Okada, Y. (1996) A one-step sandwich enzyme immunoassay for human matrix metalloproteinase 7 (matrilysin) using monoclonal antibodies. *Clin. Chim. Acta.* **244**, 181–198.

11. Matsuki, H., Fujimoto, N., Iwata, K., Knäuper, V. Okada, Y., and Hayakawa, T. (1996) A one-step sandwich enzyme immunoassay for human matrix metalloproteinase 8 (neutrophil collagenase) using monoclonal antibodies. *Clin. Chim. Acta.* **244**, 129–143.

12. Fujimoto, N., Hosokawa, N., Iwata, K., Shinya, T., Okada, Y., and Hayakawa, T. (1994) A one-step sandwich enzyme immunoassay for inactive precursor and complexed forms of human matrix metalloproteinase 9 (92 kDa gelatinase/type IV collagenase, gelatinase B) using monoclonal antibodies. *Clin. Chim. Acta.* **231**, 79–88.

13. Tamei, H., Azumano, I., Iwata, K., Yoshihara, Y., López-Otín, C., Vizoso, F., Knäuper, V., and Murphy, G. (1998) One-step sandwich enzyme immunoassays

for human matrix metalloproteinase 13 (collagenase-3) using monoclonal antibodies. *Connect. Tissue* **30**, 15–22.

14. Kodama, S., Iwata, K., Iwata, H., Yamashita, K., and Hayakawa, T. (1990) Rapid one-step sandwich enzyme immunoassay for tissue inhibitor of metalloproteinases: an application of rheumatoid arthritis serum and plasma. *J. Immunol. Methods* **127**, 103–108.

15. Fujimoto, N., Zhang, J., Iwata, K., Shinya, T., Okada, Y., and Hayakawa, T. (1993) A one-step sandwich enzyme immunoassay for tissue inhibitor of metalloproteinase-2 using monoclonal antibodies. *Clin. Chim. Acta.* **220,** 31–45.

16. Kodama, S., Kishi, J., Obata, K., Iwata, K., and Hayakawa, T. (1987) Monoclonal antibodies to bovine collagenase inhibitor. *Col. Rel. Res.* **7**, 341–350.

17. Ishikawa, E., Imagawa, M., Hashida, S., Yoshitake, S., Hamaguchi, Y., and Ueno, T. (1983) Enzyme-labeling of antibodies and their fragments for enzyme immunoassay and immunohistochemical staining. *J. Immunoassay* **4**, 209–327.

18. Hashida, S., Imagawa, M., Inoue, S., Ruan, K.-h., and Ishikawa, E. (1984) More useful maleimide compounds for the conjugation of Fab' to horseradish peroxidase through thiol group in the hinge. *J. Appl. Biochem.* **6**, 56–63.

19. LaemmLi, U. K. and Favre, M. (1973) Maturation of the head of bacteriophage T4. 1. DNA packaging events. *J. Mol. Biol.* **80**, 597–599.

20. Cooper, T. W., Eisen, A. Z., Stricklin, G. P., and Welgus, H. G. (1985) Platelet derived collagenase inhibitor: characterization and subcellular localization. *Proc. Natl. Acad. Sci. USA* **82**, 2779–2783.

19

Immunohistochemistry of MMPs and TIMPs

Yasunori Okada

1. Introduction

Immunohistochemical techniques are a convenient method to identify the cells responsible for the production of MMPs and TIMPs in local tissues under pathophysiological conditions. Direct and indirect methods are presented for immunohistochemistry, but the latter is usually utilized for the staining of MMPs and TIMPs because it is more convenient and sensitive than the former. Specific recognition of the MMP and TIMP species by the primary antibody which has no cross-reactivity with other molecules is essential for immunohistochemistry, and thus monoclonal antibodies are considered to be suitable for the purpose. In this chapter, immunohistochemistry using commercially available primary monoclonal antibodies against human MMPs and TIMPs is described. Human tissue samples obtained at surgery are fixed with periodate-lysine-paraformaldehyde (PLP) fixative, embedded in paraffin wax, and paraffin sections are immunostained according to the avidin-biotin-peroxidase complex (ABC) and/or immunogold-silver staining (IGSS) methods. Treatment of the tissues with monensin prior to fixation may be necessary to augment intracytoplasmic staining especially in mesenchymal tissues, but not in carcinoma tissues. Common problems during immunostaining are also given with a practical guide.

2. Materials

2.1. Antibodies against MMPs and TIMPs

Polyclonal antibodies were previously utilized for the immunohistochemistry for MMPs and TIMPs *(1–4)*. However, since more than 19 different MMPs and 4 TIMPs have been identified so far, it becomes difficult to raise polyclonal antibodies which are monospecific to a single species of MMP or TIMP. Thus,

From: *Methods in Molecular Biology, vol. 151: Matrix Metalloproteinase Protocols*
Edited by: I. Clark © Humana Press Inc., Totowa, NJ

monoclonal antibodies are recommended to be used for immunohistochemistry. We have developed and characterized monoclonal antibodies against MMPs and TIMPs by collaboration with Fuji Chemical Industries, Ltd. *(5–11)*. The following antibodies applicable to immunohistochemistry are now commercially available from Fuji Chemical Industries, Ltd., (530 Chokeiji, Takaoka, Toyama 933-8511, Japan; URL, http://www.fuji-chemi.co.jp):

1. Antihuman MMP-1 (41-1E5, 2 µg/mL).
2. Antihuman MMP-2 (42-5D1, 2 µg/mL).
3. Antihuman MMP-3 (55-2A4, 8 µg/mL).
4. Antihuman MMP-7 (141-7B2, 10 µg/mL).
5. Antihuman MMP-8 (115-13D2, 10 µg/mL).
6. Antihuman MMP–9 (56-2A4, 2 µg/mL).
7. Antihuman MMP-13 (181-15A12, 15 µg/mL).
8. Antihuman MT1-MMP (114-6G6, 10 µg/mL).
9. Antimouse MT2-MMP cross-reactive with human MT2-MMP (162–22G5, 30 µg/mL).
10. Antihuman TIMP-1 (147-6D11, 10 µg/mL).
11. Antihuman TIMP-2 (67-4H11, 1 µg/mL).
12. Antihuman TIMP-3 (136-13H4, 30 µg/mL).

Concentrations (µg/mL) suitable for immunohistochemistry using the ABC method are described in the parenthesis. Also, *see* **Note 1** for storage of antibodies.

2.2. Tissue Preparation

1. Because preservation of the structure and antigens in the tissues is important for immunohistochemistry, the tissues must be fixed before being immunostained. However, standard fixation processes with formaldehyde or especially glutaraldehyde sometimes cross-link proteins so strongly that antigenic sites become obscure and the immunoreaction cannot take place. Thus, a compromise between good tissue preservation and antigen availability has to be reached for each antigen. As we have screened the monoclonal antibodies by immunostaining tissues fixed with PLP fixative *(12)*, all the antibodies listed in **Subheading 2.1.** can be used for immunohistochemistry in tissues fixed with PLP fixative or paraformaldehyde fixative. However, the formaldehyde-fixed tissues should be checked for each antibody.
2. Human tissues have to be sliced or cut into blocks approx 3–5 mm thick with a razor blade if possible and then fixed for 12–24 h. To preserve good structure, the tissues should be carefully handled without being damaged and processed at the latest within 1–3 h after surgery.

2.3. Reagents

1. Stock solution of 10 m*M* monensin (Sigma, St. Louis, MO) dissolved in 100% ethanol can be stored at –20°C for 2–3 mo.

2. PLP fixative *(12)* is freshly prepared.
3. Poly(L-lysine)-coated slides are prepared by soaking slides in a 0.1% solution in distilled water and dry at 37°C overnight. Store the poly(L-lysine) solution at –20°C and refreeze after use. Slides coated with 3-aminopropyl-triethoxysilane are commercially available from Matsunami Glass IND., Ltd. (Tokyo, Japan).
4. Biotinylated secondary antibody (horse antimouse IgG) (Vector Laboratories, Burlingame, CA).
5. Avidin-biotin-peroxidase complex kit (Vector Laboratories).
6. Diaminobenzidine (DAB)-hydrogen peroxidase reaction solution should be freshly prepared and H_2O_2 added just before reaction. As DAB is difficult to dissolve, DAB solution (0.03% 3,3'-diaminobenzidine-tetrahydrochloride in 50 m*M* Tris-HCl buffer pH 7.6) is made by stirring and then 30% H_2O_2 is added at a final concentration of 0.006%.
7. 10 m*M* PBS pH 7.6 phosphate buffered saline (PBS): 29*g* Na_2HPO_4, 2*g* KH_2PO_4, 80*g* NaCl, 2g KCl in 10 L of distilled water.
8. Normal serum (from same species as secondary antibody used).
9. IGSS washing buffer: 10 m*M* PBS containing 0.8% bovine serum albumin, 0.1% gelatin and 0.02% NaN_3.
10. Colloidal gold-labeled secondary antibody (AuroProbe One GAM, Amersham, Buckinghamshire, UK).
11. A silver enhancement kit (Intense M, Amersham).
12. 25% Glutaraldehyde (TAAB Laboratories Equipment Limited, Berkshire, UK).
13. Mayer's hematoxylin and 1% methyl green solution in 20 m*M* Veronal acetate buffer pH 4.0.

3. Methods

There are two methods for immunohistochemistry: direct and indirect methods. A labeled primary antibody is applied directly to the tissue preparation in the direct method. On the other hand, in the indirect method the primary antibody is unlabeled and is identified by a labeled secondary antibody raised against the immunoglobulin of the species providing the primary antibody. For the detection of MMPs and TIMPs in human tissues, the indirect method is usually recommended, since it is more convenient and sensitive than the direct method. I focus on immunohistochemistry on paraffin sections of human pathological tissues by the ABC and IGSS methods, which are the most popular and sensitive to detect small amounts of antigens such as MMPs and TIMPs. Monensin treatment of the tissues is sometimes necessary especially if a less sensitive immunostaining method such as immunofluorescent microscopy is used *(1–3)*, since MMPs such as MMP-1 and MMP-3 are secreted without intracytoplasmic storage immediately after synthesis in fibroblasts and chondrocytes *(1,4,13)* (*see* **Note 2**).

3.1. Preparation of Paraffin Sections

1. Fix the tissue slices or blocks with PLP fixative for 12–24 h. Other fixatives including 4% paraformaldehyde can be used instead of PLP fixative. Prior to the fixation, some samples may be cultured for 3 h in the presence of 2–5 μM monensin in culture medium containing 10% fetal calf serum in a CO_2 incubator.
2. Dehydrate in graded ethanols, and embed in paraffin wax.
3. Make 4–6 μm sections in thickness and mount on the poly(L-lysine)-coated slides. Slides coated with 3-aminopropyl-triethoxysilane can be used for the ABC method, and should be used for the IGSS method since the charge interferes with silver intensification reaction.

3.2. Immunostaining by Avidin-Biotin-Peroxidase Complex Method

1. Dewax the sections in Histoclear or xylene, 3 changes, 5 min each.
2. Rehydrate in graded ethanols for 5 min in 100%, for 5 min in 95% and then for 1 min in 75%, and bring to distilled water (*see* **Note 3**).
3. Block endogenous peroxidase with 0.3% H_2O_2 and 0.1% NaN_3 in distilled water for 30 min, wash in tap water for 5 min and then in distilled water for 5 min.
4. Rinse in 10 mM phosphate-buffered-saline pH 7.6 (PBS), 3 changes, 5 min each.
5. Incubate with 10% normal serum (from the same animal species as the biotinylated antibody) for 30 min at room temperature (*see* **Notes 4** and **5**).
6. Incubate with primary monoclonal antibody to MMP or TIMP species in 10 mM PBS containing 5% normal serum and 0.02% NaN_3 for 90 min at room temperature or overnight at 4°C (*see* **Note 6**).
7. Rinse in PBS, 3 changes, 5 min each.
8. Incubate with biotinylated secondary antibody (1:200 dilution) in 10 mM PBS containing 10% normal serum and 0.02% NaN_3 for 30 min at room temperature.
9. Rinse in PBS, 3 changes, 5 min each.
10. React with preformed ABC solution for 30 min at room temperature. The ABC solution should be made by mixing A and B solutions and incubated for 30 min prior to the reaction.
11. Rinse in PBS, 3 changes, 5 min each.
12. React with DAB-hydrogen peroxide solution for 3–20 min and stop reaction by transferring to distilled water. Check the reaction (brown color) by observing under microscope.
13. Rinse in distilled water and then with running tap water for 5–10 min.
14. Counterstain the sections with hematoxylin for 30 s or methyl green for 10–30 min and wash in running tap water.
15. Dehydrate with 75%, 95% and 100% ethanol for 3 min each and then with Histoclear or xylene for 5 min.
16. Mount sections in a permanent medium with cover glasses and observe under microscope (*see* **Notes 7** and **8**).

3.3. Immunostaining by Immunogold-Silver Staining Method

1. Dewax sections on aminosilane-coated slides in Histoclear and rehydrate through graded ethanols, 3 changes, 5 min each (*see* **Note 3**).
2. Rinse in 10 m*M* PBS, 3 changes, for 5 min each.
3. Block nonspecific reactions with 5% normal serum (from the same animal species as the secondary antibody) in IGSS washing buffer for 30 min at room temperature (*see* **Notes 4** and **5**).
4. React with primary antibody in IGSS washing buffer for 90 min at room temperature or overnight at 4°C (*see* **Note 6**).
5. Rinse in the washing buffer, 3 changes, for 10 min each.
6. Incubate with colloidal gold-labeled secondary antibody (1:50 dilution in 10 m*M* PBS containing 1% bovine serum albumin) for 60 min.
7. Wash in the washing buffer, 3 changes, for 15 min each.
8. Rinse in PBS, 3 changes, for 5 min each.
9. Postfix with 2% glutaraldehyde in PBS for 10 min at room temperature.
10. Wash in distilled water, 5 changes, for 5 min each.
11. Treat by silver enhancement reaction with a silver enhancement kit (Intense M, Amersham) for 15–25 min at room temperature. The reaction solution is prepared by mixing the same volume of A and B solutions just before the reaction.
12. Wash in distilled water, 5 changes, for 5 min each.
13. Counterstain with hematoxylin or methyl green.
14. Dehydrate, clear and mount in a permanent medium (*see* **Notes 7** and **8**).

4. Notes

1. Primary monoclonal antibodies may be stored frozen and should not be repeatedly thawed and refrozen. Thus, they should be divided into aliquots for several staining and stored frozen. When stored at 4°C, sodium azide should be added to the antibody solution to a final concentration of 0.02% to prevent bacterial growth. Since azide inhibits peroxidase activity, it should not be included in the buffers used to dilute peroxidase-conjugated antibody.
2. Since most MMPs except for membrane-type MMPs are considered to be secreted immediately after synthesis into the extracellular spaces without intracytoplasmic storage, intracellular accumulation by monensin treatment is sometimes necessary for the definite detection. This is true in the case of many mesenchymal cells such as fibroblasts, chondrocytes and osteoblasts *(1–6,10)*. However, weaker staining can be obtained in the cells of the tissues without monensin treatment when sensitive immunostaining methods such as the ABC method are used *(4–6)*. MMP-9 present in neutrophils, macrophages, and osteoclasts as well as MMP-8 in neutrophils are detectable in the tissue sections without monensin treatment *(5–7,14)* since these MMPs are stored within the secretory granules. MMPs in carcinoma cells can also be stained in the carcinoma tissues without the pretreatment *(8,11)*. The secretion pathways of the MMPs in the carcinoma cells may be different from the normal cells, being stored intracellularly.

3. It is important that paraffin sections are not allowed to become dry at any time during the immunostaining, as drying results in a poor final preparation.

4. Treatment of the sections with 0.04% trypsin (Sigma) in Tris-HCl buffer pH 7.8 for 5–15 min at 37°C may be necessary for the detection of TIMP-1. Microwave treatment of the sections at 500W in 10 m*M* citrate buffer pH 6.0 (for 4 min, 3 times) may be useful for the detection of TIMP-3. Aminosilane-coated slides should be used for the micowave treatment to avoid detachment of the sections during the treatment. These treatments are carried out at the step just before the incubation with normal serum (**steps 5** and **3** for the ABC and IGSS methods, respectively).

5. Nonspecific immunostaining is sometimes obtained by the ABC method. The isoelectric point of avidin (10.5) can cause the protein to react with negatively charged tissue components, such as cell membranes and nuclei or proteoglycans in the extracellular spaces, producing unwanted staining. Other unwanted binding can be seen between avidin and endogenous biotin which is naturally present in normal human tissues such as liver, breast, adipose tissue and kidney, and also seen between avidin and mast cell granules. Thus, although extracellular matrix such as cartilage extracellular matrix is sometimes stained with antibodies against MMPs *(4,10)*, we should carefully examine if the staining is specific. Use of streptavidin or blocking the reaction with free avidin prior to the ABC reaction is useful to avoid such a problem. In addition, staining using the antibody absorbed with corresponding antigen, with nonimmune IgG and/or by different immunostaining methods is essential *(4–7)*.

6. To avoid misunderstanding of false positive and negative staining, immunoreactions with nonimmune IgG, absorption test and/or by two different immunohistochemical methods are recommended *(4–7)*. Immunoblotting and immunoassays using homogenates or culture media of the tissues are useful methods to confirm the immunostaining data *(4–11)*.

7. For the preparation of black-and-white photos by light microscope from the sections counterstained with hematoxylin, a contrast filter (Nikon, MXA 20170/ MF45IF440) is useful to abolish blue color of hematoxylin and enhance the brown color of the immunoreaction.

8. Discrepancy in the cellular localization of the protein and mRNA for MT1-MMP and MMP-1 is reported in human carcinoma tissues *(9,15)*. The reason is not clear at this moment, but it may be related to difference in stability of the proteins and mRNA in the carcinoma cells *(9)*.

References

1. Hembry, R. M., Murphy, G., Cawston, T. E., Dingle, J. T., and Reynolds, J. J. (1986) Characterization of a specific antiserum for mammalian collagenase from several species:Immunolocalisation of collagenase in rabbit chondrocytes and uterus. *J. Cell Sci.* **81**, 105–123.

2. Okada, Y., Takeuchi, N., Tomita, K., Nakanishi, I., and Nagase, H. (1989) Immunolocalisation of matrix metalloproteinase 3 (stromelysin) in rheumatoid synovioblasts (B cells):correlation with rheumatoid arthritis. *Ann. Rheum. Dis.* **48**, 645–653.

3. Okada, Y., Gonoji, Y., Nakanishi, I., Nagase, H., and Hayakawa, T. (1990) Immunohistochemical demonstration of collagenase and tissue inhibitor of metalloproteinases (TIMP) in synovial lining cells of rheumatoid synovium. *Virchows Archiv. B Cell Pathol.* **59**, 305–312.

4. Okada, Y., Shinmei, M., Tanaka, O., Naka, K., Kimura, A., Nakanishi, I., Bayliss, M. T., Iwata, K., and Nagase, H. (1992) Localization of matrix metalloproteinase 3 (stromelysin) in osteoarthritic cartilage and synovium. *Lab. Invest.* **66**, 680–690.

5. Okada, Y., Naka, K., Kawamura, K., Matsumoto, T., Nakanishi, I., Fujimoto, N., Sato, H., and Seiki, M. (1995) Localization of matrix metalloproteinase 9 (92-kilodalton gelatinase/type collagenase-gelatinase B) in osteoclasts:Implications for bone resorption. *Lab. Invest.* **72**, 311–322.

6. Yokohama, Y., Matsumoto, T., Hirakawa, M., Kuroki, Y., Fujimoto, N., Imai, K., and Okada, Y. (1995) Production of matrix metalloproteinases at the bone-implant interface in loose total hip replacements. *Lab. Invest.* **72**, 899–911.

7. Ueda, Y., Imai, K., Tsuchiya, H., Fujimoto, N., Nakanishi, I., Katsuda, S., Seiki, M., and Okada, Y. (1996) *Am. J. Pathol.* **148**, 611–622.

8. Nomura, H., Fujimoto, N., Seiki, M., Mai, M., and Okada, Y. (1996) Enhanced production of matrix metalloproteinases and activation of matrix metalloproteinase 2 (gelatinase A) in human gastric carcinomas. *Int. J. Cancer* **69**, 9–16.

9. Ueno, H., Nakamura, H., Inoue, M., Imai, K., Noguchi, M., Sato, H., Seiki, M., and Okada, Y. (1997) Expression and tissue localization of membrane-types 1, 2, and 3 matrix metalloproteinases in human invasive breast carcinomas. *Cancer Res.* **57**, 2055–2060.

10. Ohta, S., Imai, K., Yamashita, K., Matsumoto, T., Azumano, I., and Okada, Y. (1998) Expression of matrix metalloproteinase 7 (matrilysin) in human osteoarthritic cartilage. *Lab. Invest.* **78**, 79–87.

11. Yamashita, K., Azumano, I., Mai, M., and Okada, Y. (1998) Expression and tissue localization of matrix metalloproteinase 7 (matrilysin) in human gastric carcinomas. Implications for vessel invasion and metastasis. *Int. J. Cancer* **79**, 187–194.

12. McLean, I. W. and Nakane, P. K. (1974) Periodate-lysine-paraformaldehyde fixative: A new fixative for immunoelectron microscopy. *J. Histochem. Cytochem.* **22**, 1077–1083.

13. Nagase, H., Brinckerhoff, C. E., Vater, C. A., and Harris, E. D., Jr. (1983) Biosythesis and secretion of procollagenase by rabbit synovial fibroblasts. Inhibition of procollagenase secretion by monensin and evidence for glycosylation of procollagenase. *Biochem. J.* **214**, 281–288.

14. Matsuki, H., Fujimoto, N., Iwata, K., Knuper, V., Okada, Y., and Hayakawa, T. (1996) A one-step sandwich enzyme immunoassay for human matrix metalloproteinase 8 (neutrophil collagenase) using monoclonal antibodies. *Clin. Chim. Acta.* **244**, 129–143.

15. Okada, A., Bellocq, J. P., Rouyer, N., Chenard, M., Rio, M., Chambon, P., and Basset, P. (1995) Membrane-type matrix metalloproteinase (MT-MMP) gene is expressed in stromal cells of human colon, breast, and head and neck carcinomas. *Proc. Natl. Acad. Sci. USA.* **92**, 2730–2734.

20

Detecting Polymorphisms in MMP Genes

Shu Ye and Adriano M. Henney

1. Introduction

The essential role of genes is in encoding structural proteins and enzymes which enable the cell or organism to maintain homeostasis in the face of the environmental challenges experienced *(1)*. DNA containing such genetic information varies from one species to another. Even within a species, there are interindividual differences in DNA sequence, giving rise, for example, to variation in fitness and appearance. Naturally occurring sequence variants, many of which are neutral sequence changes, have been estimated to exist approximately every 200–300 bp throughout the human genome *(2)*. However, there are some sequence variants which affect the control of gene expression, the biosynthesis and transport of the gene product, or the function of the gene product itself *(2)*. Consequently, individuals within a population will have different abilities to maintain homeostasis, and the failure of some individuals to do so will lead to the development of disease.

Naturally occurring variations in the genes encoding matrix metalloproteinases (MMPs) and tissue inhibitors of matrix metalloproteinases (TIMPs) may lead to an imbalance in connective tissue turnover, through effects on gene expression or the function of gene product. A mutation in the gene for TIMP-3 has been found to cause Sorsby's fundus dystrophy, a genetic disease in the eye *(3)*. Using the polymerase chain reaction-single-strand conformation polymorphism (PCR-SSCP) analysis technique, we have detected a polymorphism in the promoter region of the MMP-3 (stromelysin) gene, which exerts allele-specific effects on promoter activity through differential binding of a transcription repressor protein, and is associated with progression of coronary atherosclerosis *(4,5)*.

A number of molecular biology techniques have been applied, or specifically developed, to detect mutations *(6)* (**Table 1**). Of these different techniques, poly-

From: *Methods in Molecular Biology, vol. 151: Matrix Metalloproteinase Protocols*
Edited by: I. Clark © Humana Press Inc., Totowa, NJ

Table 1
Mutation Detection Methods

Technique	Comments
Cytogenetic techniques	Detecting large deletions, insertions, and rearrangements that change chromosome structures
Pulsed field gel electrophoresis (PFGE)	Detecting large deletions, insertions, and reiterations that change gene structures
Polymerase chain reaction (PCR)	Detecting large deletions, insertions, and reiterations that change gene structures
Northern analysis	Detecting mutations that alter gene expression levels or splicing
Southern analysis	Detecting point mutations in restriction sites
Restriction fragment length polymorphism (RFLP) analysis	
Denaturing gradient gel electrophoresis (DGGE)	Detecting point mutations, and small deletions and insertions. Difficult to set up and requiring special equipment
Temperature gradient gel electrophoresis (TGGE)	Detecting point mutations, and small deletions and insertions. Difficult to set up and requiring special equipment
Chemical cleavage analysis	Detecting point mutations, and small deletions and insertions. Providing some indication of the location of mutations within the tested DNA fragment. Using toxic chemicals

Table 1 (*continued*)
Mutation Detection Methods

Technique	Comments
Ribonuclease A cleavage analysis	Detecting point mutations, and small deletions and insertions Providing some indication of the location of mutations within the tested DNA fragment
Heteroduplex analysis	Detecting point mutations, and small deletions and insertions
Single strand conformation polymorphism (SSCP) analysis	Detecting point mutations, and small deletions and insertions Simple, fast and cost-effective Providing no information of the location of mutations within the tested DNA fragment
Dideoxy fingerprinting (Dideoxy SSCP analysis)	Detecting point mutations, and small deletions and insertions Providing indication of the location of mutations with the tested DNA fragment Slower and more expensive than conventional SSCP analysis, but more sensitive
DNA sequencing	Detecting point mutations, and small deletions and insertions Providing accurate information regarding the location and nature of mutations Slow procedure and expensive

merase chain reaction-single strand conformation polymorphism (PCR-SSCP) analysis is most widely used *(7,8)*. The principle behind this technique is that, under nondenaturing conditions, single-stranded DNA has a folded structure that is maintained by intra-molecular interactions. Because a DNA fragment with a mutation may have a different conformation as compared with the wild type, the former and the latter can have a different electrophoretic mobility in a non-denaturing polyacrylamide gel. To search for mutations in a given DNA sequence, PCR is first carried out to amplify the target sequence using DNA templates from different individuals to be screened for mutations. The PCR products are then denatured to separate the two single strands, and subjected to native polyacrylamide gel electrophoresis. The different mobility patterns are detected by autoradiography. In this chapter, the procedure is divided and described in four Subheadings: PCR (*see* **Subheading 3.1.**), preparation of polyacrylamide gel (*see* **Subheading 3.2.**), electrophoresis (*see* **Subheading 3.3.**), and autoradiography (*see* **Subheading 3.4.**).

2. Materials

2.1. PCR Reagents

1. PCR buffer (10×): 500 mM KCl, 100 mM Tris-HCl (pH 8.3).
2. 25 mM MgCl$_2$.
3. PCR primers (*see* **Note 1**): 20–30mer oligonucleotides dissolved in distilled water or TE (1 µg/µL), store at –20°C.
4. dNTP mix: 2 mM each of dATP, dCTP, dGTP and dTTP, store at –20°C.
5. Radioisotope: 10 mCi/mL [α-^{33}P]dCTP. **Caution:** follow local rules for handling, storage and disposal of radioactivity.
6. *Taq* DNA polymerase (5 U/µL), store at –20°C.
7. DNA samples.
8. Mineral oil or paraffin.

2.2. Reagents for Polyacrylamide Gel

1. 2% dimethyldichlororosilane.
2. 49% (w/v) acrylamide stock solution (acrylamide: *bis*-acrylamide = 49:1), store at 4°C. **Caution:** unpolymerized acrylamide is a neurotoxin.
3. 10× TBE buffer: 89 mM Tris-borate, 2 mM EDTA, pH 8.3.
4. 20% (w/v) ammonium persulphate, freshly prepared with distilled water.
5. N,N,N',N'-tetramethylethylenediamine (TEMED).
6. Glycerol.

2.3. Additional Reagents and Materials

1. Formamide loading buffer (2×): 95% formamide, 20 mM EDTA, 0.05% (w/v) bromophenol blue, 0.05% (w/v) xylene cyanol, store at –20°C.
2. Whatman 3MM filter paper.
3. Saran Wrap.

4. X-ray films.
5. Thermal cycler.
6. Polyacrylamide gel electrophoresis apparatus with approx 30 cm (width) by 40 cm (length) glass plates, and 0.4 mm (thick) spacers and shark's-tooth comb.
7. Power supply.
8. Gel dryer.

3. Methods

3.1. Polymerase Chain Reaction (PCR)

1. Prepare a PCR premix by combining the following reagents (the amount of each reagent given is for one PCR. They should be scaled up appropriately according to the number of PCRs being set up):

    15.6 μL sterile distilled water
    2.5 μL 10× PCR buffer
    1.5 μL 25 mM magnesium chloride (*see* **Note 2**)
    0.2 μL forward primer (1 μg/μL)
    0.2 μL reverse primer (1 μg/μL)
    2.5 μL 2 mM 4dNTP
    0.3 μL α-[^{33}P] dCTP (10 mCi/mL)
    0.2 μL 5 U/μL *Taq* DNA polymerase

   Dispense a 23 μL aliquot of the PCR premix into a microcentrifuge tube (or microtiterplate) containing 2 μL of DNA (approx 50–150 ng/μL) sample.
2. Overlay the solution with 30 μL of mineral oil (or paraffin).
3. Place the tubes in a thermal cycler. Suggested cycling conditions (*see* **Note 2**): denaturing at 94°C for 3 min, followed by 30 cycles of 94°C for 30 s (denaturation), 55°C for 1 min (annealing), and 72°C for 1 min (extension).

3.2. Preparation of Polyacrylamide Gel

1. Thoroughly clean two glass plates (30 cm × 40 cm) with detergent and tap water, then rinse with distilled water and wipe with absolute ethanol.
2. Coat one of the plates with dimethyldichlorosilane by evenly spreading 5 mL of 2% dimethyldichlorosilane on the plate surface (one side only) with a Kimwipe tissue, and leaving it to dry in a fume hood cabinet.
3. Lay the untreated glass plate on bench. Place two 0.4 mm thick spacers onto the plate, one on each side. Place the other plate on the top, with the dimethyldichlorsilane-treated surface facing downward. Seal the sides and bottom with tape.
4. Prepare a 4.5% acrylamide gel mix by combining the following reagents:

    9 mL 49% acrylamide stock solution (acrylamide: *bis*-acrylamide 49:1)
    10 mL 10× TBE buffer
    91 mL distilled water
    100 μL freshly prepared 20% ammonium persulphate
    100 μL TEMED
    5% or 10% glycerol may be included in the gel mix (*see* **Note 3**)

5. Mix well. With the plate tilted from the horizontal, slowly inject the acrylamide mix into the space between the plates using a 50 mL syringe without forming air bubbles. Insert Shark's tooth comb with the flat side facing downward, and clipped in place to form a flat surface at the top of the gel.

3.3. Electrophoresis

1. Between 2 and 24 h after the gel is poured, remove the clips, tape and comb. Fix the plates onto an electrophoresis apparatus. Add 1× TBE buffer to the top and bottom chambers. Using a pipet, flush the flat gel surface with 1× TBE buffer. Reinsert the comb with teeth downward and just in contact with the gel surface.
2. Mix a 5 μL aliquot of PCR products with 5 μL of 2× formamide loading buffer.
3. Heat the mixture at 95°C for 3min, then place the tubes (or microtiterplate) on ice.
4. Load 3 μL onto the polyacrylamide gel.
5. Connect the electrophoresis apparatus to a power supply, and carry out electrophoresis at a constant current of 30 mA at 4°C for 3–6 h (for gels without glycerol) or 15 mA at room temperature for 12–16 h (for gels containing glycerol).

3.4. Autoradiography

1. Disconnect power and detach plates from the electrophoresis apparatus. Using a spatula, pry the plates apart. Lay a sheet of Whatman 3MM paper on the gel, press gently, and carefully lift up the 3MM paper to which the gel has adhered. Turn the 3MM paper over and cover the gel with a plastic wrap.
2. Dry the gel in a gel drier at 80°C for 1–2 h.
3. Expose an X-ray film to the gel for 72 h at room temperature without intensifying screens.

4. Notes

1. Single-strand length: the optimal length of single strand for SSCP analysis appears to be between 100–300bp. In this size range, approx 90% of single base substitutions can be detected *(9–12)*. The sensitivity decreases with increasing fragment length (67% for 300–450bp fragments) *(9–12)*. When designing PCR primers, this should be taken into consideration. If the DNA sequence to be screened for mutations is over this size range, it can be amplified in turn in smaller fragments using more than one primer pair. Alternatively, the entire sequence can be amplified by one PCR, and the amplicon cleaved into smaller fragments with suitable restriction endonucleases prior to denaturation and polyacrylamide gel electrophoresis *(13)*.
2. Optimization of PCR conditions: the fidelity and efficiency of PCR reactions are affected by a number of factors such as the amount of template DNA, the amount and melting temperature (Tm) of the primers, Mg^{2+} concentration, annealing temperature, cycling number, and so on *(14)*. PCR conditions should therefore be optimized individually for each set of primers, and the conditions described in **Subheading 3.1.** can be used as a starting point for optimization. When setting up PCR using new primer sets, it may be desirable to do a Mg^{2+} titration (e.g.,

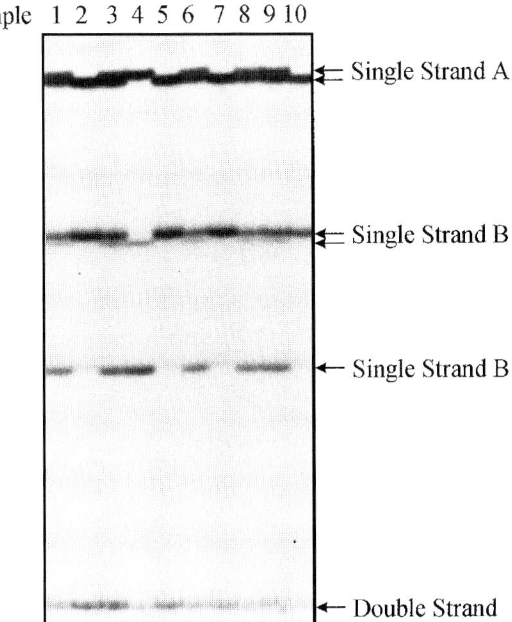

Fig. 1. Autoradiograph of SSCP analysis. A 194 bp sequence of the gelatinase B gene was PCR amplified. The amplicon was denatured and then subjected to native polyacrylamide gel electrophoresis. Note that there are three different mobility patterns: pattern 1 is seen in lanes 2, 5, 7, and 10; pattern 2 in lane 4; and pattern 3 in lanes 1, 3, 6, 8, and 9. Pattern 1 consists of three bands respectively representing the two different single strands (A and B), and double-stranded DNA. Compared with pattern 1, pattern 2 has an additional band representing a second conformation of single strand B. In addition, the mobility of single-stranded DNA is different between patterns 1 and 2 (single-strand A migrates faster in pattern 1 than in pattern 2. The opposite is seen for single-strand B). Pattern 3 consists of all bands present in both patterns 1 and 2. DNA sequencing revealed that the mobility differences between different samples in SSCP analysis were due to a guanosine (G) to adenosine (A) substitution. Samples 2, 5, 7, and 10 were G/G homozygotes; samples 1, 3, 6, 8, and 9 were heterozygotes (G/A); and sample 4 was an A/A homozygote.

setting up multiple PCR reactions respectively with 0.5 m*M*, 1 m*M*, 1.5 m*M*, 2 m*M*, 2.5 m*M*, 3 m*M*, 4 m*M*, and 5 m*M* of magnesium chloride). The annealing temperature is another important factor and should also be optimized empirically for each primer set.

3. Characteristics of the gel: the ratio of acrylamide to bisacrylamide crosslinker, and the total acrylamide concentration, determines the sieving properties of the gel. A ratio of 49 to 1 is commonly used for SSCP *(15)*. In some cases, the addition of 5% or 10% glycerol in the gel increases mobility shift. It is suggested that

this effect is due to reactions between glycerol and borate resulting in a lower pH value in the electrophoresis buffer *(16)*. It has also been reported that using low-pH buffer increases the mutation detection rate when analyzing large DNA fragments (400–800 bp) *(16)*. Gel temperature also influences SSCP mobility. Therefore, each DNA fragment is usually analyzed at different temperatures, e.g., a gel containing glycerol is run at room temperature and a gel without glycerol run at 4°C *(7,8)*.

4. Detection methods for SSCP single strands: in addition to autoradiography, other methods, such as silver staining *(17,18)*, ethidium bromide staining *(19)*, and fluorescence labeling *(20–22)*, have been applied successfully to detect DNA bands in SSCP analysis. Smaller and thicker gels are preferable when using silver or ethidium bromide staining, as larger and thinner gels are fragile and tend to break during staining. Reported SSCP gels range between 5 and 50 cm in length.

5. Anticipated results: each DNA fragment deriving from a wild-type or mutant homozygous sample commonly produces three bands, two corresponding to the two different single-stranded DNA molecules and the remainder corresponding to the double-stranded. Usually the fastest migrating band represents the double-stranded DNA, but there are exceptions. Corunning a nondenatured sample helps to identify the position of the double-stranded DNA. In some cases, there are more than three bands for each fragment because each single-stranded DNA molecule can adopt more than one conformation. In addition, double-stranded DNA from a heterozygous sample sometimes produces two or three bands, respectively representing the fast migrating homoduplex and one or two slowly migrating heteroduplexes. **Fig. 1** shows a typical SSCP autoradiograph. Note that both single strands show mobility variation between different samples, indicating sequence variation within the tested DNA fragment.

References

1. Humphries, S. E. (1995) Genetic regulation of fibrinogen. *Eur. Heart J.* **16** (**Suppl. A**), 16–20.
2. Cooper, D. N. and Krawczak, M. (1993) *Human Gene Mutation.* BIOS Scientific publishers, Oxford.
3. Weber, B. H., Vogt, G., Pruett, R. C., Stohr, H., and Felbor, U. (1994) Mutation in the tissue inhibitor of metalloproteinases-3 (TIMP3) in patients with Sorsby's fundus dystrophy. *Nat. Genet.* **8**, 352–356.
4. Ye, S., Watts, G. F., Mandalia, S., Humphries, S. E., and Henney, A. M. (1995) Genetic variation in the human stromylysin promoter is associated with progression of coronary atherosclerosis. *Br. Heart J.* **73**, 209–215.
5. Ye, S., Eriksson, P., Hamsten, A., Kurkinen, M., Humphries, S. E., and Henney, A. M. (1996) Progression of coronary atherosclerosis is associated with a common genetic variant of the human stromelysin-1 promoter which result in reduced gene expression. *J. Biol. Chem.* **271**, 13,055–13,060.
6. Spanakis, E. and Day, I. N. M. (1997) The molecular basis of genetic variation: mutation detection methodologies and limitations, in *Genetics of Common Dis-*

eases (Day, I. N. M. and Humphries, S. E., eds.) BIOS Scientific publishers, Oxford, pp. 33–74.

7. Orita, M., Iwahana, H., Kanazawa, H., Hayashi, K., and Sekiya, T. (1989) Detection of polymorphisms of human DNA by gel electrophoresis as single-strand conformation polymorphisms. *Proc. Natl. Acad. Sci. USA* **86**, 2766–2770.

8. Orita, M., Suzuki, Y., Sekiya, T., and Hayashi, K. (1989) Rapid and sensitive detection of point mutations and DNA polymorphisms using the polymerase chain reaction. *Genomics* **5**, 874–879.

9. Hayashi, K. (1991) PCR-SSCP: a simple and sensitive method for detection of mutations in the genomic DNA. *PCR Methods Appl.* **1**, 34–38.

10. Hayashi, K. and Yandell, D.W. (1993) How sensitive is PCR-SSCP? *Hum. Mutat.* **2**, 338–346.

11. Sheffield, V. C., Beck, J. S., Kwitek, A. E., Sandstrom, D. W., and Stone, E. M. (1993) the sensitivity of single-strand conformation polymorphism analysis for the detection of single base substitutions. *Genomics* **16**, 325–332.

12. Liu, Q., Feng, J., and Sommer, S. S. (1996) Bi-directional dideoxy fingerprinting (Bi-ddF): a rapid method for quantitative detection of mutations in genomic regions of 300–600 bp. *Hum. Mol. Genet.* **5**, 107–114.

13. Liu, Q. and Sommer, S. S. (1995) Restriction endonuclease fingerprinting (REF): a sensitive method for screening mutations in long, contiguous segment of DNA. *Biotechniques* **18**, 470–477.

14. Erlich, H. A. (1989) *PCR Technology. Principles and Applications for DNA Amplification.* Stockton Press, New York.

15. Glavac, D. and Dean, M. (1993) Optimization of the single-strand conformation polymorphism (SSCP) technique for detection of point mutations. *Hum. Mutat.* **2**, 404–414.

16. Kukita. Y., Tahira, T., Sommer, S. S., and Hayashi, K. (1997) SSCP analysis of long DNA fragments in low pH gel. *Hum. Mutat.* **10**, 400–407.

17. Oto, M., Miyake, S., and Yuasa, Y. (1993) Optimization of nonradioisotopic single strand conformation polymorphism analysis with a conventional minislab gel electrophoresis apparatus. *Anal. Biochem.* **213**, 19–22.

18. Calvert, R. J. (1995) PCR amplification of silver-stained SSCP bands from cold SSCP gels. *Biotechniques* **18**, 782–784.

19. Hongyo, T., Buzard, G. S., Calvert, R. J., and Weghorst, C. M. (1993) 'Cold SSCP': a simple, rapid and non-radioactive method for optimized single-strand conformation polymorphism analyses. *Nucleic Acids Res.* **21**, 3637–3642.

20. Makino, R., Yazyu, H, Kishimoto, Y., Sekiya, T., and Hayashi, K. (1992) F-SSCP: A fluorescent polymerase chain reaction-single strand conformation polymorphism (PCR-SSCP) analysis. *PCR Methods Appl.* **2**, 10–13.

21. Takahashi-Fujii, A., Ishino, Y., Shimada, A., and Kato, I. (1993) Practical application of fluorescence-based image analyzer for PCR single-stranded conformation polymorphism analysis used in detection of multiple point mutations. *PCR Methods Appl.* **2**, 323–327.

22. Iwahana, H., Yoshimoto, K., Mizusawa, N., Kudo, E., and Itakura, M. (1994) Multiple fluorescence-based PCR-SSCP analysis. *Biotechnique* **16**, 296–305.

IV

ASSAY OF MMP AND TIMP ACTIVITIES

21

Methods for Studying Activation
of Matrix Metalloproteinases

Vera Knäuper and Gillian Murphy

1. Introduction

The degradation of the extracellular matrix during development and in disease is thought to result from the combined action of several proteolytic enzyme systems, including the matrix metalloproteinases (MMPs), serine proteinases, and cysteine proteinases. The majority of the soluble MMPs are synthesized as proenzymes which require extracellular activation in order to gain proteolytic activity and the analysis of their activation mechanism is a prerequisite to understanding MMP-mediated proteolysis.

The crystal structure of the C-truncated prostromelysin-1 (MMP-3) has been solved and is a prerequisite to the understanding of the features of promatrix metalloproteinases involved in the proteolytic events of proenzyme activation *(1)*. The propeptide domain of MMP-3 is a separate folding entity, being comprised of three alpha helices. There is notable lack of electron density for the residues 1–15 and 31–39, which suggests that these residues are solvent exposed and therefore highly flexible within the structure. In the proenzyme, the groove of the catalytic site is occupied by the extended peptide chain of the propeptide, forming β-structure hydrogen bonds with the enzyme and providing Cys75 as the fourth ligand for the catalytic zinc ion. The direction of this polypeptide chain is opposite to that observed when peptide hydroxamate inhibitors of MMPs occupy the active site, and the S1' pocket is empty in the proenzyme structure. The prostromelysin structure provides evidence for the cysteine-switch model of activation *(2)*, since the active site of the proenzyme is filled by the propeptide and Cys75 directly interacts with the catalytic zinc. Activation by serine proteinases involves a stepwise mechanism, with early

From: *Methods in Molecular Biology, vol. 151: Matrix Metalloproteinase Protocols*
Edited by: I. Clark © Humana Press Inc., Totowa, NJ

cleavages taking place around residues 34–39 in the solvent accessible part of the proenzyme which has therefore been depicted as the "bait region". Subsequent proteolytic cleavages are observed, involving residues 56–59 in the loop between helices 2 and 3, which might also be readily accessible without a prior conformational rearrangement of the propeptide domain. The other cleavage sites observed are however located within the helices of the propeptide domain. It is most likely, therefore, that early cleavage events in the bait region destabilizes the proenzyme structure, leading to a structural rearrangement which must precede reactions at these sites. Final activation releases the rest of the propeptide domain and involves a structural rearrangement of Tyr83 by 17 Å.

The emphasis of this chapter is the provision of the experimental tools to study MMP activation in vitro and in cellular model systems. Hence, we will use the activation of procollagenase-3 (proMMP-13) and progelatinase A (proMMP-2) as examples of the methods used.

1.1. Expression of Matrix Metalloproteinases

The proenzymes used in this chapter were expressed using a mammalian expression system *(3,4)* and were purified using standard procedures. The purified proenzymes were aliquoted and stored at –80°C prior to the activation experiments. Latency was established by SDS-PAGE and N-terminal sequence determination.

2. Materials
2.1. Activation Reagents

1. 4-aminophenyl mercuric acetate (APMA) stock solution: a 10 mM APMA stock solution is freshly prepared by dissolving 70.4 mg APMA in 400 µL dimethyl sulphoxide followed by dilution with 19.6 mL of 50 mM Tris-HCl pH 8.8. This solution is stable for about a month at room temperature if stored in the dark.
2. Trypsin (TPCK-treated preferred): a 100 µg/mL stock is prepared in 1mM HCl and stored at –20°C.
3. Activation buffer: 100 mM Tris-HCl pH 7.6, 10 mM CaCl$_2$.

2.2. Tissue Culture of HT1080 Cells Stably Transfected with MT1-MMP

1. Growth medium for transfected HT1080 cells: DMEM (Gibco-BRL 042-90132 M), 10% fetal calf serum, 4 mM glutamine, 100 U/mL penicillin, 100 µg/mL streptomycin, HT-supplement (Gibco-BRL, 41065-012), 20 µM mycophenolic acid (Sigma M5255) and 2 mM xanthine (Sigma X-2001).
2. Wash medium: Serum free DMEM pH 9.0, 100 µM phenylmethanesulfonyl flouride (PMSF), 1 µg/mL pepstatin A, 1 µg/mL trans-epoxysuccinyl-1-leucylamid (4-guanidino)-butane (E64).
3. Lysis buffer: 5 mM Tris-HCl pH 7.6, 100 µM PMSF, 1 µg/mL pepstatin A, 1 µg/mL trans-epoxysuccinyl-1-leucylamid (4-guanidino)-butane (E64), 0.02% NaN$_3$.

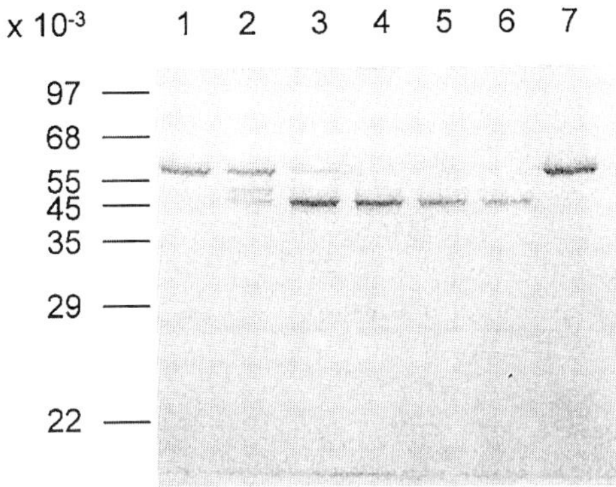

Fig. 1. Activation of procollagenase-3 by APMA. Lane 1, procollagenase-3 in the presence of APMA for 0 min; lane 2, as lane 1 for 5 min; lane 3, as lane 1 for 21 min; lane 4, as lane 1 for 43 min; lane 5, as lane 1 for 69 min; lane 6, as lane 1 for 115 min; lane 7, procollagenase-3 incubated for 115 min in the presence of buffer.

4. Reaction buffer: 20 mM Tris-HCl pH 7.8, 10 mM CaCl$_2$, 0.05% Brij 35, 100 µM PMSF, 1 mg/mL pepstatin A and 1 µg/mL trans-epoxysuccinyl-1-leucylamid (4-guanidino)-butane (E64), 0.02% NaN$_3$.
5. Serum free medium: DMEM (Gibco-BRL 042-90132 M), Insulin-Transferrin-Sodium Selenite Media Supplement (Sigma I-1884).
6. Glass homogenizer.
7. 25G and 26G needles with syringes.
8. Ultracentrifuge and rotor.
9. Broad spectrum synthetic MMP inhibitor e.g., CT-1746.

2.3. Tissue Culture of SW1353 Chondrosarcoma Cells

1. Growth medium for SW1353 chondrosarcoma cells: DMEM/NUT MIX F-12 (1:1) (Gibco-BRL 21331-020), 10 % fetal calf serum, 4 mM glutamine, 100 U/mL penicillin, 100 µg/mL streptomycin.
2. Serum-free medium for SW1353 chondrosarcoma cells: DMEM/NUT MIX F-12 (1:1) (Gibco-BRL 21331-020), 10 ng/mL interleukin-1β, 10 ng/mL oncostatin M, 0.2% lactalbumin hydrolysate (Sigma L-4379), 4 mM glutamine, 100 U/mL penicillin, 100 µg/mL streptomycin.

3. Methods

3.1. Activation In Vitro

3.1.1. Activation of Procollagenase-3 by APMA

1. Incubate full length procollagenase-3 (200–300 µg/mL) in an appropriate buffer (e.g., activation buffer) with APMA solution at a final concentration of 1 mM at 37°C and withdraw aliquots of the reaction mixture at intervals from 5 min to approx 2 h (*see* **Note 1**).

2. Analyse the samples by sodium dodecyl sulfate polyacrylamide gel electrophoresis (SDS-PAGE) as shown in **Fig. 1** *(4)*. Enzymatic activity can also be determined simultaneously using either macromolecular or fluorogenic substrates and the detailed methods for these assays are described in Chapters 22 and 31 by Cawston and Fields, respectively.

3. Where desired, N-terminal amino acid sequence determination of partially or fully activated enzymes can be performed following separation of the reaction products by reverse-phase high performance liquid chromatography (HPLC) on a Vydac C_{18} column, eluting with a linear gradient of 0–80% acetonitrile.

3.1.2. Endoproteinases

Serine proteinases, such as trypsin, chymotrypsin, or plasmin can be used to activate procollagenase-3 or other proMMPs, here we use trypsin as an example (*see* **Note 2**).

1. Incubate fu -length procollagenase-3 (200–300 µg/mL) in an appropriate buffer (e.g., activation buffer) with trypsin at a final concentration of 1 µg/mL at 37°C and withdraw aliquots of the reaction mixture at intervals from 5 min to approximately 30 min.

2. Terminate the reaction by adding a serine proteinase inhibitor at high molar excess e.g., 10 µg/mL soybean trypsin inhibitor, or 10 µ*M* diisopropylfluorophosphate (DFP) or (PMSF).

3. Analyze the samples by SDS-PAGE. Enzymatic activity can also be determined simultaneously using either macromolecular or fluorogenic substrates and the detailed methods for these assays are described in Chapters 22 and 31 by Cawston and Fields, respectively.

3.1.3. Activation of Procollagenase-3 and Progelatinase A by Cell-Membrane Preparations Containing MT1-MMP

Membrane-associated active MT1-MMP was prepared by culturing HT1080 cells stably transfected with MT1-MMP in the presence of 1 µ*M* CT1746, a peptide hydroxamate inhibitor of MMPs in order to deplete endogenous TIMP-2 binding to MT1-MMP *(5)* (*see* **Note 3**).

3.1.3.1. Day 1

Use batches of eight 175cm² flasks, and for each flask:

1. Wash the HT1080/MT1-MMP cell monolayers twice with 5 mL of ice-cold serum free DMEM pH 7.4 per 175 cm² tissue culture flask.

2. Wash the cells with an additional aliquot of 5 mL ice-cold wash medium (**Subheading 2.2., step 2**.) and scrape cells off the flask mechanically into 5 mL of ice cold wash medium.

3. Transfer the cell suspension into a 50 mL Falcon tube and centrifuge for 5 min at 2,500*g* at 4°C.

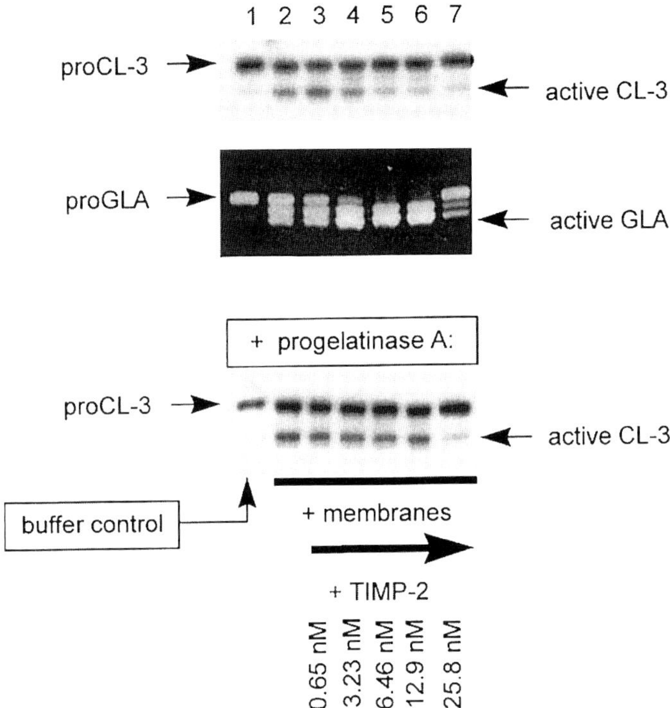

Fig. 2. Activation of procollagenase-3 and progelatinase A by membranes containing MT1-MMP: Inhibition of procollagenase-3 activation by TIMP-2 and potentiation by progelatinase A. *Upper panel:* activation of procollagenase-3 (10 ng) by membrane associated MT1-MMP (Western blot). *Mid panel:* activation of progelatinase A (10 ng) by membrane associated MT1-MMP (Zymogram). *Lower panel:* activation of procollagenase-3 (10 ng) in the presence of progelatinase A (10 ng) by membrane associated MT1-MMP (Western Blot). Lane 1, buffer control; lane 2, processing by 36.7 n*M* membrane associated MT1-MMP; lane 3–7, processing by 36.7 n*M* membrane associated MT1-MMP in the presence of 0.65–25.8 n*M* TIMP-2.

4. Carefully remove the supernatant and wash the pellet twice with 25 mL wash medium by pipetting up and down followed by a further centrifugation.
5. Remove the final supernatant and combine the cell pellets from 8 × 175 cm^2 tissue culture flasks. Store frozen at –80°C prior to further processing.

3.1.3.2. DAY 2

1. Thaw the cell pellet on ice and resuspend in 4–6 mL of lysis buffer.
2. Incubate for 10–15 min on ice.
3. Homogenize using a glass homogenizer.
4. Pass the suspension 10 times through a 25G microlance needle with a syringe to achieve further homogenization.

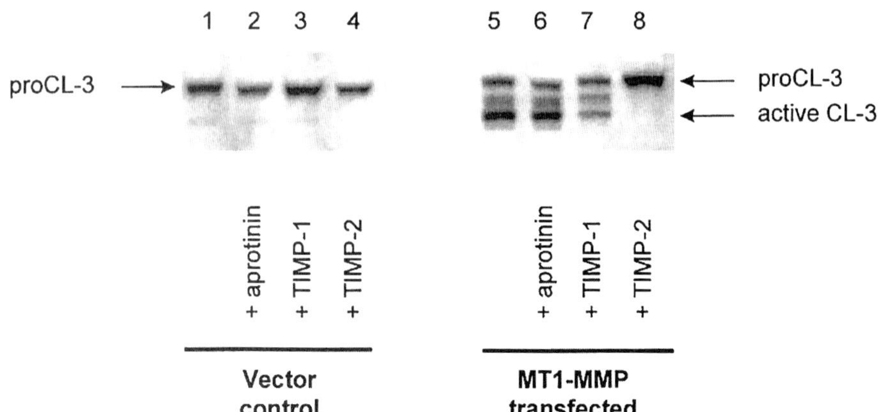

Fig. 3. Activation of procollagenase-3 by HT1080 cells transfected with MT1-MMP. Lane 1, HT1080 cells transfected with vector and supplemented with 100 ng procollagenase-3; lane 2, as lane 1 in the presence of 1 µg aprotinin; lane 3, as lane 1 in the presence of 50 ng TIMP-21; lane 4, as lane 1 in the presence of 50 ng TIMP-2. Please note that procollagenase-3 is not activated under these conditions. Lane 5, HT1080 cells transfected with MT1-MMP and supplemented with 100 ng procollagenase-3; lane 6, as lane 5 in the presence of 1 µg aprotinin; lane 7, as lane 5 in the presence of 50 ng TIMP-1; lane 8, as lane 5 in the presence of 50 ng TIMP-2. Please note that only TIMP-2 is an efficient inhibitor of procollagenase-3 activation.

5. Repeat this procedure 20 times using a 26G microlance needle.
6. Centrifuge the homogenized cell suspension for 10 min at 28,000g and 4°C using sterile centrifuge tubes and e.g., a Sorvall SS34 rotor.
7. Divide the clear supernatant (4 mL) into four centrifuge tubes and centrifuge in an ultracentrifuge using a precooled TFT 80.4 rotor (Beckmann) for 1 h at 100,000g at 4°C.
8. Resuspend each translucent pellet in 200 µL reaction buffer and disperse by passing through a 26G microlance needle.
9. Determine the concentration of active MT1-MMP in the membrane preparation by fluorogenic substrate hydrolysis using the k_{cat}/K_M value for MT1-MMP (2.6×10^5 M^{-1}s^{-1}). In this example preparation the concentration was estimated to be 36.7 nM *(6)*.
10. The membrane preparation containing active MT1-MMP is not very stable and should be stored in 50–100 µL aliquots at –80°C. Repeated freeze-thawing should be avoided and activation experiments should be performed as fast as possible.

3.1.3.3. Day 3

1. Incubate procollagenase-3 (10 ng) or progelatinase A (10 ng) either alone or in combination with 9 µL of membrane-associated MT1-MMP and in the absence

or presence of increasing amounts of TIMP-2 (from approx 0.5–30 n*M*) for approx 12 h at 37°C to allow activation (**Fig. 2**).

2. Analyze the reaction products by Western blotting (procollagenase-3) and zymography (progelatinase A).

As demonstrated in **Fig. 2** of this example, procollagenase-3 was partially activated by membrane-associated MT1-MMP under these conditions and increasing amounts of TIMP-2 inhibited activation (*upper panel*). Progelatinase A was partially converted to an intermediate form and to a lesser extent to the fully active form by MT1-MMP (lane 2, *mid panel*). Increasing amounts of TIMP-2 (0.65–12.9 nM) promoted conversion of the proenzyme form to the intermediate and active form, whereas high concentrations (25.8 n*M*) were found to be inhibitory (lane 7, *mid panel*). This finding is in agreement with the "receptor model" of TIMP-2 mediated binding and activation of progelatinase A to a partially inhibited membrane-associated MT1-MMP preparation. In this model progelatinase A binds to an "MT1-MMP/TIMP-2 receptor" via C-terminal domain interactions and is activated by an adjacent inhibitor free MT1-MMP molecule. Activation of procollagenase-3 by membrane-associated MT1-MMP was found to be more efficient in the presence of progelatinase A (**Fig. 3**, *lower panel*). This is due to the fact that gelatinase A contributes to procollagenase-3 activation *(7)*.

3.2. Activation using Cellular Model Systems

3.2.1. Activation of Procollagenase-3 by HT1080 Cells Transfected with MT1-MMP

3.2.1.1. DAY 1

1. Trypsinise HT1080 cells transfected with MT1-MMP or vector control cells and seed into a 24-well plate at a density of 1×10^5 cells/well.
2. Grow the cells in growth medium for an additional 24 h.

3.2.1.2. DAY 2

1. Wash the wells twice with 1 mL serum-free medium.
2. Supplement each well with 100 ng purified procollagenase-3 in 300 μL serum-free medium (*see* **Note 4**). Where desired some wells may be supplemented with TIMPs or serine proteinase inhibitors and so on in order to analyze their role in activation.
3. Incubate the cells in 5% humidified CO_2 for either 24 or 48 h to allow the reaction to proceed.

3.2.1.3. DAY 3 OR 4

1. Remove the supernatant and centrifuge for 5 min at 2,500*g* to remove any floating cells.
2. Remove a 20 μL aliquot from each experiment and separate on a 10% SDS-PAGE followed by Western blotting.

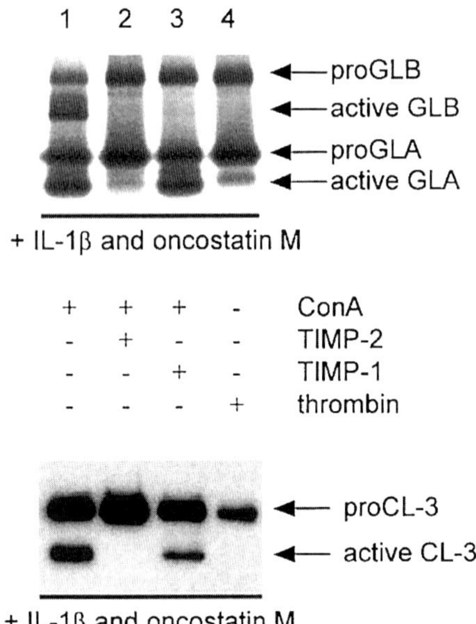

Fig. 4. Activation of endogenous progelatinase A and progelatinase B (*upper panel,* inverted zymogram) and procollagenase-3 (*lower panel,* Western blot) by concanavalin A stimulated SW1353 chondrosarcoma cells. Lane 1, activation of progelatinase A, progelatinase B and procollagenase-3 by MT1-MMP; lane 2, inhibition of progelatinase A, progelatinase B and procollagenase-3 activation by TIMP-2; lane 3, TIMP-1 does not inhibit activation of progelatinase A or procollagenase-3 by MT1-MMP but inhibits activation of progelatinase B; lane 4, thrombin does not mediate activation of progelatinase A, progelatinase B and procollagenase-3. Please note that TIMP-1 is an efficient inhibitor of progelatinase B activation, which indicates that this is a secondary event which is mediated by active gelatinase A and collagenase-3. TIMP-2 is an efficient inhibitor of activation of all three proenzymes.

3. Incubate the Western blot with a specific antiserum to procollagenase-3 and develope using a peroxidase-conjugated secondary antibody and a chemiluminescent substrate. A typical experiment is depicted in **Fig. 3**.

3.2.2. Activation by SW1353 Chondrosarcoma Cells

The human chondrosarcoma cell line SW1353 was chosen to investigate the activation of endogenous procollagenase-3, progelatinase A, and progelatinase B by MT1-MMP *(8).* SW1353 cells were maintained in growth medium in a humid atmosphere containing 5% CO_2 at 37°C, and passaged with trypsin/ EDTA as necessary.

3.2.2.1. Day 1

1. Trypsinise cells and seed at 1×10^5 cells/well in a 24-well plate.
2. Grow cells for 24 h in growth medium.

3.2.2.2. Day 2

1. After 24 h, wash the cell monolayers twice with serum-free medium
2. Culture for a further 24–72 h in 300 μL serum-free medium to induce pro-collagenase-3 synthesis. In some wells add concanavalin A (50 μg/mL) to induce MT1-MMP synthesis and study activation of procollagenase-3, progelatinase A and progelatinase B simultaneously.

3.2.2.3. Day 3–5

1. Remove the conditioned medium and centrifuged for 5 min at 2,500*g* to remove any floating cells prior to analysis.
2. Analyze the processing of the proenzymes in this system by Western blotting (procollagenase-3) and zymography (progelatinases). The results of a typical experiment are shown in **Fig. 4**.

4. Notes

1. It is noteworthy that APMA activation has to be optimized for each proenzyme preparation, since the rate of activation is dependent on the proenzyme concentration and the time intervals required to achieve conversion to the fully active form can therefore vary. Furthermore, full-length MMPs are prone to auto-proteolysis in the hinge region, which connects the N-terminal and C-terminal domains. In the case of the collagenases it is particularly important to monitor the loss of the C-terminal domain by SDS-PAGE, since the C-terminal domain determines the substrate specificity versus interstitial collagens (*9,10*) and active forms which have lost the C-terminal domain are not able to hydrolyze these substrates.

   In addition, some active enzymes are not stable for prolonged periods of time at 37°C. Therefore, the activated enzymes can be diluted (1:10) and then stored at 4°C for up to one week. Alternatively, the active stock solution can be frozen in small aliquots at –80°C, where they should be stable for up to one month. However, repeated freeze-thawing should be avoided.

   Furthermore, proMT1-MMP is very poorly activated by organomercurial compounds. A good explanation for this behavior might be that this enzyme contains a free cysteine residue within its catalytic domain which can react with these compounds leading to denaturation of the respective catalytic domain and to loss of activity.

2. Activation of proMMPs by serine proteinases has to be optimized from case to case to achieve optimal conditions and full activity. The ability of the different serine proteinases to activate proenzymes varies considerably. For example, stromelysin-1 (MMP-3) is activated by 5 μg/mL trypsin within 30 min at 25°C,

whereas 10 µg/mL plasmin are only efficient after 3 h at 37°C. Furthermore, we often see C-terminal processing due to proteolysis in the hinge region. In a number of cases this is due to the amino acid sequence of the hinge region of the respective proenzyme and for example recombinant proMT1-MMP is particularly sensitive to trypsin. In case of the collagenases again loss of the C-terminal domain should be avoided as discussed above. Thus each experiment needs careful monitoring by SDS-PAGE and is not necessary straightforward.

3. In order for activation experiments using membranes containing MT1-MMP to be performed successfully, the membrane preparations should be as fresh as possible, since long-term storage of the active MT1-MMP in these membranes is not advisable. Activity loss is even observed at –80°C. Any broad spectrum synthetic MMP inhibitor can be substituted for CT-1746 used here.

4. If exogenous procollagenase-3 or progelatinase A are added to cell monolayers of HT1080 cells expressing MT1-MMP the concentration should be kept at 100 ng/300 µL since these have been found to be optimal for these studies.

Acknowledgments

The authors would like to thank Dr. C. López-Otín, Dr. A. M. Pendas, Dr. S. Cowell, Dr. M. Balbin, Dr. G. Velasco and M. L. Stewart for their contribution to our work during the years.

References

1. Becker, J. W., Marcy, A. I., Rokosz, L. L., Axel, M. G., Burbaum, J. J., Fitzgerald, P. M. D., Cameron, P. M., Esser, C. K., Hagmann, W. K., Hermes, J. D., and Springer, J. P. (1995) Stromelysin-1: Three-dimensional structure of the inhibited catalytic domain and of the C-truncated proenzyme. *Protein Science* **4**, 1966–1976.
2. Van Wart, H. and Birkedal-Hansen, H. (1990) The cysteine switch: a principle of regulation of metalloproteinase activity with potential applicability to the entire matrix metalloproteinase gene family. *Proc. Natl. Acad. Sci. USA* **87**, 5578–5582.
3. Murphy, G., Willenbrock, F., Ward, R. V., Cockett, M. I., Eaton, D., and Docherty, A. J. P. (1992) The C-terminal domain of 72 kDa gelatinase A is not required for catalysis, but is essential for membrane activation and modulates interactions with tissue inhibitors of metalloproteinases. *Biochem. J.* **283**, 637–641.
4. Knäuper, V., López-Otín, C., Smith, B., Knight, G., and Murphy, G. (1996) Biochemical characterization of human collagenase-3. *J. Biol. Chem.* **271**, 1544–1550.
5. Butler, G. S., Butler, M. J., Atkinson, S. J., Will, H., Tamura, T., Van Westrum, S. S., Crabbe, T., Clements, J., D'Ortho, M. P., and Murphy, G. (1998) The TIMP2 membrane type 1 metalloproteinase "receptor" regulates the concentration and efficient activation of progelatinase A. *J. Biol. Chem.* **273**, 871–880.
6. Will, H., Atkinson, S. J., Butler, G. S., Smith, B., and Murphy, G. (1996) The soluble catalytic domain of membrane type 1 matrix metalloproteinase cleaves the propeptide of progelatinase A and initiates autoproteolytic activation - Regulation by TIMP-2 and TIMP- 3. *J. Biol. Chem.* **271**, 17,119–17,123.

7. Knäuper, V., Will, H., López-Otín, C., Smith, B., Atkinson, S. J., Stanton, H., Hembry, R. M., and Murphy, G. (1996) Cellular mechanisms for human pro-collagenase-3 (MMP-13) activation - Evidence that MT1-MMP (MMP-14) and gelatinase A (MMP-2) are able to generate active enzyme. *J. Biol. Chem.* **271**, 17,124–17,131.

8. Cowell, S., Knäuper, V., Stewart, M. L., D'ortho, M. P., Stanton, H., Hembry, R. M., López-Otín, C., Reynolds, J. J., and Murphy, G. (1998) Induction of matrix metalloproteinase activation cascades based on membrane-type 1 matrix metalloproteinase: associated activation of gelatinase A, gelatinase B and collagenase 3. *Biochem. J.* **331**, 453–458.

9. Murphy, G., Allan, J. A., Willenbrock, F., Cockett, M. I., O'Connell, J. P., and Docherty, A. J. P. (1992) The role of the C-terminal domain in collagenase and stromelysin specificity. *J. Biol. Chem.* **267**, 9612–9618.

10. Knäuper, V., Cowell, S., Smith, B., López-Otín, C., O'Shea, M., Morris, H., Zardi, L., and Murphy, G. (1997) The role of the c-terminal domain of human collagenase-3 (MMP- 13) in the activation of procollagenase-3, substrate specificity, and tissue inhibitor of metalloproteinase interaction. *J. Biol. Chem.* **272**, 7608–7616.

22

Assay of Matrix Metalloproteinases Against Matrix Substrates

Timothy E. Cawston, Paul Koshy, and Andrew D. Rowan

1. Introduction

Mammalian collagenases cleave all three polypeptide chains of the triple helical collagen molecule at a specific site to give characteristic one-quarter and three-quarter fragments. These denature at 37°C becoming susceptible to digestion by less specific proteinases.

The most widely used forms of collagenase assay depend on the measurement of fragments released from radiolabeled collagen *(1,2)*. Collagen is extracted from skin or tail tendons and labeled with [1-^{3}H] acetic anhydride. This labeled collagen is incubated at neutral pH and at 37°C to form collagen fibrils. Collagenases degrade this fibrillar substrate and at the end of the assay period cleaved fragments are separated from undigested collagen by centrifugation. Other methods for the separation of uncleaved substrate from the products of digestion have included precipitation with dioxane *(3)* or the use of radiolabeled collagen immobilized in microtiter plates *(4)*.

These assays for collagenases using radiolabeled substrate have the advantage of measuring enzyme activity but do not distinguish between the different collagenases (MMP-1, MMP-8, and MMP-13) or indeed other enzymes that can also cleave triple helical collagen (MMP-2 and MT1-MMP). They also have the disadvantage that measurement of activity reflects the overall balance between enzymes that can degrade collagen at neutral pH and the level of the TIMPs, and other inhibitors, that are present in the sample. Different procedures have been recommended in an attempt to overcome these problems *(5,6)*. Immunological based assays (e.g., ELISA) can accurately measure the amount

From: *Methods in Molecular Biology, vol. 151: Matrix Metalloproteinase Protocols*
Edited by: I. Clark © Humana Press Inc., Totowa, NJ

of an individual MMP present but often do not distinguish between proenzyme, active enzyme or inhibitor-complexed enzyme.

The use of radiolabeled substrates has been extended to the measurement of gelatinolytic activity *(7)* using radiolabeled gelatin (denatured collagen) and also radiolabeled casein *(8)* which can act as a general substrate for the stromelysins. These assays cannot distinguish between different enzymes which also cleave the same substrate.

2. Materials

1. Toluene or NaN_3.
2. 0.9% (w/v) NaCl.
3. 5% (w/v) NaCl.
4. 20% (w/v) NaCl.
5. Cheesecloth or fine mesh beer-making bag.
6. 0.1 *M* acetic acid.
7. 0.2 *M* acetic acid.
8. 0.5 *M* acetic acid.
9. 5% (w/v) NaCl in 0.1 *M* acetic acid.
10. 50 m*M* Tris-HCl, pH 7.6, containing 0.2 *M* NaCl, 5 m*M* calcium acetate and 0.03% (v/v) toluene (or alternatively 0.02% NaN_3).
11. Brij35 (Sigma).
12. 20 m*M* disodium hydrogen phosphate.
13. 10 m*M* NaOH.
14. 18% (w/v) trichloroacetic acid.
15. 90% (w/v) trichloroacetic acid.
16. [^{3}H] acetic anhydride (TRK-2) (Amersham Life Sciences Ltd).
17. Borate buffer: 10 m*M* disodium tetraborate pH 9.0 (adjust with NaOH), 0.2 *M* $CaCl_2$.
18. Dry isopropanol: Add molecular sieve, pore diameter 4Å to isopropanol, shake and allow to settle.
19. Dry dioxane: Add molecular sieve, pore diameter 4Å to dioxane, shake and allow to settle.
20. Buffer A: 25 m*M* sodium cacodylate, 0.05% (v/v) Brij35, pH 7.6, 0.02% NaN_3.
21. Buffer B: 100 m*M* Tris-HCl pH 7.6, 15 m*M* $CaCl_2$, 0.02% NaN_3.
22. Trypsin (type III, bovine pancreas): make up stock at 100 mg/mL in 1 m*M* HCl
23. Soybean trypsin inhibitor: make up stock at 100 mg/mL in 100 m*M* Tris-HCl, pH 7.6.
24. Bacterial collagenase (Type I, Clostridium histolyticum; Sigma): make up at 100 mg/mL in Buffer A.
25. Diisopropylphosphofluoridate (DFP): make up as 200 m*M* stock in dry isopropanol (NB: Care, DFP is a potent neurotoxin).
26. Casein: dissolve Hammmerstein grade casein (Merck) at 30 mg/mL in 10 m*M* NaOH (500 mL) and slowly stir overnight in the cold until dissolved.
27. Aminophenylmercuric acetate (APMA): make up at 10 m*M* by dissolving 35.2 mg of APMA in 200 μL of dimethylsulfoxide (DMSO) and diluting to 10 mL with 20 m*M* Tris-HCl, pH 8.0 –9.5.

28. Mistral 3000i centrifuge with a four place swing out rotor (43124–129; MSE Scientific Instruments).
29. Optiphase Supermix scintillation fluid (Wallac, Crowhill, UK).
30. Flexible 96-well sample plate (1450–401; Wallac).
31. No-crosstalk cassette (1450–101; Wallac).
32. 1450 Microbeta Trilux liquid scintillation and luminescence counter (Wallac).

3. Methods

3.1. Preparation of Substrates

3.1.1. Type I Collagen

1. Freeze new-born calf skin (30 cm × 30 cm) onto a wooden tray. Remove hair first with scissors and then with a large scalpel blade. Allow to thaw, cut skin into small pieces (approx 50 mm × 50 mm) and maintain a temperature of 4°C throughout this procedure (*see* **Note 1**).
2. Grind skin in electric mincer (e.g., Lynx Asco meat grinder) with chips of dry ice to prevent heating of the tissue during mincing. Allow minced skin to thaw and perform all subsequent steps at 0–4°C. Add toluene (0.03%; v/v), or an alternative preservative, at each stage of the procedures below.
3. Extract minced skin three times with 1.5 L of 0.9% (w/v) NaCl for 30 min with stirring followed by two extractions with ice-cold water. Filter after each extraction through cheesecloth or through a fine mesh beer-making bag (Boots Chemist). Discard supernatants.
4. Resuspend the extracted skin in 1.5 L of 0.5 M acetic acid and stir slowly overnight at 4°C, filter and retain supernatant. Repeat.
5. Combine acetic acid extracts from **step 4** and centrifuge at 7,500g for 2 h at 4°C, discard pellet.
6. Dialyze supernatant against 2 changes of 12 L of 5% (w/v) NaCl in 0.1 M acetic acid until collagen precipitates. Centrifuge at 7,500g for 30min and retain pellet.
7. Resuspend pellets in 2–3 L of 0.5 M acetic acid and stir, dialyze against 0.5 M acetic acid (5 L) overnight (or until pellet dissolves)
8. Dialyze against at least 4 changes of 20 mM disodium hydrogen phosphate (20 L) until collagen precipitates and centrifuge at 7,500g for 30 min. Retain pellets.
9. Resuspend pellets in 2–3 L of 0.5 M acetic acid, stir until dissolved and then slowly add 5% (w/v) NaCl to precipitate the collagen. Centrifuge at 7,500g for 30 min, wash pellet in 1 L 20% (w/v) NaCl and collect the collagen by centrifugation (7,500g for 30 min).
10. Resuspend pellets in 0.5 M acetic acid and stir slowly overnight until dissolved. Centrifuge at 30,000g for 1 h, retain supernatant and dialyze against 0.1 M acetic acid (10 L). Freeze-dry thoroughly and store desiccated at –20°C until required.

3.1.2. Casein

1. Dissolve Hammmerstein grade casein at 30 mg/mL in 10 mM NaOH (500 mL) and slowly stir overnight in the cold until dissolved.

2. Add glacial acetic acid until the pH is 4.6; the casein will precipitate. Centrifuge at 1,000*g* for 15 min and retain pellet (gentle centrifugation is required otherwise the pellet is very difficult to dissolve).
3. Redissolve pellet in 500 mL 10 m*M* NaOH. Repeat **steps 2** and **3**.
4. Redissolve pellet in 10 m*M* NaOH and store at –20°C at 10mg/mL.

3.1.3. Gelatin

Gelatin is prepared from collagen by incubation at 56°C for 30 min. It is important to add diisopropylphosphofluoridate (DFP) to a final concentration of 2 m*M* before incubation and prepare this solution immediately before use *(7)*.

3.2. Labeling of Substrates

3.2.1. Collagen

1. Dissolve 250 mg of freeze-dried collagen in 50 mL of 0.2 *M* acetic acid at 4°C and dialyze against 10 m*M* disodium tetraborate adjusted to pH 9.0 with NaOH and containing 0.2 *M* CaCl$_2$ (2 L, one change, *see* **Note 2**). If the collagen precipitates redialyze against acetic acid and repeat.
2. Place collagen in a conical flask with large stirrer bar and stir very slowly.
3. Place the bottom of tube containing [^{3}H]-acetic anhydride (25 mCi; TRK-2; Amersham Life Sciences Ltd) into dry ice. Open tube and add 1 mL dry dioxane into tube; it will freeze in the bottom of the tube. Thaw dioxane as quickly as possible and add dioxane containing [^{3}H] acetic anhydride into the collagen; wash tube with further 1 mL of dry dioxane. Stir for 30 min at 4°C.
4. Dialyze collagen against 50 m*M* Tris-HCl, pH 7.6, containing 0.2 *M* NaCl, 5 m*M* calcium acetate and 0.03% (v/v) toluene until the radioactivity in the diffusate falls to background levels. Dilute collagen to a final concentration of 1 mg/mL and add unlabeled collagen (1 mg/mL) until the [^{3}H] content equals 200,000 dpm/mg. This solution is dialyzed against 0.2 *M* acetic acid and stored at –20°C until required.

Specific assays can be made for type II and III collagen if these substrates are labeled *(9)*.

3.2.2. Casein

1. Casein can be labeled using the same method as for collagen but is labeled as a 10–20 mg/mL solution and calcium must be omitted from the borate buffer (*see* **Note 2**) *(8)*. Store at –20°C.

3.3. Assays

3.3.1. Collagenase Assay

1. Thaw collagen at 1 mg/mL in 0.2 *M* acetic acid and dialyze against 50 m*M* Tris-HCl buffer, pH 7.6, containing 0.2 *M* NaCl and 0.03% (v/v) toluene.

2. Set up control 400 µL microfuge tubes (tube number 1 and 2) with 100 µL of buffer A; tubes number 3 and 4 with 10 µL of trypsin (100 µg/mL) in 1 m*M* HCl + 90 µL of buffer A; tubes number 5 and 6 with 100 µL of bacterial collagenase at 100 µg/mL in buffer A. If samples contain appreciable amounts of salt then appropriate blanks should be used (*see* **Note 3**).

3. Add test samples in duplicate or triplicate to tube 7 onwards and make up volume to 100 µL with buffer A. Add 100 µL of buffer B to all tubes.

4. Add 100 µL of [³H]-labeled collagen (1 mg/mL) to all tubes. Cap and incubate in water bath at 37°C for 1–20 h. At the end of the assay period centrifuge at 13,000*g* for 10 min to remove the undigested collagen (*see* **Note 4**). Remove 200 µL of the supernatant and combine with scintillation fluid and count for [³H] in a scintillation counter.

5. Subtract mean blank values (tubes 1–2) from all test results. If trypsin digests substantial amounts of collagen (tubes 3–4) it means the collagen is denatured and assay results should be discarded (*see* **Notes 5** and **9**). The mean value obtained for tubes 5–6 represent the total lysis figure and correspond to 100 µg of collagen.

6. Results are expressed as units/mL where one unit of activity represents the amount of enzyme that degrades 1 µg of collagen/min at 37°C. Thus, to obtain results in units/mL use the following formula:

$$\text{units / mL} = \frac{[\text{test mean - mean tubes 1–2}] \times 100 \text{ (µg of collagen)} \times 1000}{[\text{mean tubes 5–6} \times \text{time (mins)} \times \text{volume of sample (µL)}]}$$

If test samples have been diluted then the dilution factor needs to be included in this formula (*see* **Notes 6** and **7**).

3.3.2. Collagenase:96-Well Plate

1. Thaw collagen at 1 mg/mL in 0.2 *M* acetic acid and dialyze against 50 m*M* Tris-HCl buffer, pH 7.6, containing 0.2 *M* NaCl and 0.03% (v/v) toluene with one change.

2. In a 96-well plate set up control wells A1 and A2 with 50 µL of buffer A; wells A3 and A4 with 5 µL of trypsin (100 µg/mL) in 1 m*M* HCl + 45 µL of buffer A; wells A5 and A6 with 50 µL of bacterial collagenase at 100 µg/mL in buffer A. If samples contain appreciable amounts of salt then appropriate blanks should be used (*see* **Note 3**).

3. Add test samples in duplicate or triplicate to well A7 onwards and make up volume to 50 µL with buffer A. Add 50 µL of buffer B to all wells.

4. Add 50 µL of [³H]-labeled collagen (1 mg/mL) to all wells. Cover with a plate sealer (ICN Pharmaceuticals Ltd, Thame, UK) and incubate in water bath at 37°C for 1–20 h. At the end of the assay period centrifuge at 1,300*g* for 30 min in a Mistral 3000I centrifuge (*see* **Note 4**) using a four place swing out rotor (43124–129; MSE Scientific Instruments). Remove 50 µL of each supernatant and combine with 200 mL of Supermix scintillation fluid (Wallac) in a flexible 96-well sample

plate (1450–401; Wallac) placed in a no-crosstalk cassette (1450–101; Wallac) and count for [³H] in a 1450 Microbeta Trilux liquid scintillation counter (Wallac).

5. Subtract mean blank values (wells A1–2) from all other results. If trypsin digests substantial amounts of collagen (wells A3–4) it means the collagen is denatured and assay results should be discarded (*see* **Notes 5** and **9**). The mean values obtained for wells A5-6 represent the total lysis figure and correspond to the total counts released from 50 µg of collagen.

6. Results are expressed as units/mL where one unit of activity represents the amount of enzyme that degrades 1 µg of collagen/min at 37°C. Thus, to obtain results in units/mL use the following formula:

$$\text{units / mL} = \frac{[\text{test mean - mean wells A1-2}] \times 50 \ (\mu g \ of \ collagen) \times 1000}{[\text{mean wells A5–6} \times \text{time (mins)} \times \text{vol of sample } (\mu L)]}$$

If test samples have been diluted then the dilution factor needs to be included in this formula (*see* **Notes 6** and **7**). Other assays have been described for collagenases and these are reviewed in **refs. *9–11***.

3.3.3. TIMP

The collagenase assay can easily be adapted to allow for the measurement of samples containing TIMPs. Extra control tubes are set up (tubes number 7 and 8) containing a known amount (approx 0.06 units) of active interstitial collagenase, MMP-1, (bacterial collagenase is not suitable as it is not inhibited by TIMPs) which is known to digest approx 70–80% of the collagen over the assay period. This amount of collagenase is added to all subsequent tubes (tube 9 onwards) along with the test samples. TIMP activity is then measured as the reduction of released collagen fragments in test samples compared to the active enzyme control. The formula for expressing the results in units/mL becomes:

$$\text{units / mL} = \frac{[\text{active collagenase mean (tubes 7–8) - Test mean}] \times 100 \ (\mu g \ of \ collagen) \times 1000}{[\text{mean tubes 5–6} \times \text{time (mins)} \times \text{volume of sample } (\mu L)]}$$

This assay can also be adapted to the 96-well plate format.

3.3.4. Gelatinase

1. Add DFP to a final concentration of 2 m*M* (*see* **Note 10**) to the required amount of [³H]-labeled gelatin and heat at 55°C for 30 min. Cool the gelatin to 4°C and use the same day.

2. Add 50 µL of buffer A to tubes number 1 and 2 (blanks) and of trypsin (100 µg/mL) in 1 m*M* HCl to tubes number 3 and 4 (total lysis). Add samples to tubes 5 onwards and make all tubes up to a volume of 50 µL with buffer A. Samples may require activation (*see* **Subheading 3.4.**).

3. Add 100 µL of buffer B to all tubes followed by 100 µL of [³H]-labeled gelatin.
4. Incubate at 37°C for 1–20 h. At the end of the assay place tubes on ice and add 50 µL of 90% (w/v) trichloroacetic acid to all tubes and leave on ice for 20 min.
5. Centrifuge tubes at 13,000*g* for 10 min. to pellet undigested gelatin and count 200 µL of the supernatant in a scintillation counter for 1 min.
6. Subtract the mean blank values from all test results and calculate units per mL as for the collagenase assay. 1 unit of gelatinase degrades 1 µg of gelatin/min at 37°C.

3.3.5. Caseinase

1. Add DFP to a final concentration of 2 m*M* (*see* **Note 10**) to the required amount of [³H] labeled casein, cool to 4°C and use the same day.
2. Add 50 µL of buffer A to tubes number 1 and 2 (blanks) and 50 µL of trypsin (100 µg/mL) in 1 m*M* HCl to tubes number 3 and 4 (total lysis). Add test samples to tube 5 onwards and make all tubes up to a vol of 50 µL with buffer A. Samples may require activation (*see* **Subheading 3.4.**).
3. Add 100 µL of buffer B to all tubes followed by 100 µL of [³H] labeled casein.
4. Incubate at 37°C for 1–20 h. At the end of the assay place tubes on ice and add 50 µL of 18% (w/v) trichloroacetic acid to all tubes and leave on ice for 20 min.
5. Centrifuge tubes at 13,000*g* for 10 min to pellet undigested casein and count 200 µL of the supernatant in a scintillation counter for 1 min.
6. Subtract mean blank values from all test results and calculate units per mL as for the collagenase assay. 1 unit of enzyme degrades 1 µg of casein/min at 37°C.

3.4. Activation of proMMPs with APMA or Trypsin

Many samples of conditioned culture medium contain proMMPs that require activation (*see* **Note 8**) and this can be accomplished in two ways.

3.4.1. APMA Activation

1. Replace buffer B in the collagenase, gelatinase, or caseinase assays with a mixture of buffer B (4 parts) to 10 m*M* APMA (1 part). The inclusion of APMA throughout the assay period is sufficient to activate proMMPs. If short assays are required (less than 3 h) then trypsin activation (*see* below) should be used or the sample should be preincubated with APMA at 37°C for 1 h before the substrate is added.

3.4.2. Trypsin Activation

1. Add an equal volume of trypsin (20 µg/mL) to each sample in the assay tube, mix and incubate at room temperature for 15 min. Then add the same volume of soybean trypsin inhibitor (100 µg/mL), mix and make up to 100 µL with buffer A. Proceed with assay by adding buffer B and so on.

4. Notes

1. Collagen preparations cannot be hurried and extra dialysis steps may be required if a flocculent white precipitate is not seen when expected (**steps 6, 8,** and **9**). Since large diameter dialysis tubing is used it is essential to ensure that each dialysis step, especially the first, is fully equilibrated before proceeding with subsequent steps. The temperature MUST be maintained at 4°C throughout the procedure so large volumes of precooled buffers are required throughout the preparation.

2. Calcium is included in the buffer when labeling collagen at high pH to prevent precipitation. It must not be included when labeling casein.

3. Blank values in the collagenase assay are affected by high salt, high serum, high calcium, other chaotrophic ions, and some metal ions. High blank values can also be caused by some collagen preparations not forming good fibrils. There is substantial variation between collagen preparations in their ability to form fibrils and variation of the levels of calcium ions can often control these differences. MMPs require calcium for thermal stability so it must be included in the assay.

4. Collagen can be difficult to pellet in polypropylene tubes since the fibrils tend to adhere to the tube at the surface of the liquid. Lower centrifugation speeds are used in the 96-well plates since this problem does not appear to occur in polystyrene plates.

5. Trypsin blanks are higher in acetylated collagen since some labeled lysine groups are located in the telopeptide region which are susceptible to trypsin. Excessive labeling with acetic anhydride increases trypsin blanks and can retard fibril formation (*see* **ref. 7**). It has been reported that trypsin blanks should be no higher than 5% of the total counts above the blank. This fictional figure is only achieved if trypsin is stored for long periods at neutral pH when it degrades itself. Trypsin should be stored frozen in small aliquots in 1 mM HCl and used immediately upon thawing.

6 The linear portion of the collagenase assay lies between 10–80% lysis and results that fall outside this range should be repeated at higher or lower dilutions.

7. Other enzymes can cleave collagen and it is theoretically possible that esterases could remove the [^{3}H] from the labeled collagen. Confirmation of the 3/4 and 1/4 products produced by collagenase can be confirmed by incubating enzyme with collagen at 23°C in the presence of 1 M glucose (to prevent fibril formation) followed by SDS PAGE to demonstrate the 3/4 and 1/4 cleavages (*see* **ref. 2**).

8. Cell culture medium with serum contains α_2-macroglobulin which inhibits MMPs. Activation of the proMMPs in the presence of α_2-macroglobulin leads to the formation of an enzyme:inhibitor complex such that activity cannot be detected. To avoid this problem serum should be treated by lowering the pH to pH 3 for 90 min and returning the pH to neutral by the addition of NaOH prior to adding to cells in culture. This destroys α_2-macroglobulin activity.

9. If blank values and trypsin values are high then the temperature can be reduced to, say, 35°C. However, this will result in a substantial loss in sensitivity of the assay.

10. Gelatin and casein are routinely treated with DFP immediately prior to the assay since low level contamination with bacteria producing serine proteinases (which can cleave these substrates) often occurs.

References

1. Gisslow, M. T. and McBride, B. C. (1975) A rapid sensitive collagenase assay. *Anal. Biochem.* **68**, 70–78.
2. Cawston, T. E. and Barrett, A. J. (1979) A rapid and reproducible assay for collagenase using [1-^{14}C]acetylated collagen. *Anal. Biochem.* **99**, 340–345.
3. Terato, K., Nagai, Y., Kawaninski, K., and Shinro, Y. (1976) A rapid assay method of collagenase activity using 14C-labeled soluble collagen as substrate. *Biochim. Biophys. Acta.* **445**, 753–762.
4. Johnson-Wint, B. and Gross, J. (1980) A quantitative collagen film collagenase assay for large numbers of samples. *Anal. Biochem.* **104**, 175–181.
5. Lefebvre, V. and Vaes, G. (1989) Enzymatic evaluation of procollagenase and collagenase inhibitors in crude biological media. *Biochim. Biophys. Acta* **992**, 355–361.
6. Murphy, G., Koklitis, P., and Carne, A. F. (1989) Dissociation of TIMP from enzyme complexes yields fully active inhibitor. *Biochem. J.* **261**, 1031–1034.
7. Sellers, A., Cartwright, E., Murphy, G., and Reynolds, J. J. (1977) Evidence that latent collagenases are enzyme-inhibitor complexes *Biochem. J.* **163**, 303–307.
8. Cawston, T. E., Galloway, W. A., Mercer, E., Murphy, G., and Reynolds, J. J. (1981) Purification of rabbit bone inhibitor of collagenase. *Biochem. J.* **195**, 159–165.
9. Harris, E. D. and Vater, C. A. (1982) Vertebrate collagenases. *Meth. Enzymol.* **82**, 423–458.
10. Cawston, T. E. and Murphy, G. (1981) Mammalian collagenases. *Meth. Enzymol.* **80**, 711–722.
11. Dioszegi, M., Cannon, P., and Van Wart, H. E. (1995) Vertebrate collagenases. *Meth. Enzymol.* **248**, 413–431

23

Zymography and Reverse Zymography for Detecting MMPs, and TIMPs

Susan P. Hawkes, Hongxia Li, and Gary T. Taniguchi

1. Introduction

Zymography and reverse zymography are techniques used to analyze the activities of matrix metalloproteinases (MMPs) and tissue inhibitors of metalloproteinases (TIMPs) in complex biological samples. The two methods are technically similar. Zymography involves the electrophoretic separation of proteins under denaturing (SDS) but nonreducing conditions through a polyacrylamide gel containing gelatin. The resolved proteins are renatured by exchange of the SDS with a nonionic detergent, such as Triton X-100 and the gel is incubated in an appropriate buffer for the particular proteinases under study. The gel is stained with Coomassie Blue and proteolytic activities are detected as clear bands against a blue background of undegraded gelatin. The success of this technique, which was originally developed for the study of serine proteases, is based on the following observations: First, gelatin is retained in the gel during electrophoresis when it is incorporated into the gel at the time of polymerization *(1)*; it can therefore function as an *in situ* protease substrate. Second, proteolytic activity can be reversibly inhibited by SDS during electrophoresis and recovered by incubating the gel in aqueous Triton X-100 *(2)*; proteolysis is thus postponed until the sample proteins have been resolved into bands of concentrated activity. Finally, the separation of MMP:TIMP complexes by SDS polyacrylamide gel electrophoresis enables their activities to be determined independently of one another, which is not possible in solution assays. A particular advantage of this system is that both the proenzyme and active forms of MMPs, which can be distinguished on the basis of molecular weight, can be detected. This is possible because the proen-

From: *Methods in Molecular Biology, vol. 151: Matrix Metalloproteinase Protocols*
Edited by: I. Clark © Humana Press Inc., Totowa, NJ

zymes are activated *in situ* presumably by the denaturation/renaturation process and autocatalytic cleavage *(3)*.

Reverse zymography was developed as a modification of zymography to detect TIMPs rather than MMPs. Initially, this was accomplished by incubating the "zymogram" with conditioned media (CM) containing MMP activity for a few hours following the Triton X-100 step. This resulted in partial degradation of the gelatin except in regions protected by TIMP activity *(4)*. Coomassie Blue staining allowed simultaneous visualization of MMP activity as clear bands against a pale blue background of partially degraded gelatin and dark blue bands locating TIMP activity. Although this allowed simultaneous visualization of both MMPs and TIMPs, we were unable to obtain a consistent level of background clearing of the gelatin using this procedure. We assumed that this was due to the presence of TIMPs in the CM and to limited diffusion of the MMP into the gel. In order to overcome these problems we modified the procedure to incorporate the CM into the gel matrix along with the gelatin during polymerization *(5)*. The CM that we use contains both MMP-2 and TIMP-2 but due to their differential mobilities during electrophoresis, much of the TIMP-2 activity migrates out of the gel leaving behind a TIMP-free, MMP-2-rich section of the gel that provides a superior matrix for the assay of TIMP activities in the samples.

This latter modification is really the only substantive difference between a zymogram and a reverse zymogram described in this chapter. Other differences depend upon the particular MMPs under study and relate to the choice of acrylamide concentration most useful for resolving their activities. Recipes for a 10% polyacrylamide zymogram and a 15% polyacrylamide reverse zymogram are included in this chapter together with a 15% regular SDS gel which is used as a control to confirm TIMP activities. These all incorporate the familiar Laemmli-SDS buffer system and provide a template that can be adjusted according to the needs of the individual study.

2. Materials

2.1. Equipment and Supplies

1. Mini-gel apparatus (for example Protean II Mini-cell [Bio-Rad] for gels) 8 cm × 7.3 cm × 0.75 mm.
2. Power supply (200 V, 500 mA).
3. Vacuum source.
4. 37°C incubator.
5. Rocking platform or rotary shaker.
6. Cellophane membrane backing (Bio-Rad).
7. Glass containers with covers, for gel incubation.

2.2. Reagents for Zymography and Reverse Zymography

Reagents 2–8 should be ultra pure or electrophoresis grade.

1. Distilled, deionized water (dd H_2O) for all buffers and solutions.
2. Acrylamide (*see* **Note 1**).
3. Bisacrylamide (*N,N'*-methylenebisacrylamide).
4. Tris [tris(hydroxymethyl)aminomethane].
5. SDS (sodium dodecyl sulfate).
6. Glycine.
7. Ammonium persulfate (APS).
8. TEMED (*N,N,N',N'*-tetramethylethylenediamine).
9. Gelatin - Type A, from porcine skin, bloom 175 (Sigma, St. Louis, MO) (*see* **Note 2**).
10. Hydrochloric acid.
11. Glycerol.
12. Potassium hydroxide.
13. Ethanol.
14. Triton X-100.
15. Sodium azide.
16. Calcium chloride.
17. Glacial acetic acid.
18. Methanol.
19. Coomassie brilliant blue R-250.
20. Molecular weight and activity standards (*see* **Note 3**).
21. Cleaning detergent (for example, Contrad 70, Fisher Scientific, Pittsburgh, PA).
22. Source of MMP-2 activity (we use conditioned media from various cell cultures, *see* **Note 4**).

2.3. Stock Solutions

1. Glass cleaning solution: 1 *M* KOH in ethanol. Store in a plastic bottle at room temperature.
2. Stock acrylamide (30% total monomer concentration): 29.2% (w/v) acrylamide, 0.8% (w/v) bisacrylamide in dd H_2O. Filter through a 0.45 μm membrane and store in dark glass bottle at 4°C for one month.
3. 4X separating gel buffer: 1.5 *M* Tris-HCl, pH 8.8. Filter and store at 4°C.
4. 4X stacking gel buffer: 0.5 *M* Tris-HCl, pH 6.8. Filter and store at 4°C.
5. 10% (w/v) SDS. Filter and store at room temperature.
6. 100X gelatin: 10% (w/v) gelatin in dd H_2O. Warm in a water bath (~ 55°C) to dissolve gelatin or heat gently in microwave oven. Store at –20°C in 1 mL aliquots.
7. 10% (w/v) APS in dd H_2O. Make fresh.
8. TEMED. Use directly from the bottle and store at 4°C, protected from light.
9. 2X (Laemmli) sample buffer: 0.125 *M* Tris-HCl, pH 6.8, 4% (w/v) SDS, 20% (v/v). glycerol, 0.04% (w/v) bromophenol blue. Do not add reducing agent (*see* **Note 5**).

Table 1

| Reagents[a] | Type of gel | | | |
	10% Zymogram	15% Reverse Zymogram	15% Regular SDS	5% Stacking[b]
Acrylamide/bis (30%)	3.33 mL	5.0 mL	5.0 mL	1.67 mL
0.5 M Tris-HCl, pH 6.8	–	–	–	2.5 mL
1.5 M Tris-HCl, pH 8.8	2.5 mL	2.5 mL	2.5 mL	–
Gelatin (100X)	100 μL	100 μL		
Conditioned media, 1% SDS	–	2.35 mL[c]	–	–
SDS (10%)	100 μL	–	235 μL[c]	100 μL
ddH$_2$O	3.92 mL	–	2.215 mL	5.68 mL
APS (10%)	50 μL	50 μL	50 μL	50 μL
TEMED	5 μL	5 μL	5 μL	5 μL

[a] Combine all ingredients except APS and TEMED, degas under vacuum for 15 min and initiate polymerization by adding APS and TEMED. Recipes are for 10 mL volume and can be adjusted according to the number and dimensions of the gels required. For example, two minigels (8.0 cm × 7.3 cm × 75 mm) can be prepared from 10 mL of separating gel solution and 5 mL of stacking gel solution.

[b] A 5 % stacking gel is suitable for all three separating gels listed.

[c] The concentration of SDS in the reverse zymogram and the control, regular SDS gel, is 0.235%.

10. 10X electrophoresis running buffer: 0.25 M Tris, 1.92 M glycine, 1% (w/v) SDS, pH 8.3 (no need to pH). Store in glass bottle at room temperature. Dilute to 1X with dd H$_2$O before use.

11. Triton X-100: 2.5% (v/v) in dd H$_2$O. Prepare fresh.

12. Development buffer: 0.05 M Tris-HCl pH 8.8, 5 mM CaCl$_2$, 0.02% NaN$_3$). Prepare fresh (*see* **Note 6**).

13. Fixing/destaining solution: methanol:acetic acid:water (4.5:1:4.5, v:v:v). Store at room temperature.

14. Staining solution: 0.1% Coomassie brilliant blue R-250 (w/v) in fixing/destaining solution. Store at room temperature.

3. Methods

3.1. Preparation of Gels
(Zymograms, Reverse Zymograms and Regular SDS Gels)

1. Thoroughly clean the gel apparatus (upper reservoir), spacers and combs by soaking in detergent solution (for example, ~0.2% [v/v] Contrad 70), rinsing well with warm tap water and then dd H$_2$O. Place the glass plates in a plastic tray and spread a small volume (few mL) of KOH/ethanol evenly with a Pasteur pipet over each surface and leave for 3–5 min. Rinse thoroughly with hot tap water and finally dd H$_2$O and let air-dry (*see* **Note 7**).

2. With clean glass surfaces inwards (if only one side treated with KOH/ethanol) assemble gel apparatus according to the manufacturer's instructions. Determine the height to which the separating gel is to be poured by inserting a comb and marking the outer plate 1–2 cm beneath the teeth of the comb.

3. Determine the gel volume from the manufacturer's instructions or by calculation. Prepare the monomer solution for the appropriate separating gel by combining all the ingredients (**Table 1**), with the exception of the ammonium persulfate (APS) and TEMED, in a clean side-arm flask. Degas the solution by applying a vacuum for approx 15 min.

4. Add the APS and TEMED (**Table 1**), gently swirl the contents of the flask and carefully pipet the monomer solution between the glass plates up to the mark determined above. Gently overlay the monomer solution with degassed dd H_2O to exclude oxygen from the surface of the polymerizing gel.

5. When the separating gel has polymerized (at least 30 min) decant the liquid above the gel and rinse with a small vol (500 µL) of 1X stacking gel buffer. Combine the stacking gel ingredients (**Table 1**), degas the solution and add APS and TEMED as described above. Pipet the stacking gel solution on top of the separating gel and insert the appropriate comb, taking care not to trap bubbles beneath the teeth. Allow the gel to polymerize for at least 30 min. Remove the comb, rinse the wells with 1X electrophoresis running buffer (*see* **Note 8**), assemble the apparatus and fill upper and lower reservoir with 1X electrophoresis running buffer.

3.2. Preparation of Samples and Electrophoresis

1. Dilute samples 1:1 with 2X sample buffer. Do not boil, but warm in a water bath (~55°C) for 3–5 min.

2. Load samples into the bottom of each well using gel loading pipet tips (new tip for each sample), taking care not to introduce bubbles or remove the pipet tip from the well until all the sample has been transferred. These assays are very sensitive and it is important not to contaminate adjacent wells.

3. Electrophorese samples at constant voltage as determined by trial with colored molecular weight standards and the tracking dye, bromophenol blue (*see* **Note 9** for guidelines).

3.3. Development and Staining of Gels

1. After electrophoresis, discard the stacking gels and transfer the separating gels to clean glass containers (1 gel per container) and wash twice for 15 min in approx 100 mL of Triton X-100 (2.5%) by gentle agitation on a rocking platform or a rotary shaker at room temperature (*see* **Note 10**).

2. After the second wash, remove all but 2–3 mL of the Triton X-100 (*see* **Note 10**) and add development buffer (100 mL). Agitate for 15 min at room temperature and then transfer to a 37°C incubator for 16–20 h.

3. After incubation in development buffer, the gels can be transferred to small containers (we use the plastic boxes in which disposable pipet tips are packaged) to

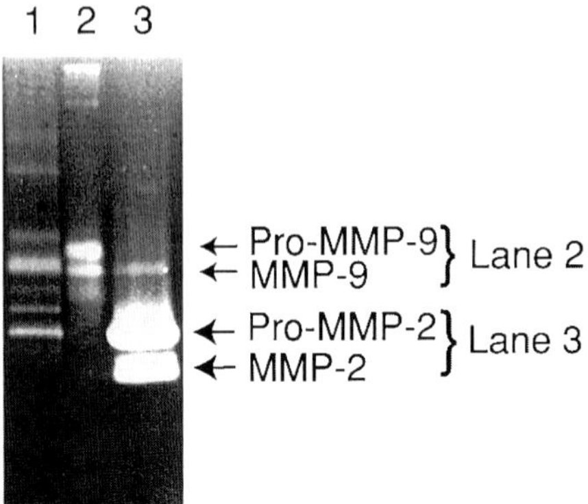

Fig. 1. Zymogram comprising 10% polyacrylamide, 1 mg/mL gelatin. The gel was cast with a 5% polyacrylamide stacking gel. Samples were electrophoresed at a constant voltage of 165 V for 1 h. Lane 1, extracellular matrix (ECM) from chicken embryo fibroblasts infected with a temperature-sensitive mutant of Rous Sarcoma Virus (RSV); lane 2, conditioned media from cultured human cytotrophoblasts; lane 3, conditioned media from the human choriocarcinoma cell line, JAR. The gel was incubated and stained with Coomassie blue as described in **Subheading 3.**

    conserve reagents. Immerse the gels in fixing/destaining solution and agitate for 15 min at room temperature.

4. Stain the gels in Coomassie Blue solution for at least 2 h (we usually stain overnight) and destain in fixing/destaining solution until the bands are clearly visible and contrast well with the background. At this stage, transfer the gels to dd H_2O, wash twice and leave in dd H_2O to stop further destaining.

5. The gels can be dried between two sheets of cellophane (for example, cellophane membrane backing from Bio-Rad). Immerse the cellophane in dd H_2O for a few minutes. Lay one piece of cellophane on a clean sheet of glass, arrange 1–2 gels on the cellophane making sure that there are no bubbles beneath them, cover with a second piece of cellophane, smooth to eliminate bubbles and clamp the edges of the cellophane to the glass with metal binder clips. Air dry overnight. If you encounter problems with cracking, equilibrate the gels and immerse the cellophane in 15% ethanol/5% glycerol instead of dd H_2O for 15 min before drying. Gels can be scanned wet or dry.

3.4. Examples and Interpretation of Results

Figure 1 shows a Coomassie blue-stained zymogram used to analyze gelatinolytic activities in various samples from cultured cells. Regions of gelatin degradation are evident as clear bands against a background of stained gelatin. In order to determine that these activities can be attributed to MMPs, it is necessary to show that they can be inhibited by EDTA or 1, 10 phenanthroline

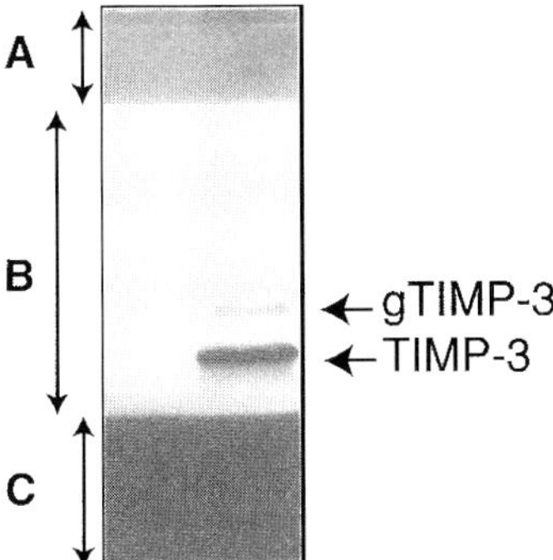

Fig. 2. Reverse zymogram comprising 15% polyacrylamide, 1 mg/mL gelatin and 23.5% (v/v) conditioned media from Rous sarcoma virus-infected chicken embryo fibroblasts. The gel was cast with a 5% stacking gel and the sample was electrophoresed at a constant voltage of 165 V for 1 h 20 min. The gel was incubated and stained with Coomassie Blue as described in **Subheading 3**. The sample contains ECM from transforming chicken embryo fibroblasts. gTIMP-3, glycosylated TIMP-3. For explanation of A, B, and C, see text.

(for example, *see* **ref. 5**). Incubation with EDTA (5 m*M*) inhibits the major activities in lanes 2 and 3 (data not shown); in particular the doublets (lane 2) which are pro-MMP-9 and activated MMP-9 and (lane 3) pro-MMP-2 and activated MMP-2, respectively. As indicated earlier, the proenzymes can be detected in this system because they are activated *in situ*. Lane 1 contains several unidentified activities, some of which are not inhibited by EDTA and are therefore not MMPs.

The total concentration of acrylamide (%T) in this zymogram is 10% where %T is the weight percentage of total monomer (acrylamide + bisacrylamide). This gel effectively resolves the zymogen and activated forms of both MMP-9 and MMP-2. For detection of lower molecular weight activities, higher % gels can be used (for optimum resolution of proteins 15–60 kDa use a 12.5% gel and for proteins 15–45 kDa, a 15% gel). Molecular weight standards are not easily detected against the background staining of the gelatin in a Coomassie Blue stained zymogram but are more readily visible if the concentration of gelatin is decreased. For controls, we use samples of conditioned media from cultured cells whose activities we have characterized (for example, the samples shown in lanes 2 and 3). Commercial standards are also available (*see* **Note 3**).

Figure 2 shows a typical reverse zymogram showing inhibitory activities of glycosylated and unglycosylated TIMP-3 in a sample of extracellular matrix (ECM) from transforming chicken embryo fibroblasts. Three distinct sections (**A–C**) are visible on the Coomassie blue stained gel. The conditioned

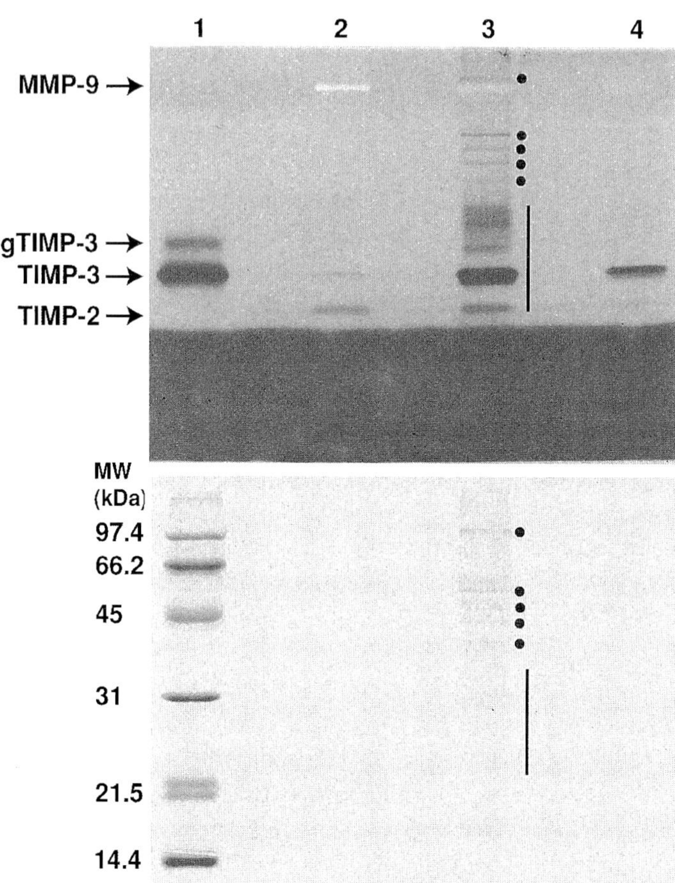

Fig. 3. Reverse zymogram (*upper panel*) and regular SDS polyacrylamide gel (*lower panel*). Both are 15% polyacrylamide separating gels (5% stacking gels) and were electrophoresed side by side at 165 V for 1 h 20 min. Lane 1 (*upper gel*), ECM from RSV-infected chicken embryo fibroblasts (TIMP-3 control), lane 1 (*lower gel*) MW standards (Bio-Rad, low range); lanes 2–4, conditioned media, total cell lysates with ECM, and ECM, respectively, from cultured human cytotrophoblasts.

media used to cast this gel contains MMP-2 and TIMP-2 activities. During electrophoresis, in the presence of SDS, the MMP:TIMP complex in the CM is disrupted and the MMP-2 and TIMP-2 migrate through the gel according to their molecular sizes (~72 kDa and 20 kDa, respectively). As a result, section A contains neither MMP-2 nor TIMP-2, B contains only MMP-2 (which migrates slower than TIMP-2) and C contains MMP-2 and the remaining TIMP-2 which has not yet migrated from the gel. The partial degradation in A is attributed to an MMP activity that we have detected in the CM, that migrates much slower than the MMP-2.

The analytical section of the gel (B) is used for detecting TIMP activities. Its effective size can be increased by increasing the electrophoresis time. In addition, a longer stacking gel will increase the distance between any TIMP-2 in the sample and the TIMP-2 in the gel. From the discussion above, it follows that if there are TIMPs whose electrophoretic mobilities are slower than that of TIMP-2 (for example, TIMP-1) in the CM incorporated in the reverse zymogram then the analytical region of the gel will be shorter. For this reason we recommend the use of CM whose major activities are MMP-2:TIMP-2, particularly as all known TIMPs effectively inhibit MMP-2. Alternatively, one can purify MMPs for inclusion in reverse zymograms *(6)* but this requires additional time and expense and is not necessary for routine qualitative analysis of TIMPs.

An important consideration for the analysis of reverse zymograms is to discriminate between stained bands of undegraded gelatin, protected by TIMP activity, and those proteins which are present in the sample at sufficiently high concentration to be stained by the Coomassie Blue. Samples extracted from tissues and total cell lysates from cells in culture can be particularly problematic in this regard. To control for this, samples must be electrophoresed at dilutions that do not allow their detection on a regular SDS gel under identical conditions. For example, lanes 2 and 4 in **Fig. 3** (reverse zymogram, upper panel) contain Coomassie Blue-stained bands which are not visible in a regular SDS gel electrophoresed under identical conditions (lower panel). These activities are TIMP-2 and TIMP-3 (lane 2) and TIMP-3 (lane 4). In lane 3, the bands indicated by closed circles in the reverse zymogram, are also visible in the SDS gel and, therefore, cannot be attributed to TIMP activities. In contrast, the bands indicated by a vertical line in the reverse zymogram are clearly TIMP activities.

The methods described in this chapter provide qualitative analyses of MMPs and TIMPs in complex biological samples. Activities can be identified by comparison with known standards and new TIMPs and MMPs may be detected. "Relative" activities among different samples can be determined by preparing a dilution series for each sample and a standard curve with a pure preparation of the protein under study. The data can be quantified by densitometric scanning of the gels. However, the use of zymography and reverse zymography for quantitative measurements of MMP and TIMP activities requires careful attention to several important factors that influence the accuracy of the data obtained. Additional information on this topic can be found in **refs. *3* and *6*.**

It is possible to perform identical techniques using other substrates, the most common of which is casein. Substrate concentration, conditioned medium, running and developing conditions must all be optimized for each substrate used.

4. Notes

1. Acrylamide monomer is a neurotoxin and can be absorbed through unbroken skin. Wear a mask while weighing out powder and wear gloves when handling acrylamide powder and solutions. Cover container until all the acrylamide is dissolved. Do not dispose of acrylamide solutions in the sewer system but add bisacrylamide (if none present), polymerize with an appropriate amount of 10% ammonium persulfate and TEMED and discard as solid waste.

2. We have tested gelatin from different sources and found this to be slightly superior in performance, particularly in the reverse zymograms.

3. We have found considerable variation in migration of protein molecular weight standards from commercial sources. For convenience, we use the low range MW standards (Bio-Rad) in regular SDS gels and the MultiMark Multi-Colored MW standards (Novex, San Diego, CA) for tracking electrophoresis in reverse zymograms. The latter can also be used for estimates of molecular size, but bear in mind that these will not be very accurate when determined under nonreducing conditions. We have also reduced and alkylated and dialyzed a set of proteins for use as standards in reverse zymography *(7)*. A number of purified MMPs and TIMPs are available commercially (Chemicon, Temecula, CA) for use as activity standards. We routinely prepare our own standards from ECM and CM samples that we have characterized from cultured cells. For example, a mix of ECM (containing TIMP-3 and glycosylated TIMP-3) and CM (containing MMP-2, TIMP-2, and TIMP-1) from cultured human cells such as FHs 173We provides a good general standard for reverse zymograms. Essentially, confluent cultures are reseeded 1:2 to 1:3 in appropriate medium without serum and cultured for 48 h before harvest. Alternatively, cells can be seeded in serum-containing medium, but washed extensively and maintained without serum for a similar time period prior to harvest. Conditioned media are collected, centrifuged at 10,000g for 15–20 min to remove cellular debris and concentrated, before adjusting to a final concentration of 1X Laemmli sample buffer without reducing agent. Extracellular matrix is prepared from the cell cultures by detachment of cell monolayers with 5 m*M* EGTA, then solubilization of the remaining matrix in Laemmli sample buffer without reducing agent. For more information on suitable growth conditions and ECM and CM collection procedures for a number of human cells, *see* **ref. 7**. We no longer dialyze the CM as described *(7)* but use it directly as above.

4. We routinely use CM from Rous Sarcoma Virus infected chicken embryo fibroblasts as a source of MMP-2 for reverse zymograms although CM from mouse Balb 3T3 and BHK cells transfected with MMP-2 cDNA is also quite satisfactory. Superior CM are those which contain MMP-2 and little or no TIMPs other than TIMP-2, whose rapid migration out of the gel minimizes any interference with the development of the reverse zymogram. MMP-2 is effectively inhibited by TIMPs-1–4 which can thus be detected in reverse zymograms containing this activity. Large batches of media are conditioned by cell cultures in the absence of

serum. They are pooled; SDS is added to a final concentration of 1% (w/v) and the CM are stored in aliquots at –70°C.

5. Mercaptoethanol, which is a component of Laemmli sample buffer, destroys MMP and TIMP activities and therefore is omitted from the sample buffer. Be aware that some commercial preparations of molecular weight standards contain reducing agents that interfere with the development of reverse zymograms and prevent visualization of the standards by Coomassie blue staining.

6. We had previously reported the use of 50 m*M* Tris-HCl at pH 8.0 in the development buffer *(5)* but have found that a 30-fold dilution of the stock separating gel buffer (1.5 *M* Tris-HCl, pH 8.8) is convenient and allows excellent development of the reverse zymograms.

7. We have routinely used KOH/ethanol to clean the glass electrophoresis plates but in the interest of safety and to avoid the possibility of etching the glass surfaces we have found that undiluted detergent, Contrad 70, is a satisfactory substitute.

8. If the gel is not going to be electrophoresed for a few hours it is better to overlay with 1X stacking gel buffer until you are ready to run the samples.

9. For 15% minigels (8 cm × 7.3 cm × 0.75 mm) we routinely electrophorese at 165 V for 1 h 20 min or as determined by trial using colored molecular weight standards (such as the 17 kDa lysozyme marker included in the MultiMark Multi-Colored MW standards [Novex] and the bromophenol blue tracking dye). For a first trial, electrophorese until the lysozyme reaches the bottom of the gel or approx 15 min after the tracking dye has run off the bottom of the gel. Less time is required for a 10% zymogram with the dimensions listed above. For an initial trial, 1 h is recommended.

10. In our experience, the following factors, in particular, improve the quality of reverse zymograms. These are:
    a. The use of scrupulously clean glass containers for the Triton X-100 washes and the development buffer incubation;
    b. The limit of one gel per container; and
    c. The inclusion of a small volume of Triton X-100 in the development buffer.

Acknowledgments

This work was supported by NIH Grant CA 39919 and the Human Frontier Science Program (to SPH). We thank Dr. Tom Meehan for reviewing the manuscript.

References

1. Heussen, C. and Dowdle, E. B. (1980) Electrophoretic analysis of plasminogen activators in polyacrylamide gels containing sodium dodecyl sulfate and copolymerized substrates. *Anal. Biochem.* **102**, 196–202.

2. Granelli-Piperno, A. and Reich, E. (1978) A study of proteases and protease-inhibitor complexes in biological fluids. *J. Exp. Med.* **148**, 223–234.

3. Kleiner, D. E. and Stetler-Stevenson, W. G. (1994) Quantitative zymography: detection of picogram quantities of gelatinases. *Anal. Biochem.* **218**, 325–329.
4. Herron, G. S., Banda, M. J., Clark, E. J., Gavrilovic, J., and Werb, Z. (1986) Secretion of metalloproteinases by stimulated capillary endothelial cells. *J. Biol. Chem.* **261**, 2814–2818.
5. Staskus, P. W., Masiarz, F. R., Pallanck, L. J., and Hawkes, S. P. (1991) The 21-kDa protein is a transformation-sensitive metalloproteinase inhibitor of chicken fibroblasts. *J. Biol. Chem.* **266**, 449–454.
6. Oliver, G. W., Leferson, J. D. Stetler-Stevenson, W. G., and Kleiner, D. E. (1997) Quantitative reverse zymography: analysis of picogram amounts of metalloproteinase inhibitors using gelatinase A and B reverse zymograms. *Anal. Biochem.* **244**, 161–166.
7. Kishnani, N. S., Staskus, P. W., Yang, T.-T., Masiarz, F. R., and Hawkes, S. P. (1994) Identification and characterization of human tissue inhibitor of metalloproteinase-3 and detection of three additional metalloproteinase inhibitor activities in extracellular matrix. *Matrix Biol.* **14**, 479–488.

24

In Situ Zymography

Sarah J. George and Jason L. Johnson

1. Introduction

The techniques of Western blotting and immunocytochemistry are suitable for the quantification and localization, respectively, of matrix metalloproteinase (MMP) and tissue inhibitor of metalloproteinase (TIMP) expression. However, they do not indicate the endogenous balance of matrix degrading activity and inhibition because: 1) at present very few antibodies distinguish between the precursor and proteolytically processed forms of MMPs; and 2) TIMPs present in the tissue sample can prevent matrix degradation by MMPs even if the enzymes are in the active form. Although biochemical studies can determine net MMP activity they do not provide any localizational information and activation of MMPs may occur during disruption of the tissue sample. *In situ* zymography not only enables the estimation of net MMP activity but also allows the localization of this activity in tissue sections. This technique was previously used for detection of enzymatic activity released by explants of developing amphibian tissue *(1)*, allows preservation of the histology and does not require species-specific reagents *(2)*. It has been recently used to detect MMP activity in human atherosclerotic plaques *(3,4)* and saphenous vein organ cultures *(5,6)*.

Using a fluorescently-labelled casein substrate, the caseinolytic activity within the tissue can be visualised as black holes within a fluorescent background. Using photographic emulsion containing silver grains, gelatinolytic activity can be detected as white holes of lysis within a black background. If other matrix proteins are of interest it is possible to label them with fluorescein isothiocyanate (FITC) prior to use in *in situ* zymography *(7)*.

From: *Methods in Molecular Biology, vol. 151: Matrix Metalloproteinase Protocols*
Edited by: I. Clark © Humana Press Inc., Totowa, NJ

2. Materials

2.1. Reagents

1. Good quality distilled water (high performance liquid chromatography [HPLC]-grade) water must be used throughout this procedure.
2. All reagents must be high quality grade.
3. Universal protease substrate (Boehringer Mannheim, Lewes, East Sussex, UK. cat no. 1 080 733).
4. MMP activity buffer: 100 mM NaCl, 100 mM Tris-HCl, pH7.5, 10 mM CaCl$_2$, 20 μM ZnCl$_2$, 0.05% (w/v) Brij 35.
5. Photographic emulsion intended for use in autoradiography. (Amersham International, Little Chalfont, Bucks, UK. cat no. LM-1).
6. Kodak D-19 developer and Unifix fixative solution (Kodak, Bridgend, UK).

2.2. Equipment

1. Water-bath or oven.
2. Glass microscope slides.
3. 22 × 40 mm glass coverslips (caseinolytic activity only).
4. Dark plastic slide boxes with 100 slide capacity (Raymond A Lamb, London, UK, cat no. E/39).

3. Methods

3.1. Preparation of Tissue Samples

1. Chill a bath of isopentane in liquid nitrogen until the first signs of freezing are observed as white spots on the base of the bath (*see* **Note 1**).
2. Drop tissue sample into the isopentane bath (*see* **Note 2**).
3. When tissue sample is solid remove with forceps and place in a suitable container.
4. Store in a –80°C freezer or under liquid nitrogen until section cutting is required.
5. On the day of cutting remove tissue samples from place of storage and embed tissue sample onto cryostat chuck using OCT (optimum cutting temperature) solution (*see* **Notes 3** and **4**).

3.2. Preparation of Slides

3.2.1. Caseinolytic Activity

3.2.1.1. PREPARATION OF SUBSTRATE

1. Add 0.1% (w/v) universal protease substrate and 0.1% (w/v) agarose to a small glass bottle containing MMP activity buffer (*see* **Notes 5** and **7**).
2. Wrap bottle in foil to protect from light.
3. Place bottle in boiling water-bath for approx 20 min until universal protease substrate and agarose are completely in solution (*see* **Note 6**).

3.2.1.2. COATING OF SLIDES

1. Warm glass microscope slides in 37°C oven (*see* **Note 8**).
2. Remove two slides from oven.
3. Pipet 50 μL of universal protease substrate solution onto one slide.
4. Place the other slide on top of the first slide to form a sandwich.
5. Gently pull the slides apart in the same manner as when making a blood smear.
6. Place slides on bench, substrate side uppermost and allow to dry.
7. Repeat coating process for the remainder of slides.
8. Coated slides must be examined under a fluorescent microscope to check that the coating is uniform (*see* **Note 9**). Discard any slides which do not have a uniform coating.
9. Using a cryostat, cut 7 μm frozen sections and place on coated slide.
10. Pipet 50 μL of MMP activity buffer onto frozen sections.
11. Carefully place a 22 × 40 mm glass coverslip on top of section.

3.2.2. Gelatinolytic Activity

3.2.2.1. PREPARATION OF GELATIN SUBSTRATE

1. Place a jar of photographic emulsion in a 37°C water-bath or oven for at least 30 min to liquefy the emulsion (*see* **Note 10**).
2. Dilute the emulsion 1:2 with MMP activity buffer.
3. Keep the diluted emulsion in the 37°C oven until required.

3.2.2.2. COATING OF SLIDES

1. Using a cryostat cut 7 μm frozen sections and place on a clean glass microscope slide.
2. Using a plastic Pasteur pipet place 3–4 drops of the diluted emulsion onto the glass slide.
3. Spread the emulsion first by tipping the slide and then by using the side of the Pasteur pipet to achieve complete coverage of the slide (*see* **Notes 11** and **12**).
4. Place the slide on the bench and the emulsion to set (*see* **Note 13**).

3.3. Incubation of Slides

1. Place 3–4 tissues on the bottom of a dark color slide box.
2. Dampen the tissue with water (*see* **Note 14**).
3. Place up to 8 slides horizontally in the box.
4. Replace the slide box lid.
5. Place in a 37°C oven overnight (*see* **Note 15**).

3.4. Visualization of MMP Activity

3.4.1. Caseinolytic Activity

1. After incubation slides can be visualized under fluorescent light (excitation 574 nm, emission 584 nm).
2. MMP activity will appear as black holes in the red fluorescent background.

3.4.2. Gelatinolytic Activity

1. After incubation, remove the lid from the slide boxes and then return the slide boxes to the 37°C oven, to allow the emulsion to dry.
2. Place slides into a slide staining rack.
3. Place the slide rack into a staining trough containing Kodak D19 developer for 8 min.
4. Wash slides in water for 1 min.
5. Place the slide rack into another staining trough containing Kodak Unifix fixative for 8 min.
6. After the slides have dried, examine them under a light microscope.

3.5. Use of Controls

1. To examine the nature of the enzymatic activity it is essential to carry out controls on serial sections which include inhibitors of each protease class (i.e., serine, cysteine, aspartic). An inhibitor of MMPs, for example EDTA, a synthetic MMP inhibitor or recombinant TIMP protein must always be included. The inhibitor should be included in the universal protease substrate solution, diluted emulsion, and MMP buffer.
2. *In situ* zymography must be carried out in duplicate to confirm the presence of MMP activity

4. Notes

1. It is essential that tissue samples are frozen in isopentane to avoid the formation of ice crystals and preserve the tissue structure.
2. The tissue must be dropped into the isopentane to avoid freezing of the tissue to the forceps.
3. On removal of the tissue from the place of storage warming of the tissue must be avoided at all times.
4. Tissue samples can be used to cut sections several times if warming of the tissue is avoided and the sample is returned to −80°C between cutting sessions.
5. The universal protease substrate is very difficult to weigh and therefore small amounts should be purchased (e.g., 15 mg) and the entire amount used.
6. It must be ensured that both the agarose and universal protease substrate are completely in solution before use as failure to do this leads to poor coverage of the slides.
7. The universal protease substrate solution can be stored at 4°C for several weeks if the solution is protected from light.
8. Slides must be warmed to avoid the solidification of the substrate solution on the slide prior to spreading.
9. It is essential to examine the slides under fluorescent light to check that the coverage is even. If holes of any size are seen these slides must be discarded.
10. The gelatin substrate can be prepared under normal light unless the emulsion is to be used for autoradiography also.

11. When spreading the diluted emulsion it should be ensured that it covers the entire unfrosted area of the slide.
12. Care must be taken to avoid damaging the tissue section when using the side of the Pasteur pipet to spread the emulsion.
13. The emulsion must be allowed to set prior to placing in the humidified chamber. This technique will not work if the emulsion is still liquid during the incubation.
14. The tissues which are placed in the bottom of the slide rack should not be made too wet as this leads to the formation of condensation on the lid of the box that can fall on to the slides during the incubation period.
15. The incubation period may be shortened or lengthened depending on the amount of MMP activity present. In the case of caseinolytic activity it is possible to examine the slides under fluorescent light and if there are no signs of MMP activity the slides can be incubated further. It may be necessary to add more MMP buffer to the slides if long incubations are carried out to avoid drying out of the tissue sections. This can be achieved by pipeting small amounts of the buffer close to the edge of the coverslip and allowing capillary action to pull it under the coverslip. If slides are allowed to dry out they must be discarded. In the case of gelatinolytic activity it is possible to examine the slides using a light microscope prior to development to get an indication of how much digestion and lysis has occurred.

References

1. Gross, J. and Lapiere, C. M. (1962) Collagenolytic activity in amphibian tissues: a tissue culture assay. *Proc. Natl. Acad. Sci. USA.* **48,** 1014–1022.
2. Sappino, A. P., Huarte, J., Vassalli, J. D., and Belin, D. (1991) Sites of synthesis of urokinase and tissue-type plasminogen activators in the murine kidney. *J. Clin. Invest.* **87,** 962–970.
3. Galis, Z. S., Sukhova, G. K., Lark, M. W., and Libby, P. (1994) Increased expression of matrix metalloproteinase and matrix degrading activity in vulnerable regions of atherosclerotic plaques. *J. Clin. Invest.* **94,** 2493–2503.
4. Johnson, J. L., Jackson, C. L., Angelini, G. D., and George, S. J. (1998) Activation of matrix-degrading metalloproteinases by mast cell proteases in atherosclerotic plaques. *Arterioscler. Thromb. Vasc. Biol.* **18,** 1707–1715.
5. George, S. J., Johnson, J. L., Angelini, G. D., Newby, A. C., and Baker, A. H. (1998) Adenovirus-mediated gene transfer of the human TIMP-1 gene inhibits SMC migration and neointima formation in human saphenous vein. *Human Gene Therapy* **9,** 867–877.
6. George, S. J., Baker, A. H., Angelini, G. D., and Newby, A. C. (1998) Gene transfer of tissue inhibitor of metalloproteinase-2 inhibits metalloproteinase activity and neointima formation in human saphenous veins. *Gene Therapy* **5,** 1552–2560.
7. Galis, Z. S., Sukhova, G. K., and Libby, P. (1995) Microscopic localization of active proteases by *in situ* zymography: detection of matrix metalloproteinase activity in vascular tissue. *FASEB J.* **9,** 974–980.

25

Detection of Focal Proteolysis
Using Texas-Red-Gelatin

Rosalind M. Hembry

1. Introduction

How do cells degrade their surrounding matrix? Constitutive and induced cellular secretion of several classes of proteinases have been implicated in extracellular degradation *(1)* and many of these proteinases have the ability to degrade extracellular matrix proteins in vitro. However, clear evidence of focal matrix degradation adjacent to cells in which enzyme synthesis occurs is sparse. In this chapter we describe a method we have developed that allows unambiguous detection of focal extracellular matrix (ECM) degradation below cells, although at the same time localizing cellular expression of enzyme: we illustrate the use of this method to demonstrate focal proteolysis by cells overexpressing the membrane-type matrix metalloproteinase-1 (MT1-MMP) on the cell surface.

Membrane-type matrix metalloproteinases (MT-MMPs), a subclass of the matrix metalloproteinase (MMP) family, uniquely possess a C-terminal transmembrane domain and have the ability to act as initiators of the activation of progelatinase A (MMP2). Most studies have concentrated on MT1-MMP, which has been shown to be present on the surface of activated cells *(2)*. Studies to define the mechanism of progelatinase A activation have shown that this is due to the formation of an MT1-MMP-TIMP-2 complex "receptor" for progelatinase A, with bound proenzyme being sequentially cleaved by active MT1-MMP and active gelatinase A *(3–5)*. The structural similarity of the catalytic domain of MT1-MMP with that of other MMPs suggests it also should be able to degrade ECM components, and the ability of MT1-MMP to digest several ECM macromolecules has been confirmed biochemically *(6–9)*. This

From: *Methods in Molecular Biology, vol. 151: Matrix Metalloproteinase Protocols*
Edited by: I. Clark © Humana Press Inc., Totowa, NJ

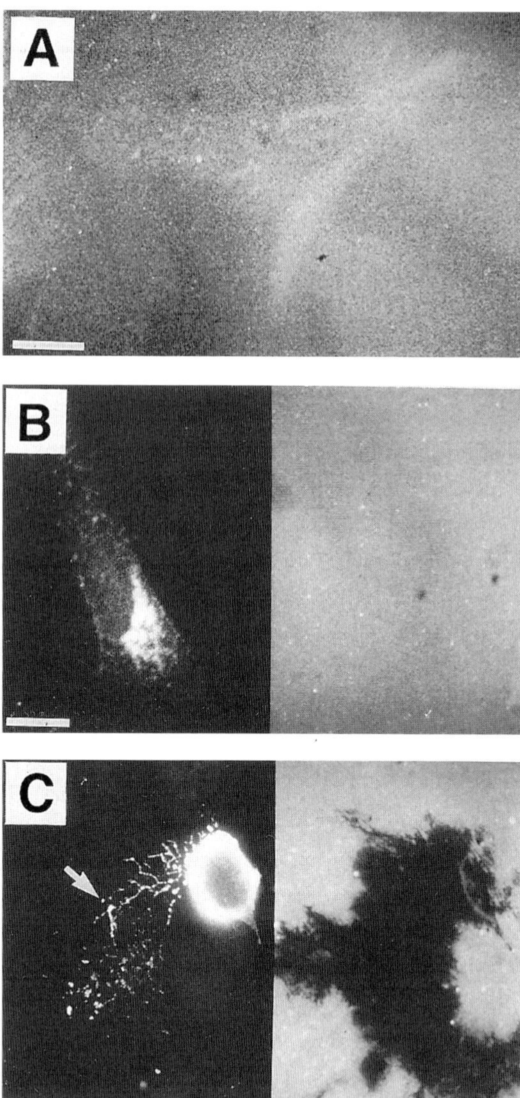

Fig. 1. (**A**) Normal HIFFs cultured for 48 h do not degrade the gelatin-TR film. Gelatin-TR films were made as described. Human infant foreskin fibroblasts were seeded at low density and cultured for 48 h. On microscopy the fibroblast is seen as a halo with the intact film below, indicating no degradation. Bar = 25 μm. (**B**) HIFFs stimulated with PMA show intracellular Golgi staining of MT1-MMP but neither membrane staining nor film degradation. Gelatin-TR films were prepared, HIFFs were added and allowed to attach for 2 h in serum containing medium, then the medium was changed to DMEM with PMA (10 ng/mL) and cells cultured for a further 24 h. Monensin (5 μ*M*) was added for the last 18 h of culture. Films were then fixed and

raised the possibility that active gelatinase A and MT1-MMP when expressed at the cell surface may both be functionally gelatinolytic.

In order to define this mechanism further, we needed to develop a method that allowed unambiguous detection of focal ECM degradation below cells, although at the same time localizing cellular expression of gelatinase A and/or MT1-MMP. We chose to do this by dual channel confocal fluorescence microscopy in order to get precise visualization of small areas of film clearance, without interference by flare from surrounding intact film. We used gelatin as a substrate for both MT1-MMP and gelatinase A, labeled it with Texas Red®, and produced a thin homogeneous film so that areas of substrate degradation would be sharply demarcated and seen as black under the microscope. We cultured Chinese Hamster ovary (CHO) cells on this film and used a transient transfection system to over-express MT1-MMP on the cell surface. This was localized by indirect immunofluorescence using fluorescein to maximize a potentially low signal. Described below is a summary of our results (*see* **ref. *10*** for original report), followed by a detailed protocol subdivided into:

1. Film preparation.
2. Addition of cells to gelatin-TR films and transfection.
3. Immunolocalization by indirect immunofluorescence.
4. Confocal microscopy.

The method is similar to those used in other studies *(11,12)* but, by mild fixation of the gelatin to the glass slide, removal of the film by cell traction is prevented. Nontransfected CHO cells that synthesize low levels of gelatinase A, or human infant foreskin fibroblasts (HIFFs) that synthesize both pro-gelatinase A and MT1-MMP constitutively, do not degrade the film when cultured on it for up to 48 h (**Fig. 1A**). HIFFs cultured on gelatin films and stimulated with phorbol myristate acetate (PMA) express MT1-MMP intracel-

---

stained as described with sheep anti MT1-MMP IgG. Simultaneous scans were collected at 488 nm (FITC, *left*) and 568 nm (Texas Red, *right*). Intracellular staining of the vesiculated Golgi apparatus and secretory vesicles are clearly visible indicating MT1-MMP synthesis but there is no cell membrane staining (*left*), and no film degradation (*right*). Bar = 25 μm. (**C**) CHO cells transiently transfected with MT1-MMP express MT1-MMP on the cell surface and degrade the gelatin film. CHO cells were grown on gelatin-TR films, transiently transfected with MT1-MMP cDNA and cultured for 19 h following glycerol shock. Films were then fixed and stained by indirect immunofluorescence with sheep anti MT1-MMP IgG. Simultaneous scans were collected at 488 nm (FITC, *left*) and 568 nm (Texas Red, *right*). MT1-MMP transfected cell has MT1-MMP immunofluorescence on cell body and on membrane remaining on the coverslip (*arrow*), and has degraded a pathway across the gelatin film during the culture period. Bar = 25 μm.

lularly but not on the surface *(10)* and no film degradation is observed (**Fig. 1B**), indicating a requirement for enzyme at the cell surface for proteolysis to occur. In contrast, when CHO cells are transfected with MT1-MMP and analyzed as above, on simultaneously scanning both channels (**Fig. 1C**) rings of FITC immunofluorescence at the level of the film are seen on 20–30% of cells (**Fig. 1C**, *left*), indicating cell surface staining of MT1-MMP on transfected cells and, by collection of 1 μm sections through cells, staining is seen to cover the entire cell surface area (data not shown). In addition, at film level, traces of MT1-MMP immunofluorescence are seen separate from the cells (**Fig. 1C**, *arrow*), presumably membrane deposits on the coverslip, detached as a result of cell movement during the culture period. A substantial fraction of the cellular integrins is known to be left behind on the substratum as the cell detaches and locomotes *(13)*, and these data suggest that cell surface bound MT1-MMP may also remain attached to the substratum. The gelatin-TR film has been degraded by these cells, seen as dark feathery patches on an otherwise homogeneous film (**Fig. 1C**, *right*). The film degradation caused by MT1-MMP expressing cells is subjacent to the cell and sharply demarcated, similar to the localized proteolysis of opsonized fibronectin films by polymorphonuclear leukocytes in the presence of proteinase inhibitors *(14)*. Tracks of film clearance are frequently observed, similar to phagokinetic tracks *(15)*, illustrating proteolysis during cell migration. Areas of film degradation are always associated with fluorescein isothiocyanate (FITC) immunofluorescence of cells and/or membrane deposits. Some cells have no cell surface staining for MT1-MMP and no adjacent film clearance; since degradation of the film is only seen beneath cells with cell membrane staining we conclude that these cells are not transfected and not actively producing cell membrane bound MT1-MMP. MT1-MMP transfected cells stained with normal sheep serum IgG show tracts of film clearance but no staining on the FITC channel *(10)* indicating that the FITC fluorescence is not due to non-specific uptake of sheep IgG by CHO cells.

2. Materials

2.1. Film Preparation

1. Either 8 well Labtek® slides (ICN Biomedicals) or glass coverslips (22 × 22 mm) washed with tissue culture detergent, well rinsed in distilled water, sterilized, and placed in 1.5 mL tissue culture dishes.
2. Sephadex G25 column (30 cm × 15 cm) equilibrated with phosphate buffered saline (PBS, pH 7.4)
3. Gelatin (Sigma).
4. Texas Red® sulphonyl chloride (Molecular Probes, Oregon, USA).
5. Dimethyl formamide dried over molecular sieve 4A 8–12 mesh beads (Sigma, M-1760)

6. Poly-L-lysine hydrobromide (Sigma P-9155) 50 µg/mL dissolved in sterile water.
7. 0.1 M Sodium bicarbonate, pH.9.0.
8. Phosphate buffered saline (PBS, pH 7.4), sterilised.
9. 4% formaldehyde freshly prepared from paraformaldehyde dissolved in PBS (pH 7.4).
10. Ammonium chloride (50 µM in PBS), sterilized.
11. Dulbecco's modified Eagle's Medium (DMEM) with 10% foetal bovine serum (FBS).

2.2. Addition of Cells to Gelatin-TR Films and Transfection

For full details of transfection, *see* Chapter 11.
1. Plasmid cDNAs.
2. 2X HEPES buffered saline (2X HBS): 8.18 g NaCl, 5.59 g HEPES, 0.2g Na_2HPO_4 (anhydrous), pH to 7.1 with NaOH and adjust final volume to 500 mL.
3. 2 M $CaCl_2$.
4. 15% v/v glycerol in DMEM.
5. DMEM.
6. DMEM with 10% FBS.
7. Enzyme inhibitors dissolved in DMEM.

2.3. Immunolocalization by Indirect Immunofluorescence

1. 4% formaldehyde freshly prepared from paraformaldehyde dissolved in PBS (pH 7.4).
2. Phosphate buffered saline.
3. Primary antibody IgG and control serum IgG in PBS.
4. Secondary antibody labeled with fluorescein isothiocyanate.
5. Vectashield mountant (Vector Laboratories Inc, Burlingame CA).

3. Methods
3.1. Preparation of Gelatin-TR Films

1. Dissolve 10 mg gelatin in 1 mL 0.1 M sodium bicarbonate, pH.9.0.
2. Dissolve Texas Red® sulphonyl chloride (1 mg) in 100 µL dry dimethyl formamide, add dropwise to the gelatin and stir (30 min, 4°C).
3. Separate the conjugated gelatin-Texas Red® (gelatin-TR, the first red fraction) from unconjugated dye (retained on column) by passing through a Sephadex G25 column equilibrated with phosphate buffered saline (PBS, pH 7.4). Add a preservative (e.g., butanol, but not sodium azide) and store at 4°C until use.
4. Coat 8 well Labtek® slides or glass coverslips with poly-L-lysine (10 µg in 200 µL H_2O per well/coverslip, 1 h, room temperature) and remove excess.
5. Add gelatin-TR (10 µL in 200 µL H_2O per well, 2 h, room temperature), then remove the excess.
6. Fix the coating to the glass well with 4% formaldehyde freshly prepared from paraformaldehyde (pH 7.4, 5 min, room temperature), remove fixative by washing three times with sterile PBS (5 min each).

7. Block unreacted aldehyde groups with sterile ammonium chloride (50 μ*M* in PBS, 10 min, room temperature).
8. Wash wells repeatedly in sterile PBS, then add 400 μL DMEM with 10% FBS and incubate overnight (37°C, 5% CO_2/air) prior to plating cells.
9. The film formed is uniform and approx 2 μm thick, measured by confocal microscopy.

3.2. Addition of Cells to Gelatin-TR Films and Transfection with cDNA

1. Plate cells (e.g., CHO-L761H at 3.2 × 10⁴ cells per well) onto Labtek® slides and culture in DMEM supplemented with 10% FBS at 37°C, with 5% CO_2 for 24 h. This is a lower cell density than normally used for transfection, but allows cells to spread out and adhere well to the substrate and facilitates identification of individual cells responsible for focal film degradation. Other cell types may be used but, again, a low cell density should be chosen such that about 50% confluence only is reached by the end of the chosen incubation time.
2. Perform transfection essentially as described by M. Butler and S. Atkinson in this volume, based on the method of Chen and Okayama *(16)*. Either pEE14 MT1-MMP or pEE14 vector control (2.5 μg DNA/cm²/well) are co-precipitated with calcium phosphate and incubated with the cells for 4 h at 37°C.
3. Remove the media and expose the cells to 15% glycerol in DMEM for 1 min.
4. Wash the cells in warm DMEM and culture for 3 h in DMEM with 10% FBS (*see* **Note 1**), 37°C, 5% CO_2 to restore cell membrane integrity.
5. Media are then replaced by DMEM without serum for a further 16 h (*see* **Note 2**). Film degradation takes place in the presence of serum proteins but their removal allows subsequent analysis of the culture medium by gelatin zymography.
6. Enzyme inhibitors may be added to the wells, if required, immediately after the glycerol shock and also included in the serum-free incubations.

3.3. Immunolocalization by Indirect Immunofluorescence

1. Wash cells cultured on gelatin-TR films briefly with warmed PBS then fix with 4% formaldehyde in PBS, freshly prepared from paraformaldehyde (5 min, room temperature), wash wells with PBS to remove fixative (*see* **Notes 3** and **4**).
2. Incubate (1 h, room temperature) with either normal sheep serum IgG (NSS), sheep antihuman MT1-MMP IgG (*[10]*, 50 μg/mL in PBS, or another appropriate anti-MT1-MMP antibody), together with 5% normal serum (from the species in which the secondary antibody is raised) to block nonspecific protein binding.
3. Remove primary antibody with 3 washes of PBS.
4. Incubate with appropriate secondary antibody with 5% normal blocking serum, 30 min, room temperature. We use an antiserum raised in a pig to sheep Fab' fragments, and prepared Fab' fragments from this labeled with FITC (pig-FITC, *[17]*, 1:200 in PBS)

5. Wash wells repeatedly in PBS, remove plastic well sides and gasket and coverslip films with Vectashield mountant (Vector Laboratories Inc, Burlingame, CA).

3.4. Confocal Microscopy

1. View on a suitable confocal microscope: we use a Bio-Rad MRC 600 with a Krypton/Argon laser.
2. Using the 568 nm laser line first with a wide confocal aperture and a low power objective, bring film into view and reduce gain and confocal aperture until film is in focus. Focussing on the (very bright) gelatin-TR film first avoids unnecessary photobleaching of the weaker FITC signal.
3. Switch to scanning on both channels and adjust 488 nm line according to brightness of signal.
4. Data can be collected in two ways: a) by simultaneously scanning both the 488 nm (FITC, *left*) and 568 nm (Texas Red, *right*) laser lines with a confocal aperture of 0.5 and kalman averaging over 10 scans, at slow scan speed; and b) by collecting serial 0.5–1 μm sections through cells and films (Z series) and merging images with Bio-Rad Comos software using false color.
5. To assess the area of film degradation the Bio-Rad Comos software can be used to outline on the screen the degradation below at least 10 separate cells in duplicate wells (>20 cells in all) and areas averaged. However, variations in local cell densities and individual cell movement may give rise to large standard errors.

4. Notes

1. Sodium butyrate was not included in the medium because preliminary experiments using this low cell density showed that it had little effect on the transfection rate.
2. Monensin (Sigma, 5 μ*M*) may be added to cell cultures for the final 3–24 h of culture to increase detectable levels of intracellular proteins (e.g., **Fig. 1B**).
3. Intracellular proteins may be exposed to antibody IgGs by permeabilizing cells with 0.1% Triton X-100 (5 min, room temperature) after fixation, followed by 3 PBS washes before application of primary antibody (e.g., **Fig. 1B**).
4. Fixatives such as acetone and methanol give poor cellular structure and much reduced staining for most MMPs.

References

1. Murphy, G. and Reynolds, J. J. (1993) Extracellular matrix degradation, in *Connective Tissue and its Heritable Disorders* (Royce, P. M. and Steinmann, B., eds.), Wiley-Liss, Inc, New York, pp. 287–316.
2. Sato, H., Takino, T., Okada, Y., Cao, J., Shinagawa, A., Yamamoto, E., and Seiki, M. (1994) A matrix metalloproteinase expressed on the surface of invasive tumour cells. *Nature* **370**, 61–65.
3. Strongin, A. Y., Collier, I., Bannikov, G., Marmer, B. L., Grant, G. A., and Goldberg, G. I. (1995) Mechanism of cell surface activation of 72-kDa type IV collagenase. Isolation of the activated form of the membrane metalloprotease. *J. Biol. Chem.* **270**, 5331–5338.

4. Atkinson, S. J., Crabbe, T., Cowell, S., Ward, R. V., Butler, M. J., Sato, H., Seiki, M., Reynolds, J. J., and Murphy, G. (1995) Intermolecular autolytic cleavage can contribute to the activation of progelatinase A by cell membranes. *J. Biol. Chem.* **270**, 30,479–30,485.

5. Butler, G. S., Butler, M. J., Atkinson, S. J., Will, H., Tamura, T., Van Westrum, S. S., Crabbe, T., Clements, J., D'Ortho, M. P., and Murphy, G. (1998) The TIMP2 membrane type 1 metalloproteinase "receptor" regulates the concentration and efficient activation of progelatinase A. *J. Biol. Chem.* **273**, 871–880.

6. Imai, K., Ohuchi, E., Aoki, T., Nomura, H., Fujii, Y., Sato, H., Seiki, M., and Okada, Y. (1996) Membrane-type matrix metalloproteinase 1 is a gelatinolytic enzyme and is secreted in a complex with tissue inhibitor of metalloproteinases 2. *Cancer Res.* **56**, 2707–2710.

7. Pei, D. Q. and Weiss, S. J. (1996) Transmembrane-deletion mutants of the membrane-type matrix metalloproteinase-1 process progelatinase A and express intrinsic matrix-degrading activity. *J. Biol. Chem.* **271**, 9135–9140.

8. Ohuchi, E., Imai, K., Fujii, Y., Sato, H., Seiki, M., and Okada, Y. (1997) Membrane type 1 matrix metalloproteinase digests interstitial collagens and other extracellular matrix macromolecules. *J. Biol. Chem.* **272**, 2446–2451.

9. D'Ortho, M.-P., Will, H., Atkinson, S., Butler, G. S., Messant, A., Gavrilovic, J., Smith, B., Timpl, R., Zardi, L., and Murphy, G. (1997) Membrane-type matrix metalloproteinases 1 and 2 exhibit broad-spectrum proteolytic capacities comparable to many matrix metalloproteinases. *Eur. J. Biochem.* **250**, 751–757.

10. D'Ortho, M.-P., Stanton, H., Butler, M., Atkinson, S. J., Murphy, G., and Hembry, R. M. (1998) MT1-MMP on the cell surface causes focal degradation of gelatin films. *FEBS Lett.* **421**, 159–164.

11. Nakahara, H., Howard, L., Thompson, E. W., Sato, H., Seiki, M., Yeh, Y. Y., and Chen, W. T. (1997) Transmembrane/cytoplasmic domain-mediated membrane type 1-matrix metalloprotease docking to invadopodia is required for cell invasion. *Proc. Natl. Acad. Sci. USA* **94**, 7959–7964.

12. Pilcher, B. K., Dumin, J. A., Sudbeck, B. D., Krane, S. M., Welgus, H. G., and Parks, W. C. (1997) The activity of collagenase-1 is required for keratinocyte migration on a type I collagen matrix. *J. Cell Biol.* **137**, 1445–1457.

13. Palecek, S. P., Schmidt, C. E., Lauffenburger, D. A., and Horwitz, A. F. (1996) Integrin dynamics on the tail region of migrating fibroblasts. *J. Cell Sci.* **109**, 941–952.

14. Owen, C. A. and Campbell, E. J. (1995) Neutrophil proteinases and matrix degradation. The cell biology of pericellular proteolysis. *Seminars in Cell Biology* **6**, 367–376.

15. Albrecht-Buehler, G. (1977)The phagokinetic tracks of 3T3 cells. *Cell* **11**, 395–404.

16. Chen, D. and Okayama, H. (1987) High efficiency transformation of mammalian cells by plasmid DNA. *Mol. Cell. Biol.* **7**, 2745–2752.

17. Hembry, R. M., Murphy, G., and Reynolds, J. J. (1985) Immunolocalization of tissue inhibitor of metalloproteinases (TIMP) in human cells. Characterization and use of a specific antiserum. *J. Cell Sci.* **73**, 105–119.

26

Antibodies to MMP-Cleaved Aggrecan

Amanda J. Fosang, Karena Last, David C. Jackson,
and Lorena Brown

1. Introduction

The matrix metalloproteinases (MMPs) have a pivotal role in both normal and pathological turnover of the extracellular matrix. Whereas MMP protein can easily be detected by immunolocalization or Western blot analysis, the determination of whether or not an MMP is active and acting on a particular substrate has been more difficult. If the primary sequence and unique cleavage sites within the substrate are known, one means of unambiguously identifying MMP activity is to use antibodies with a unique specificity for antigenic determinants on the newly created N- or C- termini of the degradation products. By definition, these "neoepitope" antibodies recognize a terminal sequence exclusively and do not recognize the same sequence of amino acids located internally as part of the intact protein. This approach has been extremely successful for detecting MMP-derived fragments of aggrecan (the large cartilage proteoglycan) *(1–5)* and collagen *(6)*, since aggrecan and collagen are abundant in cartilage matrix, and their neoepitopes are correspondingly abundant. Often MMPs are not the only proteinases involved in tissue remodeling and in this situation neoepitope antibodies allow fragments derived from MMPs and other proteinases to be distinguished from each other and compared. In cartilage, aggrecan degradation is mediated by both MMPs, and aggrecanase *(7,8)* which are members of the ADAMTS family of proteinases *(9)*. The production and use of neoepitope antibodies for investigating cartilage catabolism was first described by Hughes et al. *(10)*. Thereafter neoepitope antibodies were quickly recognized as ideal tools for resolving the products of separate degradative pathways involved in aggrecanolysis.

From: *Methods in Molecular Biology, vol. 151: Matrix Metalloproteinase Protocols*
Edited by: I. Clark © Humana Press Inc., Totowa, NJ

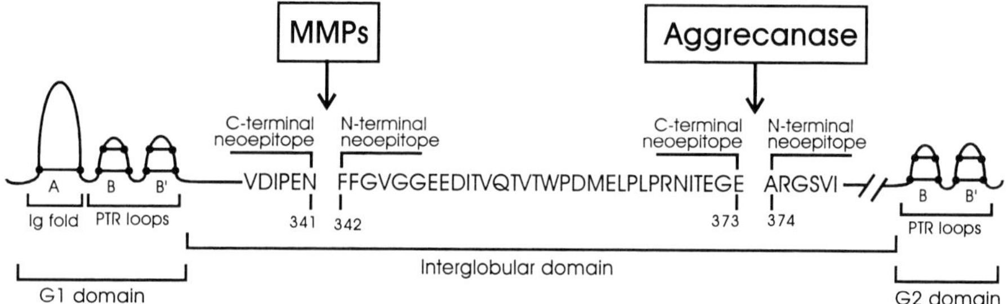

Fig. 1. Cleavage sites and neoepitopes generated by MMPs and aggrecanase in the aggrecan interglobular domain. Expansion of the G1-IGD-G2 region showing the immunoglobulin (Ig) fold motif (A loop) in G1 and the proteoglycan tandem repeat motif (PTR; B and B' loops) in G1 and G2. Cysteine residues and disulphide bonds in G1 and G2 are shown. Amino acids in the human sequence *(25)* flanking and bridging the MMP and aggrecanase cleavage sites, and the neoepitope sequences generated by cleavage are included.

It is inevitable that MMPs will cleave aggrecan at a number of sites along its 2316* amino acid core protein. However so far, precise MMP cleavage sites have only been identified within the interglobular domain (IGD) between the N-terminal G1 and G2 globular domains. The predominant site at which MMPs cleave in the IGD is ...DIPEN$_{341}$ ↓ F$_{342}$FGVG–giving rise to the amino terminal neoepitope F$_{342}$ FGVG... and the carboxy terminal neoepitope ...DIPEN$_{341}$ (**Fig. 1**). Aggrecanase cleaves at ...ITEGE$_{373}$ ↓ A$_{374}$RGSV... in the IGD, producing the amino terminal neoepitope A$_{374}$RGSV... and the carboxy terminal neoepitope ...ITEGE$_{373}$ (**Fig. 1**).

One impediment to taking full advantage of neoepitope technology is the ease with which the antibodies can be made, as some neoepitope sequences have proven to be more antigenic than others. When the Jameson-Wolf algorithm for predicting antigenicity *(11)* is applied to the ...ITEGE$_{373}$ and ...DIPEN$_{341}$ sequences, it yields a reasonably good score for the antigenic potential of these C-terminal neoepitopes. The prediction appears to have been borne out since several polyclonal antisera against these neoepitopes have been raised *(2,3,12–15)*. A polyclonal antisera recognising the ...DIPES$_{341}$ neoepitope from bovine aggrecan has also been made *(16)*. Although generation of antibodies recognising the N-terminal neoepitopes has been more challenging, with the F$_{342}$FGVG... predicted to be a difficult sequence and the A$_{374}$RGSV... N-terminal sequence predicted to be the least antigenic, antibodies to both these neoepitope sequences have been made *(1,4,17)*.

*Human aggrecan sequence *(25)*

In order to produce an antibody that exclusively recognizes a terminal sequence it is essential that the immunogen be kept short to avoid the possibility of raising antibodies against antigenic determinants on internal sequences. Purified proteins are therefore not suitable immunogens for raising neoepitope antibodies. We have used 10mer synthetic peptides as immunogens. The methodologies we used for generating the monoclonal anti-FFGVG antibody (AF-28) *(4)* were quite different from those used to produce the polyclonal anti-DIPEN antisera *(15)* and we describe both methods in this chapter. The steps described for antibody production are: 1) assembly of synthetic peptides; 2) conjugation of peptides to carrier proteins; 3) immunization of rabbits and production of polyclonal antisera; 4) immunization of mice and production of monoclonal antibodies; and 5) characterization of antibody specificity.

2. Materials

2.1. Assembly of Synthetic Peptides

Chemicals should be of analytical grade or its equivalent; some companies offer "peptide synthesis grade" solvents which may be an advantage.

1. Side arm flask supporting a glass manifold and funnels containing glass sinters (*see* **Fig. 2**).
2. *N,N'*-dimethylformamide (DMF).
3. Piperidine.
4. Dichloromethane (DCM).
5. Diethylether.
6. Diisopropylethylamine (DIPEA).
7. Methanol.
8. Acetonitrile.
9. Trifluoroacetic acid (TFA).
10. Amino acids are obtained as the fluorenylmethoxycarbonyl (Fmoc) derivatives.
11. O'benzotriazole-*N,N,N',N'*-tetra methyl-uronium-hexafluorophosphate (HBTU).
12. 1-hydroxybenzotriazole (HOBt).
13. Triisopropylsilane (TIPS).
14. Thioanisole.
15. Anisole.
16. Ethane dithiol (EDT).
17. Phenol.
18. Trinitrobenzylsulphonic acid (TNBSA).
19. 1,8-diazabicyclo-[5.4.0]undec-7-ene (DBU).
20. 0.6 *M* HOBt/0.55 *M* HBTU in DMF: This is an almost saturated solution and dissolution should be complete and the solution filtered before use. It is best to calculate the volume required before making this solution so that it is made fresh but the solution can be stored at 4°C for a week in a dark bottle with no apparent deterioration.

21. DMF, 2.5% DBU in DMF or, 20% piperidine in DMF: These can all be used in wash bottles, which allows easy washing of the peptide-support in the glass sinter funnels.
22. 1% TNBSA in DMF.
23. 0.9 M DIPEA in DMF.

2.2. Conjugation of Peptides to Carrier Proteins

2.2.1. Method A: Conjugation of FFGVGGEEDC to Keyhole Limpet Haemocyanin Using Maleimidohexanoic Acid N-Hydroxysuccinimide Ester

1. Maleimidohexanoic acid N-hydroxysuccinimide ester (MHS).
2. DMF.
3. 100 mM phosphate buffer pH 8.0.
4. 500 mM Tris-HCl buffer pH 8.0.
5. Phosphate buffered saline (PBS) pH 7.4.
6. Keyhole limpet haemocyanin (KLH).
7. Dithiothreitol.

2.2.2. Method B: Conjugation of CGGFVDIPEN to Ovalbumin and BSA Using N-Hydroxysuccinimidylbromoacetate

1. N-hydroxysuccinimidylbromoacetate (HSBA)(Sigma B-8271).
2. 0.1 M K_2HPO_4 (17.42 g/L).
3. 0.1 M KH_2PO_4 (13.61g/L).
4. 0.1 M potassium phosphate/EDTA buffer: To make 500 mL of buffer, dissolve 186 mg EDTA in a mix of 410 mL 0.1 M K_2HPO_4 and 90 mL KH_2PO_4.

2.3. Production of Polyclonal Antibodies

1. Freund's Complete Adjuvant (containing 0.5 mg/mL heat-killed *M. tuberculosis* organism).
2. Freund's Incomplete Adjuvant (without the *M. tuberculosis* organism).
3. Conjugated peptide in PBS.
4. 2 mL glass syringes.
5. Double hub connector (metal).

2.4. Production of Monoclonal Antibodies

1. Freund's Complete Adjuvant.
2. 50 μM $NaHCO_3$, pH 8.5.
3. Phosphate buffered saline (PBS): 20 mM Na_2HPO_4 in 0.15 M NaCl, pH 7.4.
4. Blocking solution: 10 mg/mL w/v bovine serum albumin in PBS.
5. PBST: PBS containing 0.05% Tween 20.
6. BSA5PBST: 5 mg/mL w/v bovine serum albumin in PBST.
7. Rabbit antimouse immunoglobulin-peroxidase conjugate (antimouse-HRP) (DAKO P0161).
8. Culture medium: RPMI 1640 supplemented with 10% heat inactivated fetal calf serum [FCS], 2 mM L-glutamine, 2 mM sodium pyruvate, 100 μM 2-mercaptoethanol, 100 IU/mL penicillin, 100 μg/mL streptomycin and 30 μg/mL gentamicin.

9. Serum-free medium: culture medium without serum.
10. Fusion medium: culture medium with a total of 15% FCS.
11. ATC buffer: 0.14 M NH$_4$Cl in 17 mM Tris-HCl, pH 7.2.
12. SP$_2$O murine myeloma cells.
13. PEG: 50% polyethylene glycol 1500; Hybri-Max, Sigma P7181.
14. 50X HT: sterile solution of 5 mM hypoxanthine and 0.8 mM thymidine; Hybri-Max, Sigma H0137.
15. 50X A: sterile solution of 20 µM aminopterin; Hybri-Max, Sigma A5159.
16. Saline: 0.85% w/v NaCl.
17. Eosin: 0.2% w/v solution of eosin in normal saline.
18. Dimethyl-sulfoxide (DMSO).
19. 2, 6, 10, 14-tetramethyl-pentadecane (Pristane).
20. BALB/c mice.
21. Needles: 19 gauge × 38 mm, 23 gauge × 32 mm, 26 gauge × 13 mm.
22. Plastic 60 × 15 mm Petri dishes.
23. TC6 plates (6-well multidish, Nunclon).
24. TC24 plates (24-well multidish, Nunclon).
25. 80 cm^2 tissue culture flasks (Nunclon).
26. Sterile flat-bottom 96-well culture plates with lids (TC microwell, Nunclon).
27. Wide-necked vessel: autoclaved 250 mL glass bottle (Schott).
28. 2 mL disposable syringe for emulsification (Once, Asik, Denmark).
29. 1 mL cryotubes (Nunc).
30. Capillary tubes: haematocrit capillaries 75 mm/75 µL, 1.5 mm external diameter.
31. Wire mesh sieves.
32. Humidified box.

2.5. Characterization of Specific Antibodies

1. 96-well ELISA plates (Immulon 4, Dynatech, 0110103855).
2. Swine antirabbit immunoglobulin-peroxidase conjugate (antirabbit-HRP) (DAKO P0399).
3. 2,2'-Azino di-ethylbenzothiazoline-sulfonic acid (ABTS).
4. Coating buffer: 20 mM sodium carbonate, pH 9.6.
5. Tris incubation buffer (Tris IB): 0.1% BSA, 0.1% Nonidet NP40, 0.15 M NaCl, 10 mM Tris-HCl, pH 7.4.
6. Citric acid buffer: 61 mM citric acid, 77 mM Na$_2$HPO$_4$.2H$_2$O, pH 4.0.
7. ABTS solution: 0.1g ABTS, 100 µL 30% H$_2$O$_2$, made to 100 mL in water.
8. Adhesive plate sealers (ICN).

3. Methods

3.1. Assembly of Synthetic Peptides

The assembly of synthetic peptides, although not usually considered to be a routine procedure, is sufficiently straightforward that access to a laboratory equipped with a fume hood and simple glassware, allows appropriately

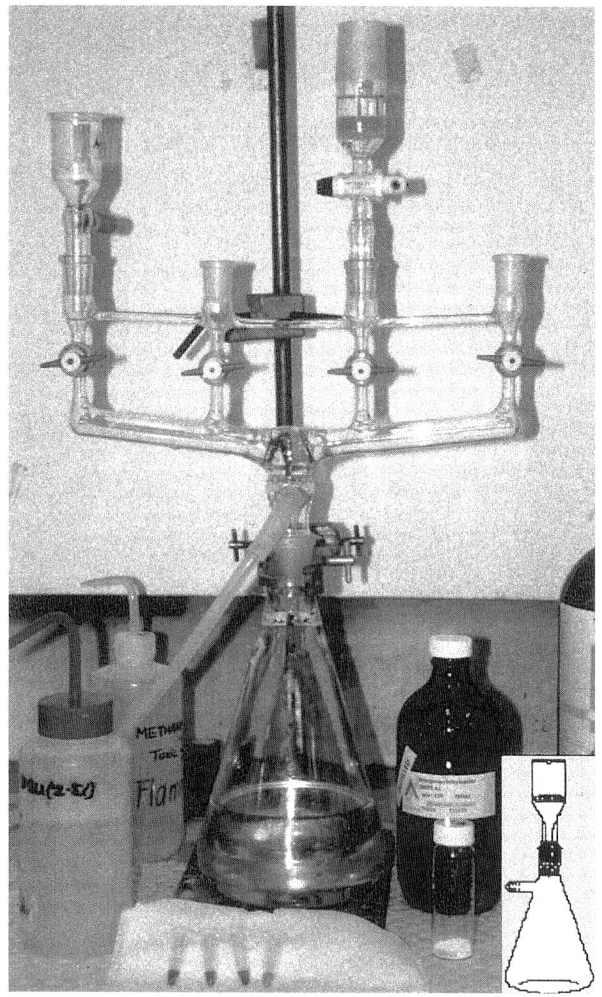

Fig. 2. Side arm flask apparatus used for peptide synthesis.

trained personnel to undertake their assembly. We have routinely used 9-fluor-
enylmethoxycarbonyl (Fmoc) chemistry over the last decade and have found
that manual synthesis of peptides is more flexible than use of automatic
synthesisers. For manual synthesis, an apparatus consisting of a flask supporting
a glass manifold and funnels containing glass sinters (**Fig. 2**) permits the simultaneous assembly of up to four peptides whereas individual synthetic peptides
can be made using a simpler arrangement with a single funnel (**Fig. 2** *insert*). In
each arrangement, the side arm of the flask is attached to a vacuum pump and a

gentle vacuum applied to aspirate solvents from the funnel. Manual synthesis using such simple apparatus permits quick and easy intervention at any point in the process allowing assessment of reaction progress and assessment of the purity of product; such issues can be important in the case of difficult sequences.

Peptide synthesis is usually carried out in the solid phase and there are a very large number of choices of solid phase supports available for peptide synthesis; the prospective peptide chemist should spend a little time familiarising themselves with the possibilities. For the peptide FFGVGGEEDC, PAL PEG PS resin (PerSeptive Biosystems GmbH Hamburg, Germany) was used.

Whereas use of sophisticated instrumentation is not necessary for assembly of the occasional peptide, access to high performance liquid chromatography (HPLC) or fast protein liquid chromatography (FPLC) are requisites. The usual method for assessing the fidelity of peptide synthesis is by using a reverse phase chromatography column installed in an HPLC or an FPLC system equipped with an appropriate detector. Subsequent downstream methods of analysis such as mass spectrometry can also be used and may often be obtained "off site" but chromatographic analysis is an essential on site requirement.

The process of peptide synthesis involves the stepwise addition of activated, protected amino acids to a solid support bearing the growing peptide chain. When this acylation reaction is complete, the Fmoc group is removed either with a solution of piperidine in DMF or of DBU in DMF. This exposes the α-NH$_2$ group in readiness for addition of the next activated amino acid. In this way, peptides are assembled from the C- to the N-terminus.

For synthesis of 0.2 mmol of peptide (e.g., 200 mg of peptide FFGVGGEEDC with a molecular weight of approx 1,000 Da)

1. Weigh 1 g PAL PEG PS resin into a sintered funnel installed in the apparatus shown in **Fig. 2** and swell in DMF at room temperature for at least 30 min. This resin that is used for assembly of the amide form of the peptide has the Fmoc group attached and needs to be swollen and then treated with either piperidine or DBU in DMF to expose the NH$_2$ group.

2. Weigh 0.8 mmol Fmoc-amino acid (i.e., a fourfold excess of amino acid relative to the substitution level of the support) into a clean and dry glass tube. McCartney tubes with a vol of 40 mL are useful containers for this purpose, as they are sufficiently large to allow direct weighing and complete dissolution of amino acid powders but small enough to allow subsequent recovery of the solution with a Pasteur pipet. Add 1.33 mL of HOBt/HBTU stock solution (i.e., an equimolar amount of HOBt and a slightly less than molar equivalent of HBTU relative to the amino acid) followed by 1.33 mL DIPEA working solution, which represents a sixfold excess over the substitution level of the solid phase support, and dissolve by gentle swirling (*see* **Note 1**). Most amino acids will dissolve but for some addition of a few drops of DMF may be necessary.

3. Remove DMF from the swollen resin in the glass sinter filter funnel by aspiration using the vacuum pump (*see* **Note 2**) and add the activated amino acid solution. Stir with a glass rod and allow to stand with occasional stirring.

4. After 30 min, remove a few beads of resin with a Pasteur pipet into an Eppendorf tube containing 1 mL of DMF. Allow the beads to settle, remove the supernatant and repeat the washing procedure twice. Add two drops of DIPEA to the beads followed by five drops of TNBSA solution. If the beads are colorless then acylation is complete and the next step can be carried out. Any trace of orange color in the beads indicates the presence of free amino groups and incomplete coupling. In these cases **step 2** should be repeated until a negative TNBSA test is returned.

5. Wash the peptide-support with DMF (3×10 mL) and then remove the N-α Fmoc group by washing either with the working DBU or the working piperidine solution (*see* **Note 3**). A total of 3 washes of 5 mL is usually sufficient for complete removal of the Fmoc group and the last wash should be allowed to stay in contact with the support for at least 5 min.

6. Wash the support well with DMF (5×10 mL) to remove any traces of DBU or piperidine.

7. Repeat **step 2** with the next amino acid.

8. At completion of peptide assembly, supports with peptide attached are washed sequentially in DMF, DCM, and diethyl ether or methanol, dried under vacuum and stored in a desiccated atmosphere at room temperature until cleaved (*see* **Notes 4** and **5**).

9. Weigh 50–100 mg of the peptide-support into a clean dry McCartney bottle and add 2 mL cleavage reagent (88% TFA, 5% phenol, 5% water, and 2% TIPS) (*see* **Note 6**). Gently flush with nitrogen and leave under N_2 for 3 h with occasional mixing (*see* **Note 7**). This procedure simultaneously cleaves the peptide from the solid support and removes side chain protecting groups from those amino acids that have them.

11. Remove the slurry into a 2.5 mL syringe barrel into which has been introduced a small plug of nonabsorbent cotton wool. Replace the syringe plunger and compress to drive all the peptide-containing supernatant into a clean dry 50 mL centrifuge tube.

12. Evaporate the solution to a volume of 200–300 µL under a gentle stream of nitrogen.

13. Add 50 mL cold diethyl ether to the peptide solution, shake vigorously and hold on ice.

14. Seal the cap and tube using paraffin wax film and centrifuge, preferably in the cold, to sediment the peptide material.

15. Wash the precipitate twice with cold ether and dissolve in 0.1% aqueous TFA (*see* **Notes 8** and **9**).

3.2. Conjugation of Peptides to Carrier Proteins

Short peptides are poorly immunogenic and need to be conjugated to carrier proteins that provide a source of T-cell help thereby allowing antibodies to be produced which are directed to the peptide epitope. Traditionally, peptide epitopes have been coupled to a source of T-cell help using relatively nonspecific reagents such as glutaraldehyde. Conjugates prepared in this way do elicit antibodies but the products of conjugation are heterogeneous and composed of homopolymers, heteropolymers as well as the desired conjugate of peptide and carrier. The use of heterobifunctional reagents permits a more controlled method for coupling peptide epitopes to protein carriers. These methods can be sequential and allow site-directed substitution of peptide and carrier to be made.

We have made extensive use of *N*-hydroxysuccinimide (NHS) esters that yield stable products following reaction with primary and secondary amines (*see* **Note 10**). If the NHS is in ester linkage with the maleimide group then controlled conjugation of the maleimide-derivatized protein carrier can be carried out through the thiol group of a cysteine residue incorporated at the N- or C-terminus of the peptide. This results in formation of a stable thioether link between peptide and carrier. The reaction is carried out in two steps and results in the site directed conjugation of epitope and carrier. Availability of *N*-hydroxy succinimide ester maleimide cross linkers with a variety of spacer groups, such as caproic, hexanoic, butyric, benzoic acid, separating the two functional groups permits the epitope to be spaced away from the carrier to a greater or lesser extent.

3.2.1. Method A: Conjugation of FFGVGGEEDC to Keyhole Limpet Hemocyanin Using Maleimidohexanoic Acid N-Hydroxysuccinimide Ester

The derivatization or acylation of amino groups of proteins by NHS esters occurs in competition with the hydrolysis of the NHS ester. Hydrolysis occurs more rapidly with increasing pH and in dilute protein solutions. We have found that good substitution levels are achieved when the reaction is carried out at pH 8.0 with protein concentrations of 10 mg/mL

3.2.1.1. DERIVATISATION OF CARRIER PROTEIN

1. Just before use, weigh and dissolve MHS in dry DMF. A typical experiment would use 5 mg MHS/100 µL DMF for 10 mg protein carrier (*see* **Note 11**).
2. Add the MHS to 1 mL protein solution (10 mg/mL) while mixing on a vortex stirrer. Some cloudiness may become apparent but on standing this usually disappears.
3. Let stand at room temperature with occasional mixing for 30 min–2 h.
4. Add a vol of 500 m*M* Tris buffer equivalent to a fivefold excess over the amount of MHS added. This will act as a scavenger for unreacted MHS.
5. Dialyze against phosphate buffer, pH 8.0.

3.2.1.2. CONJUGATION OF PEPTIDE EPITOPE TO DERIVATIZED CARRIER PROTEIN

The cysteine-containing peptide needs to be in a reduced state to ensure reaction with maleimide groups. For this reason, the peptide is first reduced using dithiothreitol before subsequent coupling to the derivatized protein.

1. Dissolve the FFGVGGEEDC peptide in 6 *M* guanidine hydrochloride in 500 m*M* Tris-HCl, pH 8.5. Add dithiothreitol at a 25-fold excess over the amount of cysteine present. Hold under N_2 overnight.

2. Separate reduced peptide from other solutes by gel permeation chromatography through Sephadex G-15, eluting with 50 m*M* ammonium bicarbonate (*see* **Note 12**). Collect the reduced peptide eluting in the void volume and lyophilise.

3. Dissolve the FFGVGGEEDC peptide in a minimum of phosphate buffer pH 7.0 (*see* **Note 13**). The amount of peptide used should represent a five- to tenfold excess of cysteine over the amount of maleimide groups present in the substituted protein carrier to ensure complete reaction. This means that 0.5–1.0 mg reduced peptide is added to 1 mg MHS-derivatized protein carrier.

4. Immediately add the peptide solution to the maleimide-derivatized protein with mixing.

5. Stand for 3 h at room temperature with occasional stirring and then overnight at 4°C.

6. Dialyze against an appropriate buffer such as phosphate buffered saline to remove nonconjugated peptide (*see* **Note 14**).

3.2.2. Method B: Conjugation of CGGFVDIPEN to Ovalbumin and BSA Using N-Hydroxysuccinimidylbromoacetate

For production of the anti-DIPEN antisera the CGGFVDIPEN peptide was conjugated to both derivatized ovalbumin and derivatized BSA. The rationale for making two different peptide-conjugates was that the ovalbumin conjugate could be used as the immunogen, whereas the BSA conjugate could be used for screening and specificity tests. Antiovalbumin antibodies were therefore not detected during screening. For large peptide immunogens, it is often sufficient to test sera against unconjugated peptide coated directly on to ELISA plates. However, the 10-mer peptides may have bound poorly to the plate, and the BSA-peptide conjugate was used to circumvent this possibility. The method for conjugation of CGGFVDIPEN to ovalbumin is described below, and conjugation to BSA was done in the same way. The method is essentially that described by Bernatowicz and Matsueda *(20)* and has been used successfully by several investigators for the production of neoepitope antibodies *(1,3,10,15)*.

3.2.2.1. DERIVATIZATION OF CARRIER PROTEINS

1. Prepare a 22 mg/mL stock of ovalbumin (*see* **Note 15**) by dissolving 154 mg ovalbumin in 7 mL of 0.1 *M* potassium phosphate buffer pH 7.5, containing 1 m*M* EDTA.

2. Dissolve HSBA in dimethyl formamide at 65 mg/mL in a glass vial. Do not use plastic.

3. Cool 1.125 mL of 22 mg/mL ovalbumin to 4°C in a 5 mL tube. Add 100 µL HSBA dropwise with stirring at 4°C, then allow to warm to room temperature for 30 min.

4. Apply the activated ovalbumin to a 1 cm × 30 cm BioGel P6-DG column (or other suitable desalting column) equilibrated in 0.1 *M* potassium phosphate buf-

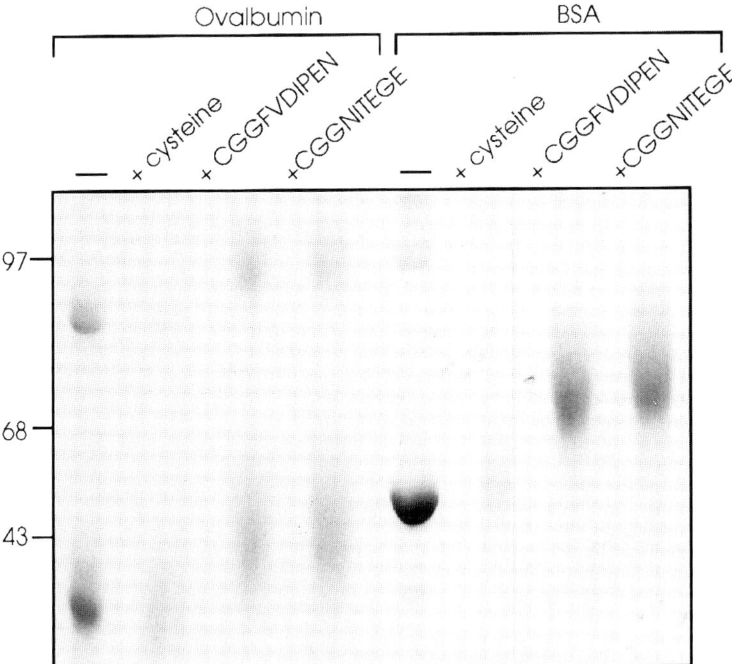

Fig. 3. Assessing conjugation of peptide to derivatized carrier protein by SDS-PAGE. Synthetic peptides CGGFVDIPEN (lanes 3 and 7), CGGNITEGE (lanes 4 and 8), or cysteine (lanes 2 and 6) were conjugated to derivatized ovalbumin (lanes 1–4) or BSA (lanes 5–8) and the final products analyzed by SDS-PAGE and staining with Coomassie blue.

fer, pH 7.5 containing 1 m*M* EDTA. Pump the column at 0.4 mL/min and collect 1 mL fractions. Monitor the absorbance at 278 nm to identify the void fractions. Pool the void fractions (usually in fractions 11–14) and concentrate to 2 mL.

3.2.2.2 Conjugation of Peptide to Derivatized Carrier Proteins

1. Dissolve the CGGFVDIPEN peptide (7.5 mg) in 1.19 mL water, bringing it to 6 m*M*.
2. Place 0.4 mL of the HSBA-activated ovalbumin solution (*see* **Note 16**) into a 10 mL hydrolysis tube (*see* **Note 17**). Add 0.4 mL 6 m*M* peptide.
3. Sparge gently with nitrogen for 5 min then seal the tube. Shake gently under nitrogen for 2 h at room temperature then stand overnight at 4°C.
4. Add 1 μL β-mercaptoethanol to block any unreacted bromoacetate groups.
5. Dialyze exhaustively against PBS.
6. The concentration of protein in the ovalbumin-peptide conjugate was estimated by assuming that (a) all the ovalbumin (22.5 mg) was recovered in the void fractions of the desalting column and, (b) no protein was lost during dialysis. On this basis, the final yield of peptide-conjugate was 700 μL, at approx 6 mg/mL.

7. Check the success of the coupling reaction on SDS-PAGE by comparing the electrophoretic mobility of the conjugated and unconjugated proteins (**Fig. 3**). In our hands, the activated ovalbumin migrated with a molecular mass of approx 37,000, while the CGGFVDIPEN-ovalbumin conjugate migrated as a broad smear from 38,000 to 55,000. The activated BSA migrated with a molecular mass of approx 48,000 and the CGGFVDIPEN-BSA conjugate migrated as a smear from 61,000 to 81,000.

3.3. Production of Anti-DIPEN Polyclonal Antibodies

3.3.1. Immunization of Rabbits

1. Before injecting the rabbits, take a prebleed of 10 mL blood. Incubate the blood for 1 h at 37°C, then stand overnight at 4°C. Remove the clot and centrifuge the remainder at 900*g* for 5 min. Transfer the serum supernatant to a fresh tube and store at –20°C.
2. Dilute the ovalbumin-peptide conjugate to 4 mg/1.2 mL in PBS. This is sufficient peptide-conjugate for 1 priming injection and 4 booster injections of 400 µg into each of 2 rabbits (*see* **Note 18**).
3. Prepare the priming injection by emulsifying 240 µL (800 µg) of the ovalbumin-peptide conjugate with 720 µL of Freund's Complete Adjuvant (*see* **Note 19**). Preparation of the emulsion is as described by Goding *(21)*. Take the aqueous immunogen into a 1 or 2 mL glass syringe and fill a second 2 mL glass syringe with the adjuvant. Remove any excess air and fit the syringes together with a metal double-hub connector. Inject the aqueous phase into the oil (not vice versa), then rapidly pass the mixture back and forth between the syringes. The emulsion is ready to inject when a single drop placed in a beaker of iced water remains as a discrete creamy globule. If the drop disperses on the surface of the water, it is an oil-in-water emulsion, unsuitable for immunization (*see* **Note 21**).
4. Use half the final emulsion for each of 2 rabbits (*see* **Note 20**). Inject into several *subcutaneous* sites.
5. Thereafter, boost the rabbits by *intramuscular* injection with 400 µg of ovalbumin-peptide conjugate emulsified in Freund's Incomplete Adjuvant, with the same ratio of aqueous immunogen:adjuvant (*see* **Note 21**). Booster injections can be given at d 14, 28, 42, and 96. Four booster injections is generally sufficient to obtain a high titre serum. After the second booster injection 10 mL of blood can be taken and the serum tested for immunoreactivity by ELISA or Western blotting (*see* below). The final bleed from each rabbit should provide about 50 mL serum (*see* **Note 22**).

3.4. Production of an Anti-FFGVG Monoclonal Antibody

3.4.1. Immunization of Mice

1. Dilute the FFGVGGEED peptide-conjugate to 750 µg/500 µL. Add dropwise to an equal volume of Freund's Complete Adjuvant while vortexing. Emulsify using the double syringe technique described above or, if connectors not available, use

repeated drawing up and expulsion from a 2 mL syringe with 23 gauge needle attached (*see* **Note 23**). When mixture becomes stiff, check to determine if completely emulsified by centrifugation at approx 400*g* in a bench top centrifuge for 3–5 min. If clear layer on top is minimal, emulsion is ready to inject. Take up into 1 mL syringe using a 19-gauge needle, drawing up a few times to suspend mycobacteria. Remove air bubbles by tapping syringe and replace needle with a 26-gauge needle for injection.

2. Inoculate each of 5 BALB/c mice (6–8 wk-old) into the peritoneal cavity with 100 μL of the emulsion, which is the equivalent of 75 μg peptide per dose.
3. After 1 mo, take a small amount of blood from the retro-orbital plexus of each mouse into a microtube using a nonheparinized capillary. Allow the blood to clot at room temperature then transfer the tubes to 4°C overnight.
4. Remove blood clots by skewering with a 19-gauge needle, spin remainder and collect serum.
5. Test serum by ELISA as in **Subheading 3.4.2.**
6. Choose 2 mice giving the best serum antibody titers and boost these, approximately seven weeks after the first inoculation, with an equivalent dose of peptide-conjugate emulsion to that used for the primary immunization (as above). The remainder of the mice should be kept as a backup in case the fusion needs to be repeated; a longer duration between priming and boosting of the mice is not a disadvantage.

3.4.2. Screening of Mouse Sera

1. Coat wells of a 96-well ELISA plate with 50 μL of a solution of FFGVGGEEDC peptide at 5 μg/mL (*see* **Note 24**). The peptide has a pI of 3.3 and dissolves well in 50 μ*M* NaHCO$_3$. Incubate plate at room temperature in a humidified box (*see* **Note 25**) for this and all subsequent steps.
2. After overnight incubation, remove peptide solution and add 100 μL of blocking solution.
3. After 1 h wash plates three times in PBST.
4. Perform twofold serial dilutions of mouse serum, starting at a 1 in 10 dilution, across 12 wells in 50 μL BSA5PBST. Use serum from an unimmunized mouse as a negative control.
5. Leave the serum dilutions for at least 4 h or overnight then wash plates 3 times in PBST.
6. Add 50 μL/well rabbit antimouse immunoglobulin-peroxidase conjugate (1 in 400 dilution in BSA5PBST).
7. After 1 h, wash plates 4 times in PBST and add 100 μL of substrate solution to each well.
8. After 15 min when color has developed monitor the absorbance at 405 nm using a microplate reader.

3.4.3. Fusion Procedure

The principle behind this technique is that the spleen is used as a source of primed B cells which make the desired antibody. B cells have only a limited lifespan in vitro but can be fused with a myeloma cell line which is capable of continuous growth. The resulting hybridomas have the ability to survive and secrete specific antibody in culture. The fusion step employs polyethylene glycol which nonspecifically joins membranes together. There will be many nonviable fusion products and these, together with any unfused spleen cells, will die in the first few days after fusion. A selection procedure must be used, however, to inhibit the growth of unfused myeloma cells. The most common of these is the "HAT" selection technique which employs the use of a mutant myeloma cell line, such as SP_2O cells, which are defective in the enzyme hypoxanthinephosphoribosyltransferase (HGPRT). Aminopterin blocks the main pathway of purine and pyrimidine synthesis for DNA replication. A salvage pathway exists that utilizes exogenous hypoxanthine for purine synthesis but requires the HGPRT enzyme. Hybridoma cells derive the HGPRT enzyme from the B cell and in the presence of aminopterin can utilize hypoxanthine to produce purines; thymidine is the source of pyrimidines. Unfused myeloma cells cannot use either the main or salvage pathways and die. Thymocytes are provided to support the growth of remaining hybridoma cells.

3.4.3.1. PREPARATION OF SPLEEN CELLS AND THYMOCYTES

1. Four days after the boost, sacrifice the two immunized mice by cervical dislocation, swab the fur with alcohol and remove the spleens aseptically into a 10 mL centrifuge tube containing culture medium in preparation for the fusion. Cut the spleens a few times to allow media access to the cells.
2. Remove the thymus from each of five 4-wk-old BALB/c mice into a 10 mL centrifuge tube containing culture medium.
3. Using sterile technique, ideally in a laminar flow hood, prepare a single cell suspension from the spleens by forcing through a wire mesh sieve into the base of a plastic Petri dish using the plunger from a 2 mL disposable syringe. Return the sieved suspension to the centrifuge tube.
4. Repeat the procedure for the thymocytes, pellet these (approx 1400g in bench top centrifuge) and wash once in 10 mL culture medium.
5. Pellet the spleen cells and resuspend in 5 mL pre-warmed (37°C) ATC buffer to lyse red cells. Incubate for 5 min exactly in a 37°C waterbath, add 5 mL of culture medium to neutralize the activity of the ATC and pellet the cells. Resuspend in culture medium. Discard half of the cells (*see* **Note 26**).

3.4.3.2. PREPARATION OF SP_2O CELLS

1. SP_2O cells are maintained in culture medium in plastic Petri dishes or TC6 plates and are passaged twice a week by transfer of 5 drops from a Pasteur pipet to 5 mL of fresh medium. Two to three days prior to the fusion, seed 1 mL of cells from a confluent dish or well into each of four 80 cm^2 flasks containing 25 mL of culture

medium. Incubate flasks on their side with lids loosened in a humidified 5% CO^2 incubator at 37°C.

2. Monitor the growth of the cells in the days leading up to the fusion; cells should be in log phase when used. An additional 25 mL of culture medium will probably be required the day before use. Cells should be further passaged if medium becomes too acidic or the percentage of FCS raised to 15% if cells are not growing rapidly enough.

3. On the day of fusion the cells should be loosened from the flask by a vigorous tap to the plastic surface and the contents dispensed into 50 mL centrifuge tubes.

4. Pellet the cells and resuspend in medium without FCS, pooling the suspended cells into a single 10 mL centrifuge tube.

3.4.3.3. FUSION PROTOCOL

1. Wash both SP_2O cells and spleen cells in 10 mL medium without FCS. Repeat a further two times and on last wash resuspend in 5 mL medium without FCS.

2. Meanwhile set up heating block in working area and adjust to 37°C (*see* **Note 27**). Place 50% PEG in 37°C waterbath to dissolve. When heating block is at the correct temperature, transfer PEG to this.

3. Count cells using a haemocytometer by diluting 10 μL of the suspension in 100 μL saline, then adding 10 μL of this with an equal volume of a vital dye, such as eosin. Calculate the number of SP_2O cells to be used, aiming at a ratio of spleen:SP_2O of between 10:1 and 2:1.

4. Add the desired number of SP_2O to the spleen cells and pellet together.

5. Remove all fluid from above cells and put tube containing pellet in heating block for 2 min.

6. Meanwhile heat a Pasteur pipet over a Bunsen burner flame to seal the end and further heat the end to curl it over to form a small hook or loop, the size of which can fit right to the bottom of the centrifuge tube.

7. Add 20 mL of serum-free medium to a 25 mL centrifuge tube. Have a timing device in plain sight and a 1 mL and 10 mL sterile pipet at hand.

8. Paying strict attention to time and keeping the cells in the heating block, add 1 mL of PEG slowly over 30 s with one hand while stirring the cells with the Pasteur pipet loop with the other. Continue stirring for a further 30 s.

9. Immediately remove 7 mL of serum-free medium from the 25 mL centrifuge tube and reset timer. Add medium back to cells while stirring as follows: 1 mL at 15 s, 2 mL at 30 s, 4 mL at 45 s. At 60 s tip contents into 25 mL tube.

10. Pellet cells gently.

11. Remove supernatant and resuspend cells gently in 10 mL fusion mix (*see* below).

3.4.3.4. PREPARATION OF FUSION MIX AND SETTING UP SELECTION CULTURES

1. Count thymocytes and transfer 5×10^8 cells to a sterile wide-necked bottle.

2. Add 2 mL 50X HT and 2 mL 50X A and make up to 100 mL total with fusion medium.

3. Use this to resuspend pellet of fused cells then transfer cell suspension to remainder of 100 mL of fusion mix.

4. Dispense into five 96-well culture trays (200 μL/well) shaking suspension regularly to ensure cells do not settle.

5. Incubate undisturbed at 37°C in 5% CO_2 for 7 d.

6. Examine wells for colonies from day 7 onwards and when sufficiently large, i.e., occupying approximately one-eighth of a well, take 50 μL of culture medium for screening. Replace with fresh medium.

7. Move the contents of positive wells into TC24 wells in 0.5 mL culture medium containing HT but not A. If cells are not looking healthy at this stage, 5×10^6 thymocytes can be added. Hybridomas should be cloned as soon as they are growing well in TC24 wells.

3.4.4. Screening of Culture Supernatants

1. Coat wells of a 96-well ELISA plate with the peptide-KLH conjugate (*see* **Note 24**) at a dilution of 5 μg peptide/mL of PBS. Leave overnight.

2. Remove contents from wells and add 100 μL blocking solution.

3. Wash 3 times with PBST and add 50 μL neat culture fluid (*see* **Note 28**). Negative control is culture fluid from a well containing no colonies. Positive control is serum from immunized mouse.

4. Incubate the plates overnight then continue the procedure as in **Subheading 3.4.2.** (*see* **Note 29**).

5. Move the contents of each positive well to a TC24 well.

6. Prior to cloning, retest culture fluid from TC24 wells on the conjugate to confirm their positive status and also on the unconjugated monomer and discard any that only bind to the conjugate.

3.4.5. Cloning by Limit Dilution

1. Prepare thymocytes as above. Allow for one mouse per plate and two plates per hybridoma to be cloned.

2. Dispense thymocytes in TC96 well plates to give 10^6 cells/100 μL/well in culture medium containing HT.

3. Resuspend the contents of a confluent TC24 well containing the hybridoma to be cloned. Take 10 μL of the suspension into 1 mL of culture medium with HT and a second 10 μL aliquot into a tube containing 10 μL eosin for counting. Calculate the cell concentration in the 1 mL vol.

4. Cultures are established at 1 cell/well in one plate and 0.5 cells/well in another plate. From the 1 mL suspension take a volume containing 100 cells and add to 10 mL medium plus HT (for 1 cell/well). Take a second aliquot containing 50 cells and add this to 10 mL medium plus HT (for 0.5 cells/well).

5. Dispense 100 μL of these into the thymocyte-containing wells and incubate undisturbed at 37°C in 5% CO_2.

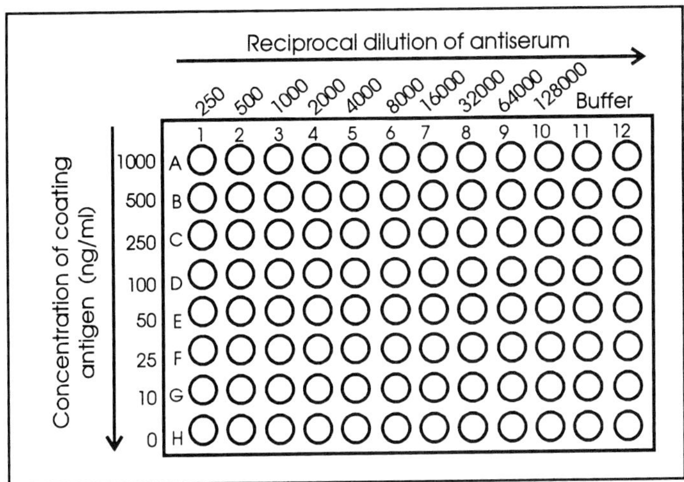

Fig. 4. Titration of sera or ascites fluid. Dilutions of antigen are coated in one direction in wells of an ELISA plate. Dilutions of sera or ascites fluid are added in the other direction in a matrix-type assay in a 96-well plate.

6. After 5 d start examining the plates for colonies, mark the number of colonies in each well when they are small, and when these are of sufficient size screen the culture fluid for antibody production as above.
7. Choose wells containing only a single colony and ideally from a plate containing no more than 30% wells positive for growth. Move the contents to a TC24 well containing 0.5 mL medium plus HT.
8. Carry out a second cycle of cloning at this stage if necessary (*see* **Note 30**).
9. When the twice cloned cells have been moved up into TC24 wells and are growing well add 0.5 mL of medium containing HT at half strength. Gradually wean the cells off HT by passage into medium with half strength HT then no HT.
10. Expand cells by passage into TC6 plates then 80 cm^2 flasks.
11. Freeze several ampoules of two subclones of each hybridoma. This can be accomplished by resuspending cells in culture medium containing 10% DMSO and an additional 10% FCS to give 2×10^6 cells per mL. Place 1 mL aliquots in labeled cryotubes and put into cell freezer at –70°C for between 5 and 24 h. Transfer to liquid nitrogen storage.

3.4.6. Production of Ascites Fluid

1. Inoculate 5 BALB/c mice with 1 mL of pristane by the intraperitoneal route. Mice should be at least 12 wk of age; retired breeding mice are ideal (*see* **Note 31**).
2. Ten days later, choose a flask with healthy cloned hybridoma cells in log phase and pellet cells. Resuspend in 2 mL medium without FCS and count cells.

3. Adjust cell concentration to 4×10^6 cells/mL and inoculate 0.5 mL per mouse by the intraperitoneal route. At the same time inoculate 0.5 mL pristane.

4. Check mice every day from d 5 onward. When the peritoneum is quite distended mice must be "milked;" holding mouse firmly in one hand, puncture the peritoneum with a 19-gauge needle held into the neck of a 10 mL centrifuge tube in the other hand. Collect the contents from each mouse into the tube then sacrifice the mice. Not all mice may be ready to milk at the same time.

5. Remove cells from the ascitic fluid by centrifugation and store the fluid at –20°C.

6. Later confirm the presence of specific antibody in the ascitic fluid by ELISA. Expect titers to be about 100- to 1000-fold greater than those in the equivalent culture fluid.

3.5. Characterization of Specific Antibodies

3.5.1. Titration of Serum and Ascites Fluid

Serum or ascites fluid is titred using a matrix-type assay in a 96-well plate, where the coating antigen (peptide-conjugate) is diluted in one direction on the plate and the antiserum is diluted in the other direction (**Fig. 4**). Use the BSA-peptide and not the ovalbumin-peptide for screening polyclonal rabbit sera. The immunizing conjugate may be used for screening monoclonal antibodies.

1. Coat wells of a 96-well ELISA plate by adding 200 µL of peptide-conjugate in coating buffer. Coat with doubling dilutions, over the range 10 ng/mL to 1 µg/mL and include one row of 12 wells with coating buffer alone to determine non-specific binding to the plate (**Fig. 4**). Cover the plate with an adhesive plate sealer and incubate overnight at 4°C (*see* **Note 32**).

2. Wash wells thoroughly with Tris incubation buffer (Tris IB)(*see* **Note 33**). Block the plate by adding 250 µL 1% BSA in Tris IB. Incubate for 1 h at 37°C.

3. Wash wells thoroughly with Tris IB. Add 200 µL of doubling dilutions of rabbit serum diluted in Tris IB, starting at a dilution of 1:250 in column 1, 1:500 in column 2, and so on to 1:128,000 in column 10 (**Fig. 4**). Add 200 µL of Tris IB to columns 11 and 12, to determine background. Incubate for 1 h at 37°C.

4. Wash wells thoroughly with Tris IB. Add 200 µL of anti-rabbit-HRP diluted 1:1000 in Tris IB (*see* **Note 34**). Incubate for 1 h at 37°C.

5. Wash wells thoroughly with Tris IB. Add 200 µL of ABTS solution (*see* **Notes 35** and **36**) and allow the color to develop at room temperature. Monitor the absorbance at 405 nm on a microplate reader.

3.5.2. Competition ELISA Assay

By definition, the antigenic determinant for a neoepitope antibody must be the N- or C- terminal sequence, and specific neoepitope antibodies must not recognize the same sequence of amino acids in the intact or undegraded protein. To confirm the specificity of the antibody, competition experiments can be done using peptides containing either extensions or truncations of one or more amino acids. For testing of the anti-DIPEN antisera, 10mer peptides representing extensions of +1

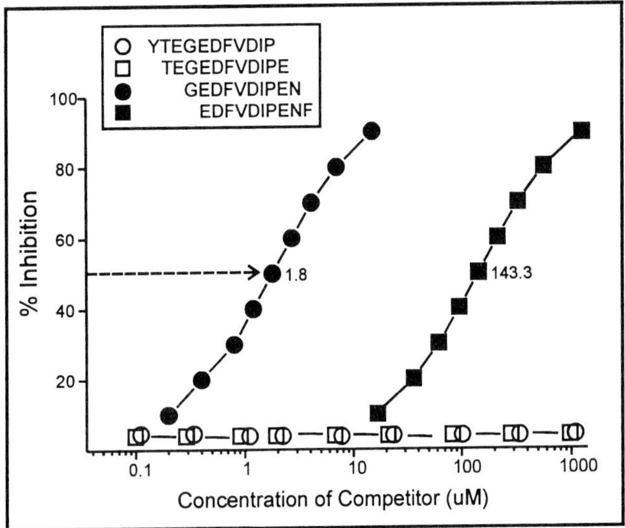

Fig. 5. Competition ELISA assay. 10mer peptides extended by one amino acid (EDFVDIPENF)(■), truncated by one amino acid (TEGEDFVDIPE)(□), truncated by two amino acids (YTEGEDFVDIP)(○) or corresponding to the true neoepitope sequence (GEDFVDIPEN)(●), were used as competitors for binding of the anti-DIPEN antisera to BSA-CGGFVDIPEN antigen coated on ELISA plates. No competition was seen with the truncated peptides and the peptide extended by one amino acid was almost 2 orders of magnitude less competitive.

(EDFVDIPENF), –1 (TGEDFVDIPE) or –2 (YTGEDFVDIP) amino acids were tested alongside the immunizing peptide (**Fig. 5**). The 10-mer peptide with an extension of one amino acid was almost two orders of magnitude less competitive than the true neoepitope peptide, while the 10-mer peptides truncated by one or two amino acids gave no competition at all.

1. Coat 96-well ELISA plates with 200 µL of appropriate peptide-conjugate diluted in coating buffer (*see* **Note 37**). Include one row or column with coating buffer alone to determine nonspecific binding to the plate. Incubate overnight at 4°C.
2. Wash wells thoroughly with Tris IB. Block the plate by adding 250 µL 1% BSA in Tris IB. Incubate for 1 h at 37°C.
3. Wash the wells thoroughly with Tris IB. Allocate one row for each peptide competitor to be tested. Add 100 µL of peptide competitor in Tris IB to cover the range 0.03–1000 µ*M*. Include several wells with no competitor to determine maximum binding. Add 100 µL of appropriately diluted sera or ascites fluid (*see* **Note 37**) diluted in Tris IB. Incubate for 1 h at 37°C.
4. Wash wells thoroughly with Tris IB. Add 200 µL of antirabbit-HRP or antimouse-HRP diluted 1:1000 in Tris IB. Incubate for 1 h at 37°C.

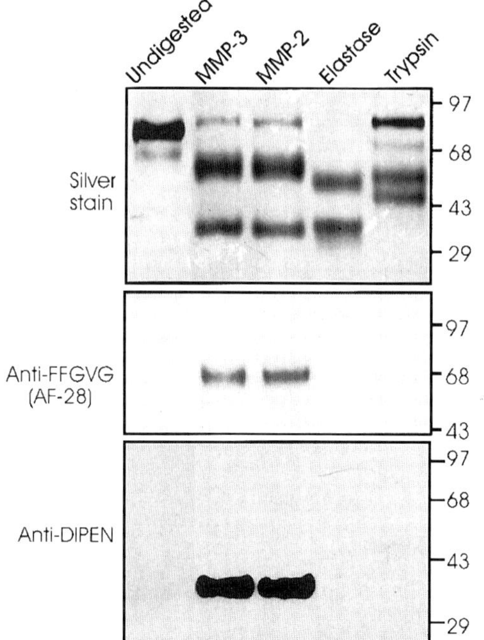

Fig. 6. Western blot analysis. A G1-G2 fragment isolated from pig laryngeal aggrecan was digested for 18 h at 37°C with 10 µg/mL MMP-3 (lane 2), 10.4 µg/mL MMP-2 (lane 3), 0.1 U/mL elastase (lane 4) or 10 ng/mL trypsin (lane 5) and the digestion products analysed by SDS-PAGE and (*top*) silver stain (*middle*) anti-FFGVG Western blotting or (*lower*) anti-DIPEN Western blotting.

5. Wash wells thoroughly with Tris IB. Add 200 µL of ABTS solution and allow the color to develop at room temperature. Monitor the absorbance at 405 nm on a microplate reader.

3.5.3. Western Blot Analysis

Further validation of the specificity of neoepitope antibodies can be obtained by Western blot analysis. In this case digest aggrecan or other suitable substrate such as the native G1-G2 fragment of aggrecan *(23)* or recombinant substrates *(15,24)* with various MMP and non-MMP enzymes. No signal should be seen in the undigested sample (**Fig. 6**, lane 1) or samples digested with non-MMP enzymes (**Fig. 6**, lanes 4 and 5). MMP digestion of native G1-G2 will produce a single G1 band (Mr ~ 50,000) that can be detected with anti-DIPEN antibody (**Fig. 6**, *lower*, lanes 3 and 4) and one or more G2-containing bands can be detected with anti-FFGVG antibodies (**Fig. 6**, *middle*, lanes 2 and 3 and **ref. *4***).

4. Notes

1. Amino acids are incorporated into the peptide chain as free acids with HOBt present in equimolar amounts, a slight deficiency of HBTU and DIPEA present at a 1.5-fold molar excess.
2. Aspiration of solvents from the peptide-support by application of a vacuum removes most of the reagent at each step and allows concentrated conditions to be achieved aiding complete reaction conditions.
3. Piperidine has a strong odor and many prefer the use of the odorless DBU reagent *(18)*.
4. Some of the polyethylene glycol-polystyrene-based resins become very sticky after exposure to these solvents; this condition is avoided if DMF and methanol are used to wash and dry the final product.
5. Ethers are a potential source of peroxides and to prevent oxidation of methionine and cysteine-containing peptides, DMF/methanol washes may be preferred.
6. Cleavage reagent should be prepared fresh. We have given here the details of only one of many cleavage reagent cocktails which have been described; we have found this to be an excellent reagent for most applications but for those cases where difficulties are encountered, the reader is referred elsewhere *(19)*.
7. Longer time periods for cleavage may be necessary for some sequences particularly those containing multiple arginine residues.
8. In those cases where peptides do not dissolve easily in 0.1% aqueous TFA, addition of acetonitrile may be necessary.
9. Assess peptide purity using reversed phase chromatography and fidelity of the target sequence by mass spectrometry.
10. NHS-maleimide esters are light and moisture sensitive and should be stored in the dark at less than 0°C in a desiccated container.
11. The carrier protein is dissolved in 100 mM phosphate buffer pH 8.0 at a concentration of 10 mg/mL and any undissolved material removed by sedimentation. As well as KLH, other proteins can be used. Popular alternatives include tetanus toxoid, thyroglobulin, BSA, and ovalbumin. In the case of toxoided proteins, the toxoiding process involves formaldehyde which substitutes side chain amino groups of lysine residues making them unavailable for derivatization. The degree of substitution due to this process may vary and be as high as 50%.
12. Chromatography on Sephadex G-15 should be carried out in a buffer which is volatile allowing reduced peptide to be freeze dried. If the pI of the peptide is above 7 elution can be done in 50 mM acetic acid; if the pI is below 7 elution can be done in 50 mM ammonium bicarbonate.
13. When the pH of the reaction is held between pH 6.5 and 7.5 the maleimide group present on the carrier protein will react specifically with the peptide thiol group. At higher pH, the reaction of maleimides with amines becomes significant.
14. The extent of the derivatization reaction can be checked by comparing electrophoretic mobilities of conjugated and nonconjugated material (*see* **Subheading 3.2.2.2.** and **Fig. 3**).

15. Be aware that the carrier protein will also elicit an immune response. If the sera is to be used for Western blotting it will be necessary to purchase or prepare ovalbumin-free molecular weight markers for gel electrophoresis; or remove the antiovalbumin antibodies by ovalbumin-Sepharose affinity chromatography.

16. Use freshly prepared bromoacetylated protein. The derivatized protein cannot be frozen.

17. Remove any possible traces of protein from the hydrolysis tubes with an acid wash. Soak the tubes overnight in 0.1 M HCl. Rinse thoroughly and dry.

18. The amount of carrier conjugate made will depend on the immunization schedule to be followed but the amount of peptide does not normally exceed 50–400 μg per dose, consequently relatively small amounts of material are used.

19. Provided the ethics committees will allow it, it is useful to immunize 2 rabbits per antigen, in case one is a poor responder.

20. Be sure to mix the Freund's Complete Adjuvant thoroughly, immediately before use, so that the bacilli organisms that settle to the bottom are resuspended well.

21. It is important to make a water-in-oil emulsion rather than an oil-in-water emulsion. The standard ratio for preparing adjuvant emulsions is 1:1, however to ensure the phases combine as water-in-oil, use 2–4 vol of adjuvant to 1 vol of aqueous immunogen *(22)*. The oil-rich emulsion does not compromise the efficacy of the standard 1:1 ratio, and makes mixing and injection much easier.

22. The exact timing of the third and fourth booster injections is not critical and could be extended for much longer periods if necessary.

23. Emulsification by this method should only be done with safety glasses and gloves ensuring that there are no unprotected workers in the vicinity. We find the indicated brand of disposable syringe is far superior to other brands tested for this purpose.

24. The unconjugated FFGVGGEEDC peptide bound well to ELISA plates however as noted above this may not be the case for all small peptides. Another option is to screen sera against both KLH and FFGVGGEEDC-KLH and discard sera with immunoreactivity against KLH alone.

25. A plastic box with a tight fitting lid lined with moist kitchen sponges is suitable. Plates coated with antigen can be kept for several days without losing antibody-binding capacity. However, it is advisable to move the box to 4°C if conditions are warm or incubation periods are long to prevent contaminants growing in the BSA.

26. One spleen is sufficient for the fusion procedure, however, we often pool two spleens in case the mice respond differently to the boost. All the splenocytes from the two mice could be used in the fusion but it would be necessary to culture the fused cells in twice as many plates.

27. If heating block is not available perform incubations in a beaker of water in a 37°C water bath. For the actual fusion, move the beaker to the working area.

28. If a manageable number of colonies develop, it is possible to screen the culture fluid from individual wells when each becomes of sufficient size. If the fusion

efficiency is high and colonies are observed in virtually all wells then screening is most easily accomplished by testing all wells simultaneously by transferring culture fluid with a multichannel pipet directly into corresponding wells of an ELISA tray. Medium should be replaced and the cultures tested again when the slower growing colonies have attained sufficient size.

29. On occasion, screening directly from unpassaged cultures will yield high backgrounds due to the presence of antibody produced by the unfused B cells prior to their death. In this case it is necessary to remove 150 μL from every well and replace with fresh culture medium containing HT. After 3 d wells can be rescreened. If the problem remains, it may be necessary to passage the wells containing colonies to fresh medium and allow to grow up before retesting.

30. Cloning should always be performed twice if, on the first cloning, the frequency of wells with growth is greater than 30% even if microscopically it appeared that only a single colony was present.

31. For ethical reasons this procedure is being phased out in many countries and the alternative strategies of purifying antibody from large volumes of tissue culture fluid or growing cells at high density in serum free synthetic medium are being employed.

32. If it is more convenient, ELISA plates can be coated for 2 h at RT, instead of overnight at 4°C.

33. Washing wells is not a trivial step in the ELISA protocol. The best way to wash the plates is with an automated plate washer, as this eliminates any cross well contamination that will inevitably occur with nonautomated washing procedures such as the flick-o-the-wrist and squirter bottle techniques. We routinely do 6 washes on an automated plate washer.

34. Sodium azide is a potent inhibitor of horseradish peroxidase and must not be included in any buffers used with the enzyme or its substrate.

35. Warning: ABTS is mutagenic.

36. ABTS solution should be stored at 4°C in the dark. Prepare fresh ABTS if the absorbance at 405 nm exceeds 0.25.

37. The appropriate concentration of coating antigen and the dilution of sera (or ascites) is determined from the titration experiments. Choose concentrations that give an OD reading of between 1.0–1.5 in about 10 min.

Acknowledgments

The CGGFVDIPEN and CGGNITEGE peptides used for the production of the anti-DIPEN and anti-NITEGE antisera were a generous gift from Drs. Peter Roughley and John Mort, Shriners Hospital, Montreal, Canada. We thank Georgia Deliyannis for the monoclonal antibody work. We gratefully acknowledge financial support from the National Health and Medical Research Council, the Royal Children's Hospital Research Institute, the Victorian Health Promotion Foundation and the Arthritis Foundation of Australia.

References

1. Hughes, C. E., Caterson, B., Fosang, A. J., Roughley, P. J., and Mort, J. S. (1995) Monoclonal antibodies that specifically recognize neo-epitope sequences generated by "aggrecanase" and matrix metalloproteinase cleavage of aggrecan: application to catabolism in situ and in vitro. *Biochem. J.* **305**, 799–804.
2. Lark, M. W., Williams, H., Hoerrner, L. A., Weidner, J., Ayala, J. M., Harper, C. F., Christen, A., Olszewski, J., Konteatis, Z., Webber, R., and Mumford, R. A. (1995) Quantification of a matrix metalloproteinase-generated aggrecan G1 fragment using monospecific anti-peptide serum. *Biochem. J.* **307**, 245–252.
3. Sztrolovics, R., Alini, M., Roughley, P. J., and Mort, J. S. (1997) Aggrecan degradation in human intervertebral disc and articular cartilage. *Biochem. J.* **326**, 235–241.
4. Fosang, A. J., Last, K., Gardiner, P., Jackson, D. C., and Brown, L. (1995) Development of a cleavage site-specific monoclonal antibody for detecting metalloproteinase-derived aggrecan fragments: detection of fragments in human synovial fluids. *Biochem. J.* **310**, 337–343.
5. Büttner, F. H., Hughes, C. E., Margerie, D., Lichte, A., Tschesche, H., Caterson, B., and Bartnik, E. (1998) Membrane type 1 matrix metalloproteinase (MT1-MMP) cleaves the recombinant aggrecan substrate rAgg1mut at the 'aggrecanase' and the MMP sites. Characterization of MT1-MMP catabolic activities on the interglobular domain of aggrecan. *Biochem. J.* **333**, 1–65.
6. Billinghurst, R. C., Dahlberg, L., Ionescu, M., Reiner, A., Bourne, R., Rorabeck, C., Mitchell, P., Hambor, J., Diekmann, O., Tschesche, H., Chen, J., Van Wart, H., and Poole, A. R. (1997) Enhanced cleavage of type II collagen by collagenases in osteoarthritic articular cartilage. *J. Clin. Invest.* **99**, 1534–1545.
7. Tortorella, M. D., Burn, T. C., Pratta, M. A., Abbaszade, I., Hollis, J. M., Liu, R., Rosenfeld, S. A., Copeland, R. A., Decicco, C. P., Wynn, R., Rockwell, A., Yang, F., Duke, J. L., Solomon, K., George, H., Bruckner, R., Nagase, H., Itoh, Y., Ellis, D. M., Ross, H., Wiswall, B. H., Murphy, K., Hillman, M. C. J., Hollis, G. F., Newton, R. C., Magolda, R. L., Trzaskos, J. M., and Arner, E. C. (1999) Purification and cloning of aggrecanase-1: A member of the ADAMTS family of proteins. *Science* **284**, 1664–1666.
8. Abbaszade, I., Liu, R. Q., Yang, F., Rosenfeld, S. A., Ross, O. H., Link, J. R., Ellis, D. M., Tortorella, M. D., Pratta, M. A., Hollis, J. M., Wynn, R., Duke, J. L., George, H. J. Hillman, M. C. J., Murphy, K., Wiswall, B. H., Copeland, R. A., Decicco, C. P., Bruckner, R., Nagase, H., Itoh, Y., Newton, R. C., Magolda, R. L., Trzaskos, J. M., Hollis, G. F., Arner, E. C., and Burn, T. C. (1999) Cloning and characterization of ADAMTS11, an aggrecanase from the ADAMTS family. *J. Biol. Chem.* **274**, 23,443–23,450.
9. Tang, B. L. and Hong, W. (1999) ADAMTS: a novel family of proteases with an ADAM protease domain and thrombospondin repeats. *FEBS LETT.* **445**, 223–225.
10. Hughes, C., Caterson, B., White, R. J., Roughley, P. J., and Mort, J. S. (1992) Monoclonal antibodies recognizing protease-generated neoepitopes from cartilage proteoglycan degradation. *J. Biol. Chem.* **267**, 16,011–16,014.

11. Jameson, B. A. and Wolf, H. (1988) The antigenic index: a novel algorithm for predicting antigenic determination. *CABIOS* **4**, 181–186.
12. Lark, M. W., Gordy, J. T., Weidner, J. R., Ayala, J., Kimura, J. H., Williams, H. R., Mumford, R. A., Flannery, C. R., Carlson, S. S., Iwata, M., and Sandy, J. D. (1995) Cell-mediated catabolism of aggrecan. Evidence that cleavage at the "aggrecanase" site (Glu^{373}-Ala^{374}) is a primary event in proteolysis of the interglobular domain. *J. Biol. Chem.* **270**, 2550–2556.
13. Hutton, S., Hayward, J., Maciewicz, R. A., and Bayliss, M. (1996) Age-related and zonal distribution of aggrecanase activity in normal and osteoarthritic human articular cartilage. *Trans. Orthop. Res. Soc.* **21**, 150.
14. Chambers, M. G., Cox, L. J., Chong, L., Maciewicz, R., Bayliss, M. T., and Mason, R. M. (1998) Localization of neoepitopes for "aggrecanase" and general metalloproteinases in normal and osteoarthritic murine articular cartilage. *Trans. Orthop. Res. Soc* **23**, 436.
15. Mercuri, F. A., Doege, K. J., Arner, E. C., Pratta, M. A., Last, K., and Fosang, A. J. (1999) Recombinant human aggrecan G1-G2 exhibits native binding properties and substrates specificity for matrix metalloproteinases and aggrecanase. *J. Biol. Chem.* **274**, 32,387–32,395.
16. Arner, E. C., Pratta, M. A., Newton, R. C., Trzaskos, J., Magolda, R., and Tortorella, M. D. (1998) Comparison of cleavage efficiency of aggrecanase and stromelysin for the aggrecan core protein. *Trans. Orthop. Res. Soc.* **23**, 922.
17. Billington, C. J., Clark, I. M., and Cawston, T. E. (1998) An aggrecan-degrading activity associated with chondrocyte membranes. *Biochem.J.* **336**, 1–212.
18. Wade, J. D., Bedford, J., Sheppard, R. C., and Tregear, G. W. (1991) DBU as an N alpha-deprotecting reagent for the fluorenylmethoxycarbonyl group in continuous flow solid-phase peptide synthesis. *Pept. Res.* **4**, 194–199.
19. Pennington, M. W. and Dunn, B. M. (eds.) (1994) *Methods in Molecular Biology, vol. 35: Peptide Synthesis Protocols.* Humana Press, Totowa, NJ.
20. Bernatowicz, M. S. and Matsueda, G. R. (1986) Preparation of peptide-protein immunogens using N-succinimidyl bromoacetate as a heterobifunctional cross-linking reagent. *Anal. Biochem.* **155**, 95–102.
21. Goding, J. W. (1986) *Monoclonal antibodies: Principles and Practice.* Academic Press, Sydney, Australia.
22. Hurn, B. A. and Chantler, S. M. (1980) Production of reagent antibodies. *Methods Enzymol.* **70**, 104–142.
23. Fosang, A. J. and Hardingham, T. E. (1989) Isolation of the N-terminal globular domains from cartilage proteoglycans. Identification of G2 domain and its lack of interaction with hyaluronate and link protein. *Biochem. J.* **261**, 801–809.
24. Hughes, C. E., Büttner, F. H., Eidenmuller, B., Caterson, B., and Bartnik, E. (1997) Utilization of a recombinant substrate rAgg1 to study the biochemical properties of aggrecanase in cell culture systems. *J. Biol. Chem.* **272**, 20,269–20,274.
25. Doege, K. J., Sasaki, M., Kimura, T., and Yamada, Y. (1991) Complete coding sequence and deduced primary structure of the human cartilage large aggregating proteoglycan, aggrecan. Human specific repeats and additional alternatively spliced forms. *J. Biol. Chem.* **266**, 894–902.

27

Cartilage Proteoglycan Release Assay

Caron J. Billington

1. Introduction

The proteoglycans are a group of macromolecules characterized by the covalent attachment of one or more sulfated glycosaminoglycan (GAG) chains to a protein core, and are an important constituent of the extracellular matrix. Cartilage contains a number of proteoglycans, the most abundant of which is aggrecan. This proteoglycan is characterized by its association with hyaluronic acid and link protein to form large macromolecular aggregates within the tissue. The large negative charge associated with the sulfated GAGs of these aggregates serves to draw water molecules into the tissue, thus giving cartilage its ability to resist compression under load *(1)*.

Proteoglycan turnover is mediated by proteolytic enzymes synthesized by the cells embedded within the cartilage, which initiate the degradation of the molecule by cleaving specific sites within the protein core *(2)*. In certain joint diseases, such as rheumatoid and osteoarthritis, an increased catabolism of proteoglycan molecules is associated with loss of joint function *(2,3)*, and thus in vitro assays to study the effects of various growth factors and inflammatory cytokines on this upregulation of proteoglycan degradation have been developed.

The method described below relies on the use of cartilage explants as both a source of proteoglycan, and the proteolytic enzymes required to degrade it. Cartilage explants may then be incubated with appropriate factors, and their effects monitored by use of a simple colorimetric assay to measure the release of sulfated GAGs into the surrounding media. This system has the advantage of maintaining the cartilage components in a three dimensional matrix, thus approximating the in vivo conditions as closely as possible. The colorimetric

From: *Methods in Molecular Biology, vol. 151: Matrix Metalloproteinase Protocols*
Edited by: I. Clark © Humana Press Inc., Totowa, NJ

assay *(4)* used to quantitate proteoglycan release is based on the binding of the basic dye 1,9 dimethylmethylene blue to the negatively charged sulfated GAGs, resulting in a metachromatic shift. The absorbance of the solution may then be read at 525 nm, and proteoglycan concentration determined by use of a standard curve.

2. Materials
2.1. Cartilage Explant Preparation and Culture

1. Cartilage source, e.g., bovine nasal septum cartilage.
2. Dulbecco's phosphate buffered saline-PBS (Sigma).
3. Dulbecco's Modified Eagle Medium-DMEM (high glucose) (Gibco-BRL).
4. L-Glutamine 200 mM (Gibco-BRL).
5. Penicillin/Streptomycin solution containing 10,000 U penicillin and 10,000 μg streptomycin/mL (Gibco-BRL).
6. Gentamicin solution 10 mg/mL (Sigma).
7. Square sterile Petri dishes and 24-well sterile tissue culture plate.
8. Post mortem knife and blade, scalpel and sterile blades (number 22), and leather punch.
9. 1.5 mL microfuge tubes.
10. 2% w/v sodium azide solution.
11. LP3 tubes with lids.
12. Phosphate buffer: 137 mL of 0.1 M NaH$_2$PO$_4$ and 63 mL of 0.1 M Na$_2$HPO$_4$, pH 6.5.
13. Papain solution: 0.25g/10 mL phosphate buffer. Prepare fresh on day of use.
14. Cysteine HCl: 0.078g/10 mL phosphate buffer. Prepare fresh on day of use.
15. EDTA: 0.19g/10 mL phosphate buffer. Prepare fresh on day of use.

2.2. Assay for Sulfated Proteoglycan

1. 1,9-Dimethylmethylene blue (DMB) from Aldrich.
2. Chondroitin sulfate from Sigma.
3. Phosphate buffer as above.
4. 96-well microtiter plate.

3. Methods
3.1. Cartilage Explant Preparation and Culture

1. Prepare the medium to be used under sterile conditions. To 500 mL of the DMEM add 5 mL of the L-glutamine (2 mM final concentration), 5 mL of the penicillin/streptomycin solution (100 U/mL penicillin and 100 μg/mL streptomycin final concentration), 1 mL nystatin (20 U/mL final concentration) and 2.5 mL gentamicin (50 μg/mL final concentration). Similarly to 500 mL of the PBS add the same amounts of nystatin, gentamicin, and penicillin/streptomycin. This PBS is used as a wash solution. Store the DMEM and PBS at 4°C once made up.

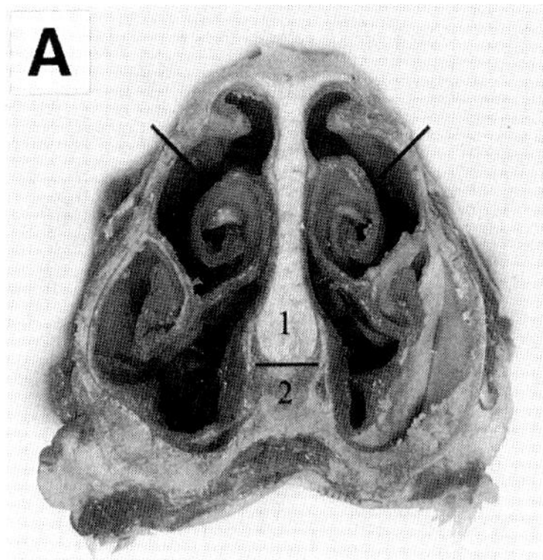

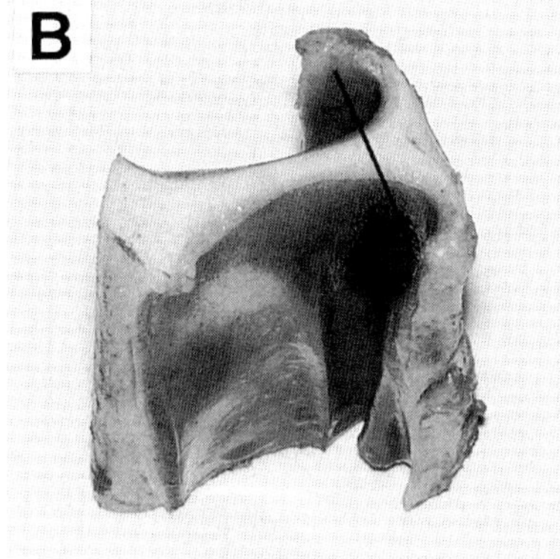

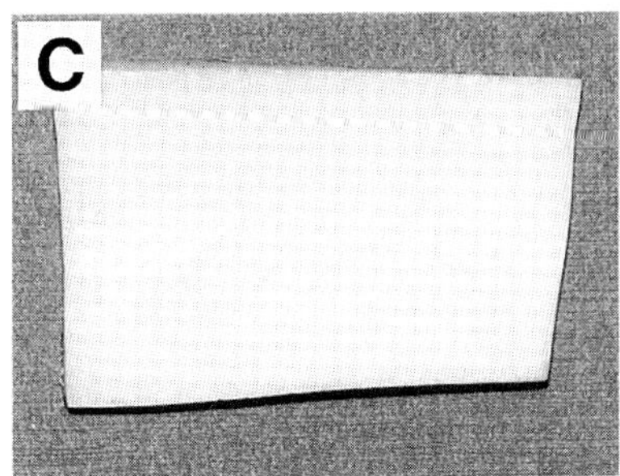

Fig. 1. Dissection of bovine nasal septum cartilage. Lines show position of incisions. (**A**) Removal of cartilage from nose. Cut through the nose at positions shown, and detach cartilage from the subchondral bone. (**B**) Following excisal from nose remove extraneous tissue by cutting at site shown. (**C**) Skinned cartilage ready for slicing.

2. Extract the nasal septum cartilage from the nose (**Fig. 1**) using a post mortem knife and place in a sterile square Petri dish. Wash with a little of the PBS, and remove the skin covering the cartilage.
3. Cut the cartilage block into three sections of equal size, and then slice each section to give cartilage pieces of an equal thickness of ~1–2 mm.
4. Transfer the slices into a clean Petri dish and cover with PBS.
5. Using an autoclaved leather punch, punch out disks from the cartilage slices and collect in a dish containing PBS. Leave in this wash buffer for 30 min with one change of buffer after 15 min.
6. Remove most of the PBS from the dish and then using the edge of a scalpel blade collect three cartilage explants and place in the first well of a 24-well plate. When each well has three explants add 1 mL of the DMEM to each well, and then incubate the plate at 37°C in an humidified atmosphere of 5% CO_2 for 24 h to allow equilibration of the explants.
7. Following equilibration remove the media from each well and replace with 600 µL of fresh media containing the appropriate stimulus. Each test condition should consist of four wells. Thus for each plate 5 different conditions may be set up together with a control for measuring basal release of proteoglycan under the experimental conditions.
8. Following addition of the media, incubate the plate for 7 d, changing the media on d 3 and 5 or as required. Collected media should be stored in microfuge tubes containing 6 µL of a 2% sodium azide solution at 4°C prior to assay for PG release.
9. On completion of the experiment, the cartilage explants must be collected for digestion and assay of the remaining PG. Carefully remove the disks from the wells into an appropriately labeled LP3 tube.
10. To each LP3 tube add 350 µL phosphate buffer, 100 µL papain solution, 50 µL cysteine-HCl and 50 µL EDTA. Cap the tubes and vortex mix to ensure the cartilage explants are in the digestion mixture. Pierce the caps with a syringe needle, and incubate at 65°C overnight.
11. Following digestion add a futher 450 µL phosphate buffer to each tube, increasing the final volume to 1 mL, and then store at 4°C prior to assay.

3.2. Assay for Sulfated Proteoglycan

1. Prepare a stock solution of the dye required for the assay by dissolving 16 mg of DMB in 5 mL of ethanol. Add to this 2 mL of formic acid and 2 g of sodium formate, and make up to 1 L with distilled water. The solution should have an absorbance of approx 0.3 at 525 nm. Store the solution at room temperature in a brown bottle (*see* **Note 1**).
2. Prepare a series of GAG standards in the phosphate assay buffer. Make up a 1mg/mL stock solution of chondroitin sulfate which can be stored in aliquots at –20°C. From this stock prepare standards of 0, 5, 10, 15, 20, 25, 30, 35, and 40 µg/mL, and add 40 µL to duplicate wells of a microtiter plate.
3. Dilute the media and digest samples to be assayed in the phosphate buffer, and add 40 µL to duplicate wells of the microtiter plate (*see* **Note 2**).

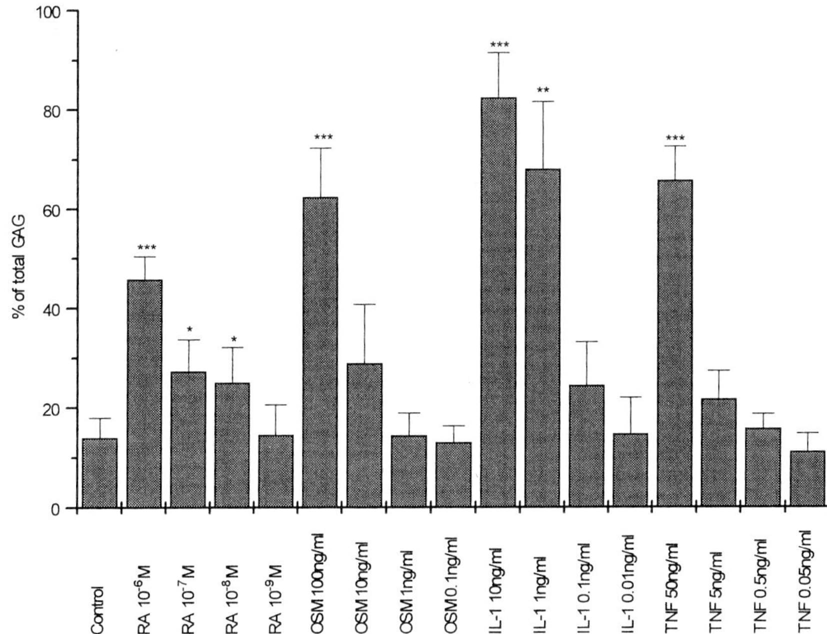

Fig. 2. Release of sulfated GAGs from cartilage explants following stimulation with retinoic acid (RA), oncostatin M (OSM), interleukin 1α (IL-1) and tumor necrosis factor (TNF) for 3 d. Control represents release from unstimulated explants. The results are expressed as the mean percentage release ± standard deviation from 4 wells, and statistical analysis was carried out to determine in each case whether the effects seen were significantly different from the control, using the Student's T-test. Levels of significance are indicated as * where p is less than 0.05, ** where p is less than 0.01, and *** where p is less than 0.001.

4. Add 250 μL of the DMB solution to each well, and read the absorbance at 525 nm within a fixed time of 60 s or less (*see* **Note 3**).

5. Determine the GAG concentration of the unknowns in μg/mL from the standard curve, correcting for any dilutions used.

6. To calculate GAG release from the explant culture first determine the amount of GAG present in each media and digest sample by correcting for the total volume of the samples (i.e., 0.6 mL for each media sample, and 1 mL for the papain digests). Determine the total amount of GAG in each well by addition of results for all media samples taken and the papain digests, and express release of GAG under the various conditions assayed as a percentage of the total GAG:

$$\% \text{ GAG Release} = [\text{GAG in medium} / \text{total GAG}] \times 100$$
$$(\text{where total GAG} = \text{GAG in medium} + \text{GAG in papain digest})$$

four wells are used for each condition, therefore the results may be expressed as a mean ± standard deviation. An example of the results obtained following a three day incubation of explants with various cytokines is shown in **Fig. 2**.

4. Notes

1. The DMB reagent is stable when stored at room temperature in a brown bottle for at least 3 mo, however storage at 4°C results in a reversible color loss.
2. To estimate roughly the dilution factor required prior to assay, dilute 1 sample from each set of 4 wells 1/10, 1/20, 1/50, and 1/100 (or higher as required), incubate with the DMB as described together with the lowest and highest (0 and 40 µg/mL) standard. It should then be possible to determine by eye which dilution will ensure the result fits on the standard curve.
3. Plates should be read as soon as possible following addition of the DMB, and excessive mixing of the well contents should be avoided as the dye-GAG complexes begin to aggregate and will eventually precipitate leading to a decrease in the absorbance.
4. Further modifications of the DMB assay have been described *(5)*, which enable quantitation of the individual sulfated GAGs by use of specific polysaccharide degrading enzymes.

Acknowledgments

I would like to thank Prof. Tim Cawston and Mr. Paul Koshy for their help in preparation of this chapter. Support was received from Chirosciences Ltd., Cambridge, UK.

References

1. Neame, P. J. (1993) Extracellular matrix of cartilage: Proteoglycans, in *Joint Cartilage Degradation* (Woessner, J. F. and Howell, D. S., ed.), Marcel Dekker, Inc. pp. 109–138.
2. Sandy, J. D., Flannery, C. R., Neame, P. J., and Lohmander, L. S. (1992) The structure of aggrecan fragments in human synovial fluid. *J. Clin. Invest.* **89**, 1512–1516.
3. Lohmander, L. S., Neame, P. J., and Sandy, J. D. (1993) The structure of aggrecan fragments in human synovial fluid. *Arth. Rheum.* **9**, 1214–1222.
4. Farndale, R. W., Sayers, C. A., and Barrett, A. J. (1982) A direct spectrophotometric microassay for sulpfated glycosaminoglycans in cartilage cultures. *Conn. Tiss. Res.* **9**, 247–248.
5. Farndale, R. W., Buttle, D. J., and Barrett, A. J. (1986) Improved quantitation and discrimination of sulphated glycosaminoglycans by use of dimethylmethylene blue. *Biochim. Biophys. Acta.* **883**, 173–177.

28

Immunoassay for Collagenase-Mediated Cleavage of Types I and II Collagens

R. Clark Billinghurst, Mirela Ionescu, and A. Robin Poole

1. Introduction

Collagen is the most abundant protein in the mammalian body. Collagen types I, II, and III are the major structural components of skin, bone, cartilage, and connective tissues. They exist as fibrils of crosslinked helical molecules composed of three α chains of approx 1000 amino acids each, with nonhelical extensions called telopeptides at both ends of each chain. During turnover in health and disease, after these fibrils are depolymerized, they are degraded, at neutral pH, by the proteolytic attack of enzymes, appropriately called mammalian collagenases, at a specific locus within the native triple-helical structure of each collagen molecule. These collagenases belong to the matrix metalloproteinase (MMP) enzyme family and include MMP-1 (collagenase-1 or fibroblast/interstitial collagenase), MMP-8 (collagenase-2 or neutrophil collagenase), and MMP-13 (collagenase-3). All three, in addition to a membrane-type MMP (MT-MMP) designated MT1-MMP (MMP-14), cleave the collagen molecules to yield characteristic 3/4 (TC^A) and 1/4 (TC^B) fragments.

The newly created C- and N-termini of the 3/4 and 1/4 fragments, respectively, represent neoepitopes that can allow for the specific identification of collagenase-cleaved triple helical collagen molecules. Cleavage site neoepitopes have already proven valuable in defining MMP activity in aggrecan degradation (*see* Chapter 26). This chapter will describe the procedures used to develop an assay utilizing an antibody, COL2-3/4C$_{short}$, produced to recognize the C-terminal neoepitope of the individual 3/4 α chain fragment created by the cleavage of triple-helical collagen types I and II with the MMPs, identified

From: *Methods in Molecular Biology, vol. 151: Matrix Metalloproteinase Protocols*
Edited by: I. Clark © Humana Press Inc., Totowa, NJ

above, that exhibit collagenase activity. This will include the production and characterization of the antibody, along with the development, validation, and experimental use of the immunoassay.

2. Materials

2.1. Antibody Production

2.1.1. Preparation of the Immunogen

1. 0.1 *M* phosphate buffer: 5.28 g/L KH_2PO_4, 16.4 g/L $Na_2HPO_4 \cdot 7H_2O$, 0.372 g/L EDTA- Na_2, 0.2 g/L NaN_3, pH 7.0.
2. Ovalbumin (OVA): 25 mg of Grade V chicken egg albumin/mL of 0.1 *M* phosphate buffer.
3. Coupling reagent: 56.5 mg N-hydroxy-succinimidylbromoacetate/mL of chilled *N,N*-dimethylformamide (Sigma).
4. Azide-free phosphate-buffered saline: 8.5 g/L NaCl, 1.61 g/L $Na_2HPO \cdot 7H_2O$, 0.55 g/L KH_2PO_4.
5. Sephadex G-25 Column (27 × 2.2 cm) with 103 mL bed volume.
6. Synthetic peptide: 5 mg/100 μL of 0.1 *M* phosphate buffer.
7. UV spectrophotometer and quartz cuvets.

2.1.2. Immunization

1. Peptide-OVA conjugate prepared in **Subheading 2.1.1.**
2. Azide-free PBS.
3. 5 mL glass syringe and 20 G needle.
4. Complete and incomplete Freund's adjuvant (Difco).
5. Two female New Zealand White rabbits weighing 2.5–3.0 kg each.

2.1.3. Antibody Purification

1. Saturated ammonium sulfate solution: 761 g/L $(NH_4)_2SO_4$, pH 7.0.
2. PBS: azide-free PBS + 0.5 g/L NaN_3.
3. 0.8 *M* sodium acetate buffer: 65.6 g/L $C_2H_3O_2Na$, 4.0 g/L NaN_3, pH 4.0.
4. 2% (w/w) pepsin: 20 mg 2X crystallized pepsin A (Worthington Biochemical)/*g* IgG.
5. 0.14 *M* phosphate buffer: 1.05 g/L KH_2PO_4, 35.5 g/L $Na_2HPO_4 \cdot 7H_2O$, 0.5 g/L NaN_3, pH 8.0.
6. Econo-Pac 10 DG columns packed with Bio-Gel P-6 desalting gel.
7. HiTrap affinity column packed with 5 mL Protein A Sepharose High Performance gel.
8. Tris-HCl buffers: 121.1 g/L (1 *M*); 12.1 g/L (100 m*M*); 1.21 g/L (10 m*M*), pH 8.0.
9. 100 m*M* glycine: 7.51 g/L, pH 3.0.
10. SulfoLink coupling gel and column (Pierce).
11. Coupling buffer: 50 m*M* Tris-HCl (6.06 g/L), 5 m*M* EDTA-Na_2 (1.86 g/L) pH 8.5.
12. Blocking reagent: 50 m*M* L-cysteine (6.1 g/L) in coupling buffer.

13. Washing solution: 58.44 g/L NaCl (1 *M*).
14. Dialysis tubing (molecular weight cut off: 12–14,000).
15. UV spectrophotometer and quartz cuvets.

2.1.4. Antibody Characterization

1. 0.1 *M* carbonate buffer: add 200 mL of 0.1 *M* Na_2CO_3 (10.6 g/L) to 0.1 *M* $NaHCO_3$ (8.4 g/L) until pH = 9.2.
2. PBS-Tween: PBS + 0.1 % (v/v) polyoxyethylenesorbitan monolaurate (Tween 20).
3. PBS-BSA: PBS + 1 % (w/v) radioimmunoassay grade bovine serum albumin (10 g/L).
4. PBS-BSA-Tween: PBS + BSA + 0.1% (v/v) Tween 20.
5. Secondary antibody: alkaline phosphatase-conjugated goat antirabbit IgG (F[ab']$_2$ fragment of goat antibody).
6. High binding, irradiated polystyrene, 96-well flat-bottom, microtiter plates (Immulon 2 HB; Dynex).
7. Polystyrene, 96-well U-bottom, microtiter plates (Dynex).
8. Photometric microplate reader with 405 and 490 nm filters and software to determine concentrations of unknowns from standard curves.
9. Freshly prepared substrate solution: 0.5 mg/mL disodium *p*-nitrophenyl phosphate (Sigma 104 phosphatase substrate 5 mg tablet/10 mL) in 1.0 *M* diethanolamine buffer (containing 24.5 mg $MgCl_2$ in 500 mL dH_2O, pH to 9.8)–store at room temperature in amber bottle.
10. ELISA amplification system (Gibco-BRL).
11. Peptides synthesized as described above with residues added to (+1, +2, +3), deleted from (–1, –2, –3), and bridging the cleavage site terminus of the immunizing peptide.
12. Native and collagenase-cleaved types I, II, and III collagens from species of choice (*see* **Note 1**).

2.2. Immunoassay

2.2.1. COL2-3/4C$_{short}$ Enzyme-Linked Immunosorbent Assay

1. All reagents listed in **Subheading 2.1.1.**, except that in the preparation of plate coating conjugate, keyhole limpet hemocyanin (KLH) is substituted for OVA, and the synthetic peptide containing glycine spacers (COL2-3/4C$_{long}$) is substituted for COL2-3/4C$_{short}$.
2. PBS; PBS-Tween; PBS-BSA; PBS-BSA-Tween (*as* **Subheading 2.1.4.**)
3. Polypropylene 96-well U-bottom microtiter plates (Costar).
4. Secondary antibody: alkaline phosphatase-conjugated goat antirabbit IgG (F[ab']$_2$ fragment of goat antibody).
5. ELISA amplification system.

2.2.2. Assay Validation

All reagents as listed in **Subheading 2.1.4.**

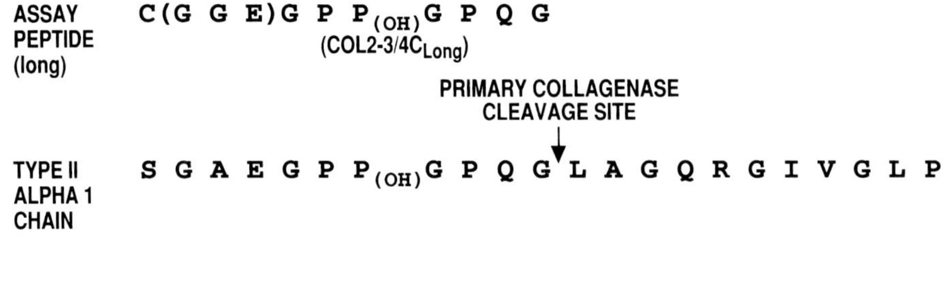

Fig. 1. Sequence of collagen α_1 (II) chain to show primary collagenase cleavage site and sequences of peptides used for immunization (COL2-3/4C$_{short}$) and plate coating (long peptide).

2.2.3. Experimental Studies Using the COL2-3/4C$_{short}$ ELISA

1. Digestion solution: 50 mg type VII α-chymotrypsin (Sigma) in 49.25 mL of 50 mM Tris-HCl (6.06 g/L, pH 7.6) with the general metallo-, cysteine and aspartic proteinase inhibitors added as 250 µL EDTA (76 mg/mL dH$_2$0), 250 µL iodoacetamide (37 mg/mL dH$_2$O), and 250 µL pepstatin A (2 mg/mL in 95% ethanol), respectively.
2. TPCK: 8 mg N-tosyl-L-phenylalanine-chloromethyl ketone in 800 µL of 95% ethanol added dropwise to 19.2 mL 50 mM Tris-HCl, pH 7.6.

3. Methods

3.1. Antibody Production

3.1.1. Preparation of the Immunogen

The amino acid sequence, Gly-Pro-Hyp-Gly-Pro-Gln-Gly, used for the synthesis of the neoepitope peptide COL2-3/4C$_{short}$, which corresponds to the C-terminus of the 3/4 (TC^A) fragment of the α chains of collagenase-cleaved, triple-helical, type II collagen was based on the published amino acid sequence for the primary MMP-1 cleavage site in type II collagen (**Fig. 1**). The five C-terminal amino acids in this peptide are identical to the five amino acids at the C-termini of collagenase-cleaved α chain 3/4 fragments of type I collagen suggesting the potential cross-reactivity of an antibody produced to the COL2-3/4C$_{short}$ peptide for both collagenase-cleaved types I and II collagens. The assignment of the third residue as a hydroxylated proline (Hyp) is based on the assumption that prolines in the Y-position of the repeating Gly-X-Y triplets in collagen α chains are potential hydroxylation sites. The

COL2-3/4C$_{short}$ peptide and those peptides used for antibody characterization (described below) were synthesized at a 0.25-mmol scale, using standard Fmoc (9-fluoroenylmethoxycarbonyl) chemistry, on a solid-phase peptide synthesizer. Crude peptides were purified by reverse phase chromatography. A cysteine was added to the N-terminus of the peptides for conjugation to the carrier proteins ovalbumin (OVA), bovine serum albumin (BSA), and keyhole limpet hemocyanin (KLH). These proteins are commonly used in hapten-protein conjugates to improve the immunogenicity of the hapten and to enhance the noncovalent binding of hapten to a solid support, such as the wells of microtiter plates used in immunoassays. The procedure for conjugation to OVA follows:

1. Slowly add (dropwise) 0.2 mL of the coupling reagent to 2.0 mL of chilled OVA, with stirring, then warm to 25°C over 30 min.
2. Centrifuge the sample and apply to a previously washed (50 mL of 0.1 *M* phosphate buffer, pH 7.0) Sephadex G-25 column, eluting with phosphate buffer at 1 mL/min and collecting 2 mL fractions.
3. Measure the absorbency at 280 nm in quartz cuvets of each fraction, pool those containing the protein peak, and measure the absorbency of the pooled sample.
4. Determine the [OVA] in mg/mL by dividing the absorbency value by 1.2 (the extinction coefficient of OVA at 1 mg/mL) and adjust [OVA] to 2–3 mg/mL.
5. Add 4 mg of the activated OVA (about 1.5 mL) to 5 mg synthetic peptide solution (100 µL) and leave at 25°C for 2 h, with occasional mixing, before transferring to 4°C overnight.
6. Dialyze against 1 L of phosphate buffer (pH 7.0), with 3 changes over 24 h, and then against 1 L azide-free PBS with one change over 24 h. Aliquot at 300 µL and store at –20°C.
7. Confirm effective incorporation by applying coupled OVA-peptide to one lane and activated OVA to another lane of a 10% SDS-PAGE gel and by showing a decrease in electrophoretic mobility of the OVA-peptide conjugate on SDS-PAGE relative to the activated OVA.

3.1.2. Immunization

The particular strain or breed of an animal species may affect the affinity or specificity of the antibodies produced in that animal. Young (but mature), disease-free, female rabbits are the principal source for raising polyclonal antibodies in the research laboratory. Polyclonals are easily raised against most antigens and haptens. The following is the protocol for the raising of the COL2-3/4C$_{short}$ polyclonal antibody:

1. Thaw out one 300 µL aliquot of peptide-OVA conjugate and add 600 µL of azide-free PBS and mix (~ 1 mg conjugate/mL).
2. Take up peptide-OVA conjugate into a glass syringe and emulsify in 900 µL of complete Freund's adjuvant (CFA) by forcing through a 20 gauge needle several

times until a drop of emulsion retains its shape when added to the surface of a
beaker of water (final concentration ~0.5 mg/mL).

3. Inject two rabbits intramuscularly (in both hind limbs) with 0.5 mL of emulsion
   (~ 200 µg of conjugate) divided between 2 different sites.
4. Give booster injections of similar quantities of peptide-OVA emulsified with incomplete Freund's adjuvant (IFA) intramuscularly every 3–4 wk.
5. Perform test bleeds and determine antibody titers by immunoassay (*see* **Subheading 3.1.4.**) after the second booster. If the titers are good, exsanguinate the
   rabbits by cardiac puncture and collect serum for antibody purification.

3.1.3. Antibody Purification

Polyclonal antibodies are usually of the IgG class and generally only 10% of
that IgG is specific for the immunizing antigen. A two-step approach is often
used to purify rabbit antibodies consisting of ammonium sulfate precipitation
followed by affinity chromatography. A 50% saturated solution of ammonium
sulfate is used as most serum components will not precipitate in this range. The
resulting crude IgG preparation is then digested with pepsin to yield antibody
fragments with the two antigen binding domains still bound together, called
$F(ab')_2$ fragments. This process removes the Fc portions of the IgG molecules
that can produce nonspecific binding in immunochemical assays. Finally the
$F(ab')_2$ fragments specific for the antigen of interest are purified by affinity
chromatography on a column containing the immunizing peptide coupled
through its added cysteine terminal end to a sulfur-linking gel.

The purification procedure used for the COL2-3/4C$_{short}$ antibody is as follows:

1. Centrifuge the volume of serum from which the antibody is to be purified for
   30 min at 3000*g* and place the supernatant in a beaker with a stirring bar.
2. With the beaker on a magnetic stirrer, slowly add an equal volume of a saturated
   solution of ammonium sulfate to the serum and incubate at 4°C overnight.
3. Centrifuge for 30 min at 3000*g*, and resuspend the pellet in 0.5 vol (of initial
   volume) of PBS; dialyze with three changes of PBS overnight.
4. Remove an aliquot of the antibody solution to determine the IgG concentration
   (mg/mL) by dividing the absorbance at 280 nm by 1.35 (at 1 mg/mL) and then
   adjust the concentration to 10–30 mg/mL by adding PBS.
5. To the crude IgG preparation add first 1/3 vol of 0.8 *M* sodium acetate buffer
   (pH 4.0), followed by 2% (wt/wt) pepsin, dissolved in a drop of distilled water,
   incubate overnight at 37°C.
6. Adjust the pH of the solution to 8.6 with NaOH to inactivate the pepsin and then
   back to neutrality with acetic acid.
7. Add an equal volume of a saturated solution of ammonium sulfate, centrifuge the
   suspension and discard the supernatant. Dissolve the precipitate in 3 mL of
   0.14 *M* phosphate buffer (pH 8.0) and desalt with 10 DG columns containing
   Bio-Gel P-6 gel, according to manufacturer's instructions (Gibco-BRL).

8. Separate the F(ab')$_2$ fragments from remaining intact antibodies and Fc fragments by passing the solution through a protein A column (10–20 mg of IgG per mL wet beads). Wash the beads with 10 column volumes of 100 m*M* Tris-HCl (pH 8.0), 10 column volumes of 10 m*M* Tris-HCl (pH 8.0) and elute the F(ab')$_2$ fragments with 0.1 *M* glycine-HCl (pH 3.0). Collect aliquots (1 mL) in tubes containing 50 μL of 1 *M* Tris-HCl (pH 8.0) and mix gently to avoid frothing. Identify the IgG-containing fractions by absorbance at 280 nm, pool and reread the absorbance at 280 nm.

9. Prepare a column with the COL2-3/4C$_{short}$ peptide coupled to 1 vol of SulfoLink coupling gel (1 mL peptide solution/mL gel) according to the manufacturer's instructions (Pierce) and wash with 3 vol of PBS.

10. Apply 1 vol of the F(ab')$_2$ solution to the column and allow to enter the gel completely before closing the bottom of the column, then add 0.4 vol of buffer and leave the column at room temperature for 1 h.

11. Wash the column with 8 vol of PBS and elute the COL2-3/4C$_{short}$ F(ab')$_2$ fragments with 4 vol of 0.1 *M* glycine-HCl (pH 3.0). Collect fractions of 1 mL into tubes containing 1 *M* Tris-HCl (pH 9.5), and read the absorbance for each at 280 nm. Pool the fractions containing the F(ab')$_2$ fragments, dialyze overnight in PBS, reread the absorbance, and then store 0.5 mL aliquots at –20°C.

3.1.4. Antibody Characterization

A true antineoepitope antibody to MMP-cleaved collagens should recognize the exact sequence of amino acids representing the newly created N- or C-termini of the collagen α chain fragments. To determine this specificity, peptides are synthesized with up to three amino acid residues either added to (+1, +2, +3) or removed from (–1, –2, –3) the end of the immunizing peptide corresponding to the cleaved terminus, and a peptide is made representing the sequence bridging the cleavage site of the native collagen α chain. Moreover, the antibody must recognize the MMP-cleaved collagen type(s) for which it was designed.

3.1.4.1. Preparing the Plates for Immunoassay

1. For the coating of the immunoassay plates to be used for all subsequent analyses, a peptide is synthesized to include two glycine spacers followed by the natural in-sequence amino acid residue between the N-terminus of the COL2-3/4C$_{short}$ peptide and the added N-terminal cysteine. This assay peptide (COL2-3/4C$_{long}$; **Fig. 1**) is conjugated to bovine serum albumin (BSA) using the methodology described in **Subheading 3.1.1.**, with the exception that the absorbance of the pooled fractions with the activated BSA is divided by 0.7 (the extinction coefficient at 1 mg/mL) to determine the [BSA].

2. Dilute the assay peptide (COL2-3/4C$_{long}$-BSA conjugate) to 20 μg/mL in 0.1 *M* carbonate buffer (pH 9.2), to promote noncovalent adsorption of the conjugate, and add 50 μL to each well (1 μg/mL) of Immulon-2 flat-bottom, 96-well microtiter plates.

3. After overnight incubation at 4°C, wash the plates three times with PBS-Tween. Block the noncoated binding sites with 150 µL/well of PBS-BSA for 30 min at room temperature and then wash once with PBS-Tween. Plates can then be stored, covered with plastic wrap, for up to 6 mo at 4°C.

3.1.4.2. ANTIBODY TITER

1. Add serial dilutions (1:10 to 1:1280) of the polyclonal antiserum and $F(ab')_2$ preparation as 50 µL aliquots, in duplicate, to wells of COL2-3/4C$_{long}$-BSA conjugate coated plates and after a 90 min incubation at 37°C, wash the plates with PBS-Tween. Seal plates in paraffin sheets for all incubations.
2. Dilute the secondary antibody conjugate 1:20,000 in PBS-BSA-Tween and add at 50 µL/well for 1 h at 37°C.
3. Wash the plates three times with PBS-Tween and once with dH$_2$O, then add the alkaline phosphatase substrate at 50 µL/well for 20–30 min at 37°C and read the absorbance at 405 nm with a microplate photometer.

3.1.4.3. ANTIBODY SPECIFICITY

1. Precoat U-bottom 96-well microtiter plates with PBS-BSA at 100 µL/well for 30 min at room temperature and wash once with PBS-Tween.
2. To act as nonspecific binding wells, add 50 µL of PBS-BSA and 50 µL of PBS-BSA-Tween to each of four wells. To all other wells add 50 µL of the appropriate dilution of polyclonal antiserum or $F(ab')_2$ preparation as determined by checkerboard analyses (*see* **Note 2**).
3. As the COL2-3/4C$_{short}$ immunoassay is an inhibition ELISA, to each of another four wells containing 50 µL antiserum, add 50 µL of PBS-BSA to determine maximum binding of the antibody in the absence of the inhibitory peptides.
4. To each of remaining test wells add, in duplicate, 50 µL of serial dilutions of the COL2-3/4C$_{short}$ peptide, the addition (+1, +2, +3) and deletion (–1, –2, –3) peptides described above, the peptide bridging the collagenase cleavage site in native type II collagen α chains, or native (triple helical), heat-denatured (80°C for 20 min) and collagenase-cleaved types I, II, and III collagens (*see* **Note 1**).
5. After 1 h at 37°C, transfer 50 µL from each well to the equivalent wells of plates precoated, as described above, with the one exception that the COL2-3/4C$_{long}$-BSA conjugate is diluted at 50 ng/well in PBS (pH 7.2).
6. Incubate the plates for 30 min at 4°C, wash three times with PBS-Tween and then add the secondary antibody conjugate, diluted 1:20,000 in PBS-BSA-Tween, for 1 h at 37°C.
7. After a further three washes with PBS-Tween and dH$_2$O, use an ELISA amplification system kit, according to manufacturer's (Gibco-BRL) instructions.
8. Read the absorbance for each well at 490 nm and subtract the mean of the nonspecific binding wells from the values for each of the other wells to obtain a corrected absorbance. Calculate the percentage binding for each peptide from their corrected absorbance values relative to the mean absorbance of the four maximum binding wells which represent 0% inhibition.

$$\% \text{ binding} = \frac{\text{corrected absorbance (absorbance} - \text{mean absorbance of nonspecific wells)} \times 100}{\text{mean absorbance of maximum binding wells}}$$

Calculate percentage inhibition by subtracting the percentage binding from 100

3.2. Immunoassay

3.2.1. COL2-3/4C$_{short}$ ELISA

The COL2-3/4C$_{short}$ enzyme-linked immunosorbent assay (ELISA) can be used to identify α chain fragments of types I and II collagene generated through the primary cleavage of the native, triple helical form of these collagen molecules by the collagenases MMP-1, MMP-8 and/or MMP-13, as well as the membrane bound MT1-MMP (MMP-14). We and others have shown that the C-termini of the 3/4 α chain fragments created by the action of these enzymes are more resistant to subsequent proteolysis than the N-termini of the 1/4 fragments, which are rapidly processed, in vitro, by these same enzymes after the initial cleavage *(1)*. Moreover, the increased levels over normal cartilage of fragments bearing the COL2-3/4C$_{short}$ neoepitope, demonstrated in α-chymotrypsin digests of osteoarthritic cartilage *(1)* and in the immunohistochemical analysis of arthritic cartilage sections (unpublished observations), supports the use of the COL2-3/4C$_{short}$ antibody for the identification of collagenase activity in tissues containing either type I or type II collagen.

The following is the procedure for the COL2-3/4C$_{short}$ ELISA which is presently used to assay samples for collagenase-cleaved type I and II collagen fragments.

1. To improve the sensitivity of the COL2-3/4C$_{short}$ ELISA, which is of particular importance when assaying biological fluids, conjugate the COL2-3/4C$_{long}$ peptide, (**Fig. 1**), to KLH as described previously for OVA and BSA, with the following exceptions. Reconstitute the KLH with 2 mL dH$_2$O and dialyze overnight against 0.1 M phosphate buffer (pH 7.0). After conjugation, determine the [KLH] by dividing the absorbance of the blue-colored pooled fractions by 1.5 (extinction coefficient at 1 mg/mL).

2. To all but the 36 perimeter wells of 96-well flat bottom Immulon 2 microtiter plates add 50 μL/well (5 ng/well) of COL2-3/4C$_{long}$-KLH conjugate (100 ng/mL) diluted in PBS (pH 7.2) and incubate the plates overnight at 4°C. The exclusion of the outer wells is due to the large standard deviations in absorbances obtained for a sample when assayed in duplicate, with one of the replicates in an outer well and the other replicate in an inner well, as compared to both replicates in the inner wells. Wash three times with PBS-Tween, then block the plates with 100 μL/well of PBS-BSA for 30 min at 25°C and wash again three times with PBS-Tween.

3. Prepare standards by diluting a 1 mg/mL stock solution of COL2-3/4C$_{short}$ peptide to 10, 2.5, 0.63, 0.16 0.04, 0.01, and 0.0025 μg/mL in PBS-BSA. Add each standard at 50 μL/well, in duplicate, to U-bottom 96-well polypropylene plates

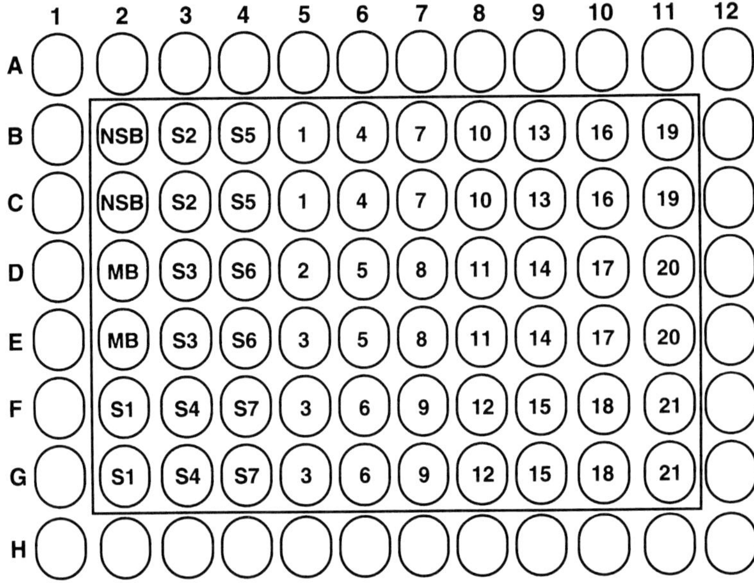

Fig. 2. Microtiter plate template for COL2-3/4C$_{short}$ ELISA. Shown is the layout of a 96-well microtiter plate for the COL2-3/4C$_{short}$ immunoassay. The perimeter wells corresponding to columns 1 and 12, and rows A and H are not used. All blanks, standards and samples are assayed in duplicate. Symbols shown represent nonspecific binding (NSB) wells with no primary antibody added, maximum binding (MB) wells with no inhibiting peptide added, standard wells (S1-S7) with 10, 2.5, 0.63, 0.16, 0.04, 0.01, and 0.0025 µg of inhibiting COL2-3/4C$_{short}$ peptide/mL of PBS/BSA, and sample wells (1–21).

as outlined in template shown as **Fig. 2**, again ignoring all outside wells. The duplicate nonspecific binding (NSB) wells consist of 50 µL PBS-BSA and 50 µL PBS-BSA-Tween, and the duplicate maximum binding (MB) wells contain 50 µL PBS-BSA and 50 µL COL2-3/4C$_{short}$ F(ab')$_2$ preparation, at a dilution in PBS (pH 7.2), as previously determined by checkerboard analysis (*see* **Note 2**).

4. Add samples in duplicate to the remaining wells, and to each of the wells (except NSB), add 50 µL/well of the diluted COL2-3/4C$_{short}$ F(ab')$_2$ preparation, before covering the plates with paraffin sheets and incubating at 37°C for 1 h.

5. Using a multichannel pipeter, transfer 50 µL from each well of the polypropylene plates to the respective well of the COL2-3/4C$_{long}$-KLH pre-coated plates. Incubate these plates for 30 min at 4°C and then wash three times with PBS-Tween.

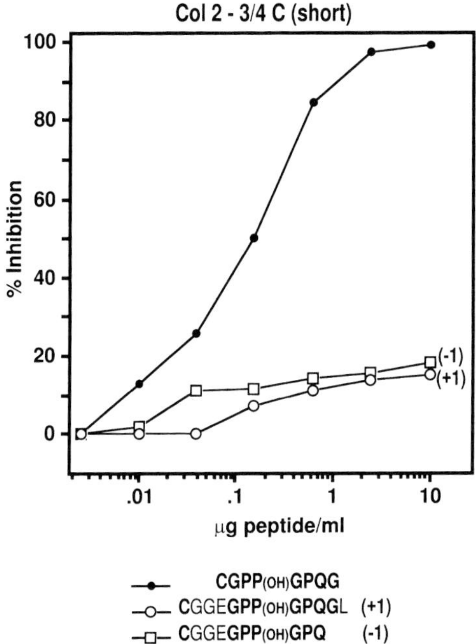

Fig. 3. Immunoassay standard curve and epitope specificity of the COL 2-3/$_4$C $_{short}$ antineoepitope antiserum determined using competing synthetic peptides in inhibition ELISAs. Varying dilutions of the nonconjugated immunizing peptide (●) were added to a 1:200 dilution of the rabbit antiserum F(ab')$_2$ preparation in 96-well round bottom microtiter plates for 1 h at 37°C. Then 50 μL was transferred from each well to the equivalent wells of Immulon-2 microtiter plates precoated with the long version of the immunizing peptide (**Fig. 1**) conjugated to keyhole limpet hemocyanin. The plates were then processed as described in Methods for the development of a standard inhibition curve. Shown are the results for the COL2-3/4C$_{short}$ ELISAS performed with the immunizing synthetic peptide and those peptides that had either 1 amino acid added to (○) or deleted from (□) the carboxyterminal end of the immunizing peptide corresponding to the cleaved termini of the alpha chain fragments. The sequences of these peptides are also shown. Reprinted with permission from **ref** *1*.

6. Dilute the secondary antibody conjugate 1:20,000 in PBS-BSA-Tween and add at 50 μL/well for 1 h at 37°C. Wash the plates three times with PBS-Tween and three times with dH$_2$O.

7. Use a two-step ELISA amplification system from Gibco-BRL according to manufacturer's instructions, stop the color development with 0.3 *M* H$_2$SO$_4$ and read the absorbances at 490 nm.

8. Determine the concentration of COL2-3/4C$_{short}$ in each sample from a log/linear plot of concentration (μg/mL) versus corrected absorbance (sample absorbance −mean of nonspecific binding absorbance) for the standards. Use nonlinear regression to estab-

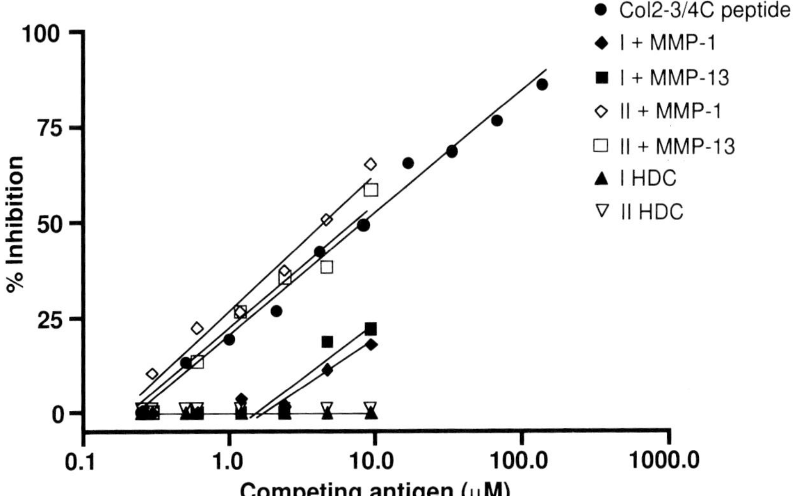

Fig. 4. ELISA to show specificity of the COL2-3/4C$_{short}$ antibody for the C-terminal neoepitope of α_1(II) collagen chains created by the digestion of triple helical type II collagen by the collagenases MMP-1 and MMP-13. Heat-denatured and collagenase-cleaved (MMP-1 and MMP-13) human type I, II, and III collagens (digestions were complete as shown by Coomassie staining of SDS-PAGE gels) were assayed in inhibition ELISAs at concentrations of 20, 58, 115, 230, 460, and 920 μg/mL of digestion buffer. A standard curve for the COL2-3/4$_{short}$ peptide (●) was also constructed using doubling dilutions from 50 to 0.1 μg of the peptide/mL of buffer in the same inhibition ELISA. Shown are typical standard curves plotted as percent inhibition against the log of the competing antigen concentrations on a μ*M* basis. Symbols show MMP-1 (◆,◇) and MMP-13 (■,□) cleaved type I and type II collagens, respectively, and heat-denatured type I (I HDC; ▲) and type II (II HDC; ▽) collagens. Reprinted with permission from **ref. *1***.

lish a sigmoidal dose response curve with variable slope for the determination of the concentrations of the unknowns. A sample standard curve is shown in **Fig. 3**.

3.2.2. Assay Validation

Assays that are developed for use in clinical sciences must be validated. There are guidelines published by the International Federation of Clinical Chemistry (IFCC).

The features of any assay that need to be validated are specificity, sensitivity, precision (within-run and between-run), analytic recovery, and identity of calibrant and unknown sample.

3.2.2.1. ANTIBODY SPECIFICITY

Antibody specificity refers to the ability of an antibody to produce a measurable response for the analyte of interest, whereas crossreactivity measures the response of the antibody to other substances. The polyclonal COL2-3/4C$_{short}$ antibody has selective, high affinity for both the immunizing COL2-3/4C$_{short}$ peptide and collagenase (MMP-1 and MMP-13)-cleaved human type II collagen 3/4 α-chain fragments, when compared on a molar basis *(1)*. The antibody demonstrates a lower affinity for similarly cleaved human type I collagen α-chains. There is negligible binding to uncleaved triple-helical and heat-denatured human types I and II collagen, and intact or cleaved type III collagen α-chains. The materials required for these procedures are listed in **Subheading 2.1.4.** and **Note 1**. The methodology is outlined in **Subheading 3.1.4.3.** and the results are shown in **Fig. 4**.

3.2.2.2. SENSITIVITY

The sensitivity of an assay is the lowest dose at which there is a significant difference in response from zero. This is generally determined by the upper 2 or 3 standard deviation (SD) limit in a zero standard study. The minimum detection limit of the COL2-3/4C$_{short}$ assay is 4.9 nM.

3.2.2.3. PRECISION

Within-run (intraassay) and between-run (interassay) precision are determined by assaying different samples that cover the range of detectable levels of epitope, in a series of runs. For within-run precision, a single sample is assayed in multiple replicates on the same plate ($n = +20$). The coefficient of variation (% CV) is determined by dividing the standard deviation (SD) by the mean of the concentration values obtained and then multiplying by 100. The between-run precision is determined from the same samples assayed on multiple plates ($n = +5$) and the % CV determined from the SD of the plate means and the overall concentration mean. Precision values obtained for collagenase-cleaved human type II collagen and culture medium from IL-1 stimulated bovine articular cartilage explants range from CV of 4.6–8.3% for within-run and CV of 5.4–6.9 % for between-run.

3.2.2.4. ANALYTIC RECOVERY

Spike recovery is determined by the addition of known quantities of the purified COL2-3/4C$_{short}$ peptide into fluid samples with different levels of endogenous COL2-3/4C$_{short}$. Typical results for synovial fluid, urine, and cartilage explant culture medium range from 89 to 110% recovery.

Linearity is shown by diluting samples serially and comparing the observed values with those that are expected. Typical recovery rates of 94–111% have been noted for cleaved type II collagen and culture medium from IL-1 stimulated articular cartilage explants.

3.2.2.5. IDENTITY OF ANALYTE AND CALIBRANT

The demonstration of parallelism between the dose response curves of the calibrant and the unknown sample proves identity. The neoepitope peptide, COL2-3/4C$_{short}$, and the sample (cleaved collagen, culture medium, serum, synovial fluid, urine, etc.) are each progressively diluted and assayed separately as described above. The results for each are plotted on the same graph of concentration (dilution) vs % inhibition and the curves for each are compared for parallelism. A typical experiment demonstrating parallelism between MMP-1 and MMP-13 cleaved types I and II collagen and the COL2-3/4C$_{short}$ peptide is shown in **Fig. 4**.

3.2.3. Experimental Studies Using COL2-3/4C$_{short}$ ELISA

The COL2-3/4C$_{short}$ ELISA can be used to assess the degree of MMP activity, specifically for the collagenases MMP-1, MMP-8, and MMP-13, in terms of cleaved product, in any system where there is type I and/or II collagen. It can be used to evaluate the effect of MMP inhibitors on the generation of collagenase-cleaved collagen products in vitro and in vivo *(1)*. To evaluate the levels of collagenase-cleaved collagen *in situ*, tissue is digested with α-chymotrypsin, which releases cleaved and denatured collagens from the extracellular matrix and preserves the recognition of the neoepitope by the COL2-3/4C$_{short}$ antibody. The protocol for digestion of articular cartilage and extraction of collagenase-cleaved collagen follows:

1. Mince articular cartilage and incubate 50–75 mg overnight at 37°C with 500 μL of α-chymotrypsin digestion solution containing proteinase inhibitors.
2. Incubate the digest for 20 min at 37°C with 200 μL (160 μg/mL final concentration) of TPCK.
3. Centrifuge the samples and remove the supernatants for assaying with the COL2-3/4C$_{short}$ ELISA as described in **Subheading 3.2.1.**

4. Notes

1. If the collagens for the species of interest cannot be purchased, they can be purified from tissues of that species. Type II collagen can be purified from articular cartilage of the species of choice, by pepsin digestion and differential salt precipitation *(2)*, and types I and III collagen can be prepared by pepsin digestion and differential denaturation/renaturation *(3)*. For cleavage by the collagenases (MMP-1, MMP-8, and/or MMP-13), these lyophilized collagens are dissolved separately in 0.5 *M* acetic acid and diluted to 2.5 mg/mL in a digestion buffer consisting of 50 m*M* Tris, 10 m*M* CaCl$_2$, 0.5 *M* NaCl, 0.01% Brij 35 and 0.02% NaN$_3$, pH 7.6. The proenzymes are activated with 2 m*M* (final concentration) aminophenyl mercuric acetate (APMA) in the same digestion buffer for 90 min at 37°C. The activated enzyme solution is added to each of the collagen solutions

at molar ratios of from 1:5 to 1:10, the pH adjusted to 7.5 with NaOH (if necessary), and the samples are incubated at 30°C for 24 h. The MMPs are inactivated with 20 mM (final concentration) EDTA and cleavage is confirmed by SDS-PAGE under denaturing conditions using 10%, 1 mm thick, mini-Protean gels (Bio-Rad). To confirm immunoreactivity of the cleavage-site neoepitope antibody, the electrophoretically separated collagenase-cleaved α chain fragments are transferred to a nitrocellulose membrane. After blocking in PBS-3% BSA, the membrane is incubated overnight with the antineoepitope antibody. After washing with PBS-BSA-Tween, secondary antibody conjugate diluted in PBS-3% BSA-Tween is added for 1 h at 25°C, the membranes washed three times in PBS-BSA-Tween and three times in dH$_2$O, and alkaline phosphatase substrate solution added (5-bromo-4-chloro-3-indolyl phosphate and nitroblue tetrazolium) from a kit (Bio-Rad). After the color has developed for the immunostained bands, and before excess background staining (10–20 min), the substrate is removed and the membranes washed with dH$_2$O before drying on filter paper.

2. Checkerboard analysis is used for titration of both the antigen and antibody to determine the optimum concentration of antigen to coat the microtiter plate wells and the optimum dilution of primary antibody to allow maximum sensitivity and minimal background in the ELISA. The COL2-3/4C-carrier conjugate is diluted across the plate from left to right, with the last column of wells containing no antigen, and allowed to adsorb to the plate. The COL2-3/4C$_{short}$ antibody is then diluted across the plate from top to bottom, thereby obtaining a checkerboard titration of antigen against antibody. After incubation at room temperature for 1 h, enzyme-conjugated secondary antibody is added, incubated at 37°C for 1 h, followed by the addition of the enzyme substrate and the absorbances read. A graph is made showing a plot for each antibody dilution of antigen dilution versus OD (optical density), and from this the background of the plate and the nonspecific adsorption of conjugate, as well as the plateau region of maximum antigen binding to the plate (plate saturation) can be determined.

Acknowledgments

Financial Support for these studies was provided by the Shriners Hospital for Children, Medical Research Council of Canada, National Institutes of Health (to A. Robin Poole) and the Fonds de la Recherche en Sante du Quebec (to R. Clark Billinghurst).

References

1. Billinghurst, R. C., Dahlberg, L., Ionescu, M., Reiner, A., Bourne, R., Rorabeck, C., Mitchell, P., Hambor, J., Diekmann, O., Tschesche, H., Chen, J., Van Wart, H., and Poole, A. R. (1997) Enhanced cleavage of type II collagen by collagenases in osteoarthritic articular cartilage. *J. Clin. Invest.* **99**, 1534–1545.
2. Miller, E. J. (1971) Isolation and characterization of a collagen from chick cartilage containing three identical α chains. *Biochemistry* **10**, 1652–1659.

3. ChandraRajan, J. (1978) Separation of type III collagen from type I collagen and pepsin by differential denaturation and renaturation. *Biochem. Biophys. Res. Commun.* **83**, 180–186.

29

Collagen Degradation Assays

Anthony P. Hollander

1. Introduction

Degradation of fibrillar collagens by matrix metalloproteinases (MMPs) is thought to be a major catabolic pathway in various connective tissues. However, the complex structure of collagen molecules and their degradation products makes specific assay of these events very challenging. In this chapter, those particular features of collagen biochemistry which contribute to the difficulty of measuring its degradation will be outlined and the principles of recently developed methodology for assaying collagen degradation *in situ* will be described. Although the most abundant collagen in the body is type I collagen (found, for example, in skin, bone, ligaments and tendons), much of this chapter will focus on type II collagen, the major collagen of hyaline cartilage, for which much of the new methodology has already been established.

1.1. The Structure of Collagen Fibrils

Fibrillar collagen molecules consist of three α-chains wound around each other into a right-handed triple helix, with short, nonhelical, telopeptides at each end *(1)*. Although all collagen α-chains are synthesized with large N- and C-terminal propeptides, these are usually removed by specific propeptidases soon after the newly synthesized molecule has been released from the cell *(1)*. In the extracellular matrix, individual collagen molecules become crosslinked to each other by the formation of stable, covalent pyridinium crosslinks between lysine and hydroxylysine residues of different α-chains *(2)*. Within a fibril, the collagen molecules are organized in a quarter-stagger array with a 67 nm banding pattern *(3)* which can be directly observed by electron microscopy (**Fig. 1**). The pyridinium cross-links help to coordinate the uniform structure of each fibril and they also stabilize the individual collagen molecules.

From: *Methods in Molecular Biology, vol. 151: Matrix Metalloproteinase Protocols*
Edited by: I. Clark © Humana Press Inc., Totowa, NJ

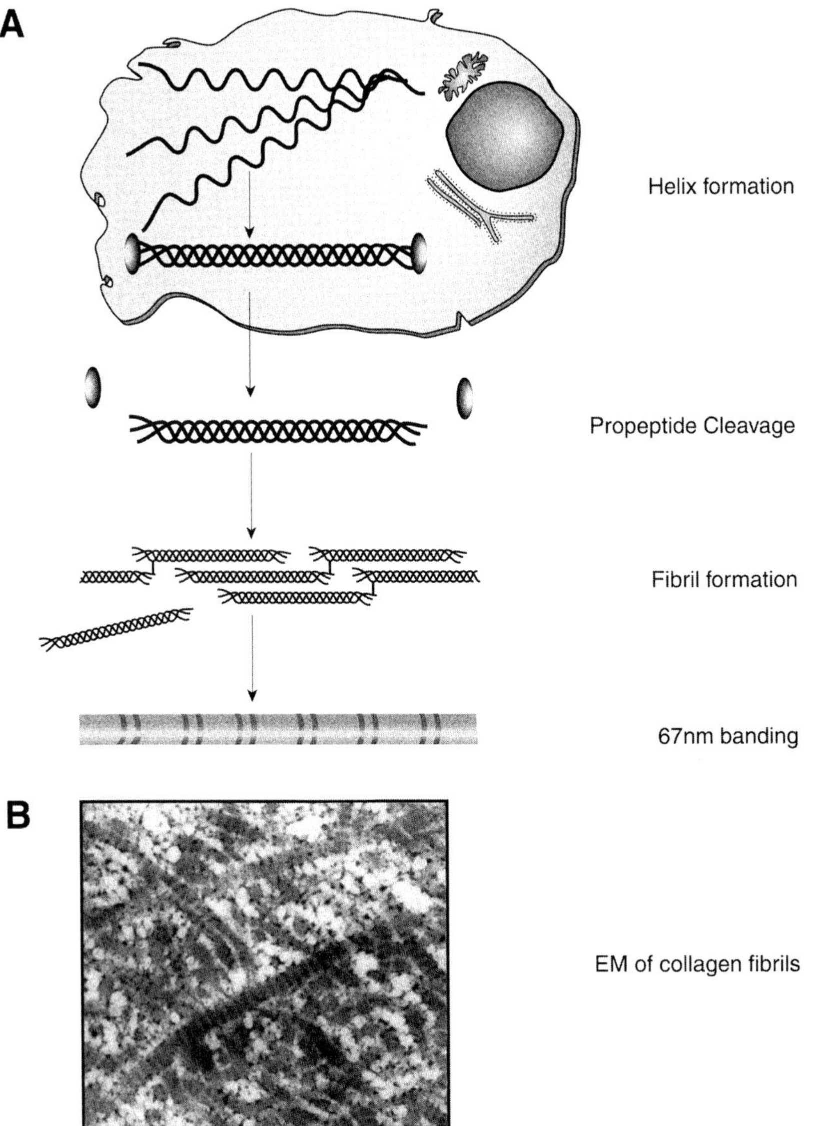

Fig. 1. Collagen fibril formation. (**A**) Collagen molecules form a triple helix within the cell. Following their release the propeptides are rapidly removed and and covalent crosslinking progresses gradually, in the extracellular matrix. This process results in the formation of fibrils with a uniform, periodic structure. (**B**) Electron micrograph of bovine articular cartilage, showing the typical banding pattern of type II collagen fibrils. This micrograph was kindly prepared by Dr. Vincent Everts, Academic Medical Centre, University of Amsterdam.

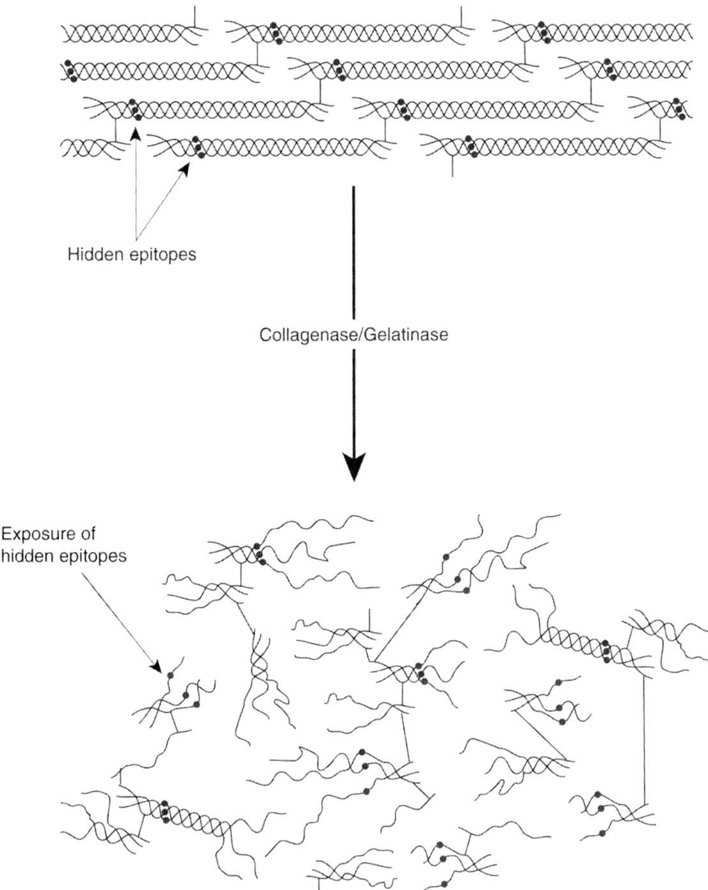

Fig. 2. Collagen fibril degradation. Cleavage of collagen molecules within the triple helix by a collagenase results in their denaturation, a process that exposes epitopes that were previously unavailable for antibody binding. Cleavage of the denatured α-chains by gelatinases leads to further degradation. However, the collagen fragments can be retained within the extracellular matrix by covalent crosslinks, allowing detection of the hidden epitopes *in situ*.

1.2. Collagen Fibril Degradation

Collagens are relatively resistant to proteolysis and this stability is increased proportionately with increasing levels of cross-link formation *(4,5)*. Theoretically, a fibrillar network could be depolymerized by cleavage of the telopeptides, within the cross-link sites. A number of enzymes can act as telopeptidases, including stromelysin (MMP-3) *(6)*, collagenase-3 (MMP-13)

(7), cathepsin B *(8)*, cathepsin K *(9)*, cathepsin G *(10)* and neutrophil elastase *(10)*. There is no direct evidence that such activity is a major route of collagen degradation in the normal turnover or pathological breakdown of adult tissues. However telopeptidase-mediated depolymerization may be important in certain embryonic and neonatal tissues *(7)*.

Cleavage within the triple helix has generally been thought to be a property of only a small number of enzymes, the collagenases. Interstitial collagenase (MMP-1), neutrophil collagenases (MMP-8), and MMP-13 all cleave fibrillar collagens at a single site, approximately three-quarters of the distance from the N-terminus of the α-chain *(11–13)*. More recently, we have found that cathepsin K can cleave collagen types I and II in the helix, close to their N-terminus *(9)*. Whereas this cysteine proteinase may play a major role in the degradation of bone type I collagen *(14)*, the MMP collagenases are likely to be the major mediators in most tissues.

At physiological temperature, helical cleavage is followed by spontaneous denaturation (unwinding) of the α-chains. The denatured α-chains are then highly susceptible to further, extensive degradation by gelatinolytic proteinases *(15)* such as gelatinase A (MMP-2) and gelatinase B (MMP-9). A particularly important point with respect to the assay of collagen degradation is that denatured α-chains and their fragments can be retained within the extracellular matrix by the covalent cross-links *(16–18)*, as shown in **Fig. 2**. The methodology described in this chapter has been designed specifically to measure those fragments that have been retained *in situ* within the extracellular matrix. In order to measure the extent of collagen degradation within a given tissue, the denatured molecules must be selectively removed and assayed. The remaining intact collagen fibrils must then be solubilized prior to their quantitation, so that the amount of degraded fragments can be determined as a proportion of total collagen in the tissue.

1.3. Hidden Epitopes

Specific assay of individual α-chains and their degradation products can be achieved using antibodies to hidden epitopes. These are linear amino acid sequences that are not normally available for antibody binding when they are contained within an intact helical structure, but they become exposed once the triple helix has been cleaved and denatured (**Fig. 2**). The earliest reference to such an approach was in studies by Rucklidge et al. *(19)* in which sheep were immunized with native type I collagen. Antibodies specific for the denatured molecule were isolated from the sheep antiserum by affinity chromatography, indicating that a portion of the intact collagen had degraded during the immunization procedure. The antibodies were used to immunolocalize denatured type I collagen in resorbing renal tubules of the rat *(20)*. Dodge et al. developed the method by immunizing rabbits with CNBr peptides of type II collagen *(16)*. The resulting antiserum could be used to immunolocalize denatured collagen in arthritic cartilage. A similar approach was used for the detection of denatured type I collagen in the guinea-pig dilating cervix *(21)*. More recently, we refined the method still further by raising a monoclonal antibody to a synthetic peptide (CB11B), based on an amino acid sequence from the type II collagen α-chain *(17)*. The monoclonal antibody, Col2-3/4m, recognized denatured type II collagen and not the native molecule. The advantage of making an antipeptide antibody was that the specific epitope could be selectively isolated from denatured or native α-chains. We showed that α-chymotrypsin can cleave denatured type II collagen without destroying the CB11B epitope itself. Like most mammalian proteinases, α-chymotrypsin cannot cleave intact, native type II collagen molecules. It can therefore be used to selectively extract the degraded collagen. A second enzyme, proteinase K, was also found to spare the CB11B epitope, but it fully solubilized cartilage pieces during an overnight incubation at 56°C. It could therefore be used to extract the intact epitope from all α-chains, native or denatured. Sequential extraction of cartilage using these proteinases, combined with assay of the extracts by inhibition ELISA using antibody Col2-3/4m, was found to provide a reliable method for assessing the proportion of denatured type II collagen in arthritic articular cartilage as well as in the lumbar intervertebral disks *(22,23)*.

1.4. Purpose of Chapter

In this chapter the assay of type II collagen degradation in cartilage will be described in detail, as an example of how selective proteolysis and hidden epitope immunoassay can be used in combination. For the purpose of this description, it will be assumed that an antibody is available for use in the assay. If alternative anti-collagen antibodies are to be used then the appropriate method of proteolytic extraction of specific epitope(s) will have to be established before the procedure can be adapted. In addition, an alternative procedure will be described which is not reliant on antibodies (*see* **Note 6**).

2. Materials

1. Tris hydroxylamine (Tris-HCl): Prepare as a large volume stock solution at 50 m*M*, including sodium azide to a final concentration of 0.05% w/v and adjust the pH to 7.6 with HCl.
2. Phosphate buffered saline (PBS): Prepare as a large volume stock solution of 0.85% w/v NaCl, 0.16% w/v Na_2HPO_4, 0.054% w/v KH_2PO_4 and 0.05% w/v sodium azide. Adjust the pH to 7.4.
3. Carbonate buffer: Prepare 1 L of 0.1 *M* $NaHCO_3$ and 200 mL of 0.1 *M* Na_2CO_3. Gradually add the Na_2CO_3 to the $NaHCO_3$, with mixing, until the pH is 9.2.

4. Proteinase Inhibitors (all from Sigma): Prepare EDTA and iodoacetamide, each as a 200 m*M* stock solution in water and store at –20°C; Prepare Pepstatin A as a 2 mg/mL stock solution in 95% v/v ethanol and store at –20°C.
5. α-Chymotrypsin (EC 3.4.21.1), Sigma: prepare as a 1 mg/mL stock solution in 50 m*M* Tris-HCl, pH 7.6, containing 1 m*M* EDTA, 1 m*M* iodoacetamide, and 10 μg/mL pepstatin A. Store in 5 mL aliquots at –20°C.
6. Proteinase K (E C 3.4.21.64), Sigma: prepare as a 1 mg/mL stock solution in 50 m*M* Tris-HCl, pH 7.6, containing proteinase inhibitors: 1 m*M* EDTA, 1 m*M* iodoacetamide, and 10 μg/mL pepstatin A. Store in 5 mL aliquots at –20°C.
7. *N*-Tosyl-L phenylanaline-chloromethyl ketone (TPCK; α-chymotrypsin inhibitor), Sigma: prepare initially as a 10 mg/mL stock solution in 100% v/v ethanol and then add it dropwise to 50 m*M* Tris-HCl, pH 7.6, to a final concentration of 400 μg/mL. Store in 2 mL aliquots at –20°C.
8. Bovine Type II Collagen, Sigma: store the lyophilized powder at –20°C (not at 4°C as suggested on the bottle label).
9. Heat denatured type II collagen (HDC), prepared immediately before use: dissolve native bovine type II collagen in carbonate buffer, pH 9.2, to a final concentration of 1 mg/mL in a screw-capped 1.5 mL centrifuge tube. At this concentration it will be partly in suspension. Place the tube in a water bath at 80°C for 20 min, but remove it for brief periods of mixing over the first 2 min, to ensure complete dissolution of the collagen. Dilute the HDC in carbonate buffer to a final concentration of 40 μg/mL.
10. Bovine serum albumin (BSA), Sigma: prepare a stock solution of 1% w/v BSA in PBS and store at 4°C.
11. PBS/Tween: add Tween-20 (Sigma) to PBS to a final concentration of 0.1% v/v.
12. Antibody Col2-3/4m or equivalent.
13. Peptide CB11B or equivalent.
14. Second antibody: goat antimouse Ig labeled with alkaline phosphatase (available from many sources).
15. Diethanolamine buffer: prepared as 8.9 m*M* diethanolamine and 0.25 m*M* MgCl$_2$. Store at ambient temperature.
16. Alkaline phosphatase substrate solution: prepared immediately before use as 0.5 mg/mL *p*-nitro-phenyl phosphate (Sigma) in diethanolamine buffer.

3. Methods

3.1. Extraction of Cartilage

1. Cut the cartilage into small pieces and distribute into 1.5 mL centrifuge tubes with a V-shaped bottom and a screw-cap lined with an o-ring. Each tube should ideally contain 10–50 mg of tissue (wet weight).
2. To each tube add 500 μL of 1 mg/mL α-chymotrypsin (including proteinase inhibitors, as described above). Put the cap on tightly and tap the tube until all the cartilage is submerged in the solution.
3. Incubate overnight in an oven or water-bath, at 37°C.

4. Add 200 µL of the 400 µg/mL TPCK solution to each tube. Mix and allow to stand at ambient temperature for 15 min.

5. Using an appropriate pipet, transfer all of the α-chymotrypsin **C**artilage-**E**xtract solution (**CE**) to a new screw-capped 1.5 mL centrifuge tube, taking care to leave all the undigested **C**artilage **R**esidue (**CR**) in the original tube. Mix the CE and then transfer 300 µL to another screw-capped 1.5 mL centrifuge tube for further **D**igestion of the **E**xtract (**DE**). Store the remaining CE at 4°C.

6. To the CR add 500 µL of 1 mg/mL proteinase K (including proteinase inhibitors, as described above). Put the cap on tightly and tap the tube until all the cartilage is submerged in the solution. To the DE add 100 µL of 1 mg/mL proteinase K and put the cap on.

7. Incubate CR and DE overnight in an oven or water-bath, at 56°C. Then increase the temperature to 100°C and allow the samples to boil for 10–15 min, to inactivate any remaining proteinase K. *See* **Note 1** below for a discussion of problems encountered with digesting cartilage residues.

8. Assay CE, DE, and CR for epitope CB11B, by inhibition ELISA (*see* below).

3.2. Inhibition ELISA for Epitope CB11B

1. Distribute HDC in carbonate buffer at 50 µL/well in all except the outermost wells of Immulon-2 ELISA plates. Wrap in cling-film and store at 4°C for 72 h.

2. Wash the plates 3 times in PBS/Tween (*see* **Note 2** for the washing method) and shake them dry. Wrap in cling-film and store at 4°C until required (the HDC-coated plates are usable even after several months of storage).

3. Immediately before assay, prepare a 96-well round-bottom microtiter plate by blocking the plastic with BSA (100 µL/well) for at least 30 min at ambient temperature. Wash it once with PBS/Tween. This is the preculture plate.

4. Add 100 µL of Tris-HCl to each of 6 non-specific binding (NSB) wells in the preculture plate. Add 50 µL of Tris-HCl to each of 6 maximum binding (MB) wells. Add 50 µL of antibody COL2-3/4m at an appropriate dilution (*see* **Note 3**) to all wells except NSB. Do not use the outermost wells as these are prone to excessive evaporation during the incubation stages. Prepare standard peptide CB11B in Tris-HCl at 9 different concentrations within the range 2–10 µg/mL. These standards can be stored at 4°C for several weeks, but they do tend to lose reactivity with time, so the standard curve must be monitored and the standards replaced as necessary. Add 50 µL of each standard to duplicate wells already containing antibody. Prepare appropriate dilutions (*see* **Note 4**) of CE, DE, and CR in Tris-HCl and add 50 µL of each one to duplicate wells already containing antibody. Note that a complete standard curve should be set up on each of the plates to be used.

5. Cover the preculture plate tightly with Parafilm and place in a humidified incubator at 37°C overnight. The humidity helps to minimize evaporation of the samples.

6. The most challenging part of the assay is to transfer all the samples from the preculture plate to the ELISA plate within 45 s. This timing is important because

the subsequent incubation time on the ELISA plate is short (30 min) and so small variations in the transfer time can lead to quite large variations in the assay result. In practice, the best method is to use a multichannel pippet to transfer from 10 wells at a time. First transfer row B from one plate to the other, then row C and so on. It is important to use just one set of pipet tips for each plate during this transfer. There is no time to change tips in-between transferring each row of the plate. As soon as the last row has been transferred, start timing the 30 min incubation time, leaving the plate at ambient temperature, uncovered. If more than one plate is being used, transfer all samples on the next one and then write down the time interval between plate 1 and plate 2 and so on. When 30 min have elapsed, immediately wash plate 1 three times in PBS/Tween and start the timer again so that subsequent plates can be washed at the appropriate time intervals. Note that if the 30 min incubation period is extended there is generally a significant loss of reactivity in the assay (presumably due to dissociation of the solution phase antigen-antibody complex in favor of antibody binding to the solid-phase antigen).

7. Prepare a commercial second antibody such as alkaline-phophatase-labeled goat-antimouse Ig, at the dilution recommended by the manufacturer. In some cases the required dilution will need to be optimized in the assay. The antibody should be diluted in PBS-Tween containing 1% w/v BSA. Add 50 µL of it to all wells of the ELISA plate being used in the assay. Incubate at 37°C for 1–2 h (humidity is not necessary at this stage).

8. Wash the plates 3 times in PBS/Tween, then once in water and shake them dry. Add 50 µL/well of alkaline phosphatase substrate solution and leave the plates at ambient temperature to allow color to develop.

9. Read the OD at 405 nm in a plate reader. The OD of the MB wells should be in the range of 0.4–1.2 for optimal calculation of % inhibition values. The mean NSB value should be subtracted from all others prior to calculation. Most modern plate-readers will be able to construct the best-fit standard curve (usually a log-log plot) and calculate specific values for each sample.

10. Calculate % denaturation of type II collagen (*see* **Note 5**).

4. Notes

1. Solubility of cartilage residues: digestion of cartilage residues by proteinase K generally results in complete solubilization of the tissue, which is necessary for an accurate estimate of the total collagen content. However in some instances proteinase K fails to solubilize the tissue. We have found two different reasons for this. The first is if too much cartilage was added to the centrifuge tube. In this case the remaining undigested residue can be transferred to fresh proteinase K. The second is that calcified cartilage is resistant to proteolysis. This problem is most common with cartilage from animals such as rabbits or guinea-pigs because the uncalcified tissue layer is very thin. The way to resolve this problem is to increase the concentration of EDTA in the α-chymotrypsin and proteinase K solutions to 400 m*M* (the tetra-sodium salt of EDTA must be used for this).

2. Plate washing: A number of techniques have been used to wash ELISA plates. In the method outlined here speed of washing is essential at the point when samples are being removed following a 30 min incubation time. Whereas plates can be rapidly immersed in a large bowl of washing buffer, this tends to leave the entire plate rather slippery from the detergent, making it difficult to handle thereafter. Automated plate-washers can be unreliable and are often rather slow. We have found the best method is to use a large reservoir (10 or 15 mL volume) with a tap at the bottom. The reservoir should be positioned above a sink (e.g., on a shelf top) and a flexible tube should be attached to the outlet and allowed to run down into the sink. When the tap is opened the washing buffer will run out onto the ELISA plate under the force of gravity. The flow-rate can be adjusted by squeezing the tube, either directly or using a clip.

3. Optimal antibody dilution: The anticollagen antibody must be used at an appropriate dilution. If it is too dilute then the MB value will be too low and inhibition will be difficult to detect. If it is too concentrated then higher concentrations of epitope will be needed to inhibit its binding to HDC on the ELISA plate. For optimal conditions, the antibody should be bound to HDC at a range of dilutions over a 30 min incubation time at ambient temperature. After washing the plate, bound antibody can be detected using the second antibody/alkaline phosphatase detection system. A plot of log-dilution against OD_{405nm} will approximate to a straight line. The dilution producing an OD_{405nm} that is about 40% of maximal OD is optimal for use in the assay. However note that when the antibody is added to the preculture plate it is diluted 1:1 with sample (or buffer) so it should be prepared initially at $2 \times$ the chosen concentration.

4. Sample dilutions: The appropriate sample dilutions will vary enormously depending on the type of cartilage, amount of tissue used, and extent of collagen degradation. But as a general guideline, the CE and DE samples may require no dilution, or they may need to be diluted to about 1:4. The CR samples will always require diluting, often by as much as 1:40 or 1:80. Clearly these values will be even more varied when other tissues are being extracted and assayed for their specific collagens.

5. Calculation of % denaturation: Total CB11B in the CE compartment represents the amount of denatured type II collagen. The digestion of CE with proteinase K to generate DE is a means of degrading any native type II collagen extracted intact by α-chymotrypsin, thus making it available for assay. The amount of CB11B in DE should be equal to or greater than CB11B in CE. Therefore total CB11B epitope in the cartilage = CB11B in DE + CB11B in CR. Calculate type II collagen degradation as:

$$\% \text{ Denaturation} = (\text{CB11B in CE/Total CB11B}) \times 100$$

6. Assaying collagen degradation *in situ* without antibodies: The use of hidden epitope antibodies in the method outlined here provides a high degree of specificity. However in some cases it may be adequate to measure the % of denatured

collagen without taking account of the collagen type. Such an approach would rely entirely on the selective proteolysis of denatured α-chains. The tissue should be extracted first with α-chymotrypsin and then with proteinase K, as described above, however there is no need to produce DE from CE in this case. The extracts should then be hydrolysed by mixing 1:1 with 12 *M* HCl and heating overnight at 110°C. The hydrolyzates can then be assayed for hydroxyproline either by a standard colourimetric method *(24,25)* or else by HPLC *(26)*. Hydroxyproline in the α-chymotrypsin extract should be expressed as a percentage of total hydroxyproline. Apart from the lack of specificity for a given collagen type, the main disadvantage of this approach is that the colorimetric assay is relatively insensitive whereas the HPLC assay is not suitable for the processing of large numbers of samples.

References

1. Kadler, K. (1994) Extracellular matrix 1: fibril-forming collagens. *Protein Profile* **1**, 519–635.
2. Eyre, D. R., Paz, M. A., and Gallop, P. M. (1984) Cross-linking in collagen and elastin. *Annu. Rev. Biochem.* **53**, 717–748.
3. Kucharz, E. J. (1992) *The Collagens: Biochemistry and Pathophysiology.* Springer-Verlag, Berlin.
4. Hamlin, C. R., Luschin, J. H., and Kohn, R. R. (1978) Partial characterization of the age-related stabilizing factor of post-mature human collagen-I. By the use of bacterial collagenase. *Exp. Geront.* **13**, 403–414.
5. Hamlin, C. R., Luschin, J. H., and Kohn, R. R. (1978) Partial characterization of the age-related stabilizing factor of post-mature human collagen-II. By the use of trypsin. *Exp. Geront.* **13**, 415–423.
6. Hui, K. Y., Haber, E., and Matsueda, G. R. (1983) Monoclonal antibodies to a synthetic fibrin-like peptide bind to human fibrin but not fibrinogen. *Science* **222**, 1129–1132.
7. Liu, X., Wu, H., Byrne, M., Jeffrey, J., Krane, S., and Jaenisch, R. (1995) A targeted mutation at the known collagenase cleavage site in mouse type I collagen impairs tissue remodelling. *J. Cell Biol.* **130**, 227–237.
8. Burleigh, M. C., Barrett, A. J., and Lazarus, G. S. (1974) Cathepsin B1. A lysosomal enzyme that degrades native collagen. *Biochem. J.* **137**, 387–398.
9. Kafienah, W., Brömme, D., Buttle, D. J., Croucher, L. J., and Hollander, A. P. (1998) Human cathepsin K cleaves native type I and II collagens at the N-terminal end of the triple helix. *Biochem. J.* **331**, 727–732.
10. Barrett, A. J. (1978) Capacity of leukocyte elastase and cathepsin G to degrade mature collagen fibers, In *Neutral Proteases of Human Polymorphonuclear Leukocytes* (Havemann, K., and Janoff, A., eds.), Urrban & Schwarzenberg, Inc., Baltimore, pp. 385–389.
11. Gross, J., Highberger, J. H., Johnson-Wint, B., and Biswas, D. (1980) Mode of action and regulation of tissue collagenase, in *Collagenase in Normal and Patho-*

logical Connective Tissues (Woolley, D. E., and Evanson, J. M., eds.) Wiley, Chichester, UK, pp. 11–35.

12. Netzel-Arnett, S., Fields, G., Birkedal-Hansen, H., and Van Wart, H. E. (1991) Sequence specificities of human fibroblast and neutrophil collagenases. *J. Biol. Chem.* **266**, 6747–6755.

13. Mitchell, P. G., Magna, H. A., Reeves, L. M., Lopresti-Morrow, L. L., Yocum, S. A., Rosner, P. J., Geoghegan, K. F., and Hambor, J. E. (1996) Cloning, expression and type II collagenolytic activity of matrix metalloproteinase-13 from human osteoarthritic cartilage. *J. Clin. Invest.* **97**, 761–768.

14. Brömme, D. and Okamoto, K. (1995) Human cathepsin O2, a novel cysteine protease highly expressed in osteoclastomas and ovary. Molecular cloning, sequencing and tissue distribution. *Biol. Chem. Hoppe-Seyler* **376**, 379–384.

15. Murphy, G. and Reynolds, J. J. (1993) Extracellular matrix degradation, in *Connective Tissue and Its Heritable Disorders. Molecular, Genetic and Medical Aspects* (Royce, P.M., and Steinmann, B., eds.), Wiley-Liss, Inc., New York, pp. 287–316.

16. Dodge, G. R. and Poole, A. R. (1989) Immunohistochemical detection and immunochemical analysis of type II collagen degradation in human normal, rheumatoid, and osteoarthritic articular cartilages and in explants of bovine articular cartilage cultured with interleukin-1. *J. Clin. Invest.* **83**, 647–661.

17. Hollander, A. P., Heathfield, T. F., Webber, C., Iwata, Y., Bourne, R., Rorabeck, C., and Poole, A. R. (1994) Increased damage to type II collagen in osteoarthritic articular cartilage detected by a new immunoassay. *J. Clin. Invest.* **93**, 1722–1732.

18. Hollander, A. P., Pidoux, I., Reiner, A., Rorabeck, C., Bourne, R., and Poole, A. R. (1995) Damage to type II collagen in aging and osteoarthritis starts at the articular surface, originates around chondrocytes, and extends into the cartilage with progressive degeneration. *J. Clin. Invest.* **96**, 2859–2869.

19. Rucklidge, G. J., Riddoch, G. I., and Robins, S. P. (1986) Immunocytochemical staining of rat tissues with antibodies to denatured type I collagen: a technique for localizing areas of collagen degradation. *Coll. Relat. Res.* **6**, 41–49.

20. Rucklidge, G. J., Milne, G., Riddoch, G. I., and Robins, S. P. (1986) Evidence for renal tubular resorption of collagen fragments from immunostaining of rat kidney with antibodies specific for denatured type I collagen. *Coll. Relat. Res.* **6**, 185–193.

21. Rajabi, M. R., Dodge, G. R., Solomon, S., and Poole, A. R. (1991) Immunochemical and immunohistochemical evidence of estrogen-mediated collagenolysis as a mechanism of cervical dilatation in the guinea pig at parturition. *Endocrinology* **128**, 371–378.

22. Hollander, A. P., Heathfield, T. F., Liu, J. J., Pidoux, I., Roughley, P.J., Mort, J. S., and Poole, A. R. (1996) Enhanced denaturation of the a1(II) chains of type-II collagen in normal adult human intervertebral discs compared with femoral articular cartilage. *J. Orthop. Res.* **14**, 61–66.

23. Antoniou, J., Steffen, T., Nelson, F., Winterbottom, N., Hollander, A. P., Poole, A. R., Aebi, M., and Alini, M. (1996) The human lumbar intervertebral disc. Evi-

dence for changes in the biosynthesis and denaturation of the extracellular matrix with growth, maturation, ageing and degeneration. *J. Clin. Invest.* **98**, 996–1003.

24. Stegmann, H. and Stadler, K. (1967) Determination of hydroxyproline. *Clin. Chim. Acta.* **18**, 267–273.

25. Reddy, G. K. and Enwemeka, C. S. (1996) A simplified method for the analysis of hydroxyproline in biological tissues. *Clin. Biochem.* **29**, 225–229.

26. Bank, R. A., Krikken, M., Beekman, B., Stoop, R., Maroudas, A., Lafeber, F. P. J. G., and Te Koppele, J. M. (1997) A simplified measurment of degraded collagen in tissues: application in healthy, fibrillated and osteoarthritic cartilage. *Matrix Biology* **16**, 233–243.

30

Invasion Assays and Matrix Metalloproteinases

Quantification of Cellular Invasion Using Propidium Iodide Labeling and Confocal Laser Scanning Microscopy

Ulrike Benbow, Kenneth A. Orndorff, and Alice L. Givan

1. Introduction

Matrix metalloproteinases are believed to play a major role in cellular invasion and tumor metastasis. Therefore, to understand the molecular mechanisms underlying invasion, a variety of in vitro assays have been developed. Most assays assess the interaction of tumor cells with the surrounding extracellular matrix *(1–4)*, although the invasive ability of other cell types could be measured. These methods involve cell invasion into a matrix coated on top of a filter. The layering of matrices requires precise thickness, because treatments are conventionally compared based on whether or not the cells have penetrated the matrix completely and appeared on the underside of the filter. In addition, until the development of methods using colorimetric, radiometric, or fluorescent dyes *(5–8)*, the counting of recovered cells from the lower chambers was slow and statistically uncertain *(9)*. More recently an improvement in the invasion assay using a confocal laser scanning system to obtain images of light reflected by cells within the matrix has been described *(10)*. In order to take full advantage of the ability of confocal laser scanning microscopy (CLSM) to examine specific fluorescence and to acquire images with high resolution, the technique presented here has modified the reflectance method by using propidium iodide (PI) and small pixel size to image nuclei in matrices on filters *(11–14)*.

Manual coating of filters with matrix material allows the use of a variety of matrices as well as easy adjustment of the matrix thickness. Cells of interest are seeded in the upper invasion chamber on the matrix (**Fig. 1A**). Che-

From: *Methods in Molecular Biology, vol. 151: Matrix Metalloproteinase Protocols*
Edited by: I. Clark © Humana Press Inc., Totowa, NJ

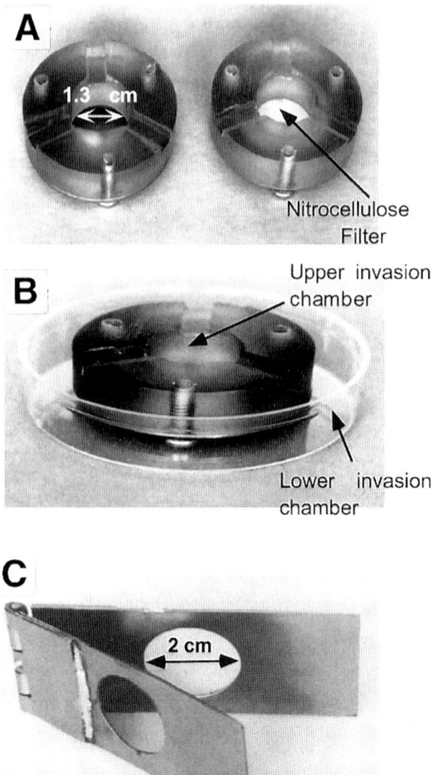

Fig. 1. Invasion chambers. Invasion chambers were custom-made using polysulfone, a heat and chemical resistant material. (**A**) The chambers have an outer diameter of 4 cm with an inner invasion chamber of 1.3 cm in diameter. A nitrocellulose filter (depicted in white) is inserted and held in place by a metal plate with the outside diameter of the chamber and a 1.3 cm opening. (**B**) The screws, holding the metal plate and filter in place, also act to elevate the invasion chamber when placed in a tissue culture dish, allowing free diffusion between the two compartments. (**C**) The holders for processing the filters are made from stainless steel. The filters are placed between the two open holes. A binder clip is used to hold the filters in place.

moattractants may be added to the lower chamber (**Fig. 1B**). For co-cultures, cells can be placed in the lower compartment. This allows free diffusion of factor(s) between the upper and lower invasion chambers without direct cell–cell contact. Invasion assays can be terminated at any given time point.

After incubation in the chambers, filters are removed and cells are fixed and treated with RNase followed by PI staining of nuclei. In the presence of RNase,

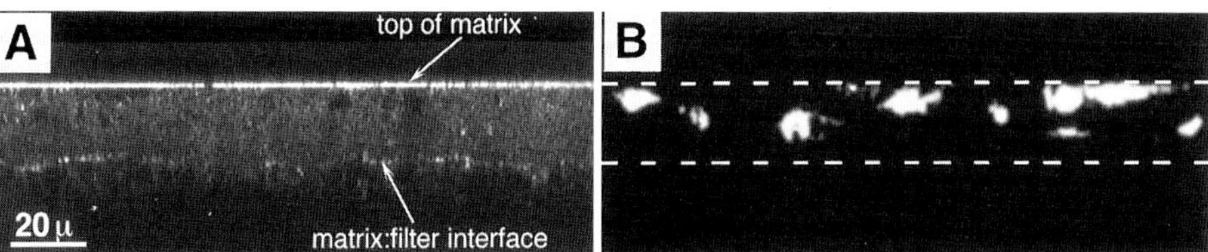

Fig. 2. X/Z image of reflectance and PI fluorescence of cells invading into Matrigel®. A2058 melanoma cells were seeded on Matrigel (20 μm thickness) and the x/z image was taken after 96 h. (**A**) is the reflectance image, indicating the high reflectance at the SlowFade®: matrix interface and the reflective granular texture of the Matrigel itself. (**B**) is the red fluorescence image of the same field, showing cell nuclei at different depths in the matrix.

the red fluorescence of PI is proportional to DNA content. A laser-scanning confocal system, in the reflection mode, is used to detect the top of the matrix *(14)* (**Fig. 2**). Images of red fluorescence from nuclei are then taken in 2 μm steps into the z-direction starting two steps above the top of the matrix (**Fig. 3A, B**). Software returns a value for the area of red fluorescence (above background) in each optical section. Results for the red fluorescent area in each section are calculated and expressed as a percent of the summed red fluorescent areas in all sections (**Fig. 3C**). In addition, the mean depth of the red fluorescence in the matrix can be calculated and this value can be used to describe the rate of cell migration under different conditions (**Fig. 4**).

2. Materials

2.1. Coating of Nitrocellulose Filters and Seeding of Cells

Diluted solutions of collagen and Matrigel® (Collaborative Biomedical Products; Bedford, MA) are stable at 4°C for up to 1 mo. The diluted collagen can be stored at –20°C for up to 4 mo. Before use, the collagen and Matrigel should be thawed at 4°C. Pipet tips and sterile water for coating are kept at 4°C. All steps are performed under sterile conditions in a tissue culture hood.

1. Serum-free medium, containing antibiotics (1% penicillin-streptomycin).
2. Chemoattractants in serum-free medium (if desired).
3. Collagens and Matrigel®: dilute in serum-free medium (*see* **Note 1**).
4. Gentamicin: dilute 1:100 in serum-free medium (prepare fresh before use).
5. Nitrocellulose filters, pore size 12 μm, diameter 25 mm (for invasion chambers with diameters of 1.3 cm).
6. Invasion chambers (*see* **Fig. 1**).
7. Tissue culture dishes (60 mm).
8. Forceps.

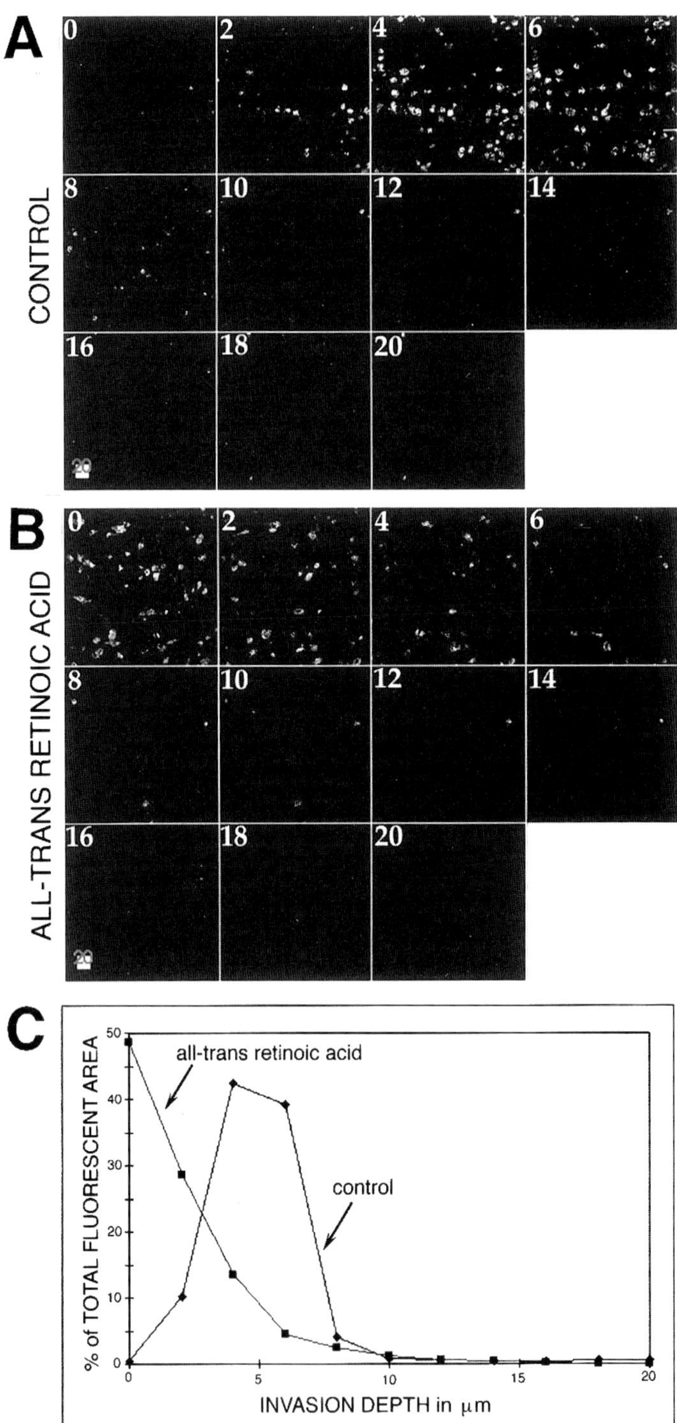

A
CONTROL
0 2 4 6
8 10 12 14
16 18 20
B
ALL-TRANS RETINOIC ACID
0 2 4 6
8 10 12 14
16 18 20
C
% of TOTAL FLUORESCENT AREA
all-trans retinoic acid
control
INVASION DEPTH in μm

2.2. Processing Filters for CLSM

1. 1XPBS: Phosphate-buffered saline, pH 7.5.
2. RNase A: Dilute to 1 mg/mL in 1 × PBS, pH 7.5.
3. Propidium iodide: Stock solution 5 mg/mL in dimethyl sulfoxide (DMSO). Store at −20°C. Working solution: 0.1–0.2 µg/mL, diluted with 10 mM Tris-HCl, pH 7.6, 150 mM NaCl.
4. Tissue culture dishes (60 mm).
5. 2-propanol.
6. Xylene.
7. Mounting medium such as SlowFade® or Canada Balsam (Molecular Probes, Eugene OR; Sigma, St. Louis, MO).
8. Filter holders (*see* **Fig. 1C**).
9. Stainless steel slide rack which fits filter holders.
10. Commercially available glass staining dishes for dehydration steps and xylene treatment.

3. Methods
3.1. Coating of Nitrocellulose Filters

The procedure and concentrations given below are for assays using the invasion chambers described in **Fig. 1A,B**. The coating is performed under sterile conditions in a tissue culture hood.

1. Fit filters into invasion chambers and autoclave (*see* **Note 2**).
2. Place chambers into 60 mm tissue culture dishes.
3. Wet filters with 200 µL of ice cold sterile water.
4. Remove water and pipett 150 µL of collagen or Matrigel solution onto the filter (*see* **Note 3**).
5. Cover dishes and place into a 37°C, CO_2 incubator for 30 min.
6. Remove and store uncovered at RT in a tissue culture hood for 30–45 min.
7. Repeat coating and drying two times (total of 450 µL).
8. Add the final coating of 450 µL collagen/Matrigel (total 900 µL, *see* **Note 3**).

Fig. 3. *(opposite page)* Optical sections of PI fluorescence of tumor cells invading a collagen type I matrix. MDA231 breast cancer cells were seeded on a type I collagen matrix (10 µm thickness) in the absence or presence of all-trans retinoic acid ($10^{-6} M$). Cells were incubated for 48 h and then the filters were processed and mounted with Canada Balsam. A montage of all optical sections is shown. The numbers correspond to the depth in µm taken from top of the matrix. (**A**) In untreated cultures, the majority of the nuclei are present between 4 and 6 µm. Only a few cells were able to invade into the nitrocellulose filter (below 10 µm). (**B**) In the presence of all-trans retinoic acid, the invasive potential of the tumor cells is greatly reduced with the majority of the nuclei located on top of the matrix and in the 2 µm section. Reprinted by permission of Wiley-Liss, Inc. *(14)*.

9. Incubate chambers at 37°C, in a CO_2 incubator overnight.
10. Remove plates, uncover and air dry for 40–60 min in a tissue culture hood.
11. Add 200 µL diluted gentamicin to the upper invasion chamber (1–3 min).
12. Remove gentamicin and place chambers into new 60 mm tissue culture dishes containing 5 mL of serum-free medium (*see* **Note 4**).
13. Seed cells at a density of $1–1.5 \times 10^5$ cells/mL into upper invasion chamber (1 mL total).
14. Cover and incubate for up to 96 h in a 37°C CO_2 incubator (*see* **Note 5**).

3.2. Processing Filters for CLSM

PI stains both DNA and double-stranded RNA *(15)*. This protocol, therefore, includes an RNase digestion step to reduce cytoplasmic background from RNA. These steps are carried out at room temperature unless otherwise indicated. Five mL of each solution is sufficient to cover the filter. Swirl dishes after each change of solution. A shaking platform is not essential.

1. Carefully remove filter from the invasion chamber with forceps. Place filter in a new tissue culture dish containing 5 mL $1 \times$ PBS for 5 min. Remove PBS.
2. Fix for 10 min in 2-propanol.
3. Wash in two changes of $1 \times$ PBS for 5 min each.
4. Incubate with RNase A (1 mg/mL in $1 \times$ PBS, pH 7.5) for 30 min (*see* **Note 6**).
5. Wash in $1 \times$ PBS for 5 min.
6. Incubate with PI (working solution) for 30–45 min.
7. Wash in H_2O for 5 min.
8. Place filters in filter holders and filter rack. Dehydrate in staining dishes successively using 30, 50, 75, 95, and 100% 2-propanol for 3 min each (*see* **Note 7**).
9. Place filter rack for 4 min in staining dish containing xylene, repeat three times in fresh xylene (*see* **Note 8**).
10. Mount filters on microscope slides (*see* **Note 9**).
11. Seal slides.
12. Image with CLSM (*see* **Note 10**).

3.3 Confocal Laser Scanning Microscopy

1. Slides are imaged with a confocal laser scanning microscope, using a 40X (NA 1.3) objective, 488 nm or 488 plus 568 nm excitation light, and detection of light signals with a quarter wave plate/polarizer filter for reflectance and a 605/32 nm band pass filter for PI fluorescence. The red detector is set to maximize the distinction between background red fluorescence and the positive signal from the nuclei (*see* **Note 10**).
2. The slide is scanned first at 0.18 µm steps in the z-direction (with a small confocal aperture) to ascertain the position of the transition between low and high reflectance, marking the top of the matrix layer (*see* **Note 11**).

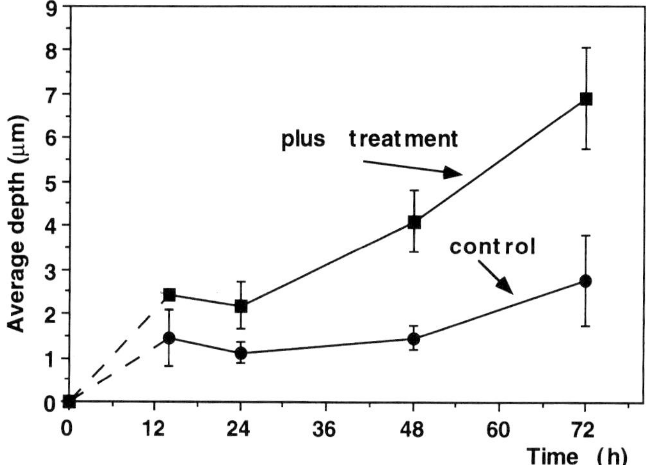

Fig. 4. Kinetics of A2058 melanoma cell invasion. Time-course data was used to determine the rate of migration of A2058 cells into a collagen type I matrix of 20 μm. The average depth of nuclei was calculated at each time point and plotted against time. Cells treated with conditioned medium (■) progressed at a nearly linear rate over 72 h when compared to control cells (●) in serum-free medium. Data are presented as ± standard error of quadruplicate determinations. Reprinted by permission of Wiley-Liss, Inc. *(14)*.

3. Sequential z-direction scans are begun at 2 steps (4 μm) above this transition position, and 512 × 512 pixel images (pixel size 0.47 μm) of the red fluorescence are acquired at z-step intervals of 2.0 μm until the filter layer is reached (*see* **Note 11**).
4. Image files of red fluorescence are stored for subsequent analysis.

3.4. Image Analysis and Quantification

1. Using image analysis software to analyze the saved 3-dimensional data files, a threshold level is set to exclude the background red fluorescence of the matrix material. This threshold level is then maintained throughout the analysis of all optical sections and all samples from a given experiment.
2. Each optical section is then analyzed for the total red fluorescent area (above background) in that section.
3. The fluorescent area above background for all sections is then summed and the area in each optical section is expressed as the % of this total summed area for all sections (**Fig. 3C**).
4. Knowing the z-position (in μm) of each fluorescence image (its distance from the high reflectivity interface at the start of the matrix), the % of total fluorescent area in each section can be plotted at known distances into the matrix (**Fig. 3C**).
5. By multiplying the % of total fluorescent area in each section by the z-distance (in μm) of that section, and then summing these values from all the sections and dividing the sum by 100, the mean distance of migration for cells can be calculated (**Fig. 4**).

4. Notes

1. We noted that commercially-available collagens, when diluted and stored, often form a precipitate. Therefore, before diluting, we routinely perform an overnight salt precipitation at 4°C, by adding 15% NaCl in 3% of 0.1 *M* acetic acid in a ratio of 1 vol of 15% NaCl to 3 vol of collagen stock. After centrifugation at 6,000*g* for 30 min at 4°C, discard the supernatant and resuspend the precipitate in 3% 0.1 *M* acetic acid. Dialyze against several changes of 20 m*M* Tris-HCl, pH 7.6, 200 m*M* NaCl, 0.02% sodium azide, at 4°C. The final concentration of matrix used for layering on the filter affects the thickness of the matrix. For example, at a concentration of 2 mg/mL, layering a volume of 1000 µL results in approx 20 µm of matrix thickness. For each batch of collagen or Matrigel®, the desired thickness should be confirmed using CLSM in the reflection mode.

2. The invasion chambers were custom-made using polysulfone, a material that is highly resistant to chemicals, heat, and steam.

3. Successive manual coating produces an even matrix surface. The thicker coating and the thickness of the nitrocellulose filter, when compared to commercially coated polycarbonate filters (less than 13 µm total), allows spatial separation of invading cells.

4. The tissue culture dish represents the lower invasion chamber. Chemoattractant agents may be added to the lower compartment. In addition, cells may be grown on the surface of the lower compartment for co-culture systems.

5. Assays can be stopped at any time point. However, we do not advise increasing the incubation time for more than 72–96 h, due to the possibility of contamination. One milliliter of serum-free culture medium should be added to the upper invasion chamber every 24 h to prevent the filters from drying.

6. If the background fluorescence is too high, the last 15 min of the incubation time in RNase should be carried out at 37°C.

7. At the end of dehydration the filters can be stored in 100% 2-propanol for up to 1 h.

8. This step is necessary for translucency of the otherwise opaque filters. Xylene is harmful with long exposure to humans and should be used under a hood. Xylene substitutes are now available that yield transparency similar to xylene.

9. Remove noncoated parts of the filter. Different mounting media can be applied. The authors prefer SlowFade (Molecular Probes, Eugene OR), due to the higher reflectance line obtained at the transition between the mounting medium and the matrix (**Fig. 2A**). However, it should be noted that slides mounted with SlowFade (a glycerol/water based medium) should be run on the confocal system within 4 h because the aqueous SlowFade® diffuses into the matrix and leads to loss of translucency and lack of a clear interface. In contrast, slides mounted with Canada Balsam are stable for weeks at 4°C. The drawback in using Canada Balsam is that this mounting medium does not give a strong reflectance line at the solution/matrix interface. It is, however, possible to use the bright granular reflectance of the matrix itself as orientation for the scanning start position.

10. In the authors' laboratory, a Bio-Rad MRC-1024 laser scanning confocal system is used for this assay, with a 15 mw krypton/argon ion laser and a Zeiss Axioskop upright microscope with a 40X (NA 1.3) Zeiss Plan Neofluar objective. The confocal iris is set at its minimal diameter (0.7 mm). The authors routinely use both the 488 nm and 568 nm lines for excitation; however, similar results have been obtained with just the 488 nm argon line. The filter for PI fluorescence in this Bio-Rad system is a 605/32 band pass filter. For reflectance, a quarter wave plate/polarizer filter is used to minimize internal lens reflection.

11. To determine the exact position of the transition between low and high reflectance it is important to scan in the z-direction at small steps. We recommend using steps between 0.18–0.5 µm for the reflectance scan to find the starting position and then switching to 2.0 µm steps to collect images of PI fluorescence within the matrix. Due to the possibility that some of the cells remain on top of the matrix we recommend starting the fluorescence scan two steps (4 µm) above the matrix surface.

Acknowledgments

The authors would like to thank Dr. Matthias P. Schoenermark for his contribution to this work and the generous gift of invasion chambers and filter holders. We also would like to thank Mr. Wilbur Clark for the construction of additional chambers and filter holders at the Central Facility at the Dartmouth Medical School, Hanover NH 03755. Finally, we would like to thank Dr. Constance E. Brinckerhoff for her helpful suggestions and careful reading of the manuscript. The Englert Cell Analysis Laboratory was established by equipment grants from the Fannie E. Rippel Foundation and is supported, in part, by the core grant (CA23108) to the Norris Cotton Cancer Center. This work was also supported by NIH grants (AR-26599 and NRSA 1F32-AR-08437) and a grant from the RGK Foundation (Austin, TX).

References

1. Boyden, S. V. (1962) The chemotactic effect of mixtures of antibody and antigen on polymorphonuclear leukocytes. *J. Exp. Med.* **115**, 453–466.

2. Mareel, M. M. (1983) Invasion in vitro. Methods of analysis. *Cancer Metastasis Rev.* **2**, 201–218.

3. Welch, D. R , Lobl, T. J., Seftor, E. A., Wack, P. J., Aeed, P.A., Yohem, K. H., and Seftor, R. E. (1989) Use of the Membrane Invasion Culture System (MICS) as a screen for anti-invasive agents. *Int. J. Cancer* **15**, 449–457.

4. Hendrix, M. J., Seftor, E. A., Seftor, R. E., Misiorowski, R. L., Saba, P. Z., Sundareshan, P., and Welch, D. R. (1989) Comparison of tumor cell invasion assays: human amnion versus reconstituted basement membrane barriers. *Invasion Metastasis* **9**, 278–297.

5. Gohla, A., Eckert, K., and Maurer, H. R. (1996) A rapid and sensitive fluorometric screening assay using YO-PRO-1 to quantify tumour cell invasion through Matrigel. *Clin. Exp. Metastasis* **14**, 451–458.

6. Saito, K., Oku, T., Ata, N., Miyashiro, H., Hattori, M., and Saiki, I. (1997) A modified and convenient method for assessing tumor cell invasion and migration and its application to screening for inhibitors. *Biol. Pharm. Bull.* **20**, 345–348.

7. Muir, D., Sukhu, L., Johnson, J., Lahorra, M. A., and Maria, B. L. (1993) Quantitative methods for scoring cell migration and invasion in filter-based assays. *Anal. Biochem.* **215**, 104–109.

8. Repesh, L. A. (1989) A new in vitro assay for quantitating tumor cell invasion. *Invasion Metastasis* **9**, 192–208.

9. Sieuwerts, A. M., Klijn, J. G. M., and Foekens, J. A. (1997) Assessment of the invasive potential of human gynecological tumor cell lines with the in vitro Boyden chamber assay: influences of the ability of cells to migrate through the filter membrane. *Clin. Exp. Metastasis* **15**, 53–62.

10. Schoenermark, M.P., Bock, O., Steinmeier, R., Benbow, U., and Lenarz, T. (1997) Quantification of tumor cell invasion using confocal laser scan microscopy. *Nat. Med.* **3**, 167–171.

11. Benbow, U., Schoenermark, M. P., Mitchell, T. I., Rutter, J. L., Shimokawa, K., Nagase, H., and Brinckerhoff, C. E. (1999) Host/tumor cell interaction activates matrix metalloproteinase 1 and mediates invasion through type I collagen. *J. Biol. Chem.* **274**, 25,371–25,378.

12. Benbow, U., Schoenermark, M. P., Orndorff, K. A., Givan, A. L., and Brinckerhoff, C. E. (1999) Human breast cancer cells "autoactivate" collagenase-1 and invade type I collagen: invasion is inhibited by all-trans retinoic acid. *Clin. Exp. Metastasis* **17**, 231–238.

13. Benbow, U., Maitra, R., Hamilton, J. W., and Brinckerhoff, C. E. (1999) Selective modulation of collagenase 1 gene expression by the chemotherapeutic agent doxorubicin. *Clin. Cancer Res.* **5**, 203–208.

14. Benbow, U., Orndorff, K. A., Brinckerhoff, C. E., and Givan, A. L. (2000) A confocal assay for invasion: the use of propidium iodide fluorescence and laser reflectane to quantify the rate of migration of cells through a matrix. *Cytometry* (in press).

15. Suzuki, T., Fujikura, K., Higashiyama, T., and Takata, K. (1997) DNA staining for fluorescence and laser confocal microscopy. *J. Histochem. Cytochem.* **45**, 49–53.

31

Using Fluorogenic Peptide Substrates to Assay Matrix Metalloproteinases

Gregg B. Fields

1. Introduction

1.1. Development of Peptide Substrates for Matrix Metalloproteinases

Catabolism of extracellular matrix (ECM) components has been ascribed to a family of Zn^{2+} metalloenzymes. These matrix metalloproteinases (MMPs; also termed matrixins) are believed to be important in connective tissue remodeling during development and wound healing. MMPs have also been implicated in a variety of disease states, including arthritis, glomerulonephritis, periodontal disease, tissue ulcerations, and tumor cell invasion and metastasis *(1–4)*. Because of their potential involvement in pathological conditions, much research effort has focused on designing substrates for MMP family members. As early as 1976, Nagai and associates used synthetic collagen-sequence peptides to study the specificity of tadpole skin collagenase *(5,6)*. Since those initial studies, a great deal of sequence specificity studies for MMP family members have been performed with peptides based on protein sequences surrounding MMP cleavage sites *(7)*. Results of these studies were subsequently used to design fluorogenic substrates for the MMPs.

1.1.1. Collagenases (MMP-1, -8, and -13) and Gelatinases (MMP-2 and -9)

Initial MMP peptidase studies focused on substrates designed from the MMP-1/MMP-8 cleavage site in the α1 chain of types I or III collagen. The sequence utilized, Pro-Gln/Leu-Gly~Ile-Ala-Gly, encompasses the substrate P_3 through P_3' subsites according to the nomenclature of Schechter and Berger *(8)*. Collagen sequence peptides were used initially to study tadpole MMP sub-

From: *Methods in Molecular Biology, vol. 151: Matrix Metalloproteinase Protocols*
Edited by: I. Clark © Humana Press Inc., Totowa, NJ

strate specificities *(5,6,9)*. The first mammalian MMP peptidase activities demonstrated were the cleavage of Dnp-Pro-Gln-Gly~Ile-Ala-Gly-Gln-D-Arg and Dnp-Pro-Leu-Gly~Ile-Ala-Gly-D-Arg by MMP-2 *(10)*, Dnp-Pro-Leu-Gly~Ile-Ala-Gly and Dnp-Pro-Leu-Gly~Ile-Ala-Gly-Arg-NH_2 by MMP-8 *(11)*, and several peptides containing Pro-Gln-Gly~Ile-Ala by MMP-9 *(11)*. Weingarten and associates used a series of peptides, peptolides, and peptide esters to explore certain aspects of MMP-1 sequence specificity *(12)*. The specificity at subsite P_1' was further examined using the substrate Ac-Pro-Leu-Gly~X-Leu-Gly-OC_2H_5 *(13)*. MMP-1 preference at P_1' was Leu > Ile > Val > Phe > Ala. These substrates were subsequently used to examine MMP-2 sequence specificities *(13–15)*. The MMP-2 P_1' subsite preferences were Leu > Ile ~ Phe >Val > Ala.

The $\alpha1(I)$ collagen sequence Gly-Pro-Gln-Gly~Ile-Ala-Gly-Gln was used as the starting substrate for a comprehensive study of MMP-1 and MMP-8 sequence specificities *(16,17)*. Comparison of the interstitial collagenases MMP-1 and MMP-8 showed similar, but not identical, specificities. For example, substitution of the P_1 subsite Gly with Glu enhanced MMP-8 activity but diminished MMP-1 activity, whereas substitution with His enhanced MMP-1 activity but diminished MMP-8 activity. Also, substitution of the P_1' subsite Ile with Tyr enhanced MMP-8 activity but diminished MMP-1 activity, while substitution of the P_3' subsite Gly with Met enhanced MMP-1 activity but diminished MMP-8 activity. The X-ray crystal structures of the catalytic domains of MMP-1 and MMP-8 revealed that MMP-8 has a larger S_1' binding pocket than MMP-1 *(18)* and that the Arg_{195} uniquely found in the S_1' pocket of MMP-1 makes aromatic residues at the P_1' site less favorable *(19,20)*. Rat collagenase *(21)*, which has higher sequence similarity to MMP-13 (collagenese 3) *(22)* than MMP-1 or MMP-8, had a very different P_1' subsite preference than either MMP-1 or MMP-8 *(13)*.

The collagen sequence-based substrates reported by Fields *(17)* were subsequently used to study the sequence specificities of MMP-2 and MMP-9 *(23)*. The general patterns for subsite preferences were more similar between the two gelatinases (MMP-2 and MMP-9) than those between the two collagenases (MMP-1 and MMP-8). Only four significant differences were seen: MMP-2 prefers Arg over Gln in subsite P_2 and Ser over Gly in subsite P_3' and MMP-9 prefers Glu over Ala in subsite P_2' and Thr over Gln in subsite P_4'.

Relative substrate specificities of MMP-1 and MMP-9 have been evaluated by "substrate mapping" in which a mixture of substrates is hydrolyzed by MMP-1 or MMP-9 and the products analyzed by reversed-phase high performance liquid chromatography (HPLC) coupled to fast-atom-bombardment mass spectrometry (FABMS) *(24,25)*. Comparison of the initial substrate mapping for MMP-1 using Dnp-Pro-Leu-Gly~Leu-Trp-Ala-D-Arg-NH_2 *(24)* with the earlier comprehensive MMP-1 study *(17)* showed good quantitative agreement with regard to the P_1' subsite. Comparison of the substrate mapping and comprehensive MMP-9 studies *(23,25)* fell into some similar patterns as the MMP-1 studies. Common trends of MMP-9 were seen at the P_2 subsite, where Leu was preferred over Val, Arg, or Asp. Both studies agreed with respect to the P_1 subsite, as no substitution provided significantly enhanced activity compared to Gly. In similar fashion, both studies agreed on the P_1' subsite, as the only natural amino acid that provided significant rate enhancement compared to Leu was Met. Substrate mapping found enhanced rates by substitution of Gly at the P_1' position with the unnatural amino acids *S*-mercaptoethylcysteine (Emc), norleucine (Nle), norvaline (Nva), $Cys(CH_2CH_3)$, $Cys(CH_3)$, and phenylglycine (Phg).

Triple-helical peptide (THP) models of the collagenase cleavage site in type I collagen have been constructed in order to further examine the substrate specificity of MMP family members. The $\alpha1(I)772\text{-}786$ THP was hydrolyzed by MMP-1, MMP-1$(\Delta_{243\text{-}450})$, MMP-2, and MMP-13 but not by MMP-3 or MMP-3$(\Delta_{248\text{-}460})$ *(26)*. Sequence and mass spectrometric analysis of the products generated by the MMPs indicated that MMP hydrolysis occurred primarily at the Gly-Ile bond, which is analogous to the bond cleaved in native type I collagen. Thus, the $\alpha1(I)772\text{-}786$ THP contained all necessary information to direct MMP-1, MMP-2, and MMP-13 binding and proteolysis. Similar results were reported by Ottl et al. *(27)*, who found that a heterotrimeric THP incorporating the $[\alpha1(I)]_2\alpha2(I)772\text{-}783$ sequence was efficiently hydrolyzed by MMP-8.

1.1.2. Stromelysin (MMP-3)

Also amongst the first mammalian MMP peptidase activities demonstrated was the cleavage of Dnp-Pro-Leu-Gly~Ile-Ala-Gly-Arg-NH_2 by MMP-3 *(28)*. The $\alpha1(I)$ collagen sequence Gly-Pro-Gln-Gly~Ile-Ala-Gly-Gln was used as the starting substrate for a comprehensive study of MMP-3 sequence specificity *(7,17)*. The MMP-3 comprehensive sequence specificity showed similar patterns to MMP-1 and MMP-8, which include preferences for: (1) a long chain (Leu, Met) or bulky, uncharged residue in subsite P_2, (2) Ala in subsite P_1, (3) Leu and aromatics (Phe, Trp, Tyr) in subsite P_1', and (4) aromatics (Phe and Trp), Leu, or Arg in subsite P_2'. The P_1' subsite was very sensitive to substitution. The differences in MMP-3 behavior compared to MMP-1 and MMP-8 were preferences for Val over Gln in subsite P_2, Leu over Gly in subsite P_1, Trp, or Phe over Ile in subsite P_1', and Arg over Gly in subsite P_3'. A different MMP-3 sequence specificity study has been performed using substance P sequence analogs *(29,30)*. The general trends of the two MMP-3 studies *(17,30)* were remarkably similar, especially for subsites P_3, P_2, P_1', and P_2'. The only significant differences at the P_1 subsite for the substance P study were Leu being less preferred than Gly and Phe being equivalent to Gln. The identification of active substrate sequences for MMP-3 has been achieved by screening of M13 phage peptide-display libraries *(31)*. The sequence information was used to create Ac-Pro-Phe-Glu~Leu-Arg-Ala-NH$_2$ (k_{cat}/K$_M$ = 126,000 s^{-1}M^{-1} for MMP-3) *(31)*.

1.1.3. Matrilysin (MMP-7)

The collagen sequence-based substrates reported by Fields *(17)* have been used to study the sequence specificity of MMP-7 *(23)*. The MMP-7 sequence specificity had many deviations from that of MMP-2 and MMP-9, and appeared to have peptidase activities closer to MMP-1 and MMP-8. Most unique for MMP-7 was the four-fold enhancement in activity by substitution of the P$_2$ subsite Gln to Met. The identification of active substrate sequences for MMP-7 has been achieved by screening of M13 phage peptide-display libraries *(31)*. The sequence information was used to develop Ac-Pro-Leu-Glu~Leu-Arg-Ala-NH$_2$ (k_{cat}/K$_M$ = 177,000 s^{-1}M^{-1} for MMP-7) *(31)*.

1.1.4. Caveats to MMP Sequence Specificity Studies

The studies of sequence specificities of MMP substrates show various similarities and differences. Discrepancies have suggested that there may be several caveats to be considered *(7)*. First, different modes of zymogen activation can create MMPs with different specific activities *(32–34)*. MMP peptide studies have used different activation methods, and thus discrepancies in relative activities may be the result of the activation method and the subsequent MMP species produced. Also, different MMP forms can exist due to the ability of these enzymes to autodegrade. Second, peptides used in the studies can themselves degrade over time, especially if stored in aqueous solution and repeatedly freeze-thawed *(35)*. For most studies, peptide integrity has been evaluated by mass spectrometry, and thus substrate degradation is not perceived to be a great problem. Third, the starting peptide "template" can influence the individual substitution results. Single substitutions within a substrate are often assumed to be independent (noninteractive), and hence the sum of the free energy changes from single substitutions are equal to that of a substrate possessing the multiple substitutions. Exceptions to the assumption of mutation additivity for enzymatic activity, and protein–protein interactions in general, have been recognized for some time *(36)*. The work of McGeehan et al. *(25)* demonstrated definitively that sequence specificity of MMP substrates can be "template dependent."

1.2. General Characteristics of Fluorogenic Substrates

Fluorogenic substrates provide a particularly convenient enzyme assay method, as they can be monitored continuously and utilized at reasonably low concentration ranges. There are three types of fluorogenic substrates: (1) aromatic amines, (2) contact-quenched, and (3) resonance energy transfer quenched *(37)*. MMP fluorogenic substrates have been developed using resonance energy transfer quenching, which are sometimes referred to as intramolecular fluorescence energy transfer substrates (IFETS) *(38)*.

1.2.1. Resonance Energy Transfer

Resonance energy transfer may occur between a fluorescent donor group and a quenching acceptor group when the fluorophore has a high quantum yield (Φ_F) and a fluorescence emission spectrum that is exactly overlapped by a strongly absorbing acceptor (quencher) *(37)*. The efficiency of transfer will also depend upon the distance between the donor and acceptor *(37,38)*. Resonance energy transfer quenched fluorogenic substrates are thus created by incorporating the donor and acceptor on opposite sides of the scissile bond, at a distance which allows for highly efficient energy transfer. When the substrate is cleaved by the enzyme, diffusion of the donor-containing substrate fragment away from the acceptor-containing substrate fragment results in loss of energy transfer and subsequent appearance of fluorescence. The creation of an optimal MMP fluorogenic substrate thus depends upon (1) incorporation of a fluorescent donor that has a high quantum yield, (2) incorporation of an acceptor that absorbs at the donor fluorescence emission wavelength (λ_{emiss}), and (3) the number of amino acid residues between the donor and acceptor.

1.2.2. Chemical Moieties Used for Donor/Acceptor Pairs

There is a wide variety of donor and acceptor groups that have been used in fluorogenic substrates *(37,38)*. MMP substrates have used one of five different fluorophores. The first was Trp *(39)*, which has $\varepsilon_{280} = 5600\ M^{-1}cm^{-1}$ and $\Phi_F = 0.2$. The quencher group was *N*-2,4-dinitrophenyl (Dnp) *(30,39)*, which has an absorption maximum at $\lambda = 363$ nm and a prominent shoulder at $\lambda = 410$ nm *(40)*. Alternately, a fluorogenic substrate for the *Leishmania* surface metalloproteinase has used the dansyl group to quench Trp fluorescence *(41)*. It should be noted that dansyl quenching of Trp decreases rapidly with intervening residue distance *(37)*. Knight and colleagues proposed the use of (7-methoxycoumarin-4-yl)acetyl (Mca) as a fluorophore for MMP substrates *(40)*. Mca has $\varepsilon_{325} = 14,500\ M^{-1}cm^{-1}$ and $\Phi_F = 0.49$. Mca is efficiently quenched by Dnp moieties *(40,42)*, as the shoulder in the Dnp absorption spectrum overlaps the Mca fluorescence emission spectrum *(40)*. Dnp is also used as a quencher for MMP substrates containing the fluorophore *N*-methylanthranilic acid (Nma) *(43)*. 5-[(2-Aminoethyl)amino]naphthalene-1-sulfonic acid (Edans) has been used as a donor group for an MMP substrate *(44)*. Edans has $\varepsilon_{336} = 5400\ M^{-1}cm^{-1}$ and $\Phi_F = 0.13$. Edans fluorescence is quenched by the 4-(4-dimethylaminophenylazo)benzoic acid (Dabcyl) group *(44)*. Lastly, MMP substrates have been developed with Lucifer Yellow (LY) as a fluorophore *(45,46)*. LY has ε_{420} = 12,000 $M^{-1}cm^{-1}$ and is quenched by 5-carboxytetramethylrhodamine (Ctmr) *(45)*.

1.2.3. Practical Aspects of Donor/Acceptor Groups

Trp can be incorporated easily by solid-phase peptide methodology and either Dnp or dansyl is acylated to the N-terminus of the peptide. Other moieties require additional synthetic steps. For example, to use the Mca/Dnp pair, a Dnp derivative, such as *N*-3-(2,4-dinitrophenyl)-L-2,3-diaminopropionic acid (Dpa) *(40)* or Lys(Dnp) *(42)*, must be synthesized (*see* below) and incorporated by solid-phase methods, while Mca is acylated to the peptide N-terminus. Alternatively, the Mca derivative L-2-amino-3-(7-methoxy-4-coumaryl) propionic acid (Amp) can be synthesized (*see* below) and incorporated by solid-phase methods *(47)*. For Dabcyl/Edans substrates, derivatives of Glu(Edans) must be prepared (*see* **Subheading 3.1.5.**). Substrates containing Nma can be synthesized using "on-resin" reactions to create Lys(Nma) moieties (*see* **Subheading 3.1.6.** and **3.1.7.***)*.

In contrast to solid-phase assembly of fluorogenic substrates, "post-synthesis" approaches have been used for the preparation of peptides containing the LY/Ctmr pair *(45,46)*. Although they are more laborious, post-synthesis approaches do alleviate one potential purification problem of fluorogenic peptides. Typically, fluorophores and quenchers are hydrophobic, and thus the reversed-phase (RP) HPLC separation of the desired peptide from deletion peptides containing these groups can be difficult.

1.3. Fluorogenic Substrates for Matrix Metalloproteinases

Stack and Gray developed the first fluorogenic MMP substrate, Dnp-Pro-Leu-Gly~Leu-Trp-Ala-D-Arg-NH$_2$ *(39)*. This substrate was hydrolyzed at the Gly-Leu bond by MMP-1 and MMP-2 *(39)* (**Table 1**). The activities of the full length and the C-terminal-truncated MMP-1 and MMP-3 have also been compared toward the substrate Dnp-Pro-Leu-Gly~Leu-Trp-Ala-D-Arg-NH$_2$ *(48)*. The k_{cat}/K_M values for MMP-1 hydrolysis of Dnp-Pro-Leu-Gly~Leu-Trp-Ala-D-Arg-NH$_2$ do not change significantly even if the enzyme lacks the C-terminal domain (**Table 1**). C-terminal truncation also has little effect on MMP-3 hydrolysis of this peptide (**Table 1**). MMP-1 and MMP-3 peptidase specificity appears to be determined by the catalytic domain, with only a small, if any, contribution from the C-terminal hemopexin-like domain.

The fluorogenic substrate Dnp-Pro-Leu-Gly~Leu-Trp-Ala-D-Arg-NH$_2$, when dissolved in trifluoroethanol, shows a significant change in structure (as monitored by circular dichroism spectroscopy) in the presence of Ca^{2+} *(49)*. These authors have suggested a role for Ca^{2+} in the binding and hydrolysis of substrate by MMP-1 *(49)*. In general, it has been noted that several peptide substrates of MMP-1 or MMP-2 bind Ca^{2+}, resulting in structural changes in the substrate *(49,50)*.

Combining the principle of Stack and Gray's substrate with MMP-1, MMP-3, and MMP-8 specificity studies, Fields proposed Dnp-Pro-Tyr-Ala~Leu-Trp-Ala-Arg-NH$_2$ as an MMP-1 substrate, Dnp-Pro-Tyr-Ala~Tyr-Trp-Met-Arg-NH$_2$ as an MMP-3 substrate, and Dnp-Pro-Leu-Ala~Tyr-Trp-Ala-Arg-NH$_2$ as an MMP-8 substrate *(17)*. In similar fashion, MMP-2 and MMP-9 specificity studies *(23)* were used to design the substrate Dnp-Pro-Leu-Gly~Met-Trp-Ser-Arg *(51)*. With the exception of MMP-8 substrates Dnp-Pro-Leu-Ala~Leu-Trp-Ala-Arg and Dnp-Pro-Leu-Ala~Tyr-Trp-Ala-Arg, these "optimized" substrates were not very active *(51)* (**Table 1**). Niedzwiecki et al. *(30)* developed Dnp-Arg-Pro-Lys-Pro-Leu-Ala~Nva-Trp-NH$_2$ and Dnp-Arg-Pro-Lys-Pro-Leu-Ala~Phe-Trp-NH$_2$ as MMP-3 substrates (**Table 1**).

Knight and colleagues designed Mca-Pro-Leu-Gly~Leu-Dpa-Ala-Arg-NH$_2$, which proved to be the best substrate for MMP-1, MMP-2, MMP-7, and MMP-9 *(40)* (**Table 1**). Subsequent studies have shown that this substrate is hydrolyzed by MMP-8, MMP-13, and C-terminal-truncated MMP-13 *(34,52)* (**Table 1**). As was the case in the prior study of MMP-1 and MMP-3 *(48)*, the k_{cat}/K_M values for MMP-13 hydrolysis of Mca-Pro-Leu-Gly~Leu-Dpa-Ala-Arg-NH$_2$ do not change significantly even if the enzyme lacks the C-terminal domain *(52)*. The Mca/Dpa pair was subsequently used for Mca-Pro-Cha-Gly~Nva-His-Ala-Dpa-NH$_2$ (where Cha is 3-cyclohexylalanine), which proved to be an efficient substrate for MMP-1, MMP-8, and MMP-13 *(34,52)* (**Table 1**).

Nagase et al. *(42)* designed several fluorogenic substrates based on the Mca/Dnp combination (NFF-1, NFF-2, and NFF-3), one of which that would be hydrolyzed by MMP-3 exclusively. NFF-3 [Mca-Arg-Pro-Lys-Pro-Val-Glu~Nva-Trp-Arg-Lys(Dnp)-NH$_2$] was hydrolyzed rapidly by MMP-3 ($k_{cat}/K_M = 218{,}000$ s$^{-1}M^{-1}$) and very slowly by MMP-9 ($k_{cat}/K_M = 10{,}100$ s$^{-1}M^{-1}$), but there was no significant hydrolysis by MMP-1, MMP-2, and MMP-7 *(7,42)* (**Table 1**). NFF-3 was the first documented synthetic substrate hydrolyzed by only certain members of the MMP family, and thus has important application for the discrimination of MMP-3 activity from that of other MMPs.

The fluorogenic substrate Dnp-Pro-Cha-Gly~Cys(CH$_3$)-His-Ala-Lys(Nma)-NH$_2$ was found to be hydrolyzed rapidly by the 19 kDa MMP-1 and MMP-9 *(43)* (**Table 1**). Dabcyl-Gaba-Pro-Gln-Gly~Leu-Glu(Edans)-Ala-Lys-NH$_2$, where Gaba is γ-amino-n-butyric acid) is a very good substrate for MMP-1, MMP-2, MMP-3, and MMP-9 *(44)* (**Table 1**). The combination of LY as fluorophore and Ctmr as quencher has been used for the assembly of the MMP-1 substrate LY-Gly-Pro-Leu-Gly~Leu-Arg-Ala-Lys(Ctmr) *(45)* (**Table 1**).

Table 1
Hydrolysis of Fluorogenic Substrates by MMPs

Substrate	$k_{cat}{}^a$ (s^{-1})	$K_M{}^a$ (μM)	$k_{cat}/K_M{}^a$ $(s^{-1}M^{-1})$	Reference
MMP-1				
Dnp-Pro-Gln-Gly~Ile-Ala-Gly-Trp	ND	ND	230^b	**51**
Dnp-Pro-Leu-Gly~Leu-Trp-Ala-D-Arg-NH$_2$	ND	ND	827	**48**
Dnp-Pro-Leu-Gly~Leu-Trp-Ala-D-Arg-NH$_2$	ND	ND	830	**40**
Mca-Arg-Pro-Lys-Pro-Tyr-Ala ~Nva-Trp-Met-Lys(Dnp)-NH$_2$ (NFF-2)	0.15	142	1060	**42**
Dnp-Pro-Leu-Gly~Leu-Trp-Ala-D-Arg-NH$_2$	0.011^c	7.1^c	1500^c	**39**
Mca-Pro-Lys-Pro-Gln-Gln ~Phe-Phe-Gly~Leu-Lys(Dnp)-Gly (NFF-1)	0.04	16.6	2510	**42**
Dnp-Pro-Leu-Ala~Tyr-Trp-Ala-Arg	0.13^b	50^b	2700^b	**51**
Dnp-Pro-Leu-Ala~Leu-Trp-Ala-Arg	1.2^b	130^b	9400^b	**51**
Mca-Pro-Leu-Gly~Leu-Dpa-Ala-Arg-NH$_2$	ND	ND	$12,100^d$	**34**
Mca-Pro-Cha-Gly~Nva-His-Ala-Dpa-NH$_2$	ND	ND	$14,700^d$	**34**
Mca-Pro-Leu-Gly~Leu-Dpa-Ala-Arg-NH$_2$	ND	ND	14,800	**40**
Dabcyl-Gaba-Pro-Gln-Gly ~Leu-Glu(Edans)-Ala-Lys-NH$_2$	ND	ND	21,000	**44**
LY-Gly-Pro-Leu-Gly~Leu-Arg-Ala-Lys(Ctmr)	ND	3.9	ND	**45**
MMP-1(Δ243-450)				
Dnp-Pro-Leu-Gly~Leu-Trp-Ala-D-Arg-NH$_2$	ND	ND	653	**48**
MMP-1 (19 kDa)				
Dnp-Pro-Leu-Gly~Leu-Trp-Ala-D-Arg-NH$_2$	ND	ND	3350^b	**43**
Dnp-Pro-Cha-Gly~Cys(Me)-His-Ala-Lys(Nma)-NH$_2$	ND	ND	$13,000^b$	**43**
MMP-1(1-242) + MMP-3(248-460)				
Dnp-Pro-Leu-Gly~Leu-Trp-Ala-D-Arg-NH$_2$	ND	ND	511	**48**
MMP-2				
Dnp-Pro-Gln-Gly~Ile-Ala-Gly-Trp	ND	ND	4200^b	**51**
Dnp-Pro-Leu-Gly~Met-Trp-Ser-Arg	0.069^b	7.8^b	8900^b	**51**
Mca-Pro-Lys-Pro-Gln-Gln ~Phe-Phe-Gly-Leu-Lys(Dnp)-Gly (NFF-1)	1.07	100	10,800	**42**
Mca-Arg-Pro-Lys-Pro-Tyr-Ala ~Nva-Trp-Met-Lys(Dnp)-NH$_2$ (NFF-2)	2.70	50	54,400	**42**
Dnp-Pro-Leu-Gly~Leu-Trp-Ala-D-Arg-NH$_2$	MND	ND	$58,000^d$	**40**
Dnp-Pro-Leu-Gly~Leu-Trp-Ala-D-Arg-NH$_2$	0.32^c	2.6^c	$122,000^c$	**39**

Substrate	$k_{cat}{}^a$ (s⁻¹)	$K_M{}^a$ (μM)	$k_{cat}/K_M{}^a$ (s⁻¹M⁻¹)	Reference
Dabcyl-Gaba-Pro-Gln-Gly ~Leu-Glu(Edans)-Ala-Lys-NH₂	ND	ND	619,000	**44**
Mca-Pro-Leu-Gly~Leu-Dpa-Ala-Arg-NH₂	ND	ND	629,000[d]	**40**
MMP-3				
Dnp-Pro-Gln-Gly~Ile-Ala-Gly-Trp	ND	ND	61[b]	**51**
Dnp-Pro-Leu-Gly~Leu-Trp-Ala-D-Arg-NH₂	ND	ND	2200	**48**
Dnp-Pro-Leu-Gly~Leu-Trp-Ala-D-Arg-NH₂	ND	ND	2200	**40**
Dnp-Pro-Tyr-Ala~Tyr-Trp-Met-Arg-NH₂	0.24[b]	100[b]	2400[b]	**51**
Mca-Pro-Lys-Pro-Gln-Gln ~Phe-Phe-Gly-Leu-Lys(Dnp)-Gly (NFF-1)	0.53	50	10,900	**42**
Dnp-Arg-Pro-Lys-Pro-Leu-Ala~Phe-Trp-NH₂	ND	ND	17,000[d]	**30**
Mca-Pro-Leu-Gly~Leu-Dpa-Ala-Arg-NH₂	ND	ND	23,000	**40**
Dabcyl-Gaba-Pro-Gln-Gly ~Leu-Glu(Edans)-Ala-Lys-NH₂	ND	ND	40,000	**44**
Mca-Arg-Pro-Lys-Pro-Tyr-Ala ~Nva-Trp-Met-Lys(Dnp)-NH₂ (NFF-2)	3.95	66	59,400	**42**
Dnp-Arg-Pro-Lys-Pro-Leu-Ala~Nva-Trp-NH₂	ND	ND	45,000[d]	**30**
Mca-Arg-Pro-Lys-Pro-Val-Glu ~Nva-Trp-Arg-Lys(Dnp)-NH₂ (NFF-3)	1.31[d]	20[d]	65,700[d]	**42**
Mca-Arg-Pro-Lys-Pro-Val-Glu ~Nva-Trp-Arg-Lys(Dnp)-NH₂ (NFF-3)	5.40	25	218,000	**42**
Mca-Arg-Pro-Lys-Pro-Val-Glu ~Nva-Trp-Arg-Lys(Dnp)-NH₂ (NFF-3)	2.20[e]	7.5[e]	302,000[e]	**42**
MMP-3(Δ248-460)				
Dnp-Pro-Leu-Gly~Leu-Trp-Ala-D-Arg-NH₂	ND	ND	2390	**48**
MMP-3(1-233) + MMP-1(229-450)				
Dnp-Pro-Leu-Gly~Leu-Trp-Ala-D-Arg-NH₂	ND	ND	2750	**48**
MMP-7				
Dnp-Pro-Leu-Gly~Leu-Trp-Ala-D-Arg-NH₂	ND	ND	11,700	**40**
Dnp-Pro-Leu-Gly~Leu-Trp-Ala-D-Arg-NH₂	ND	ND	13,000[f]	**63**
Mca-Pro-Leu-Gly~Leu-Dpa-Ala-Arg-NH₂	ND	ND	169,000	**40**
MMP-8				
Dnp-Pro-Gln-Gly~Ile-Ala-Gly-Trp	21.0[b]	1100[b]	19,000[b]	**51**
Dnp-Pro-Leu-Ala~Leu-Trp-Ala-Arg	9.4[b]	50[b]	190,000[b]	**51**
Dnp-Pro-Leu-Ala~Tyr-Trp-Ala-Arg	3.1[b]	7.7[b]	400,000[b]	**51**

(continued)

Table 1 *(continued)*

Substrate	$k_{cat}{}^a$ (s^{-1})	$K_M{}^a$ (μM)	$k_{cat}/K_M{}^a$ (s$^{-1}M^{-1}$)	Reference
MMP-8 (Met80 or Leu81 *N*-terminus)				
Mca-Pro-Leu-Gly~Leu-Dpa-Ala-Arg-NH$_2$	ND	ND	135,000[e]	**34**
Mca-Pro-Cha-Gly~Nva-His-Ala-Dpa-NH$_2$	ND	ND	166,000[e]	**34**
MMP-8 (Phe79 *N*-terminus)				
Mca-Pro-Leu-Gly~Leu-Dpa-Ala-Arg-NH$_2$	ND	ND	193,000[d]	**34**
Mca-Pro-Cha-Gly~Nva-His-Ala-Dpa-NH$_2$	ND	ND	252,000[d]	**34**
MMP-9				
Dnp-Pro-Gln-Gly~Ile-Ala-Gly-Trp	ND	ND	1200[b]	**51**
Mca-Arg-Pro-Lys-Pro-Val-Glu ~Nva-Trp-Arg-Lys(Dnp)-NH$_2$ (NFF-3)	0.282	28	10,100	**42**
Dnp-Pro-Leu-Gly~Met-Trp-Ser-Arg	0.069[b]	5.2[b]	13,000[b]	**51**
Mca-Arg-Pro-Lys-Pro-Tyr-Ala ~Nva-Trp-Met-Lys(Dnp)-NH$_2$ (NFF-2)	1.38	25	55,300	**42**
Dnp-Pro-Leu-Gly~Leu-Trp-Ala-D-Arg-NH$_2$	ND	ND	69,400[b]	**43**
Dnp-Pro-Cha-Gly~Cys(Me)-His-Ala-Lys(Nma)-NH$_2$	ND	ND	86,600[b]	**43**
Dabcyl-Gaba-Pro-Gln-Gly ~Leu-Glu(Edans)-Ala-Lys-NH$_2$	ND	ND	206,000	**44**
MMP-13				
Mca-Pro-Leu-Gly~Leu-Dpa-Ala-Arg-NH$_2$	ND	ND	757,000[d]	**34**
Mca-Pro-Cha-Gly~Nva-His-Ala-Dpa-NH$_2$	ND	ND	1,090,000[d]	**34**
MMP-13(Δ249-451)				
Mca-Pro-Leu-Gly~Leu-Dpa-Ala-Arg-NH$_2$	ND	ND	639,000[d]	**52**
Mca-Pro-Cha-Gly~Nva-His-Ala-Dpa-NH$_2$	ND	ND	1,000,000[d]	**52**

[a]Assays were performed at 37°C, pH 7.5, except were noted. ND = not determined.

[b]Assay performed at 23°C.

[c]Porcine, not human, MMP.

[d]Assay performed at 25°C.

[e]Assay performed at pH 6.0.

[f]Assay performed at pH 6.5.

2. Materials

2.1. Synthesis of Derivatives Used in Fluorogenic Substrates

1. Fluoren-9-ylmethoxycarbonyl (Fmoc) *N*-hydroxysuccinimide ester.
2. Dimethoxyethane.

3. Lys(Dnp).
4. 10% aqueous Na_2CO_3.
5. Concentrated HCl.
6. Ethyl acetate.
7. $CHCl_3$.
8. Silica gel 60 (Merck 9385).
9. $CHCl_3$-methanol-acetic acid: 9.25:0.5:0.25, v/v.
10. Fmoc-Gly.
11. 1-Hydroxybenzotriazole (HOBt).
12. 4-Dimethylaminopyridine (DMAP).
13. Dichloromethane (DCM).
14. *N,N*-dimethylformamide (DMF).
15. HMP resin.
16. *N,N'*-diisopropylcarbodiimide (DIPCDI).
17. Piperidine.
18. 4-(2',4'-dimethoxyphenylaminomethyl)phenoxy (DMPAMP).
19. Fmoc-DMPAMP resin.
20. Fmoc-Asn.
21. Pyridine.
22. (Bis[trifluoroacetoxy]iodo)benzene.
23. Diethyl ether.
24. Solid Na_2CO_3.
25. Ethanol.
26. 1-Fluoro-2,4-dinitrobenzene.
27. (2*R*)-Borane-10,2-sultam glycinate (Oxford Asymmetry, Abingdon, UK).
28. 4-bromomethyl-7-methoxycoumarin (Sigma-Aldrich, St. Louis, MO).
29. NBu_4HSO_4.
30. $LiOH{\cdot}H_2O$.
31. $MgSO_4$.
32. Tetrahydrofuran.
33. Edans (Sigma).
34. Boc-Glu-OBzl.

2.2. Synthesis of Fluorogenic Substrates

1. Benzotriazolyl *N*-oxytrisdimethylaminophosphonium hexafluorophosphate (BOP).
2. *N,N*-Diisopropylethylamine (DIEA).
3. Glacial acetic acid.
4. $KHSO_4$.
5. Benzotriazole-1-yl-oxy-tris-pyrrolidinophosphonium hexafluorophosphate (PyBOP).
6. *N*-methylmorpholine (NMM).
7. 7-methoxycoumarin-4-acetic acid.
8. 2-(1H-benzotriazol-1-yl)-1,1,3,3-tetramethyl uronium hexafluorophosphate (HBTU).

9. *p*-(*p*-dimethylaminophenylazo)benzoic acid (Sigma).
10. *N*-methylpyrrolidone (NMP).
11. Thiophenol
12. Succinimidyl *N*-methylanthranilate (Molecular Probes, Eugene, OR).
13. Hydrofluoric acid (HF).
14. Trifluoroacetic acid (TFA).
15. 0.025 *M* sodium phosphate, pH 7.0.
16. NaOH, 1 *M*.
17. $NaIO_4$: 0.04 *M* dissolved in 0.025 *M* sodium phosphate, pH 7.0 (final pH of the periodate solution, pH 6.7, not further adjusted).
18. Lucifer Yellow CH (Molecular Probes, Eugene, OR): 0.03 *M* in 0.1 *M* sodium acetate, pH 4.5.
19. 5-Carboxytetramethylrhodamine succinimidyl ester (Molecular Probes, Eugene, OR).
20. Acetonitrile.

2.3. Assaying Matrix Metalloproteinases Using Fluorogenic Substrates

1. Dimethylsulfoxide (DMSO).
2. Assay buffer A: 0.5 *M* Tris-HCl pH, 7.7, 5 m*M* $CaCl_2$, 0.2 *M* NaCl, up to 20% DMSO.
3. Assay buffer B: 0.1 *M* HEPES, 0.01 *M* $CaCl_2$, pH 7.5.
4. Assay buffer C: 50 m*M* Tris-HCl pH 7.5, 0.15 *M* NaCl, 10 m*M* $CaCl_2$, 0.05% Brij 35, 0.02% NaN_3.
5. Assay buffer D: 50 m*M* sodium acetate buffer, pH 6.0.
6. Assay buffer E: 0.1 *M* Tris-HCl, pH 7.5, 0.1 *M* NaCl, 10 m*M* $CaCl_2$, 0.05% Brij 35.
7. Assay buffer F: 50 m*M* Tris-HCl, pH 7.6, 150 m*M* NaCl, 5 m*M* $CaCl_2$, 1 m*M* $ZnCl_2$, 0.01% Brij 35.
8. Assay buffer G: 50 m*M* Tris-HCl, pH 7.6, 200 m*M* NaCl, 5 m*M* $CaCl_2$, 20 m*M* $ZnSO_4$, 0.05% Brij 35.
9. Assay buffer H: 0.05 *M* Tris-HCl, pH 7.7, 0.15 *M* NaCl, and 0.005 *M* $CaCl_2$.

3. Methods

3.1. The Synthesis of Derivatives Used in Fluorogenic Substrates

3.1.1. Synthesis of Fmoc-Lys(Dnp) (42)

1. Dissolve fluoren-9-ylmethoxycarbonyl (Fmoc) *N*-hydroxysuccinimide ester (1.89 g; 5.60 mmol) in 30 mL of dimethoxyethane and store at 4°C.
2. Dissolve Lys(Dnp) (1.63 g; 4.67 mmol) in 10 mL of 10% aqueous Na_2CO_3 and add slowly to the dimethoxyethane solution. Allow the reaction to proceed for 2 h at 4°C or overnight at room temperature.
3. Filter the solution and acidify the filtrate to pH ~3.0 with concentrated HCl.
4. Remove the dimethoxyethane by heating the solution under reduced pressure.
5. Extract the solution with ethyl acetate, and reduce the ethyl acetate layer by heating under reduced pressure to an oil.

6. The oil (Fmoc-Lys[Dnp]) can be used without further purification. Alternatively *(53)*, dissolve the oil in CHCl$_3$ (30 mL) and apply to a column (2 × 7 cm) of silica gel 60 (Merck 9385). Elute the product with 60 mL of CHCl$_3$-methanol-acetic acid (9.25:0.5:0.25, v/v).

7. Remove the solvents in vacuo leaving an oil that crystallizes upon trituration with light petroleum. Fmoc-Lys(Dnp) can be recrystallized as a dicyclohexylamine salt.

3.1.2. Fmoc-Lys(Dnp)-Gly-4-Hydroxymethylphenoxy (HMP) Resin (42)

1. Dissolve Fmoc-Gly (5.77 g; 19.4 mmol), 1-hydroxybenzotriazole (HOBt) (2.97 g; 19.4 mmol), and 4-dimethylaminopyridine (DMAP) (0.237 g; 1.94 mmol) in 100 mL DCM-*N,N*-dimethylformamide (DMF) (1:1) and add to 5.0 g HMP resin (4.85 mmol).

2. Add *N,N'*-diisopropylcarbodiimide (DIPCDI) (3.04 mL; 19.4 mmol), and allow esterification to proceed for 3 h.

3. Add an additional 200 mL of DMF, and allow the reaction to continue for 3.5 h.

4. Wash the resin with DMF and DCM and store under vacuum overnight. Fulvene-piperidine analysis *(35)* gives a substitution level of 0.45 mmol/g for Fmoc-Gly-HMP resin.

5. Deprotect 3.0 g of Fmoc-Gly-HMP resin (1.36 mmol) by treating with 50 mL piperidine-DMF (1:4) for 0.5 h and washing 3 times with DMF.

6. Dissolve Fmoc-Lys(Dnp) (0.232 g; 4.07 mmol) and HOBt (0.623 g; 4.07 mmol) in 50 mL of DCM-DMF (1:1) and add to the resin.

7. Add DIPCDI (0.637 mL; 4.07 mmol) and allow coupling to proceed for 3.5 h.

8. Wash the resultant Fmoc-Lys(Dnp)-Gly-HMP resin with DMF and DCM and store under vacuum.

The attempted synthesis of peptide analogues using *stored* Fmoc-Lys(Dnp)-Gly-HMP resin produces peptides which, by FABMS analysis, are missing the *C*-terminal Lys(Dnp)-Gly (–352.1 Da) *(42)*. Thus, the initial base treatment of the stored Fmoc-Lys(Dnp)-Gly-HMP resin causes diketopiperazine formation *(35)*. The diketopiperazine side-reaction can be avoided by incorporating Fmoc-Lys(Dnp) onto an amide-producing peptide-resin linker 4-(2',4'-dimethoxyphenylaminomethyl)phenoxy (DMPAMP).

3.1.3. Fmoc-Lys(Dnp)-DMPAMP Resin (42)

1. Deprotect 2.0 g of Fmoc-DMPAMP resin (0.86 mmol) by treatment with 20 mL piperidine-DMF (1:1) for 0.5 h and washing 3 times with DMF.

2. Dissolve Fmoc-Lys(Dnp) (0.2 g; 3.5 mmol) and HOBt (0.536 g; 3.5 mmol) in 20 mL of DCM-DMF (1:1) and add to the resin.

3. Add DIPCDI (0.548 mL; 3.5 mmol) and allow coupling to proceed for 4.5 h.

4. Wash the resultant Fmoc-Lys(Dnp)-DMPAMP resin with DMF and DCM and store under vacuum.

3.1.4. N²-*Fmoc*-N³-*Dnp-2,3-diaminopropionic Acid (Fmoc-Dpa)* **(40)**

1. Dissolve Fmoc-Asn (1.42 g, 4.0 mmol) in 10 mL DMF.
2. Add H_2O (2.0 mL), pyridine (0.75 mL, 8.0 mmol), and (bis[trifluoroace-toxy]iodo)benzene (2.60 g, 6.0 mmol), and stir the mixture at 20°C overnight.
3. Remove the solvents at 50°C in vacuo and dissolve the remaining oil in H_2O (100 mL) and concentrated HCl (10 mL).
4. Extract the solution twice with diethyl ether (50 mL each) and bring to pH 7 with solid Na_2CO_3.
5. Add $NaHCO_3$ (0.67 g, 8.0 mmol), ethanol (100 mL), and 1-fluoro-2,4-dinitroben-zene (0.5 mL, 4.0 mmol), and stir the mixture at 20°C for 2 h.
6. Remove ethanol at 40°C in vacuo and bring the aqueous solution to pH 1.0 with HCl.
7. Recrystallize the crude product (1.59 g, 81% yield) from hot ethanol/H_2O.

3.1.5. *Fmoc-L-Amp* **(54)**

1. Suspend (2*R*)-borane-10,2-sultam glycinate (Oxford Asymmetry, Abingdon) (1.89 g, 5.0 mmol), 4-bromomethyl-7-methoxycoumarin (Fluka) (1.62 g, 6.0 mmol), NBu_4HSO_4 (2.04 g, 6.0 mmol), and $LiOH·H_2O$ (0.84 g, 20.0 mmol) in DCM (45 mL) and H_2O (7.5 mL) and stir vigorously for 18 h at 4°C.
2. Filter the mixture, dilute with DCM (50 mL), extract twice with H_2O (50 mL each), and dry over $MgSO_4$.
3. Evaporate the DCM, and isolate the product, a coumarin-sultam adduct, as a yellow foam by flash chromatography on SiO_2 eluted with $CHCl_3$-methanol (98:2).
4. Dissolve the coumarin-sultam adduct in tetrahydrofuran (100 mL) and stir with 1 *M* HCl (50 mL) at room temperature. After 18 h, hydrolysis of the bis(methyl-thio)methylidene group is complete and aminoacyl-sultam is formed.
5. Filter the reaction mixture and evaporate solvents.
6. Add H_2O (50 mL) and evaporate. Wash the residual solid twice with diethyl ether (30 mL each).
7. Dissolve the aminoacyl-sultam in tetrahydrofuran (100 mL) and stir with 0.4 *M* $LiOH·H_2O$ in H_2O at room temperature for 1 h.
8. Evaporate the solvents to 50 mL, add H_2O (50 mL), and extract the solution twice with DCM (50 mL each).
9. Filter the aqueous phase, reduce in volume to 60 mL, and neutralize with 12 *M* HCl.
10. Stir the precipitate of L-2-amino-3-(7-methoxy-4-coumaryl)-propionic acid (L-Amp) at 0°C for 24 h and recover by filtration (1.06 g, 81% yield).
11. Prepare Fmoc-L-Amp by standard methods using Fmoc *N*-hydroxysuccinimide ester (*47*).

3.1.6. *Boc-Glu(Edans)-OH* **(55)**

1. Mix Edans (Sigma) (300 mg, 1.13 mmol), Boc-Glu-OBzl (381 mg, 1.13 mmol), and benzotriazolyl *N*-oxytrisdimethylaminophosphonium hexafluorophosphate (BOP) (489 mg, 1.13 mmol) with 10 mL of dry DMF.

2. Add *N,N*-diisopropylethylamine (DIEA) (392 µL, 2.25 mmol) to the suspension, and stir the mixture at room temperature under a nitrogen atmosphere for 4 h, yielding a clear solution.

3. Remove DMF by rotary evaporation under reduced pressure, and purify the crude residue twice by flash chromatography, once in methanol (saturated with NH_3)-DCM (1:4), followed by methanol-acetic acid-ethyl acetate (6:1:43) to yield 486 mg (71%) of Boc-Glu(Edans)-OBzl as a pale gold solid after lyophilization from H_2O.

4. Dissolve the product (~120 mg) in 30 mL methanol in a Parr bottle. Add Pd/C (10%; 33 mg) and shake the reaction in a Parr apparatus under an H_2 atmosphere at 40 psi for 8 h.

5. Remove the catalyst by filtration, and remove the solvent by evaporation under reduced pressure to yield Boc-Glu(Edans)-OH.

3.1.7. Fmoc-Glu(Edans)-OH (56)

1. Mix Edans (Sigma) (1.07 g, 4 mmol), Fmoc-Glu-O*t*Bu (1.70 g, 4 mmol), and BOP (1.77 g, 4 mmol) with 10 mL of dry DMF.

2. Add DIEA (2 mL, 12 mmol) to the suspension, and stir the mixture at room temperature under a nitrogen atmosphere for 2 h, yielding a clear solution.

3. Add DCM (50 mL) and 0.5 M $KHSO_4$ (50 mL), and wash the organic layer with 0.1 M $KHSO_4$ and H_2O, dry over $MgSO_4$, and evaporate *in vacuo* to yield 2.8 g of an oil.

4. Add glacial acetic acid (10 mL) and 36% HCl (1 mL), and stir the solution for 1 h.

5. Evaporate the solvents *in vacuo*, and remove traces of acetic acid by co-evaporation with DMF (2 × 10 mL).

6. Purify the product, in 45% yield, by C_{18} RP-HPLC.

3.2. The Synthesis of Fluorogenic Substrates

3.2.1. Substrates Containing Dnp, Trp, Edans, Mca, and/or Dabsyl

1. Incorporation of individual amino acids has typically been by Fmoc solid-phase methodology *(40,42,44,56)*, although the Boc strategy has been utilized for substrates containing Dnp and Trp *(39)*. It is recommended that for Trp-containing substrates the Fmoc-Trp(Boc) derivative be utilized *(57)*. Dnp is incorporated at the N-terminus of the peptide by using the appropriate, commercially available Dnp-amino acid derivative for the final coupling *(43)*.

2. Couple Fmoc-Glu(Edans)-OH using a 10-fold excess with benzotriazole-1-yl-oxy-tris-pyrrolidinophosphonium hexafluorophosphate (PyBOP)/*N*-methylmorpholine (NMM) for 3 h *(56)*.

3. Peptide-resins may be analyzed prior to N-terminal acylation by Edman degradation analysis to evaluate the efficiency of assembly *(42)*.

4. Acylate the N-termini of peptide-resins with 7-methoxycoumarin-4-acetic acid using standard synthesis cycles with 2-(1H-benzotriazol-1-yl)-1,1,3,3-tetramethyl uronium hexafluorophosphate (HBTU) and HOBt and 2–4 h coupling times *(42)* or

multiple couplings with PyBOP and HOBt *(40)*. *N*-terminal acylation with Dabcyl is performed using 10 equiv *p*-(*p*-dimethylaminophenylazo)benzoic acid (Sigma), 10 equiv PyBOP, and 20 equiv NMM in 250 μL *N*-methylpyrrolidone (NMP) for 3 h *(56)*.

3.2.2. Substrates Containing Lys(Nma)

To create a fluorogenic substrate containing Lys(Nma), the peptide is assembled by Boc chemistry using a Boc-Lys(Fmoc) derivative *(43)*.

1. Treat the peptide-resin with thiophenol (2 mL) in NMP (15 mL) for 1 h to remove Dnp side-chain protection from His if present (**Note**: this treatment will *not* remove an N^α-Dnp group).
2. Wash the peptide-resin with DCM (6 × 20 mL) and remove Fmoc side-chain protection with two treatments of piperidine (2 mL) in NMP (15 mL) for 20 min each.
3. React the resin with succinimidyl *N*-methylanthranilate (Molecular Probes, Eugene, OR) (1 mmol) in NMP (15 mL) with DIEA (1.1 mmol) for 16 h.
4. Wash the resin with DCM (6 × 20 mL) and dry.
   This substrate could also be assembled by Fmoc chemistry, if an orthogonal Lys side-chain protecting group such as allyloxycarbonyl (Aloc) or 1-(4,4-dimethyl-2,6-dioxocyclohexylidene)ethyl (Dde) is utilized *(35,58,59)*.
5. Cleave peptides from the resin and deprotect the side-chains by treatment with HF plus appropriate scavengers *(39,43)* or TFA plus appropriate scavengers *(57,60)*.
6. Purify peptides by preparative C_{18} RP-HPLC *(40,42,43)*.
7. Characterize peptides by electrospray mass spectrometry *(42)*, FABMS *(42,43)*, or matrix-assisted laser desorption/ionization mass spectrometry *(44,56)*.

3.2.3. Substrates Containing LY and Ctmr

When the combination of LY and Ctmr are used for a substrate, both the fluorophore and quencher are incorporated after peptide synthesis and purification *(45,46)*. The peptide is synthesized with an N-terminal Ser, which is converted by periodate oxidation to an N^α-glyoxylyl group as follows.

1. Dissolve peptide (7.7 μmol) in 0.25 mL of 0.025 *M* sodium phosphate, pH 7.0.
2. Titrate the resulting solution with 1 mL drops of 1 *M* NaOH until the pH is near 7.0.
3. Treat the solution with 0.25 mL of 0.04 *M* $NaIO_4$ dissolved in 0.025 *M* sodium phosphate, pH 7.0 (final pH of the periodate solution, pH 6.7, not further adjusted), to create a solution at 22°C containing 0.02 *M* $NaIO_4$ and 0.015 *M* peptide. After 15 min the peptide should be converted to the N^α-glyoxylyl form.
4. Treat the solution with 1 mL of 0.03 *M* Lucifer Yellow CH (Molecular Probes, Eugene, OR) in 0.1 *M* sodium acetate, pH 4.5, and incubate the resulting 1.5 mL of reaction mixture at 22°C for 2 h.
5. The glyoxylyl group forms a hydrazone with the carbohydrazide LY. Isolate the product by C_{18} RP-HPLC.
6. React the free N^ε-amino group of Lys with 5-carboxytetramethylrhodamine succinimidyl ester as follows. Redissolve the LY-peptide (~11 mg) in 2.25 mL of

0.1 M NaHCO$_3$ and treat with 0.75 mL of 5-carboxytetramethylrhodamine succinimidyl ester (Molecular Probes, Eugene, OR) prepared as a 20 mM suspension in acetonitrile.

7. After 1 h at 22°C, add 0.15 mL of the suspension of 5-carboxytetramethylrhodamine succinimidyl ester to drive the reaction toward completion. Purify the LY-peptide-Ctmr product by C$_{18}$ RP-HPLC.

3.3. Assaying Matrix Metalloproteinases Using Fluorogenic Substrates (see Note 1)

3.3.1. Dnp-Pro-Leu-Gly~Leu-Trp-Ala-D-Arg-NH$_2$ (39)

1. Prepare the substrate as a stock solution in dimethylsulfoxide (DMSO) and determine the concentration spectrophotometrically using ε_{372} = 16,000 M^{-1}cm^{-1} (*see* **Note 2**).
2. Perform assays in assay buffe A, at 37°C using peptide concentrations ranging from 2.5 to 40 mM.
3. Add enzyme, and determine the initial rate of substrate hydrolysis at λ_{excit} = 280 nm and λ_{emiss} = 346 nm.
4. Calibrate the fluorimeter by adding aliquots of a standard solution of Trp directly to the reaction mixture at the conclusion of each run. Fluorescent intensities are not corrected for the inner filter effect as the maximum correction factor is less than 8% at the highest substrate concentration.

3.3.2. Dnp-Arg-Pro-Lys-Pro-Leu-Ala~Nva-Trp-NH$_2$ and Dnp-Arg-Pro-Lys-Pro-Leu-Ala~Phe-Trp-NH$_2$ (30)

1. Prepare substrates as stock solutions in DMSO.
2. Perform assays using 2.96 mL of assay buffer B and 0.020 mL substrate in a 3 mL cuvet at 25°C.
3. Initiate the reaction by the addition of 0.020 mL of enzyme solution.
4. Monitor reaction progress at λ_{excit} = 290 nm and λ_{emiss} = 340 nm. Values of k$_{cat}$/K$_M$ are calculated from first-order progress curves for fluorescence increase with [S]$_O$ = 2 μM << K$_M$.

3.3.3. NFF-1, NFF-2, and NFF-3 (42)

1. Prepare substrates as 10 mM stock solutions in DMSO.
2. Perform fluorescent assays at λ_{excit} = 325 nm and λ_{emiss} = 393 nm.
3. Run initial total hydrolysis assays at a substrate concentration of 5 μM to avoid filtering effects. The change in fluorescence at this concentration is 110 based on total hydrolysis of 1 μM substrate.
4. Incubate 100 μL of various concentrations of a substrate with 10 μL of an enzyme solution (2–20 nM) in assay buffer C at 37°C. (MMP-3 activity at pH 6.0 is measured in assay buffer D, instead of Tris-HCl buffer).
5. Stop the reaction by addition of 900 μL of 3% (v/v) glacial acetic acid.

The amount of substrate hydrolysis is calculated based on the fluorescence values of the Mca-Arg-Pro-Lys-Pro-Gln standard solution after subtraction of the reaction blank value (stopping solution added before the enzyme). Individual kinetic parameters (k_{cat} and K_M) are determined over a substrate concentration range of 2.5–75 mM and calculated by double reciprocal plots.

3.3.4. Mca-Pro-Leu-Gly~Leu-Dpa-Ala-Arg-NH$_2$ (37,40)

1. Prepare substrate as a stock solution in DMSO and determine the concentration spectrophotometrically using ε_{410} = 7500 M^{-1}cm^{-1}.
2. Perform fluorescent assays at at λ_{excit} = 328 nm and λ_{emiss} = 393 nm.
3. Zero the fluorimeter with substrate in assay buffer E and then calibrate with Mca-Pro-Leu so that the full scale deflection corresponds to between 2 and 10% hydrolysis of the substrate.

    For each MMP, the initial rate of substrate cleavage, measured over 10–15 min, was proportional to substrate concentration in the range of 1–8 μM. A concentration of 1.6 μM is used to determine k_{cat}/K_M values. Absorptive quenching at higher substrate concentrations can occur, and it is thus important to calibrate at each substrate concentration.
4. To determine the extent of quenching, zero the fluorimeter with buffer in the cuvet and measure the fluorescence of substrate solutions of increasing concentration.
5. Plot the values versus substrate concentration and extrapolate the initial linear portion of the curve. The difference between the extrapolated and experimental points at any concentration shows the extent of quenching.
    Mca-Pro-Leu-Gly~Leu-Dpa-Ala-Arg-NH$_2$ has also been used for flow injection analysis of MMP-7 activity and inhibitors (*61*).

3.3.5. Dabcyl-Gaba-Pro-Gln-Gly~Leu-Glu(Edans)-Ala-Lys-NH$_2$ (44)

1. Perform fluorescent assays at λ_{excit} = 340 nm and λ_{emiss} = 485 nm.
2. Measure purified MMP activity at enzyme concentrations of 1.5–25 nM with 1.8 μM substrate in assay buffer F at 37°C. Crude MMP activity is measured using 5 μM substrate in a total volume of 150 mL assay buffer at 37°C.

3.3.6. Dnp-Pro-Cha-Gly~Cys(CH$_3$)-His-Ala-Lys(Nma)-NH$_2$ (43)

1. Prepare substrate as a 5 mM stock solution in DMSO with the concentration determined spectrophotometrically using ε_{372} = 16,000 M^{-1}cm^{-1}.
2. Perform fluorescent assays at λ_{excit} = 340 nm and λ_{emiss} = 440 nm.
3. Determine the k_{cat}/K_M values at a substrate concentration of 0.5 μM in a quartz cuvette containing 3 mL of assay buffer G at 23°C.
4. For microtiter plates, conduct assays in a total volume of 0.3 mL assay buffer containing 3 nM MMP-1 in each well of a black 96-well plate.
5. Initiate assays by substrate addition (10 μM final concentration) and measure the product formation at λ_{excit} = 365 nm and λ_{emiss} = 450 nm after 40–60 min.

3.3.7. LY-Gly-Pro-Leu-Gly~Leu-Arg-Ala-Lys(Ctmr) **(45)**

1. Measure MMP activity at room temperature by monitoring the change in fluorescence of a sample containing various amounts of substrate (0.1–4 μM) and enzyme dissolved in assay buffer H.
2. Excite the fluorescent substrate at 430 nm (10-nm slit) and read emission at 530 nm (5-nm slit).
   Filtering effects of LY and fluorescence of Ctmr may need to be considered at higher substrate concentrations *(45,46)*.

4. Notes

1. The fluorophores exhibiting greater quantum yields allow for the most sensitive assays, provided that quenching is reasonably efficient. On this basis, substrates using Mca are preferred over Trp, Nma, or Edans *(37)*. Trp is the most problematic fluorophore, primarily because assays could suffer from high background due to MMP or other protein internal Trp residues. The fluorogenic substrate Dnp-Pro-Cha-Gly~Cys(CH$_3$)-His-Ala-Lys(Nma)-NH$_2$ was developed to be compatible with commercial 96-well plate reading systems to allow for rapid inhibitor screening *(43)*. Mca-Pro-Leu-Gly~Leu-Dpa-Ala-Arg-NH$_2$ has been used for high-density screening of a combinatorial library of potential MMP-7 inhibitors *(62)*.
2. Solubility in aqueous buffer is a problem for many fluorogenic substrate due to the hydrophobic nature of fluorescent donors and acceptors. The solubility of the Dabcyl-Gaba-Pro-Gln-Gly~Leu-Glu(Edans)-Ala-Lys-NH$_2$ substrate in assay buffer (50 mM Tris, 150 mM NaCl, 5 mM CaCl$_2$, 1 μM ZnCl$_2$, 0.01% Brij 35, pH 7.6) is 80 mM, while the solubilities of Mca-Pro-Leu-Gly~Leu-Dpa-Ala-Arg-NH$_2$ and Dnp-Pro-Cha-Gly~Cys(CH$_3$)-His-Ala-Lys(Nma)-NH$_2$ in the same buffer are 4.3 and 4.6 μM, respectively *(44)*. NFF-3 is soluble up to a concentration of 250 μM using 2.5% DMSO in water *(42)*, which appears to be better than the solubility of Mca-Pro-Leu-Gly~Leu-Dpa-Ala-Arg-NH$_2$ *(40)*.

Acknowledgments

I gratefully acknowledge support of my laboratory's research by NIH grants CA77402 and AR01929 and Pfizer Central Research.

References

1. Woessner, J. F. (1991) Matrix metalloproteinases and their inhibitors in connective tissue remodeling. *FASEB J.* **5**, 2145–2154.
2. Birkedal-Hansen, H., Moore, W. G. I., Bodden, M. K., Windsor, L. J., Birkedal-Hansen, B., DeCarlo, A., and Engler, J. A. (1993) Matrix metalloproteinases: a review. *Crit. Rev. Oral Biol. Med.* **4**, 197–250.
3. Stetler-Stevenson, W. G., Aznavoorian, S., and Liotta, L. A. (1993) Tumor cell interaction with the extracellular matrix during invasion and metastasis. *Ann. Rev. Cell Biol.* **9**, 541–573.

4. Nagase, H. (1996) Matrix metalloproteinases in *Zinc Metalloproteases in Health and Disease* (Hooper, N. M., ed.), Taylor & Francis, London, UK, pp.153–204.

5. Nagai, Y., Masui, Y., and Sakakibara, S. (1976) Substrate specificity of vertebrate collagenase. *Biochim. Biophys. Acta.* **445**, 521–524.

6. Masui, Y., Takemoto, T., Sakakibara, S., Hori, H., and Nagai, Y. (1977) Synthetic substrates for vertebrate collagenase. *Biochem. Med.* **17**, 215–221.

7. Nagase, H. and Fields, G. B. (1996) Human matrix metalloproteinase specificity studies using collagen sequence-based synthetic peptides. *Biopolymers* **40**, 399–416.

8. Schechter, I. and Berger, A. (1967) On the size of the active site in proteases 1: papain. *Biochem. Biophys. Res. Commun.* **27**, 159–162.

9. Gray, R. D. and Saneii, H. H. (1982) Characterization of vertebrate collagenase activity by high-performance liquid chromatography using a synthetic substrate. *Anal. Biochem.* **120**, 339–346.

10. Seltzer, J. L., Adams, S. A., Grant, G. A., and Eisen, A. Z. (1981) Purification and properties of a gelatin-specific neutral protease from human skin. *J. Biol. Chem.* **256**, 4662–4668.

11. Williams, H. R. and Lin, T. Y. (1984) Human polymorphonuclear leukocyte collagenase and gelatinase. *Int. J. Biochem.,* **16,** 1321–1329.

12. Weingarten, H., Martin, R., and Feder, J. (1985) Synthetic substrates of vertebrate collagenase. *Biochemistry* **24**, 6730–6734.

13. Weingarten, H. and Feder, J. (1986) Cleavage site specificty of vertebrate collagenase. *Biochem. Biophys. Res. Commun.* **139**, 1184–1187.

14. Seltzer, J. L., Weingarten, H., Akers, K. T., Eschbach, M. L., Grant, G. A., and Eisen, A. S. (1989) Cleavage specificity of type IV collagenase (gelatinase) from human skin. *J. Biol. Chem.* **264**, 19,583–19,586.

15. Seltzer, J. L., Akers, K. T., Weingarten, H., Grant, G. A., McCourt, D. W., and Eisen, A. Z. (1990) Cleavage specificity of human skin type IV collagenase (gelatinase). *J. Biol. Chem.* **265**, 20,409–20,413.

16. Fields, G. B., Van Wart, H. E., and Birkedal-Hansen, H. (1987) Sequence specificity of human skin fibroblast collagenase: Evidence for the role of collagen structure in determining the collagenase cleavage site. *J. Biol. Chem.* **262**, 6221–6226.

17. Fields, G. B. (1988) The application of solid phase peptide synthesis to the study of structure-function relationships in the collagen-collagenase system. (Ph.D. Thesis). Florida State University, Tallahassee, FL.

18. Stams, T., Spurlino, J. C., Smith, D. L., Wahl, R. C., Ho, T. F., Qoronfleh, M. W., Banks, T. M., and Rubin, B. (1994) Structure of human neutrophil collagenase reveals large S1′ specificity pocket. *Nature Struct. Biol.* **1**, 119–123.

19. Lovejoy, B., Hassell, A. M., Luther, M. A., Weigl, D., and Jordan, S. R. (1994) Crystal structures of recombinant 19-kDa human fibroblast collagenase complexed to itself. *Biochemistry* **33**, 8207–8217.

20. Grams, F., Reinemer, P., Powers, J. C., Kleine, T., Pieper, M., Tschesche, H., Huber, R., and Bode, W. (1995) X-ray structures of human neutrophil collagenase complexed with peptide hydroxamate and peptide thiol inhibitors: implications for substrate-binding and rational drug design. *Eur. J. Biochem.* **228**, 830–841.

21. Quinn, C. O., Scott, D. K., Brinckerhoff, C. E., Matrisian, L. M., Jeffrey, J. J., and Partridge, N. C. (1990) Rat collagenase: cloning, amino acid sequence comparison and parathyroid hormone regulation in osteoblastic cells. *J. Biol. Chem.* **265**, 22,342–22,347.

22. Freije, J. M. P., Diez-Itza, T., Balbin, M., Sanchez, L. M., Blasco, R., Tolivia, J., and López-Otín, C. (1994) Molecular cloning and expression of collagenase-3, a novel human matrix metalloproteinase produced by breast carcinomas. *J. Biol. Chem.* **69**, 16,766–16,773.

23. Netzel Arnett, S., Sang, Q. X., Moore, W. G. I., Navre, M., Birkedal-Hansen, B., and Van Wart, H. E. (1993) Comparative sequence specificities of human 72- and 92-kDa gelatinases (type IV collagenases) and PUMP (matrilysin). *Biochemistry* **32**, 6427–6432.

24. Berman, J., Green, M., Sugg, E., Anderegg, R., Millington, D. S., Norwood, D. L., McGeehan, G., and Wiseman, J. (1992) Rapid optimization of enzyme substrates using defined substrate mixtures. *J. Biol. Chem.* **267**, 1434–1437.

25. McGeehan, G. M., Bickett, D. M., Green, M., Kassel, D., Wiseman, J. S., and Berman, J. (1994) Characterization of the peptide substrate specificities of interstitial collagenase and 92-kDa gelatinase. *J. Biol. Chem.* **269**, 32,814–32,820.

26. Tuzinski, K. A., Nagase, H., and Fields, G. B. (1998) Matrix metalloproteinase, hydrolysis of triple-helical peptide models of interstitial collagens in *Peptides* (Tam, J. P. and Kaumaya, P. T. P., eds.), Kluwer Academic Publishers, Dordrecht, The Netherlands (In press).

27. Ottl, J., Battistuta, R., Pieper, M., Tschesche, H., Bode, W., Kuhn, K., and Moroder, L. (1996) Design and synthesis of heterotrimeric collagen peptides with a built-in cystine-knot. *FEBS Lett.* **398**, 31–36.

28. Galloway, W. A., Murphy, G., Sandy, J. D., Gavrilovic, J., Crawston, T. E., and Reynolds, J. J. (1983) Purification and characterization of a rabbit bone metalloproteinase that degrades proteoglycan and other connective-tissue components. *Biochem. J.* **209**, 741–752.

29. Teahan, J., Harrison, R., Izquierdo, M., and Stein, R. L. (1989) Substrate specificity of human fibroblast stromelysin: hydrolysis of substance P and its analogues *Biochemistry* **28**, 8497–8501.

30. Niedzwiecki, L., Teahan, J., Harrison, R. K., and Stein, R. L. (1992) Substrate specificity of the human matrix metalloproteinase stromelysin and the development of continuous fluorometric assays. *Biochemistry* **31**, 12,618–12,623.

31. Smith, M. M., Shi, L., and Navre, M. (1995) Rapid identification of highly active and selective substrates for stromelysin and matrilysin using bacteriophage peptide display libraries. *J. Biol. Chem.* **270**, 6440–6449.

32. Suzuki, K., Enghild, J. J., Morodomi, T., Salvesen, G., and Nagase, H. (1990) Mechanisms of activation of tissue procollagenase by matrix metalloproteinase 3 (stromolysin). *Biochemistry* **29**, 10,261–10,270.

33. Reinemer, P., Grams, F., Huber, R., T., K., Schnierer, S., Piper, M., Tschesche, H., and Bode, W. (1994) Structual implications for the role of the N-terminus in the superactivation of collagenases. *FEBS Lett.* **338**, 227–233.

34. Knäuper, V., López-Otín, C., Smith, B., Knight, G., and Murphy, G. (1996) Biochemical characterization of human collagenase-3. *J. Biol. Chem.* **271**, 1544–1550.
35. Fields, G. B., Tian, Z., and Barany, G. (1992) Principles and practice of solid-phase peptide synthesis, in *Synthetic Peptides: A User's Guide* (Grant, G. A., ed.), W. H. Freeman & Co., New York, pp. 77–183.
36. Wells, J. A. (1990) Additivity of mutational effects in proteins. *Biochemistry* **29**, 8509–8517.
37. Knight, C. G. (1995) Fluorometric assays of proteolytic enzymes. *Meth. Enzymol.* **248**, 18–34.
38. Gershkovich, A. A. and Kholodovych, V. V. (1996) Fluorogenic substrates for proteases based on intramolecular fluorescence energy transfer (IFETS). *J. Biochem. Biophys. Methods* **33**, 135–162.
39. Stack, M. S. and Gray, R. D. (1989) Comparison of vertebrate collagenase and gelatinase using a new fluorogenic substrate peptide. *J. Biol. Chem.* **264**, 4277–4281.
40. Knight, C. G., Willenbrock, F., and Murphy, G. (1992) A novel coumarin-labelled peptide for sensitive continuous assays of the matrix metalloproteinases. *FEBS Lett.* **296**, 263–266.
41. Bouvier, J., Schneider, P., and Malcolm, B. (1993) A fluorescent peptide substrate for the surface metalloprotease of *Leishmania*. *Exp. Parasitol.* **76**, 146–155.
42. Nagase, H., Fields, C. G., and Fields, G. B. (1994) Design and characterization of a fluorogenic substrate selectively hydrolyzed by stromelysin 1 (matrix metalloproteinase-3). *J. Biol. Chem.* **269**, 20,952–20,957.
43. Bickett, D. M., Green, M. D., Berman, J., Dezube, M., Howe, A. S., Brown, P. J., Roth, J. T., and McGeehan, G. M. (1993) A hight throughput fluorogenic substrate for interstitial collagenase (MMP-1) and gelatinase (MMP-9). *Anal. Biochem.* **212**, 58–64.
44. Beekman, B., Wouter, J., Bloemhoff, W., Ronday, K., Tak, P. P., and Te Koppele, J. M. (1996) Convenient fluorometric assay for matrix metalloproteinase activity and its application in biological media. *FEBS Lett.* **390**, 221–225.
45. Geoghegan, K. F., Emery, M. J., Martin, W. H., McColl, A. S., and Daumy, G. O. (1993) Site-directed double fluorescent tagging of human renin and collagenase (MMP-1) substrate peptides using the periodate oxidation of *N*-terminal serine: An apparently general strategy for provision of energy-transfer substrates for proteases. *Bioconjugate Chem.* **4**, 537–544.
46. Geoghegan, K. F. (1996) Improved method for converting an unmodified peptide to an energy transfer substrate for proteinase. *Bioconjugate Chem.* **7**, 385–391.
47. Knight, C. G. (1991) A quenched fluorescent substrate for thimet peptidase containing a new fluorescent amino acid, DL-2-amino-3-(7-methoxy-4-coumaryl)propionic acid. *Biochem. J.* **274**, 45–48.
48. Murphy, G., Allan, J. A., Willenbrock, F., Cockett, M. I., O'Connell, J. P., and Docherty, A. J. P. (1992) The role of the *C*-terminal domain in collagenase and stromelysin specificity. *J. Biol. Chem.* **267**, 9612–9618.
49. Upadhye, S. and Ananthanarayanan, V. S. (1995) Interaction of peptide substrates of fibroblast collagenase with divalent cations: Ca^{++} binding by substrate as a

suggested recognition signal for collagenase action. *Biochem. Biophys. Res. Comm.* **215**, 474–482.

50. Liko, Z., Botyanszki, J., Bodi, J., Vass, E., Majer, Z., Hollosi, M., and Suli-Vargha, H. 1996) Effect of Ca^{2+} on the secondary structure of linear and cyclic collagen sequence analogs. *Biochem. Biophys. Res. Comm.* **227**, 351–359.
51. Netzel-Arnett, S., Mallya, S. K., Nagase, H., Birkedahl-Hansen, H., and Van Wart, H. E. (1991) Continuously recording fluorescent assays optimized for five human matrix metalloproteinases. *Anal. Biochem.* **195**, 86–92.
52. Knäuper, V., Cowell, S., Smith, B., López-Otín, C., O'Shea, M., Morris, H., Zardi, L., and Murphy, G. (1997) The role of the *C*-terminal domain of human collagenase-3 (MMP-13) in the activation of procollagenase-3, substrate specificity, and tissue inhibitor of metalloproteinase interaction. *J. Biol. Chem.* **272**, 7608–7616.
53. Anastasi, A., Knight, C. G., and Barrett, A. J. (1993) Characterization of the bacterial metalloendopeptidase pitrilysin by use of a contunuous fluorescence assay. *Biochem. J.* **290**, 601–607.
54. Knight, C. G. (1998) Stereospecific synthesis of L-2-amino-3(7-methoxy-4-coumaryl)propionic acid, an alternative to tryptophan in quenched fluorescent substrates for peptidases. *Lett. Peptide Sci.* **5**, 1–4.
55. Maggiora, L. L., Smith, C. W., and Zhang, Z. Y. (1992) A general method for the preparation of internally quenched fluorogenic protease substrates using solid-phase synthesis. *J. Med. Chem.* **35**, 3727–3730.
56. Drijfhout, J. W., Nagel, J., Beekman, B., Te Koppele, J. M., and Bloemhoff, W. (1996) Solid phase synthesis of peptides containing the fluorescence energy transfer Dabcyl-Edans couple, in *Peptides: Chemistry, Structure and Biology* (Kaumaya, P. T. P. and Hodges, R. S., eds.), Mayflower Scientific Ltd., Kingswinford, pp. 129–131..
57. Fields, C. G. and Fields, G. B. (1993) Minimization of tryptophan alkylation following 9-fluorenylmethoxycargonyl solid-phase peptide synthesis. *Tetrahedron Lett.* **34**, 6661–6664.
58. Bycroft, B. W., Chan, W. C., Chhabra, S. R., and Hone, N. D. (1993) A novel lysine-protecting procedure for continuous flow solid phase synthesis of branched peptides. *J. Chem. Soc. Chem. Commun.*, 778–779.
59. Fields, C. G., Lovdahl, C. M., Miles, A. J., Matthias-Hagen, V. L., and Fields, G. B. (1993) Solid-phase synthesis and stability of triple-helical peptides incorporating native collagen sequences. *Biopolymers* **33**, 1695–1707.
60. King, D. S., Fields, C. G., and Fields, G. B. (1990) A cleavage method which minimizes side reactions following Fmoc solid phase peptide synthesis. *Int. J. Peptide Protein Res.* **36**, 255–266.
61. Itoh, M., Osaka, M., Chiba, T., Masuda, K., Akizawa, T., Yoshioka, M., and Seiki, M. (1997) Flow injection analysis for measurement of activity of matrix metalloproteinase-7 (MMP-7). *J. Pharmaceut. Biomed. Anal.* **15**, 1417–1426.
62. Schullek, J. R., Butler, J. H., Zhi-Jie, N., Chen, D., and Yuan, Z. (1997) A high-density screening format for encoded combinatorial libraries: assay miniturization and its application to enzymatic reactions. *Anal. Biochem.* **246**, 20–29.

63. Crabbe, T., Willenbrock, F., Eaton, D., Hynds, P., Carne, A. F., Murphy, G., and Docherty, A. J. P. (1992) Biochemical characterization of matrilysin: activation conforms to the stepwise mechanisms proposed for other matrix metalloproteinases. *Biochemistry* **31**, 8500–8507.

32

Kinetic Analysis of the Inhibition of Matrix Metalloproteinases by Tissue Inhibitor of Metalloproteinases (TIMP)

Mike Hutton and Frances Willenbrock

1. Introduction

The kinetic analysis of the inhibition of matrix metalloproteinases by tissue inhibitor of metalloproteinases (TIMP) yields valuable information on the mechanism and specificity of the TIMPs. When combined with the use of genetic engineering or chemical methods of modification to alter specifically the structure of either enzyme or inhibitor, kinetic techniques can also be used to identify the contribution of individual amino acid residues to binding and complex stabilization. Nuclear magnetic resonance (NMR) and crystallographic structures of TIMP-enzyme complexes show that the binding region is very large, covering about 1300Å^2, and consists of six separate polypeptide segments of TIMP-1 *(1)*. Thus, a systematic approach to modifying residues at the enzyme-TIMP interface can identify the important features of such a large binding site.

TIMP and active matrix metalloproteinases form a reversible tight-binding complex with a 1:1 stoichiometry. To date four members of the family have been identified, all of which share considerable higher structural similarity, and may therefore also share a common inhibitory mechanism. Similarities between the structures of complexes of the catalytic domain of stromelysin-1 with TIMP-1 and with N-TIMP-2 support this possibility *(1,2)*. TIMP binds in the active site cleft of stromelysin-1, with the α-carbonyl of Cys-1 of TIMP appearing to coordinate the catalytic zinc. Thus it is possible that TIMP competes directly with substrate as a competitive inhibitor, although we have not been able to study the type of inhibition with respect to our substrates due to limited substrate solubility.

From: *Methods in Molecular Biology, vol. 151: Matrix Metalloproteinase Protocols*
Edited by: I. Clark © Humana Press Inc., Totowa, NJ

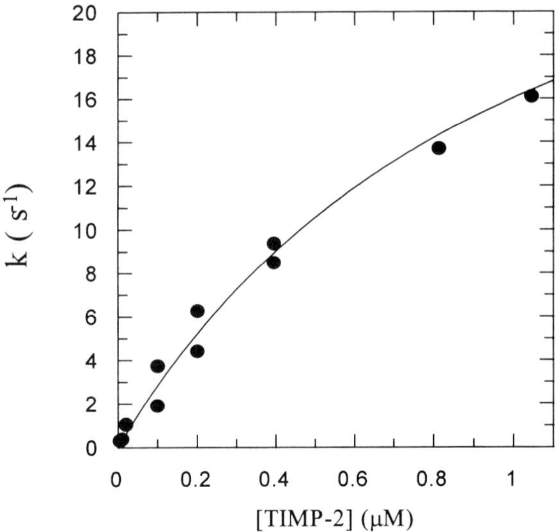

Fig. 1. The dependence of the observed first order rate constant, k, on TIMP-2 concentration with gelatinase A from Hutton et al. 1998; as described by Mechanism 2. Assays performed at 25°C in 50 mM Tris/HCl, pH 7.5, 150 mM NaCl, 10 mM CaCl$_2$, 0.05% Brij 35 using Applied Photophysics stopped flow spectrofluorimeter (slits = 10 nm) using an excitation wavelength of 328 nm and monitoring fluorescence above 360 nm. (Reprinted with permission from *Biochemistry* [1998], **37**, 10,094–10,098. Copyright by American Chemical Society.)

For most of the systems we have studied, we have been unable to obtain evidence to suggest that the mechanism is any more complex than that of a simple bimolecular collision (Mechanism 1) for which $K_i = k_{-1}/k_1$.

Mechanism 1: $\qquad\qquad\qquad\qquad E + I \underset{k_{-1}}{\overset{k_1}{\rightleftharpoons}} EI$

However, we have recently demonstrated that the interaction between gelatinase A and TIMP-2 occurs by a more complex mechanism (*3*) and it is probable that such a mechanism applies to other TIMP-enzyme interactions. This more complex mechanism occurs *via* formation of an intermediate (EI) which then rapidly isomerises to form the final complex, EI* (Mechanism 2).

Mechanism 2: $\qquad\qquad E + I \underset{k_{-1}}{\overset{k_1}{\rightleftharpoons}} EI \underset{k_{-2}}{\overset{k_2}{\rightleftharpoons}} EI*$

The overall dissociation constant for the reaction, K_i^*, is given in **Eq. 1**:

$$K_i^* = k_{-1} k_{-2} / k_1 (k_2 + k_{-2})$$ (1)

The formation of the intermediate (EI) is demonstrated by observing saturation in the dependence of the observed rate constant for inhibition with increasing inhibitor concentration, **Fig. 1.** Saturation occurs because at high inhibitor concentrations the rate-limiting step is no longer the association of enzyme and inhibitor but the isomerization step, which is independent of reagent concentration.

In most of the enzyme-TIMP interactions that we have studied, with the exception of MMP-7 (**Subheading 3.2.**), the K_i for the interaction is less than $10^{-10}M$ *(4)* and this has given rise to a number of practical problems. In order to determine the K_i of an interaction directly it is essential to work under conditions in which $[I] \sim K_i$. Under these conditions the proportion of free enzyme to bound enzyme is most sensitive to [I] and the effect of altering inhibitor concentration on the concentration of free enzyme can therefore readily be determined. Under the conditions where $[I] \ll K_i$ or $[I] \gg K_i$ the ratio of free to bound enzyme is virtually unaffected by changes in [I], and any results obtained will be highly inaccurate. For the reasons described in **Subheading 3.1.1,** we are unable to fulfill the conditions of $[I] \sim K_i$. Therefore instead of determining the K_i directly we have to analyse the individual rate constants and then from this data we can calculate the K_i of the interaction (as described in **Subheading 3.2.**).

We have established two types of methods for determining rate constants: (1) by following the inhibition of enzyme activity by TIMP and (2) by following the binding of TIMP and enzyme directly.

1. *Inhibition of enzyme activity:* the advent of quenched fluorescent substrates *(5,6)* and their subsequent enhancement by replacing the fluorescent tryptophan with derivatives of 7-methoxyycoumarin *(7)* has been an important development in the study of the interaction of TIMPs and matrix metalloproteinases. Enzyme activity is followed using these quenched fluorescent substrates that consist of a short sequence of amino acids containing the scissile peptide bond separating the fluorescent group from a dinitrophenyl group that acts as an internal quencher. The quenching effect has been shown to be almost unaffected by the sequence of amino acids, thus allowing specificity to be introduced for individual enzymes *(8)*. The substrates are highly sensitive so that enzyme concentrations of $10^{-12} - 10^{-11} M$ can be used in routine assays. The change in fluorescence can be recorded continuously and, as the wavelengths used (λ_{ex} 328 nm, λ_{em} 393 nm) are considerably higher than those at which protein fluorescence is observed, there is very little background interference. We use enzyme activity assays mainly to determine the rate constants for the association of enzyme and TIMP by following the time course of inhibition when TIMP is added to a reaction mixture of enzyme and substrate.

2. *Binding of TIMP to enzyme:* methods for studying the binding of enzyme and inhibitor directly involve radioactively labeling TIMP and following the formation, or dissociation, of labeled complex. The procedure is time-consuming and laborious as the labeled complex must be separated from free enzyme before being quantified. In order to be accurate several time points are required when following the progress of the reaction leading to a large number of samples being processed. At present, however, it is the only method available for studying the dissociation of these tight-binding complexes.

2. Materials

1. All chemicals should be of the highest grade, in particular any detergents used, as lower grade detergents have been found to interfere with both the protein interactions and product fluorescence.
2. Quenched fluorescent substrates can be purchased from a variety of sources e.g., NovaBiochem (La Jolla, CA), Bachem (Switzerland). The two most commonly used substrates are Mca-Pro-Leu-Gly-Leu-Dpa-Ala-Arg *(7)* which is a good general purpose MMP substrate and Mca-Pro-Leu-Ala-Nva-Dpa-Ala-Arg *(9)* a more specific substrate for the stromelysins. Substrate solutions are made up in dimethysulfoxide (DMSO) and can be stored for several months at 4°C.
3. ^{125}I can be purchased as Bolton and Hunter reagent from Amersham Life Science, UK.

3. Methods

3.1. Determination of the Rate Constants Characterizing the Inhibition of Matrix Metalloproteinases by TIMP

3.1.1. Inhibition of Enzyme Activity

The effectiveness of enzyme activity assays to follow enzyme-inhibitor interactions depends on both the protein and substrate concentrations that can be used in the assay. Ideally, (1) enzyme concentrations should be lower than the K_i for the enzyme-inhibitor interaction; (2) inhibitor concentrations should be used that cross the K_i ; and (3) a range of substrate concentrations should be used which crosses the K_m for the enzyme-substrate reaction.

Unfortunately in most TIMP-enzyme systems (1) TIMP has such a high affinity for the enzymes, with K_i values of 10^{-15}–10^{-12} *M*, that no substrates are sufficiently sensitive to permit assays to be performed under the condition of [E] < K_i ; (2) if inhibitor concentrations that approximate the K_i are used, the reaction is so slow that the enzymes are not sufficiently stable to permit such assays to be performed; and (3) quenching of product fluorescence at high substrate concentrations greatly reduces the sensitivity of the assay and prevents the use of substrate concentrations $\gg K_m$. We have found that when the substrate concentration is 8 – 10 µ*M*, product fluorescence is almost totally quenched. It is possible in some cases to circumvent this problem by using

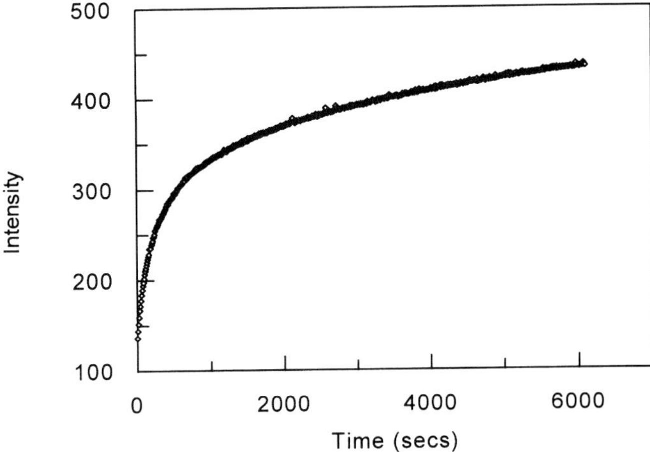

Fig. 2. An example of the gradual inhibition of gelatinase A by TIMP-2. [Gelatinase A] = 42 p*M*, [TIMP-2] = 198 p*M* in 50 m*M* Tris-HCl, pH 7.5, 10 m*M* CaCl$_2$, 100 m*M* NaCl, 0.025% Brij 35 at 25°C; wavelengths = λ_{ex} 328 nm, λ_{em} 393 nm.

stopped assays at higher substrate concentrations, so that the reaction mixture can be diluted to give a substrate concentration of 1–2 µ*M* before the fluorescence is read. Unfortunately, the substrates which are hydrolyzed most readily by the MMPs are not very soluble so we have found that, even when using the stopped assays, it is not possible to use substrate concentrations approaching the K$_m$. Increasing the organic solvent content of the assay in order to increase the substrate concentration does not help as the concomitant decrease in water concentration results in a significant decrease in enzyme activity *(10)*.

As a result of the above problems, we are therefore restricted in most cases to using conditions in which the enzyme and inhibitor concentrations are greater than the K$_i$ and the substrate concentration is much less than the K$_m$.

3.1.1.1. ASSOCIATION KINETICS

The procedure to set up the assays for both conventional and stopped-flow fluorimetry is described in **Notes 1–3**. The apparent association rate constant, k$_{on}$, has been determined for the interaction of many matrix metalloproteinases and TIMPs *(11)*. Many of these reactions are very rapid (k$_{on}$ = 10^5 – 10^6 $M^{-1}s^{-1}$) but the use of very low concentrations of enzyme and inhibitors allows the reaction to be followed by conventional fluorimetry. The optimum length of time for the assays is between 5 and 60 min. If TIMP concentrations are used which give a very long reaction time, it can be difficult to determine when the steady-state rate has been attained. When reactions take less than 5 min, too few data points are obtained in the early part of the curve for the analysis of such data to be accurate. Thus, to follow the reaction at higher TIMP concentrations it is necessary to use a stopped-flow spectrophotometer or a stopped-flow attachment to the fluorimeter. It is best to aim for as large a change in fluorescence as possible over the time course, so the enzyme concentration should be increased when the TIMP concentration is increased to allow a greater proportion of the substrate to be hydrolyzed. Care must be taken when altering the enzyme concentration, however, to keep the conditions of [E] << [I] and [E] << [S] and to keep the total change in fluorescence within the linear portion of the progress curve obtained in absence of inhibitor, as described in **Note 3**.

Initially substrate and enzyme are mixed and TIMP is then added at t = 0. The gradual inhibition of enzyme activity results in curvature in the progress curve of substrate hydrolysis. The reaction must be followed until the rate of substrate hydrolysis is once again linear, indicating that the system has reached equilibrium, **Fig. 2**. The complete progress curve can then be analyzed using a curve-fitting program such as Grafit *(12)*. Analysis is facilitated considerably by using pseudo-first-order conditions (i.e., [I] >> [E]). Under these conditions the pseudo-first-order rate constant, k, is obtained from (**Eq. 2**):

$$P = v_s t + (v_i - v_s) (1 - e^{-kt}) / k \qquad (2)$$

where P is the product concentration at time t, and v_i and v_s are the initial and steady state rates, respectively *(13)*. Several TIMP concentrations over as wide a range as possible are used and k is plotted against [TIMP]. For Mechanism 1, a simple bimolecular collision, this plot will be linear and is described by (**Eq. 3**):

$$k = k_{-1} + k_1 I \qquad (3)$$

If curvature in the dependence of k against [TIMP] is observed, with saturation at high [TIMP] and a decrease in v_i with increasing TIMP concentration, then this is evidence for a pathway ϡ—involving an intermediate as shown in mechanism 2. In this case the plot of k against [TIMP] is fitted to (**Eq. 4**):

$$k = k_{-2} + k_2 [k_1 I / (k_{-1} + k_1 I)] \qquad (4)$$

where the rate constants are as shown earlier in Mechanism 2. If k_2 is very much greater than the maximum value that can be obtained for k then curvature will not be detected. In this case, $k_1 I << k_{-1}$ and **Eq. 4** can be simplified to (**Eq. 5**):

$$k = k_{-2} + k_2 k_1 I / k_{-1} \qquad (5)$$

Therefore the gradient of a plot of k against [TIMP] will be equivalent to k_1 k_2 / k_{-1}. It is important that independent evidence for the existence of an intermediate be obtained if (**Eq. 5**) is used to analyze a linear plot.

3.1.1.2. DISSOCIATION KINETICS

In order to use enzyme activity assays to follow the dissociation of enzyme-inhibitor complex, it is essential to be able to perform activity assays at an enzyme concentration lower than the K_i of the system. The enzyme-inhibitor complex is formed at a concentration greater than the K_i using equimolar amounts of the two reagents. The complex is then diluted into assay mixture (a solution of substrate in buffer) to a give a final concentration much less than the K_i and the appearance of enzyme activity as the complex dissociates is followed until equilibrium is reached (i.e., a linear rate of hydrolysis is observed). So far, we have only been able to use this method for the matrilysin-TIMP-1 interaction which has a K_i of 0.37 nM *(14)*. In this case complexes were initially formed at a concentration of 50 nM before dilution to a concentration of 0.1 nM in the assay mixture. Recovery of enzyme activity was followed for 4 h. Curves are analysed to obtain the apparent first-order rate constant for dissociation, k, using (**Eq. 2**) as for complex association, except in this case v_s is greater than v_i. It is important that the complex has fully formed before dilution and this can readily be checked by comparing values of v_i and v_s (v_i should be no greater than 2% of v_s). If the interaction under study follows Mechanism 1, then k is equivalent to k_{-1}. If Mechanism 2 applies, then it is most probable that k_{-2} will always be the rate-limiting step for dissociation and its value will therefore be equal to k.

3.1.2. Enzyme-Inhibitor Binding Studies

In order to follow enzyme-inhibitor binding directly, the complex must be distinguishable from the free enzyme and inhibitor. Ideally the complex would have different spectroscopic characteristics, the appearance of which could readily be followed as the reaction proceeds without the need for any purification steps.

Although as yet we do not have such a convenient system, we are currently testing methods for the incorporation of spectroscopic labels into TIMP *(15)* and hope to be able to use these modified inhibitors in future studies. In the meantime, we have used radioactively labeled TIMP (*see* **Note 4**) to study the gelatinase A-TIMP-2 interaction *(3)*. Labeled complexes can be separated from free gelatinase A and free TIMP-2 by using gel filtration or affinity chromatography. These methods have the disadvantage that the separation step has to be modified for each different enzyme-TIMP system. All other aspects of the analysis, however, are generally applicable for the study of other matrix metalloproteinases and TIMPs.

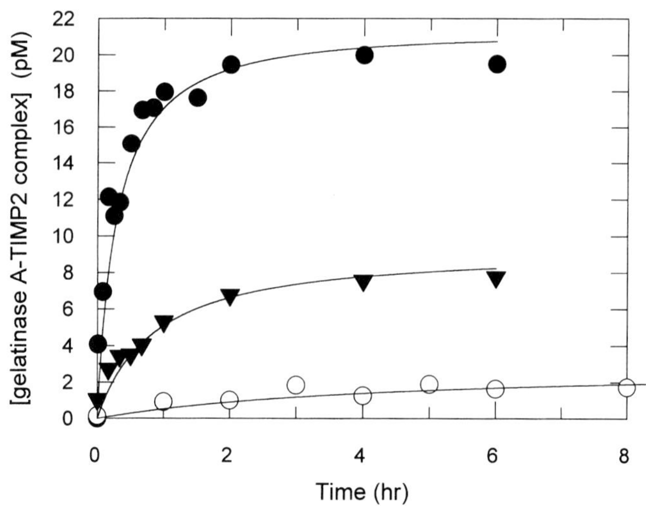

Fig. 3. The rate of binding of gelatinase A to [125]I -TIMP-2. Data were obtained at several TIMP-2 and gelatinase A concentrations and analyzed to obtain the apparent second order rate constant. Concentrations used in the figure are 4 p*M* (○), 10 p*M* (▼) and 21 p*M* (●), giving values of $4.8 \times 10^7\ M^{-1}s^{-1}$, $2.5 \times 10^7\ M^{-1}s^{-1}$ and $5.1 \times 10^7\ M^{-1}\ s^{-1}$ respectively. (Reprinted with permission from *Biochemistry* **37**, 10,094–10,098. Copyright 1998 American Chemical Society.)

3.1.2.1. ASSOCIATION KINETICS

The rate of formation of the gelatinase A-TIMP-2 complex is followed using ^{125}I-TIMP-2. Enzyme and ^{125}I-TIMP are incubated at equimolar concentrations in buffer at 25°C. At time intervals the reaction is stopped by the addition of a large excess of unlabeled TIMP-1 (approx 1000-fold) which binds to and therefore effectively removes any free enzyme from the reaction mixture. To the reaction mixture is added 1mL of gelatin-agarose which binds to free gelatinase A and gelatinase A-TIMP-2 complex, thus separating free TIMP-2 from bound TIMP-2. After mixing at 25°C for 30 min, the agarose suspension is centrifuged at 3,000*g* for 3 min and the number of counts in the pellet (i.e., gelatinase A-TIMP-2 complex) and the supernatant (i.e., free TIMP-2) quantified. The determination is performed at a number of enzyme and inhibitor concentrations (**Fig. 3**). As the reaction is effectively irreversible over the time scale of the experiment, the results can be analyzed by fitting data to the equation for a second-order irreversible reaction (**Eq. 6**).

$$1/I = k_{on}t + 1/I_0$$

(6)

in which I and I_0 are the TIMP concentrations at time t and time 0 respectively and k_{on} is the apparent second-order rate constant.

In the case of gelatinase A-TIMP-2, we could perform the binding studies at low reagent concentrations only, where the rate-limiting step is the initial association and the apparent second-order rate constant is equivalent to $k_1 k_2 / k_{-1}$. For other systems it may be possible to follow the binding using a greater range of concentrations extending to those at which isomerisation becomes rate-limiting. In these cases it would be possible to obtain estimates for each of the rate constants by using simulation programs such as KINSIM *(16)* and FITSIM *(17)* to fit the data to reaction Mechanism 2.

3.1.2.2. Dissociation Kinetics

The rate of dissociation of the gelatinase A-TIMP-2 complex is determined by following the exchange of free unlabeled TIMP-2 with complexed ^{125}I-TIMP-2 over a period of time. Complex is prepared by incubating equimolar concentrations of enzyme and inhibitor. Care must be taken to ensure that the complex has fully formed and this can be tested by checking for enzyme activity. The complex is then mixed with a 10-fold excess of unlabeled TIMP-2 at 25°C. At regular time intervals, for up to 6 wk, samples are removed and complexed TIMP-2 is separated from free TIMP-2 by gel filtration using a Superdex 75 FPLC column. The two pools of radioactive TIMP-2 are quantified and the rate of release of ^{125}I-TIMP-2 determined as the gradient of a plot of ln[complexed TIMP-2] against time. The experiment is also performed using unlabeled TIMP in the initial complex and by incubating this complex with a 10-fold excess of ^{125}I-TIMP to check whether labeling of the TIMP-2 with ^{125}I affects the stability of the complex.

The gelatinase A-TIMP-2 complex is extremely stable, with a half-life of approx 1 yr. As this is considerably longer than the duration for which the experiment can be performed, the estimated rate constant of 2×10^{-8} s^{-1} is thus subject to large error.

3.2. Determination of the Binding Constant (K_i) for the Matrix Metalloproteinase-TIMP Interaction

Where the interaction between TIMP and enzyme is very tight (i.e., K_i <200 pM) the K_i can be calculated from the rate constants obtained using the methods described in the previous sections. If the mechanism is unknown an apparent value for the K_i can be obtained simply by dividing the dissociation rate constant obtained from **Subheading 3.1.2.1.** by the association rate constant obtained from **Subheading 3.1.1.1.** In a system similar to that observed for gelatinase A-TIMP-2 in which an intermediate is formed (Mecha-

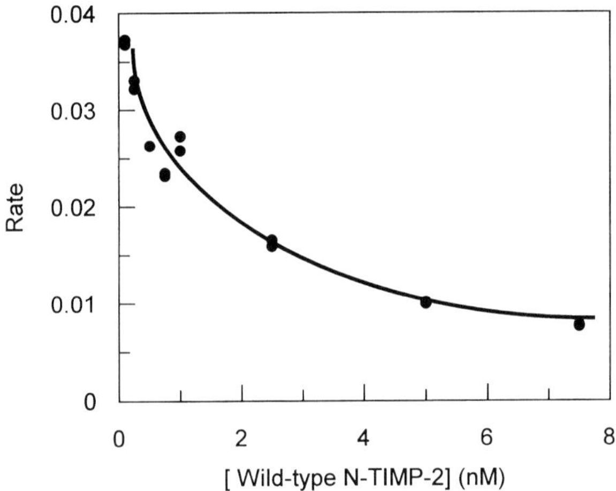

Fig. 4. An example of K_i determination for the interaction of matrilysin and N-TIMP-2 (B. Wattam and G. Murphy, unpublished data). The $K_i = 1.2$ nM, matrilysin concentration is 0.1 nM. Assays were performed at 25°C in 50 mM Tris-HCl, pH 7.5, 175 mM NaCl, 10 mM CaCl$_2$, 0.05% Brij 35. Data were analysed using a form of the tight binding equation described by Morrison and Walsh, 1988 (*13*).

nism 2) and the value for k_{-2} is very much smaller than that of the other rate constants, the dissociation rate constant will be k_{-2} and the association rate constant will be $k_1 k_2 / k_{-1}$ giving a K_i of $k_{-1} k_{-2} / k_1 k_2$, which approximates to K_i^* in (**Eq. 1**).

However for enzyme-TIMP interactions with a higher K_i value (> 200 pM) such as matrilysin and TIMP-1 or matrilysin and N-TIMP-2, the K_i can be determined without knowledge of the individual rate constants using enzyme activity assays. Assays are set up as described in **Note 1** under conditions of $[E] \leq K_i$ and $[I] \sim K_i$. For the inhibition of matrilysin by TIMP-1 the K_i of 0.37 nM was determined using an enzyme concentration of 0.1 nM and TIMP concentrations of 0.2–20 nM (*14*). Enzyme and TIMP were preincubated for 4 h and then diluted 10-fold into the assay mixture to give the enzyme concentration of 0.1 nM. Dissociation of the complex was followed by following the increase in rate of substrate hydrolysis until a linear rate of hydrolysis was observed, indicating that equilibrium had been reached. The final linear rate of substrate hydrolysis, v_s, was determined at each TIMP concentration and the plot of v_s against [TIMP] fitted to a form of the tight-binding inhibition equation described by Morrison and Walsh 1988 (*13*) (**Eq. 7**) to obtain a value for the apparent K_i (K_I').

$$v_s = (v_0 / 2E_t)\{[(K_i' + I_t - E_t)^2 + 4K_i' E_t]^{1/2} - (K_i' + I_t - E_t)\} \tag{7}$$

where v_0 is the rate in the absence of inhibitor, v_s is the steady-state rate in the presence of inhibitor, and E_t and I_t are the total enzyme and inhibitor concentrations respectively (**Fig. 4**). It is important to follow the reaction for a significant length of time to get an accurate value of v_s, and also to perform the determinations at two different enzyme concentrations to check that the condition $[E] < K_i$ is fulfilled. If results from the two determinations are not consistent, then the enzyme concentration should be decreased until two sets of consistent results are obtained.

4. Notes

1. Sample preparation: The samples used should be as pure as possible so that they do not interfere with product fluorescence. The concentration of the proteins used must be known accurately, a detailed set of methods for determining the concentration of the TIMP and MMPs is available in Murphy and Willenbrock 1995 *(4)*. The essence of this technique is the titration of active MMP against a known concentration of TIMP. TIMP concentration can be determined for a pure sample from published extinction coefficients *(18)*. We have found that, once activated, most enzymes are stable for up to 5 d at 4°C; TIMPs can be stored for longer, at 4°C, providing antibacterial agents are present e.g., NaN_3.

2. Choice of conditions: As with all kinetic assays it is important to specify the temperature and buffer conditions used. The temperature of the reaction must be chosen with care because the stability of some of the enzymes, such as gelatinase A and B is markedly temperature dependent. At 37°C, for example, gelatinase A becomes completely inactive within 10 min, whereas at 25°C it is stable for several hours, allowing the longer assays to be performed. Stability is also affected by pH and matrix metalloproteinases are stable in the pH range of 5.5–8.5. Most of the enzymes have a pH optimum of pH 6.0–8.0 and therefore assays are usually performed at pH 7.0–7.5. Stromelysin-1, however, exhibits a different pH dependence with an optimum at pH 6 so, if maximum sensitivity is required, it may be preferable to perform assays at this lower pH.

    The organic solvent content should be kept constant because it affects enzyme activity markedly due to substrate partitioning effects and sensitivity of k_{cat} / K_m to water concentration. Thus, if different concentrations of substrate are to be used, the stock solution should be diluted in DMSO and the same volume of the substrate solutions added throughout. We have found that a final concentration of 1% DMSO is acceptable. Low concentrations of detergent (0.025%), such as Brij 35, are usually added to prevent protein loss through absorption to the cuvets. Detergent can be omitted, if necessary, as long as high protein concentrations are used ($> 10^{-7}$ *M*). Contamination by residual TIMP or enzyme from previous assays sticking to the cuvettes can be a significant problem when very low concentrations are to be used in subsequent assays. TIMP can be particularly difficult to remove even by acid washing. For very sensitive assays it is advisable to use disposable cuvets. These must be of the highest quality as inferior makes give a high background fluorescence. We regularly use disposable cuvets from Hughes and Hughes Limited, Somerset, UK.

The presence of $CaCl_2$ helps to maintain the stability of the enzymes. Although NaCl is often added to assays it has no effect on the enzyme activity and, in many cases, little effect on the TIMP-enzyme interaction. However, the rate of association of TIMP-2 and gelatinase A is sensitive to the ionic strength of the buffer and it is advisable to check the effect of ionic strength on each TIMP-enzyme system and to ensure that a constant ionic strength is used throughout. A buffer commonly used in our laboratory is 50 mM Tris/HCl pH 7.5, 10 mM $CaCl_2$, 100 mM NaCl, 0.025% Brij 35.

3. Calibration of assay: It is important to be able to give an absolute value for the measured fluorescence and thus to calibrate the fluorimeter prior to use. We routinely use a solution of Mca, the fluorescent group in the quenched fluorescent peptides. The concentration of this stock solution is determined by measuring its absorbance at 410 nm in 1% DMSO and using the molar extinction coefficient of 7500 mM^{-1} cm^{-1}. The assay should be calibrated for each condition used, particularly if the substrate concentration or slit widths are altered. This is performed by adding known concentrations of the fluorescent group, Mca (either commercially available or produced by adding a high concentration of enzyme to a known concentration of substrate and incubating overnight), to a solution of buffer and determining the change in fluorescence for a given concentration of fluorophore. In order to establish that curvature in the assays described in **Subheading 3.1.1.1.** is due to inhibition by TIMP and not substrate depletion or enzyme stability, progress curves must initially be run at the appropriate substrate concentration in the absence of inhibitor to determine the fluorescence range over which the assay is linear. Once this is established, it must be ensured that assays performed in the presence of TIMP never exceed this fluorescence range. If the permissible change in fluorescence is exceeded, then the assay must be repeated using a lower enzyme concentration. Enzyme stability must also be checked in the longer assays by performing the assay in the absence of inhibitor in order to ascertain a linear progress curve is obtained.

4. Preparation of TIMP for labeling: We labeled TIMP-2 with ^{125}I-Bolton and Hunter reagent following the protocol as specified by Amersham Life Science and this resulted in a specific activity of ^{125}I-TIMP-2 of 0.094µCi/µg.

Acknowledgments

We would like to thank Augustin Amour, Mark Fox, Vera Knäuper, and Gillian Murphy for their helpful discussions and ideas in the preparation of this chapter. We thank the Arthritis Research Campaign (ARC) for their financial support of this project.

References

1. Gomis-Rüth, F.-X., Maskos, K., Betz, M., Bergner, A., Huber, R., Suzuki, K., Yoshida, N., Nagase, H., Brew, K., Bourenkov, G. P., Bartunik, H., and Bode, W.

(1997) Mechanism of inhibition of the human matrix metalloproteinase stromelysin-1 by TIMP-1. *Nature* **389**, 77–79.

2. Williamson, R. A., Carr, M. D., Frenkiel, T. A., Feeney, J., and Freedman, R. B. (1997) Mapping the binding site for matrix metalloproteinases on the N-terminal domain of the tissue inhibitor of metalloproteinaes-2 by NMR chemical shift perturbation. *Biochemistry* **36**, 13,882–13,889.

3. Hutton, M., Willenbrock, F., Brocklehurst, K., and Murphy, G. (1998) Kinetic analysis of the mechanism of interaction of full length TIMP-2 and gelatinase A: Evidence for the existence of a low affinity Intermediate. *Biochemistry* **37**, 10,094–10,098.

4. Murphy, G. and Willenbrock, F. (1995) Tissue inhibitors of matrix metalloendopeptidases. *Methods Enzymol.* **248**, 496–510.

5. Stack, M. S. and Gray, R. D. (1989) Comparison of vertebrate collagenase and gelatinase using a new fluorogenic substrate peptide. *J. Biol. Chem.* **264**, 4277–4281.

6. Netzel-Arnett, S., Mallya, S. K., Nagase, H., Birkedal-Hansen, H., and Van Wart, H. E. (1991) Continuously recording fluorescent assays optimised for five human matrix metalloproteinases. *Anal. Biochem.* **195**, 86–92.

7. Knight, C. G., Willenbrock, F., and Murphy, G. (1992) A novel coumarin-labeled peptide for sensitive continuous assays of the matrix metalloproteinases. *FEBS Lett.* **296**, 263–266.

8. Nagase, H., Fields, C. G., and Fields, G. B. (1994) Design and characterisation of a fluorogenic substrate selectively hydrolyzed by stromelysin 1 (matrix metalloproteinase-3). *J. Biol. Chem.* **269**, 20,952–20,957.

9. Murphy, G., Willenbrock, F., Crabbe, T., O'Shea, M., Ward, R., Atkinson, S., O'Connell, J., and Docherty, A. (1994) Regulation of matrix metalloproteinase activit. *Ann. NY Acad. Sci.* **732**, 31–41.

10. Willenbrock, F., Knight, C. G., Murphy, G., Phillips, I. R., and Brocklehurst, K. (1995) Evidence for the importance of weakly bound water for matrix metalloproteinase activity. *Biochemistry* **34**, 12,012–12,018.

11. Willenbrock, F. and Murphy, G. (1994) Structure-function relationships in the tissue inhibitors of metalloproteinases. *Am. J. Respir. Crit. Care Med.* **150**, S165–S170.

12. Leatherbarrow, R. J. (1992) Grafit version 3.0, Erithacus Software Ltd., Staines, U.K.

13. Morrison, J. F. and Walsh, C. T. (1988) The behaviour and significance of slow-binding enzyme inhibitors. *Adv. Enzymol. Relat. Areas Mol. Biol.* **61**, 201–301.

14. O'Shea, M., Willenbrock, F., Williamson, R. A., Cockett, M. I., Freedman, R. B., Reynolds, J. J., Docherty, A. J. P., and Murphy, G. (1992) Site-directed mutations that alter the inhibitor activity of the tissue inhibitor of metalloproteinases-1: importance of the N-terminal region between cysteine 3 and cysteine 13. *Biochemistry* **31**, 10,146–10,152.

15. Wallis, R., Leung, K., Pommer, A. J., Videler, H., Moore, G. R., James, R., and Kleanthous, C. (1995) Cognate and noncognate interactions that span the millimolar to femtomolar affinity range. *Biochemistry* **34**, 13,751–13,759.

16. Barshop, B. A., Wrenn, R. F., and Frieden, C. (1983) Analysis of numerical models for computer simulation of kinetic processes: development of KINSIM—a flexible, portable system. *Anal. Biochem.* **130**, 134–145.

17. Zimmerlle, C. T. and Frieden, C. (1989) Analysis of progress curves by simulations generated by numerical integration. *Biochem. J.* **258**, 381–387.
18. Douglas, D. A., Shi, Y. E., and Sang, Q. X. A. (1997) Computational sequence analysis of the tissue inhibitor of metalloproteinase family. *J. Protein Chem.* **16**, 237–255.

33

Assaying Growth Factor Activity of Tissue Inhibitors of Metalloproteinases

Kyoko Yamashita and Taro Hayakawa

1. Introduction

Tissue inhibitors of metalloproteinases (TIMPs) are well known as inhibitors of matrix metalloproteinases (MMPs) such as interstitial collagenase, gelatinase A and B, and stromelysin 1, and have been suggested to play an important role in the regulation of MMPs. Although MMP inhibition is the primary action of the TIMPs, these inhibitors also have cell growth-stimulating activity that seems to be independent of their MMP inhibitory activity *(1–10)*. The purpose of this chapter is to introduce three methods for assaying the growth factor activity of TIMPs, that is, the direct counting of cell number, the measurement of DNA content for cell proliferation, and the [^{3}H]thymidine incorporation assay for cell mitogen activity.

2. Materials

1. Human and bovine recombinant TIMPs (*see* **Notes 1–3**).
2. Trypsin-EDTA: 0.25% trypsin with 1 m*M* EDTA-4Na.
3. Phosphate-buffered saline (pH 7.2), PBS: 0.10 g $CaCl_2$ (anhyd.), 0.20 g KCl, 0.20 g KH_2PO_4, 0.10 g $MgCl_2 \cdot 6H_2O$, 8.0 g NaCl, 2.16 g $Na_2HPO_4 \cdot 7H_2O$ in 1 L.
4. PBS(-): as PBS but not including $MgCl_2$ and $CaCl_2$.
5. Trypan blue solution: 0.3% Trypan blue in PBS.
6. DABA: 0.4 g/mL 3,5-diaminobenzoic acid dihydrochloride (Aldrich 618-56-4) (prepare immediately prior to use, *see* **Note 4**).
7. DNA standard: 50.0 µg/mL in 0.6 *N* $HClO_4$ (sodium salt from thymus, Sigma D3664) (*see* **Note 5**).
8. TCA: trichloroacetic acid.
9. [^{3}H]Thymidine: [methyl-^{3}H]thymidine (aqueous solution containing 2% ethanol, sterilized, Amersham Life Sci. TRK758).

From: *Methods in Molecular Biology, vol. 151: Matrix Metalloproteinase Protocols*
Edited by: I. Clark © Humana Press Inc., Totowa, NJ

"

3. Methods

3.1. Cell Culture

The desired cells are cultured at 37°C in phenol red-containing medium supplemented with 10% FCS and antibiotics and buffered with sodium bicarbonate, in a humidified atmosphere containing 5% CO_2. They are passaged with trypsin-EDTA in order to maintain them in an exponential growth phase.

3.2. Cell Proliferation

3.2.1. Preparation of Cells

1. Trypsinize the cell precultures.
2. Pipet up and down to disrupt cell clumps and obtain a suspension of single cells; and then count the cells.
3. Plate 1.5×10^5 cells per well in a 6-well tissue culture plate in 2 mL of three kinds of medium fetal calf serum ((FCS)-free medium, FCS-free medium containing TIMP at test concentrations (*see* **Notes 2** and **3**), medium containing 10% FCS).
4. All cultures should be terminated when the cells in the 10% FCS medium reach subconfluence (70–80% confluence).
5. Count the cell number.

3.2.2. Measurement of Cell Number

Many assays exist to measure cell proliferation. Rapid and accurate assessment of viable cell number and cell proliferation is an important requirement in many experimental situations involving in vitro cell cultures (*see* **Note 6**). Two methods, that is, the direct counting of cell number and the measurement of DNA content are introduced here.

3.2.2.1. Direct Counting

Direct counts of cell number in a microscopic counting chamber (hemocytometer), when combined with vital staining with Trypan blue can distinguish viable and nonviable cells. This method is quick, inexpensive, and requires only a small fraction of total cells from a cell population.

1. After trypsinizing the cells, centrifuge them at 186*g* for 5 min.
2. Resuspend the cell pellet in 100 µL of PBS, and then pipet to the disrupt cell clumps.
3. Remove 50 µL of cell suspension to a tube and add the same amount of trypan blue solution.
4. Count the cell number using a hemocytometer under a microscope.

3.2.2.2. Measurement of DNA Content

This method utilizes the reaction between diaminobenzoic acid (DABA) and the deoxyribose sugars exposed after removal of the purine base by hot acid hydrolysis and is applicable to DNA in solution as well as to that in cellular suspensions *(11)*.

1. Remove culture medium and wash the cells twice with PBS(-).
2. Collect cells (*see* **Note 7**) and resuspend them in 2 mL of PBS(-).
3. Add 1 mL of 30% cold TCA (final concentration 10%) and then place the cells on ice for 10 min.
4. Centrifuge the cells (2040g, 10 min), remove the TCA, and resuspend the cells in 1 mL of 10% cold TCA; repeat three times.
5. Remove TCA as completely as possible (*see* **Note 8**) and then add 200–400 µL of 0.2 N NaOH.
6. Add the same amount of 1 N HClO$_4$ as NaOH and incubate at 70°C for 30 min (*see* **Note 9**).
7. Centrifuge at 1670g for 10 min, and collect the supernatant for DNA measurement.
8. Prepare various concentrations (2.5–25 µg/mL) of standard DNA solution and transfer 0.1 mL of each into a brown test tube.
9. Transfer 0.1 mL of DNA extract from the cells into a brown test tube.
10. Add 0.1 mL of DABA solution and then incubate sample and standards at 60°C for 30 min (*see* **Notes 9** and **10**).
11. Stop reaction by adding 0.9 mL of 0.6 N HClO$_4$.
12. Measure fluorescent intensity at 507 nm for emission using 412 nm for excitation.

3.3. Cell Mitogen Activities

During cell proliferation, the DNA has to be replicated before the cell divides into two daughter cells. This close association between DNA synthesis and cell doubling makes the measurement of DNA synthesis very attractive for assessing cell proliferation. If labeled DNA precursors are added to the cell culture, cells that are about to divide incorporate the labeled nucleotide into their DNA. A traditional assay using tritiated thymidine ([³H]thymidine) is introduced here.

3.3.1. [³H]Thymidine Incorporation Assay **(12)**

3.3.1.1. PREPARATION OF CELLS

1. Trypsinize the cell pre-cultures.
2. Pipet to disrupt cell clumps and then count the cells.
3. Plate 2×10^5 cells per well in a 6-well tissue culture plate containing 2 mL of medium supplemented with 10% FCS per well.
4. When the cells reach approx 50% confluence, remove the medium and wash the cultures three times with serum-free medium.

3.3.1.2. STIMULATION OF DNA SYNTHESIS

1. Add 1 mL of each of the three kinds of medium (FCS-free medium, FCS-free medium containing TIMP at test concentrations, medium containing 10% FCS) to separate wells.

2. Incubate the plates for 16–24 h (*see* **Note 11**) in a CO_2 incubator at 37°C.
3. Add [³H]thymidine (final concentration, 1 μCi/mL), and incubate further for 6 h in the CO_2 incubator at 37°C.

3.3.1.3. DETECTION OF [³H]THYMIDINE INCORPORATED

1. Remove the medium from the wells by aspiration.
2. Rinse the cells three times with cold PBS and add 1 mL of cold 5% TCA per well. Place the plate on ice for 20 min.
3. Collect the cells (*see* **Note 7**) and centrifuge them, and then wash them three times with 1 mL of cold 5% TCA.
4. Remove TCA as completely as possible (*see* **Note 8**), solubilize the cell pellet with 0.5 mL of 0.5 *M* NaOH at room temperature, and then neutralize with 0.5 mL of 0.5 *M* HCl.
5. Remove 0.5 mL to a scintillation vial. Add 10 ml of scintillation fluid, mix, and count the radioactivity in a β-counter.

4. Notes

1. Both human and bovine recombinant TIMP-1 and human recombinant TIMP-2 are commercially available (COSMO BIO, CNI, CALBIOCHEM, access http://www.fuji-chem.co.jp to get further information).
2. The concentrations of TIMPs should be checked by use of the appropriate one-step sandwich enzyme immunoassay (e.g., TIMP-1 ELISA kit from DAIICHI Chem. Co. or Amersham Life Sciences).
3. TIMPs should be sterilized by filtration, kept, and added to culture medium in a solution as concentrated as possible. Their activity is easily lost in the diluted condition.
4. DABA solution should be prepared just before use and shielded from light.
5. DNA from thymus is easily dissolved in a hot (50–60°C) solvent.
6. As another cell count method, a colorimetric assay based on the metabolic activity of viable cells has been developed. The principle is that a tetrazolium salt (like MTT, XTT, MTS, or WST-1) is reduced by mitochondorial succinate-tetrazolium reductase in viable cells to a colored formazan and is quantified in a conventional ELISA plate reader at 570 nm. This assay is easy to use, but one has to keep in mind that the color development may vary greatly in viable cells due to the metabolic state of the cells. Many commercial assay kits are available (Wako Pure Chem., Amersham Life Sci., Boehringer Mannheim, etc.).
7. Make sure by observation under a microscope that cells have been quantitatively removed from each well.
8. Filter paper strips are suitable to remove TCA almost completely from the inside of the test tube.
9. Tubes should be capped during the incubation to avoid evaporation.
10. After the addition of DABA, the sample should be shielded from light.
11. A time course experiment to determine the optimal incubation time with TIMPs is recommended because the incubation time for maximal thymidine uptake is different from cell type to cell type.

References

1. Bertaux, B., Hornebeck, W., Eisen, A. Z., and Dubertret, L. (1991) Growth stimulation of human keratinocytes by tissue inhibitor of metalloproteinases. *J. Invest. Dermatol.* **97**, 679–685.
2. Hayakawa, T., Yamashita, K., Tanzawa, K., Uchijima, E., and Iwata, K. (1992) Growth-promoting activity of tissue inhibitor of metalloproteinases-1 (TIMP-1) for a wide range of cells. A possible new growth factor in serum. *FEBS Lett.* **298**, 29–32.
3. Young, T.-T. and Hawkes, S. P. (1992) Role of the 21-kDa protein TIMP-3 in oncogenic transformation of cultured chicken embryo fibroblasts. *Proc. Natl. Acad. Sci. USA* **89**, 10,676–10,680.
4. Nemeth, J. A. and Goolsby, C. L. (1993) TIMP-2, a growth-stimulatory protein from SV40-transformed human fibroblasts. *Exp. Cell Res.* **207**, 376–382.
5. Hayakawa, T., Yamashita, K., Ohuchi, E., and Shinagawa, A. (1994) Cell growth-promoting activity of tissue inhibitor of metalloproteinases-2 (TIMP-2). *J. Cell Sci.* **107**, 2373–2379.
6. Corcoran, M. L. and Stetler-Stevenson, W. G. (1995) Tissue inhibitor of metalloproteinase-2 stimulates fibroblast proliferation via a cAMP-dependent mechanism. *J. Biol. Chem.* **270**, 13,453–13,459.
7. Yamashita, K., Suzuki, M., Iwata, H., Koike, T., Hamaguchi, M., Shinagawa, A., Noguchi, T., and Hayakawa, T. (1996) Tyrosine phosphorylation is crucial for growth signaling by tissue inhibitors of metalloproteinases (TIMP-1 and TIMP-2). *FEBS Lett.* **396**, 103–107.
8. Nemeth, J. A., Rafe, A., Steiner, M., and Goolsby, C. L. (1996) TIMP-2 growth-stimulatory activity: a concentration- and cell type-specific response in the presence of insulin. *Exp. Cell Res.* **224**, 110–115.
9. Kikuchi, K., Kadono, T., Furue, M., and Tamaki, K. (1997) Tissue inhibitor of metalloproteinase 1 (TIMP-1) may be an autocrine growth factor in scleroderma fibroblasts. *J. Invest. Dermatol.* **108**, 281–284.
10. Murate, T., Yamashita, K., Isogai, C., Suzuki, H., Ichihara, M., Hatano, S., Nakahara, Y., Kinoshita, T., Nagasaka, T., Yoshida, S., Komatsu, N., Miura, Y., Hotta, T., Fujimoto, N., Saito, H., and Hayakawa, T. (1997) The production of tissue inhibitors of metalloproteinases (TIMPs) in megakaryopoiesis: possible role of platelet- and megakaryocyte-derived TIMPs in bone marrow fibrosis. *Br. J. Haematol.* **99**, 181–189.
11. Heinegardner, R. T. (1971) An improved fluorometric assay for DNA. *Anal. Biochem.* **39**, 197–201.
12. Hirata, Y. and Orth, D. N. (1979) Conversion of high molecular weight human epidermal growth factor (hEGF)/urogastrone (UG) to small molecular weight hEGF/UG by mouse EGF-associated arginine esterase. *J. Clin. Endocrinol. Metab.* **49**, 481–483.

Index

A

Activation, of MMPs by, 377–386
  APMA, 395
    MMP-1, 213, 279, 280
    MMP-2, 282
    MMP-3, 200, 201, 283
    MMP-7, 285
    MMP-8, 288
    MMP-9, 290
    MMP-10, 291
    MMP-13, 379, 380
  cell membranes
    MMP-2, 380–383
    MMP-13, 380–383
  cells
    MMP-13, 383–385
  endoproteinases, 395
    MMP-3, 283
    MMP-13, 380
Aggrecan, *see* substrates
Antibodies
  characterization, 442–444, 463–465
  ELISA, 348, 442–444
  immunohistochemistry, 359–360
  monoclonal, production, 436–442
  neoepitope, 425–447, 457–471
    anti-DIPEN, 436
    anti-FFGVG, 436–442
    collagen, 457, 458
  polyclonal, production, 436,
    460–465
AP-1, *see* Transcription factors
Assays

caseinase, 395, 407
collagen degradation, 457–471,
    473–482
collagenase, 392–394
  96-well plate format, 393, 394
ELISA, 347–356, 442–444, 465–
    470, 479–480
fluorogenic peptide, 511–513
gelatinase, 394, 395, 399–409
  using TR-gelatin, 417–423
growth factor, 533–536
invasion, 485–493
  image analysis, 491
neoepitope, 425–427, 457, 458
proteoglycan release, 451–456
TIMP, 394
  reverse zymography, 399–409
  growth factor activity, 533–536
zymography, 399–409
  In situ, 411–415, 417–423

B

Baculovirus, 203–216
Bullous pemphigoid, 162, 163

C

Cancer, 155–161, 167–169
Cardiovascular disease, 154, 155
Cartilage
  explant culture, 452–454
  extraction, 478, 479
Casein, *see* Substrates
Cells
  BHK, 181, 186–188

C3H10T1/2, 181
CHOL761h, 241, 243
COS-1, 181, 182, 241
HTC-75, 244
NS0, 188
SF-9, 208–210
Cloning, 25–39
    following biochemical
        characterization, 26–31
    general methods, 26
    low stringency hybridization, 32, 33
    RT-PCR amplification, 33–37
    screening EST databases, 37, 38
Collagen, *see* Substrates
Collagenase
    -1, see MMP-1
    -2, see MMP-8
    -3, see MMP-13
    -4, see MMP-18
    assay, 213,214, 392–394, 457–471,
        473–482
    interstitial, *see* MMP-1
    neutrophil, *see* MMP-8
    peptide substrate, 495–497
    triple helicase activity, 102–106
Confocal microscopy, *see*
        Microscopy
Contact hypersensitivity, 164

D

Dimethylmethylene blue, *see* Assay,
        proteoglycan release
DMB, *see* Dimethylmethylene blue
DNA content, measurement, 534, 535

E

E. coli expression, *see* Expression
Elastase
    macrophage, *see* MMP-12
Electroporation

of *Pichia pastoris*, 229
ELISA, 347–356
    antibodies, 348
        Fab' preparation, 351
        Labeling, 350, 351
    bead method, 353
    COL2-3/4C$_{short}$, 465–470
        validation, 468–470
    collagen epitope CB11B, 479, 480
    plate method, 352, 353
    standards, preparation of, 351, 352
Enamelysin, *see* MMP-20
Envelysin, 8
Enzyme-linked immunosorbent
        assay, *see* ELISA
Erythroid potentiating factor, *see*
        TIMP-1
Expression, *see also* individual
        MMPs or TIMPs
    inducible, 241, 242
    recombinant
        Baculovirus, 207–216
        E. coli, 191–203, 240, 245,
            246, 267–272
        mammalian, 181–188, 241,242
        yeast, 219–236

G

Gelatin, *see* Substrates
Gelatinase
    A, *see* MMP-2
    Assay, 394, 395
    B, *see* MMP-9
    Peptide substrate, 495–497
Gene knockout, *see* Transgenic mice
Gene targeting, *see* Transgenic mice
Glutathione S-transferase, *see* GST
Growth factor activity, *see* Assays,
        growth factor

GST
   fusion protein, 240

H

Hidden epitope antibody, *see*
      Neoepitope antibody
His-tag, 240, 245, 246
HnRNA, *see* Transcription

I

Immunolocalization, 359–364
   antibodies, 359, 360
   DIG in ISH, 341
   MMP-14, 248, 249
   paraffin sections, 362
   staining
      ABC stain, 362
      immunogold, 363
   tissue preparation, 360
In situ hybridization, 335–345
   detection, 341
   hybridization, 340
   in vitro transcription, 339
   isotopic vs non-isotopic, 336, 337
   preparation of cDNA template,
      339
In situ zymography, *see*
      Zymography
Interstitial collagenase, *see* MMP-1
Invasion, *see* Assays, invasion

K

Kinetic analysis, TIMP, 519–530
   binding studies, 525–527
      association, 526,527
      dissociation, 527
   Ki determination, 527–529
   MP inhibition, 522–525
      association, 523–525
      dissociation, 525

Kinetics
   impact of exosites, 100–102
Knockout, *see* Transgenic mice

M

Mammary gland development, 153
Matrilysin, *see* MMP-7
Matrix metalloproteinases
   assay, *see* assays
   cancer, 155–161
   cell-type specific expression of,
      130–133
   cloning, 25–39
   domain, 53–65, 83, 84
      catalytic domain, 53–58
      haemopexin, 63–65, 97, 98
      prodomain, 61–63
   exosite, 92–110
      function, 98–110
   history, 1–23
   kinetic analysis, 519–530
   module, 84, 85
   motif, 85, 86
   posttranscriptional regulation of,
      126, 127
   purification, *see* individual
      MMPs and ELISA standards
   structural studies, 45–65
   transcriptional regulation of, 123–126
Mechanism
   catalytic, 57, 91
   inhibition by TIMP, 66–69
Membrane-type metalloproteinase,
      see MT-MMP
   expression, 239–253
Microscopy
   confocal
      to detect proteolysis of TR-
         gelatin, 422, 423

        invasion assays, 490, 491
MMP-1
   cancer, 157
   cell-type specific expression of,
        130–133
   embryonic development, 153
   expression
        Baculovirus, 207–216
   history, 2
   peptide substrate, 495–497
   pulmonary emphysema, 161, 162
   purification, 279–282
   regulation
        transcriptional, 123–130
   postranscriptional, 126, 127
   truncated, 209
   wound repair, 153, 154
MMP-2
   activation
        exosite model, 106–110
   cancer, 158–160
   fibronectin type II modules, 93–97
   history, 3,4
   peptide substrate, 495–497
   purification, 282, 283
   triple helicase activity, 102–106
MMP-3
   cancer, 158
   cardiovascular disease, 154, 155
   contact hypersensitivity, 164
   expression, 191–203
   history, 4
   mammary gland development, 153
   peptide substrate, 497, 498
   purification, 197–200, 283–285
   truncated, 195–202
MMP-4, 4
MMP-5, 4
MMP-6, 4

MMP-7
   cancer, 156,157
   history, 4–5
   peptide substrate, 498
   purification
        from culture medium, 285, 286
        from rat uterus, 286–288
   wound repair, 154
MMP-8
   cancer, 157
   embryonic development, 153
   history, 5
   peptide substrate, 495–497
   purification, 288–290
MMP-9
   bullous pemphigoid, 162, 163
   cancer, 158–160
   contact hypersensitivity, 164
   embryonic development, 151, 152
   fibronectin type II repeats, 93–97
   history, 5
   transgenic mice, 151,152
   peptide substrate, 495–497
   purification, 231, 232, 290, 291
MMP-10
   history, 5
   purification, 291, 292
MMP-11
   cancer, 158
   history, 6
MMP-12
   cancer, 160
   cardiovascular disease, 155
   history, 6
   knockout, 162
   pulmonary emphysema, 162
   purification, 293, 294
MMP-13
   cancer, 157

embryonic development, 153
History, 6
  peptide substrate, 495–497
MMP-14
  expression, 239–253
  history, 6,7
  knockout, 171
  detection, 247–249
MMP-15, -16, -17, -18, -19, -20
  history, 7
MT1-MMP, *see* MMP-14
MT2-MMP, *see* MMP-15
MT3-MMP, *see* MMP-16
MT4-MMP, *see* MMP-17

N

Neoepitope antibody, *see* Antibody
Neutrophil collagenase, *see* MMP-8
Nuclear factor kappa B, *see*
  Transcription factors
Nuclear run-off, *see* Transcription

P

PEA3, *see* Transcription factors
Peptide
  conjugation, 432–436, 460, 461
  fluorogenic, 500–504
  substrate, 201, 202
  synthesis, 429–432, 506–511
Pichia pastoris, 219–236
  expression of recombinant
    protein in, 221–223
  screening for phenotype, 229–
    231
  transformation, 227–229
PCR, *see* Polymerase chain reaction
Polymerase chain reaction,
  detection of gene polymorphisms,
    371

reverse transcriptase, 305–320
  hot start, 310
  hnRNA analysis, 329–331
  multi-competitor, 314–317
    competitor RNA, generation
      of, 314–316
  primer dropping, 308–314
    GAPDH, equalizing to, 311
    range finding, 311–314
  quantification, 311, 314–317
Polymorphisms, 367–374
  autoradiography, 372
  detection methods, 368, 369
  electrophoresis, 372
  PCR, 371
  polyacrylamide gel, 371, 372
Propidium iodide, 485–493
Protein
  recombinant (*E. coli*)
    extraction of 196, 197, 260–262,
      270
    induction of, 195,196, 270
    purification of, 197–200, 262,
      263, 271
    refolding of, 197, 198, 257–264,
      270, 271
Pulmonary emphysema, 161, 162
Pump-1, *see* MMP-7

R, S

RASI-1, *see* MMP-19
Refolding, *see* Protein, refolding of
Reverse zymography, 186, 399–409
  gels, 402, 403
  sample preparation, 403
  standards, 408, 409
RNA extraction, 308, 309
RT-PCR, *see* Polymerase chain
  reaction, reverse transcriptase

Single-strand conformation
    polymorphism, *see*
    Polymorphism
Spheroplast, 227, 228
Sorsby's Fundus Dystrophy, 10
Soybean MMP, 8
SSCP, *see* polymorphisms
Stromelysin
  -1, *see* MMP-3
  -2, *see* MMP-10
  -3, *see* MMP-11
Structure
  MMP
    catalytic domain 53–58
    haemopexin-like domain, 63–65
    prodomain, 61–63
  TIMP, 66
Substrates
  aggrecan
    neoepitopes, 425–427
    release assay, 451–456
  casein
    labeling, 392
    preparation of, 391, 392
  collagen
    labeling, 392
    neoepitopes, 457–471, 473–482
    preparation of, 391
  gelatin
    preparation, 392
    Texas-Red, 417–423
  peptide
    fluorogenic, 495, 500–504

T

Tetracycline
Tet-off, 241, 242
Texas-Red-Gelatin, 417–423
Thymidine incorporation, 535, 536

TIMP-1
  angiogenesis, 170
  cancer, 167–169
  development, 166, 167
  erythroid potentiating factor, 9
  expression
    Baculovirus, 203–216
    E. coli, 267–272
  growth factor activity, 533–536
  history, 9
  host defense, 169, 170
  knockout, 165
  purification
    recombinant, 271
    from culture medium, 294, 295
    from human plasma, 295, 296
  refolding, 270, 271
  steroidigenesis, 169
  transgenic mice, 164–166
TIMP-2
  expression
    E. coli, 257–264
  growth factor activity, 533–536
  history, 9, 10
  knockout, 165, 170
  purification, 262, 263, 296, 297
  refolding 260–262
  transgenic mice, 165
TIMP-3
  history, 10
  transgenic mice, 164, 165
TIMP-4
  history, 10
Tissue inhibitor of
    metalloproteinases, *see also*
    individual TIMPs
  assay, 213, 214, 294, 394, *see
    also* assays
  reverse zymography, 399–409

cloning, 25–27
growth factor activity, 533–536
history, 9, 10
kinetic analysis, 519–530
purification, *see* individual
TIMPs and ELISA
standards
structural studies, 66–69
TR-gelatin, *see* Texas-Red-Gelatin
Transgenic mice, 149–171
apolipoprotein E, 154, 155
collagenase, 153
MMP2, 158–160
MMP-3, 153, 158
MMP-7, 156, 157
MMP-9, 151, 152, 158–160
MMP11, 158
MMP-12, 160, 162
MMP-14, 171
plasminogen activator, 154, 155
TIMP-1, 164–170
TIMP-2, 165, 170
TIMP-3, 164–166
Transcription factors, role of, 124, 125
AP-1, 124, 129
CAAT boxes, 125
MMP-1 and -3 transcription, 124, 125
nuclear factor kappa-B, 125, 128
PEA3, 124, 125, 129
STAT, 130
Transfection
insect cells, 210
stable, 186–188, 250
transient, 184–186, 246, 247
Transin, *see* Stromelysin

Transcription, measuring
analysis of hnRNA, 323, 329–331
primer design, 329
RT-PCR, 329, 330
internal DNA standard, 330, 331
nuclear run-off, 321, 326–328
isolation of nuclei, 326, 327

V

Vectors
PBAT, 209
pEE14, 243
pET, 259, 268-269
pGEMEX, 192, 193
pGEX, 240
pIC9, 224
PLV-1393, 209
pNUT, 183
pRSET, 240
pTRE, 242

W

Western blot, 249, 444
Wound healing, 153, 154

X,Y,Z

XMMP, 7
Yeast, *see Pichia pastoris*
Zymography, 399–409
casein, 201, 213, 407
gelatin, 247, 248, 402–404
gels, 402, 403
in situ, 411–415
reverse, *see* reverse zymography
sample preparation, 403
standards, 408, 409